Intermediate
Algebra
with
Trigonometry

Intermediate Algebra with Trigonometry

Charles P. McKeague

CUESTA COLLEGE

Academic Press
(Harcourt Brace Jovanovich, Publishers)
Orlando □ San Diego
San Francisco □ New York □ London
Toronto □ Montreal □ Sydney □ Tokyo □ São Paulo

Academic Press, Inc.
Orlando, Florida 32887

United Kingdom Edition Published by
Academic Press, Inc. (London) Ltd.
24/28 Oval Road, London NW1 7DX

ISBN: 0-12-484780-3

Printed in the United States of America

Preface to the Instructor

The first twelve chapters of this book cover all the topics usually found in an intermediate algebra course, while the last three chapters cover most of the essential topics from trigonometry.

The three chapters on trigonometry have been included in order to bridge the gap that exists between intermediate algebra and precalculus or trigonometry. The material on trigonometry is intended to be an introduction to the subject. Although most of the main topics found in trigonometry are covered in these chapters, they are not covered in quite as much detail as they are in a regular trigonometry course.

There are a number of ways in which these three chapters can be used. If a simple introduction to the six trigonometric functions is all that is required, then Chapter 13 should be all that is needed. If a more detailed coverage of trigonometry is needed, then either Chapter 14 or 15, or both, can be included after Chapter 13 has been covered. It isn't necessary to cover Chapter 14 before Chapter 15.

Organization of the Text The book begins with a preface to the students explaining what study habits are necessary to ensure success in mathematics.

The rest of the book is divided into chapters. Each chapter is organized as follows:

1. **Chapter Preface** Each chapter begins with a preface that explains in a very general way what the student can expect to find in the chapter and some of the applications associated with topics in the chapter. This preface also includes a list of previous material that is used to develop the concepts in the chapter.

2. **Sections** Following the preface to each chapter, the body of the chapter is divided into sections. Each section contains explanations and examples.

   The explanations are made as simple and intuitive as possible. The ideas, properties, and definitions from Chapters 1 and 2 are used continuously throughout the book. The idea is to make as few rules and definitions as possible and then refer back to them when new situations are encountered.

   The examples are chosen to clarify the explanations and preview the problems in the problem sets.

3. **Problem Sets** Following each section of the text is a problem set. There are five main ideas incorporated into each of the problem sets.
   a. *Drill:* There are enough problems in each problem set to ensure student proficiency with the material once they have completed all the odd-numbered problems.
   b. *Progressive Difficulty:* The problems increase in difficulty as the problem set progresses.
   c. *Odd-Even Similarities:* Each pair of consecutive problems is similar. The answers to the odd problems are listed in the back of the book. This gives the student a chance to check their work and then try a similar problem.
   d. *Application Problems:* Whenever possible, I have included a few application problems toward the end of each problem set. My experience is that students are always curious about how the mathematics they are learning can be applied. They are also much more likely to put some time and effort into trying application problems if there are not an overwhelming number of them to work.
   e. *Review Problems:* Starting with Chapter 2, each problem set ends with a few of review problems. Generally, these review problems cover material that will be used in

the next section. I find that assigning the review problems helps prepare my students for the next day's lecture, and makes reviewing part of their daily routine.

4. **Chapter Summaries** Following the last problem set in each chapter is a chapter summary. Each chapter summary lists all the properties and definitions found in the chapter. In the margin of each chapter summary, next to most topics being summarized, is an example that illustrates the kind of problem associated with that topic.

5. **Chapter Test** Each chapter ends with a chapter test. These tests are designed to give the student an idea of how well he or she has mastered the material in the chapter. All answers for these chapter tests are included in the back of the book.

I think you will find this book to be very flexible and easy to use. It has been written to assist both you and your students in the classroom.

ACKNOWLEDGMENTS I want to thank Richard Christopher, the project editor, for all his hard work and attention to detail. It is a pleasure to be associated with him. I also want to thank Lori Lawson for her assistance in putting together the manuscript for this book. Her word processing ability is without equal. Thanks also to my wife Diane and my children Patrick and Amy for making me come out of my office from time to time.

Preface to the Student

I have seen a number of students over the last few years for whom this quote really makes sense. They don't enjoy math and science classes because they are always worried about whether they will understand the material. Most of them just try to get by and hope for the best. (Are you like that?)

Then there are other students who just don't worry about it. They know they can become as proficient in mathematics as they want (and get whatever grade they want, too). I have people in my class who just can't help but do well on the tests I give. Most of my other students think these people are smart. They may be, but that's not the reason they do well. These successful students have learned that topics in mathematics are not always understandable the first time around. They don't worry about understanding the material; they are successful because they work lots of problems.

That's the key to success in mathematics: working problems. The more problems you work, the better you become at working problems.

It's that simple. Don't worry about being successful or understanding the material, just work problems. Lots of them. The more problems you work, the better you will understand the material.

That's the answer to the big question of how to master this course. Here are the answers to some other questions that are often asked by students.

How much math do I need to know before taking this class?

You should have passed a beginning algebra class. If it has been a few years since you took it, you may have to put in some extra time at the beginning of the semester to get back some of the skills you have lost.

What is the best way to study?

The best way to study is consistently. You must work problems every day. The more time you spend on your homework in the beginning, the easier the sections in the rest of the book will seem. The first two chapters in the book contain the basic properties and methods that the rest of the chapters are built on. Do well on these first two chapters and you will have set a successful pace for the rest of the book.

If I understand everything that goes on in class, can I take it easy on my homework?

Not necessarily. There is a big difference between understanding a problem someone else is working and working the same problem yourself. You are watching someone else think through a problem. There is no substitute for thinking it through yourself. The concepts and properties are understandable to you only if you yourself work problems involving them.

I'm worried about not understanding it. I've passed algebra before but I'm not really sure I understood everything that went on.

There will probably be few times when you can understand absolutely everything that goes on in class. This is standard with most math classes. It doesn't mean you will never understand it. As you read through the book and try problems on your own, you will understand more and more of the material. But if you don't try the problems on your own, you are almost guaranteed to be left confused. By the way, reading the book is important, even if it seems difficult. Reading a math book isn't the same as reading a novel. (You'll see what I mean when you start to read this one.) Reading the book through once isn't going to do it. You will have to read some sections a number of times before they really sink in.

If you have decided to be successful, here is a list of things you can do that will help you attain that success.

How to Be Successful in Algebra

1. Attend class. There is only one way to find out what goes on in class and that is to be there. Missing class and then hoping to get the information from someone who was there is not the same as being there yourself.

2. Read the book and work problems every day. Remember, the key to success in mathematics is working problems. Be wild and crazy about it. Work all the problems you can get your hands on. If you have assigned problems to do, do them first. If you don't understand the concepts after finishing the assigned problems, then keep working problems until you do.

3. Do it on your own. Don't be misled into thinking someone else's work is your own.

4. Don't expect to understand a topic the first time you see it. Sometimes you will understand everything you are doing and sometimes you won't. If you are a little confused, just keep working problems. There will be times when you will not understand exactly what you are doing until after you have worked a number of problems. In any case, worrying about not understanding the material takes time. You can't worry and work problems at the same time. It is best to worry when you worry, and work problems when you work problems.

5. Spend whatever amount of time it takes to master the material. There is really no formula for the exact amount of time you have to spend on algebra to master it. You will find out as you go along what is or isn't enough time for you. If you end up having to spend 2 or 3 hours on each unit to get to the level you are interested in attaining, then that's how much time it takes. Spending less time than that will not work.

6. Relax. It's probably not as difficult as you think.

Contents

3

EXPONENTS AND POLYNOMIALS 86

4

RATIONAL EXPRESSIONS 133

5

RATIONAL EXPONENTS AND ROOTS 177

9

THE CONIC SECTIONS   338

10

RELATIONS AND FUNCTIONS   377

11

LOGARITHMS   413

12

SEQUENCES AND SERIES 444

13

INTRODUCTION TO TRIGONOMETRY 472

14

TRIGONOMETRIC IDENTITIES AND EQUATIONS 530

15

TRIANGLES 565

APPENDIX A: SYNTHETIC DIVISION A1

APPENDIX B: THE DISCRIMINANT A5

APPENDIX C: TABLE OF POWERS, ROOTS, AND PRIME FACTORS A7

APPENDIX D: TABLE OF COMMON LOGARITHMS A11

APPENDIX E: TABLE OF TRIGONOMETRIC FUNCTIONS A15

ANSWERS TO ODD-NUMBERED EXERCISES AND CHAPTER TESTS A29

INDEX I1

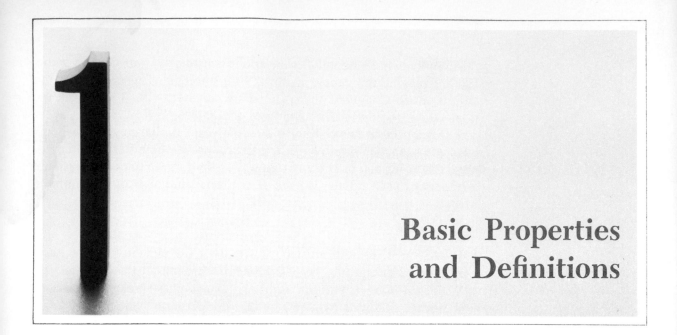

Basic Properties and Definitions

To the student:

The material in Chapter 1 is some of the most important material in the book. It is also some of the easiest material to understand. Be sure that you master it. Your success in the following chapters is directly related to how well you understand the material in Chapter 1.

Here is a list of the most essential concepts from Chapter 1 that you will need in order to be successful in the succeeding chapters:

1. You must know how to add, subtract, multiply, and divide positive and negative numbers. You must be consistently accurate in getting correct answers to simple arithmetic problems.

2. You must understand and recognize the commutative, associative, and distributive properties. These are the three most important properties of real numbers. They are used many times throughout the book to justify and explain other rules and properties.

3. You should know the major classifications of numbers. Some rules and properties hold only for specific kinds of numbers. You must therefore know the difference between whole numbers, integers, rational numbers, and real numbers.

Actually, what Chapter 1 is all about is listing the rules of the game. We are playing the game (algebra) with numbers. The properties and definitions in Chapter 1 tell us what we can start with. The rest of the book, then, is simply the playing of the game.

The material in this chapter may seem very familiar to you. You may have a tendency to skip over it lightly because it is familiar. Don't do it. Understanding algebra begins with understanding Chapter 1. Make sure you *understand* Chapter 1, even if you are familiar with the material in it.

1.1
Basic Definitions

This section is, for the most part, simply a list of many of the basic symbols and definitions we will be using throughout the book.

We begin by reviewing the symbols for comparison of numbers and expressions, followed by a review of the operation symbols.

Comparison Symbols

In symbols	*In words*
$a = b$	a is equal to b
$a \neq b$	a is not equal to b
$a < b$	a is less than b
$a \leq b$	a is less than or equal to b
$a \geq b$	a is greater than or equal to b
$a > b$	a is greater than b
$a \not> b$	a is not greater than b
$a \not< b$	a is not less than b
$a \Leftrightarrow b$	a is equivalent to b
	(This symbol is usually used when accompanying logical statements.)

Operation Symbols

Operation	*In symbols*	*In words*
Addition	$a + b$	The sum of a and b
Subtraction	$a - b$	The difference of a and b
Multiplication	ab, $a \cdot b$, $a(b)$, $(a)b$, or $(a)(b)$	The product of a and b
Division	$a \div b$, a/b, or $\frac{a}{b}$	The quotient of a and b

The key words are *sum, difference, product,* and *quotient*. They are used frequently in mathematics. For instance, we may say the product of

3 and 4 is 12. We mean both the statements $3 \cdot 4$ and 12 are called the product of 3 and 4. The important idea here is that the word *product* implies multiplication, regardless of whether it is written $3 \cdot 4$, 12, 3(4), or (3)4.

The following is an example of translating expressions written in English into expressions written in symbols.

▼ **Example 1**

In English	*In symbols*
The sum of x and 5.	$x + 5$
The product of 3 and x.	$3x$
The quotient of y and 6.	$y/6$
Twice the difference of b and 7.	$2(b - 7)$
The difference of twice b and 7.	$2b - 7$
The quotient of twice x and 3.	$2x/3$
The product of x and y is less than the sum of x and y.	$xy < x + y$
The quotient of 3 times x and 5 is greater than or equal to the difference of x and 4.	$\dfrac{3x}{5} \geq x - 4$ ▲

We now list the basic definitions and operations associated with sets.

DEFINITION A *set* is a well-defined collection of objects or things. The objects in the set are called *elements* or *members* of the set.   *Sets*

The concept of a set can be considered the starting point for all the branches of mathematics. For instance, most of the important concepts in statistics are based, in one form or another, on probability theory. The basic concept in probability theory is the idea of an event, and an event is nothing more than a set.

Sets are usually denoted by capital letters and elements of sets by lower-case letters. To show an element is contained in a set we use the symbol $\in$. That is,

$x \in A$ is read "x is an element (member) of set A"
     or   "x is contained in A."
The symbol $\notin$ is read "is *not* a member of."

DEFINITION   Set A is a subset of set B, written $A \subset B$, if every element in A is also an element of B. That is,

$$A \subset B \text{ if and only if } A \text{ is contained in } B.$$

We use the braces { } to enclose the elements of a set.
Here are some examples of sets and subsets:

▼   **Example 2**

a.   The set of numbers used to count things is $\{1, 2, 3, \ldots\}$. The dots mean the set continues indefinitely in the same manner. This is an example of an infinite set.

b.   The set of all numbers represented by the dots on the faces of a regular die is $\{1, 2, 3, 4, 5, 6\}$. This set is a subset of the set in part a. It is an example of a *finite* set, since it has a limited number of elements.

c.   The set of all Fords is a subset of the set of all cars, since every Ford is also a car.   ▲

DEFINITION   The set with no members is called the *empty* or *null set*. It is denoted by the symbol $\emptyset$. (*Note:* a mistake is sometimes made by trying to denote the empty set with the notation $\{\emptyset\}$. The set $\{\emptyset\}$ is not the empty set, since it contains one element, the empty set $\emptyset$.)
The empty set is considered a subset of every set.

Operations with Sets

There are two basic operations used to combine sets. The operations are union and intersection.

DEFINITION   The *union* of two sets A and B, written $A \cup B$, is the set of all elements that are either in A or in B, or in both A and B. The key word here is *or*. For an element to be in $A \cup B$ it must be in A or B. In symbols the definition looks like this:

$$x \in A \cup B \quad \text{if and only if} \quad x \in A \text{ or } x \in B$$

DEFINITION   The *intersection* of two sets A and B, written $A \cap B$, is the set of elements in both A and B. The key word in this definition is the word *and*. For an element to be in $A \cap B$ it must be in both A and B, or

$$x \in A \cap B \quad \text{if and only if} \quad x \in A \text{ and } x \in B$$

▼ **Example 3** Let $A = \{1, 3, 5\}$, $B = \{0, 2, 4\}$, and $C = \{1, 2, 3, \ldots\}$.
Then

a. $A \cup B = \{0, 1, 2, 3, 4, 5\}$
b. $A \cap B = \varnothing$ (A and B have no elements in common.)
c. $A \cap C = \{1, 3, 5\} = A$
d. $B \cup C = \{0, 1, 2, 3, \ldots\}$ ▲

We can represent the union and intersection of two sets A and B with
pictures. The pictures are called Venn diagrams.

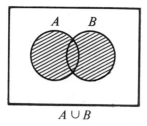

$A \cup B$

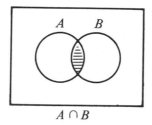

$A \cap B$

DEFINITION Two sets with no elements in common are said to be
disjoint or *mutually exclusive*. Two sets are disjoint if their intersection is
the empty set.

A and B are disjoint if and only if
$A \cap B = \varnothing$

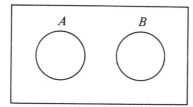

A and B are disjoint.

Along with giving a graphical representation of union and intersection, Venn diagrams can be used to test the validity of statements
involving combinations of sets and operations on sets.

▼ **Example 4** Use Venn diagrams to check the expression

$$A \cap (B \cup C) = (A \cap B) \cup (A \cap C)$$

(Assume no two sets are disjoint.)
  Solution We begin by making a Venn diagram of the left side
with $B \cup C$ shaded in with vertical lines.

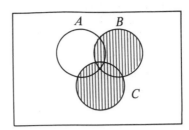

Using the same diagram we now shade in set A with horizontal lines.

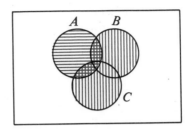

The region containing both vertical and horizontal lines is the intersection of A with $B \cup C$, or $A \cap (B \cup C)$.

We diagram the right side and shade in $A \cap B$ with horizontal lines and $A \cap C$ with vertical lines.

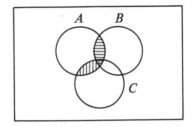

Any region containing vertical or horizontal lines is part of the union of $A \cap B$ and $A \cap C$, or $(A \cap B) \cup (A \cap C)$.

The original statement appears to be true.

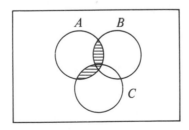

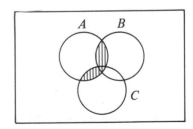

$A \cap (B \cup C)$ $(A \cap B) \cup (A \cap C)$ ▲

Up to this point we have described the sets we have encountered by listing all the elements and then enclosing them with braces, { }, or we have used Venn diagrams. There is another notation we can use to describe sets. It is called *set-builder* notation. Here is how we would write our definition for the union of two sets A and B using set-builder notation.

$$A \cup B = \{x \mid x \in A \text{ or } x \in B\}$$

The right side of this statement is read "the set of all x such that x is a member of A or x is a member of B." As you can see, the vertical line after the first x is read "such that."

▼ **Example 5** Let A and B be two intersecting sets neither of which is a subset of the other. Use a Venn diagram to illustrate the set

$$\{x \mid x \in A \text{ and } x \notin B\}.$$

Solution Using vertical lines to indicate all the elements in A and horizontal lines to show everything that is not in B we have

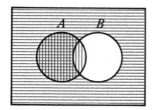

Since the connecting word is "and" we want the region that contains both vertical and horizontal lines.

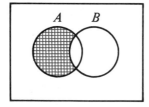

▲

▼ **Example 6** Let $A = \{1, 2, 3, 4, 5, 6\}$ and find

$$C = \{x \mid x \in A \text{ and } x \geq 4\}.$$

Solution We are looking for all the elements of A that are also

greater than or equal to 4. They are 4, 5, and 6. Using set notation we have

$$C = \{4, 5, 6\}. \qquad \blacktriangle$$

Problem Set 1.1

Translate each of the following sentences into symbols:

1. The sum of x and 5.
2. The sum of y and -3.
3. The difference of 6 and x.
4. The difference of x and 6.
5. The product of t and 2 is less than y.
6. The product of $5x$ and y is equal to z.
7. The quotient of $3x$ and $2y$ is greater than 6.
8. The quotient of $2y$ and $3x$ is not less than 7.
9. The sum of x and y is less than the difference of x and y.
10. Twice the sum of a and b is 15.
11. Three times the difference of x and 5 is more than y.
12. The product of x and y is greater than or equal to the quotient of x and y.
13. The difference of s and t is not equal to their sum.
14. The quotient of $2x$ and y is less than or equal to the sum of $2x$ and y.
15. Twice the sum of t and 3 is not greater than the difference of t and 6.
16. Three times the product of x and y is equal to the sum of $2y$ and $3z$.

For problems 17–34, let $A = \{0, 2, 4, 6\}$, $B = \{1, 2, 3, 4, 5\}$, $C = \{1, 3, 5, 7\}$, and $D = \{-2, -1, 0, 1, 2\}$ and find the following:

17. $A \cup B$	18. $A \cup C$
19. $A \cap B$	20. $A \cap C$
21. $C \cup D$	22. $B \cup D$
23. $C \cap D$	24. $B \cap D$
25. $A \cup D$	26. $B \cup C$
27. $A \cap D$	28. $B \cap C$
29. $A \cup (B \cap C)$	30. $C \cup (A \cap B)$
31. $(A \cup B) \cap (A \cup C)$	32. $(C \cup A) \cap (C \cup B)$
33. $A \cap (B \cup D)$	34. $B \cap (C \cup D)$

35. List all the subsets of $\{a, b, c\}$.
36. List all the subsets of $\{0, 1\}$.
37. Give an example, using any sets you want, of two mutually exclusive sets.
38. Give an example of two nonmutually exclusive sets.

Use Venn diagrams to show each of the following regions. Assume sets A, B, and C all intersect one another.

39. $(A \cup B) \cup C$ 40. $(A \cap B) \cap C$
41. $A \cap (B \cap C)$ 42. $A \cup (B \cup C)$
43. $A \cap (B \cup C)$ 44. $(A \cap B) \cup C$
45. $A \cup (B \cap C)$ 46. $(A \cup B) \cap C$
47. $(A \cup B) \cap (A \cup C)$ 48. $(A \cap B) \cup (A \cap C)$
49.   Use a Venn diagram to show that if $A \subset B$, then $A \cap B = A$
50.   Use a Venn diagram to show that if $A \subset B$, then $A \cup B = B$

Let A and B be two intersecting sets neither of which is a subset of the other. Use Venn diagrams to illustrate each of the following sets.

51. $\{x \mid x \in A \text{ and } x \in B\}$ 52. $\{x \mid x \in A \text{ or } x \in B\}$
53. $\{x \mid x \notin A \text{ and } x \in B\}$ 54. $\{x \mid x \notin A \text{ and } x \notin B\}$
55. $\{x \mid x \notin A \text{ or } x \notin B\}$ 56. $\{x \mid x \notin A \text{ or } x \in B\}$
57. $\{x \mid x \in A \cap B\}$ 58. $\{x \mid x \in A \cup B\}$

Let $A = \{1, 2, 3, 4, 5, 6\}$ and $B = \{2, 4, 6, 8\}$, and find

59. $\{x \mid x \in A \text{ and } x \leq 3\}$ 60. $\{x \mid x \in B \text{ and } x \geq 8\}$
61. $\{x \mid x \in A \text{ and } x \notin B\}$ 62. $\{x \mid x \in B \text{ and } x \notin A\}$
63. $\{x \mid x \in B \text{ and } x \neq 6\}$ 64. $\{x \mid x \in A \text{ and } x \neq 6\}$
65. $\{x \mid x \in A \cap B\}$ 66. $\{x \mid x \in A \cup B\}$

The notation $n(A)$ is used to denote the *number* of elements in set A. For example, if $A = \{a, b, c\}$, then $n(A) = 3$.

67.   If A and B are sets such that $n(A) = 4$, $n(B) = 5$, and $A \cap B = \emptyset$, find $n(A \cup B)$. (How many elements are in the union of A and B?)
68.   If $A \cap B = \emptyset$, what is $n(A \cap B)$?
69.   If $n(A) = 7$, $n(B) = 8$, and $n(A \cap B) = 2$, find $n(A \cup B)$.
70.   If $n(A) = 4$, $n(B), = 10$, and $n(A \cap B) = 3$, find $n(A \cup B)$.

In this section we will give a definition of real numbers in terms of the real number line. We will then classify all pairs of numbers that add to 0, and do the same for all pairs of numbers whose product is 1. We will end the section with two definitions of absolute value and a list of some special sets.

The real number line is constructed by drawing a straight line and labeling a convenient point with the number 0. Positive numbers are in increasing order to the right of 0, negative numbers are in decreasing order to the left of 0. The point on the line corresponding to 0 is called the origin.

1.2
The Real Numbers, Opposites, Reciprocals, and Absolute Value

The Real Numbers

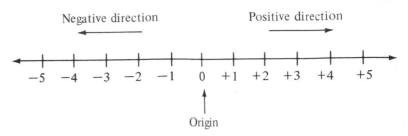

The numbers associated with the points on the line are called *coordinates* of those points. Every point on the line has a number associated with it. The set of all these numbers makes up the set of real numbers.

DEFINITION A *real number* is any number that is the coordinate of a point on the real number line.

There are many different sets of numbers contained in the real numbers. We will classify some of them later. Among those numbers contained in the real numbers are all positive and negative fractions and decimals.

▼ **Example 1** Locate the numbers -4.5, $-2\frac{1}{4}$, $-.75$, $\frac{1}{2}$, $\sqrt{2}$, π, and 4.1 on the real number line.

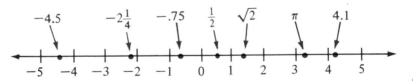

▲

Note In this book we will refer to real numbers as being on the real number line. Actually, real numbers are *not* on the line; only the points they represent are on the line. We can save some writing, however, if we simply refer to real numbers as being on the number line.

*Opposites
and Reciprocals*

DEFINITION Any two real numbers, the same distance from 0, but in opposite directions from 0 on the number line, are called *opposites* or *additive inverses*.

▼ **Example 2** The numbers -3 and 3 are opposites. So are π and $-\pi$, $\frac{3}{4}$ and $-\frac{3}{4}$, and $\sqrt{2}$ and $-\sqrt{2}$.   ▲

The negative sign in front of a number can be read in a number of different ways. It can be read as "negative" or "the opposite of." We say

-4 is the opposite of 4 or negative 4. The one we use will depend on the situation. For instance, the expression $-(-3)$ is best read "the opposite of negative 3." Since the opposite of -3 is 3, we have $-(-3) = 3$. In general, if a is any positive real number, then

$$-(-a) = a \qquad \text{(The opposite of a negative is positive.)}$$

One property all pairs of opposites have is that their sum is 0. It is for this reason they are called additive inverses.

Before we go further with our study of the number line, we need to review multiplication with fractions. Recall that for the fraction $\frac{a}{b}$, a is called the numerator and b is called the denominator. To multiply two fractions we simply multiply numerators and multiply denominators.

Review of Multiplication with Fractions

▼ **Example 3**   Multiply $\frac{3}{5} \cdot \frac{7}{8}$.

Solution   The product of the numerators is 21 and the product of the denominators is 40.

$$\frac{3}{5} \cdot \frac{7}{8} = \frac{3 \cdot 7}{5 \cdot 8} = \frac{21}{40} \qquad \blacktriangle$$

▼ **Example 4**   Multiply $8 \cdot \frac{1}{5}$.

Solution   The number 8 can be thought of as the fraction $\frac{8}{1}$.

$$8 \cdot \frac{1}{5} = \frac{8}{1} \cdot \frac{1}{5} = \frac{8 \cdot 1}{1 \cdot 5} = \frac{8}{5} \qquad \blacktriangle$$

Note   In past math classes you may have written fractions like $\frac{8}{5}$ (improper fractions) as mixed numbers, such as $1\frac{3}{5}$. In algebra it is usually better to leave them as improper fractions.

The idea of multiplication of fractions is useful in understanding the concept of the reciprocal of a number. Here is the definition.

DEFINITION   Any two real numbers whose product is 1 are called *reciprocals* or *multiplicative inverses*.

▼ **Example 5**

Number	*Reciprocal*	
a. 3	$\frac{1}{3}$	because $3 \cdot \frac{1}{3} = \frac{3}{1} \cdot \frac{1}{3} = \frac{3}{3} = 1$

Number	Reciprocal	
b. $\dfrac{1}{6}$	6	because $\dfrac{1}{6} \cdot 6 = \dfrac{1}{6} \cdot \dfrac{6}{1} = \dfrac{6}{6} = 1$
c. $\dfrac{4}{5}$	$\dfrac{5}{4}$	because $\dfrac{4}{5} \cdot \dfrac{5}{4} = \dfrac{20}{20} = 1$
d. $\sqrt{2}$	$\dfrac{1}{\sqrt{2}}$	because $\sqrt{2} \cdot \dfrac{1}{\sqrt{2}} = \dfrac{\sqrt{2}}{1} \cdot \dfrac{1}{\sqrt{2}} = \dfrac{\sqrt{2}}{\sqrt{2}} = 1$
e. a	$\dfrac{1}{a}$	because $a \cdot \dfrac{1}{a} = \dfrac{a}{1} \cdot \dfrac{1}{a} = \dfrac{a}{a} = 1 \quad (a \neq 0)$

▲

    Although we will not develop multiplication with negative numbers until later in this chapter, you should know that the reciprocal of a negative number is also a negative number. For example, the reciprocal of -5 is $-\frac{1}{5}$. Likewise, the reciprocal of $-\frac{3}{4}$ is $-\frac{4}{3}$. We should also note that 0 is the only number without a reciprocal, since multiplying by 0 always results in 0, never 1.

The Absolute Value of a Real Number

Sometimes it is convenient to consider only the distance a number is from 0 and not its direction from 0.

DEFINITION   The *absolute value* of a number (also called its *magnitude*) is the distance the number is from 0 on the number line. If x represents a real number, then the absolute value of x is written $|x|$.

    Since distances are always represented by positive numbers or 0, the absolute value of a quantity is always positive or 0; the absolute value of a number is never negative.

    The above definition of absolute value is geometric in form since it defines absolute value in terms of the number line.

    Here is an alternate definition of absolute value that is algebraic since it involves only symbols.

DEFINITION   If x represents a real number, then the *absolute value* (or *magnitude*) of x is written $|x|$, and is given by

$$|x| = \begin{cases} x & \text{if } x \geq 0 \\ -x & \text{if } x < 0 \end{cases}$$

    If the original number is positive or 0, then its absolute value is the number itself. If the number is negative, its absolute value is its opposite

(which must be positive). We see that $-x$, as written in this definition, is a positive quantity since x in this case is a negative number.

Note   It is important to recognize that if x is a real number, $-x$ is not necessarily negative. For example, if x is 5, then $-x$ is -5. On the other hand, if x were -5, then $-x$ would be $-(-5)$, which is 5. To assume that $-x$ is a negative number we must know beforehand that x is a positive number. If x is a negative number, then $-x$ is a positive number.

▼   **Example 6**

a. $|5| = 5$ e. $-|-3| = -3$
b. $|-2| = 2$ f. $-|5| = -5$
c. $|-\frac{1}{2}| = \frac{1}{2}$ g. $-|-\sqrt{2}| = -\sqrt{2}$
d. $|\frac{3}{4}| = \frac{3}{4}$ h. $-|\frac{5}{6}| = -\frac{5}{6}$

The last four parts of this example do not contradict the statement that the absolute value of a number is always positive or 0. In part e we are asked to find the *opposite* of the absolute value of -3, $-|-3|$. The absolute value of -3 is 3, the opposite of which is -3. If the absolute value of a number is always positive or 0, then the opposite of its absolute value must always be negative or 0.   ▲

Note   There is a tendency to confuse the two expressions

$$-(-3) \quad \text{and} \quad -|-3|$$

The first expression is the opposite of -3, which is 3. The second expression is the opposite of the absolute value of -3, which is -3.

$$-(-3) = 3 \quad \text{and} \quad -|-3| = -3$$

Note   It is incorrect to write $|x| = x$. This statement is true only if x is a positive number. If x were negative, then, by definition, $|x| = -x$.

We will end this section by listing the major subsets of the set of real numbers.   *Subsets of the Real Numbers*

$$\text{Counting numbers} = \{1, 2, 3, \ldots\}$$

These are also called *natural numbers* or the *set of positive integers*. If we include the number 0 in the set of counting numbers, we have the *set of whole numbers:*

$$\text{Whole numbers} = \{0, 1, 2, 3, \ldots\}$$

The whole numbers along with the opposites of all the counting numbers make up the *set of integers:*

$$Integers = \{\ldots -3, -2, -1, 0, 1, 2, 3, \ldots\}$$

The set of all numbers that can be written as the ratio of two integers is called the *set of rational numbers.* (The ratio of two numbers a and b is the same as their quotient $\frac{a}{b}$.)

$$Rational\ numbers = \{\tfrac{a}{b} \,|\, a \text{ and } b \text{ are integers, } b \neq 0\}$$

This is read "the set of all $\frac{a}{b}$ such that a and b are integers, where $b \neq 0$."

▼   **Example 7**

a. $\frac{3}{4}$ is a rational number since it is the ratio of the two integers 3 and 4.

b. $-\frac{5}{6}$ is a rational number because it can be thought of as the ratio of -5 to 6. (It can also be thought of as the ratio of 5 to -6.)

c.   5 is a rational number since it is the ratio of 5 to 1.

$$5 = \frac{5}{1}$$

d. -8 is a rational number because it is the ratio of -8 to 1.

$$-8 = \frac{-8}{1}$$

e.   0.75 is a rational number since it can be written as $\frac{75}{100}$.

f.   The number 0.333... is a rational number because it is equal to $\frac{1}{3}$. (To check this, divide 1 by 3 with long division.)   ▲

In general the set of rational numbers includes all the whole numbers and integers, as well as all terminating decimals (like 0.75) and all repeating decimals (like 0.333...).

The real numbers that cannot be written as the ratio of two integers comprise the *set of irrational numbers.* These numbers have decimal representations that are all nonrepeating, nonterminating decimals—that is, decimals with an infinite number of digits past the decimal point in which no repeating pattern of digits can be found.

Irrational numbers $= \{x \,|\, x \text{ is a nonrepeating, nonterminating decimal}\}$

Since irrational numbers cannot be written as the ratio of integers, and their decimal representations never terminate or repeat, they have to be written in other ways.

▼ **Example 8**   The following are irrational numbers:

a. $\sqrt{2}$
b. $-\sqrt{3}$
c. $4 + 2\sqrt{3}$
d. π
e. $\pi + 5\sqrt{6}$ ▲

Note   We can find decimal approximations to some irrational numbers by using a calculator. For example, on an eight-digit calculator

$$\sqrt{2} = 1.4142135.$$

This is not exactly $\sqrt{2}$ but simply an approximation to it. There is no decimal that gives $\sqrt{2}$ exactly.

In the beginning of this section we defined real numbers as numbers associated with points on the real number line. Since every point on the number line is named by either a rational number or an irrational number, we can give a second, but equivalent, definition for the set of real numbers.

Real numbers $= \{x \mid x$ is rational or x is irrational$\}$

▼ **Example 9**   For the set $\{-5, -3.5, 0, 3/4, \sqrt{3}, \sqrt{5}, 9\}$, list the numbers that are **a.** whole numbers, **b.** integers, **c.** rational numbers, **d.** irrational numbers, and **e.** real numbers.

Solution

a.   whole numbers $= \{0, 9\}$
b.   integers $= \{-5, 0, 9\}$
c.   rational numbers $= \{-5, -3.5, 0, 3/4, 9\}$
d.   irrational numbers $= \{\sqrt{3}, \sqrt{5}\}$
e.   They are all real numbers.   ▲

---

1. Locate the numbers -3, -1.75, $-\frac{1}{2}$, 0, $\frac{1}{3}$, 1, 1.3, and 4.5 on the number line.

2. Locate the numbers -3.25, -3, -2.5, -1, $-\frac{1}{3}$, 0, $\frac{3}{4}$, 1.2, and 2 on the number line.

Problem Set 1.2

Copy and complete the following table.

	Number	Opposite	Reciprocal
3.	4		
4.	-3		
5.	$-\frac{1}{2}$		
6.	$\frac{5}{6}$		
7.		-5	
8.		7	
9.		$-\frac{3}{8}$	
10.		$\frac{1}{2}$	
11.			-6
12.			-3
13.			$\frac{1}{3}$
14.			$-\frac{1}{4}$
15.		$-\sqrt{3}$	
16.		$\sqrt{5}$	
17.			$-\sqrt{2}$
18.			$-\frac{3}{5}$
19.	x		
20.	0		

21. Name two numbers that are their own reciprocals.
22. Give the number that has no reciprocal.
23. Name the number that is its own opposite.
24. The reciprocal of a negative number is negative—true or false?

Write each of the following without absolute value symbols:

25. $|-2|$ 26. $|-7|$
27. $|-\frac{3}{4}|$ 28. $|\frac{5}{6}|$
29. $|\pi|$ 30. $|-\sqrt{2}|$
31. $-|4|$ 32. $-|5|$
33. $-|-2|$ 34. $-|-10|$
35. $-|-\frac{3}{4}|$ 36. $-|\frac{7}{8}|$

Find the value of each of the following expressions:

37. $-(-2)$ 38. $-(-\frac{3}{4})$
39. $-[-(-\frac{1}{3})]$ 40. $-[-(-1)]$
41. $|2| + |3|$ 42. $|7| + |4|$
43. $|-3| + |5|$ 44. $|-1| + |0|$
45. $|-8| - |-3|$ 46. $|-6| - |-1|$
47. $|-10| - |4|$ 48. $|-3| - |2|$
49. $|-2| + |-3| - |5|$ 50. $|-6| - |-2| + |-4|$

For the set $\{-6, -5.2, -\sqrt{7}, -\pi, 0, 1, 2, 2.3, \frac{9}{2}, \sqrt{17}\}$ list all the elements that are:

51. Counting numbers   **52.** Whole numbers
53. Rational numbers   **54.** Integers
55. Irrational numbers   **56.** Real numbers
57. Nonnegative integers   **58.** Positive integers

Label the following true or false. For each false statement give a counter example (that is, an example that shows the statement is false).

59. Zero has an opposite and a reciprocal.
60. Some irrational numbers are also rational numbers.
61. All whole numbers are integers.
62. Every real number is a rational number.
63. All integers are rational numbers.
64. Some negative numbers are integers.
65. Zero is both rational and irrational.
66. Zero is not considered a real number.
67. The opposite of an integer is also an integer.
68. The reciprocal of an integer is an integer.
69. The counting numbers are a subset of the whole numbers.
70. The rational numbers are a subset of the irrational numbers.

Multiply the following.

71. $\frac{3}{5} \cdot \frac{7}{8}$ **72.** $\frac{6}{7} \cdot \frac{9}{5}$
73. $4 \cdot \frac{3}{5}$ **74.** $9 \cdot \frac{2}{7}$
75. $15 \cdot \frac{8}{17}$ **76.** $17 \cdot \frac{3}{9}$
77. $\frac{5}{3} \cdot 7 \cdot \frac{8}{3}$ **78.** $\frac{3}{5} \cdot 6 \cdot \frac{1}{7}$
79. $\frac{3}{5} \cdot \frac{4}{7} \cdot \frac{6}{11}$ **80.** $\frac{4}{5} \cdot \frac{6}{7} \cdot \frac{3}{11}$
81. $\frac{4}{3} \cdot \frac{3}{4}$ **82.** $\frac{5}{8} \cdot \frac{8}{5}$

83. $\sqrt{2} \cdot \dfrac{1}{\sqrt{2}}$ **84.** $\sqrt{3} \cdot \dfrac{1}{\sqrt{3}}$

85. $2 \cdot \dfrac{1}{\sqrt{5}} \cdot \sqrt{5}$ **86.** $3 \cdot \sqrt{7} \cdot \dfrac{1}{\sqrt{7}}$

87. $\pi \cdot \dfrac{1}{\pi}$ **88.** $e \cdot \dfrac{1}{e}$

89. Name two numbers that are 5 units from 2 on the number line.
90. Name two numbers that are 6 units from -3 on the number line.
91. If the rational number $\frac{1}{3}$ can be written as the repeating decimal $0.333\ldots$, what rational number can be written as the repeating decimal $0.999\ldots$?
92. Write the rational number $\frac{7}{9}$ as a repeating decimal. (*Hint:* Divide 7 by 9 using long division.)
93. Write the rational number $\frac{28}{99}$ as a repeating decimal. (Divide 28 by 99 using long division.)

94. Write the rational number $\frac{123}{999}$ as a repeating decimal.

95. A person has a balance of \$25 in their checking account. If they write a check for \$40, what negative number will give the new balance in the persons account?

96. A man wins \$37 on one hand of cards. On the next hand he loses \$57. What negative number can be used to represent his net gain?

97. Name two numbers whose absolute value is 6.

98. Name two numbers whose absolute value is 3/4.

1.3
Graphing Simple and Compound Inequalities

In this section we will use some of the ideas developed in the first two sections to graph inequalities. The graph of an inequality uses the real number line to give a visual representation of an algebraic expression.

▼   **Example 1**   Graph $\{x \mid x \leq 3\}$.

Solution   We want to graph all the real numbers less than or equal to 3—that is, all the real numbers below 3 and including 3. We label 0 on the number line for reference as well as 3 since the latter is what we call the end point. The graph is as follows.

We use a solid circle at 3 since 3 is included in the graph.   ▲

▼   **Example 2**   Graph $\{x \mid x < 3\}$.

Solution   The graph will be identical to the graph in Example 1 except at the end point 3. In this case we will use an open circle since 3 is not included in the graph.

▲

In Section 1.1 we defined the *union* of two sets A and B to be the set of all elements that are in either A or B. The word *or* is the key word in the definition. The *intersection* of two sets A and B is the set of all elements contained in both A and B, the key word here being *and*. We can put the words *and* and *or* together with our methods of graphing inequalities to graph some compound inequalities.

▼ **Example 3** Graph $\{x \mid x \leq -2 \text{ or } x > 3\}$.

Solution The two inequalities connected by the word *or* are re-
ferred to as a *compound inequality*. We begin by graphing each
inequality separately.

$x \leq -2$
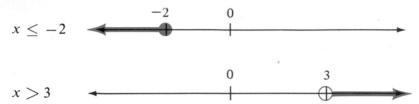

$x > 3$

Since the two are connected by the word *or,* we graph their union.
That is, we graph all points on either graph.

▲

▼ **Example 4** Graph $\{x \mid x > -1 \text{ and } x < 2\}$.

Solution We first graph each inequality separately.

$x > -1$
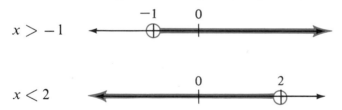

$x < 2$

Since the two inequalities are connected by the word *and,* we graph
their intersection—the part they have in common.

▲

NOTATION Sometimes compound inequalities that use the word *and* as
the connecting word can be written in a shorter form. For example, the
compound inequality $-3 \leq x$ and $x \leq 4$ can be written $-3 \leq x \leq 4$.
The word *and* does not appear when an inequality is written in this
form. It is implied. Inequalities of the form $-3 \leq x \leq 4$ are called
continued inequalities. This new notation is useful because is takes fewer
symbols to write it and because if $-3 \leq x$ and $x \leq 4$, then x must be

between -3 and 4. With the notation $-3 \leq x \leq 4$, it "looks" as though x is between -3 and 4. The graph of $-3 \leq x \leq 4$ is

▼   **Example 5**   Graph $\{x \mid 1 \leq x < 2\}$.

Solution   The word *and* is implied in the continued inequality $1 \leq x < 2$. We graph all the numbers between 1 and 2 on the number line, including 1 but not including 2.

   ▲

Problem Set 1.3

Graph the following on the real number line:

1. $\{x \mid x < 1\}$
2. $\{x \mid x > -2\}$
3. $\{x \mid x \leq 1\}$
4. $\{x \mid x \geq -2\}$
5. $\{x \mid x \geq 4\}$
6. $\{x \mid x \leq -3\}$
7. $\{x \mid x > 4\}$
8. $\{x \mid x < -3\}$
9. $\{x \mid x > 0\}$
10. $\{x \mid x < 0\}$
11. $\{x \mid -2 < x\}$
12. $\{x \mid 3 \geq x\}$
13. $\{x \mid 4 \leq x\}$
14. $\{x \mid 2 > x\}$

Graph the following compound inequalities:

15. $\{x \mid x < -3 \text{ or } x > 1\}$
16. $\{x \mid x \leq 1 \text{ or } x \geq 4\}$
17. $\{x \mid x \leq -3 \text{ or } x \geq 1\}$
18. $\{x \mid x < 1 \text{ or } x > 4\}$
19. $\{x \mid -3 \leq x \text{ and } x \leq 1\}$
20. $\{x \mid 1 < x \text{ and } x < 4\}$
21. $\{x \mid -3 < x \text{ and } x < 1\}$
22. $\{x \mid 1 \leq x \text{ and } x \leq 4\}$
23. $\{x \mid -1 \leq x \text{ and } 2 \leq x\}$
24. $\{x \mid 3 \leq x \text{ and } 4 \leq x\}$
25. $\{x \mid -1 \leq x \text{ or } 2 \leq x\}$
26. $\{x \mid 3 \leq x \text{ or } 4 \leq x\}$
27. $\{x \mid x < -1 \text{ or } x \geq 3\}$
28. $\{x \mid x < 0 \text{ or } x \geq 3\}$
29. $\{x \mid x \leq -1 \text{ and } x \geq 3\}$
30. $\{x \mid x \leq 0 \text{ and } x \geq 3\}$
31. $\{x \mid x > -4 \text{ and } x < 2\}$
32. $\{x \mid x > -3 \text{ and } x < 0\}$

Graph the following continued inequalities:

33. $\{x \mid -1 \leq x \leq 2\}$
34. $\{x \mid -2 \leq x \leq 1\}$
35. $\{x \mid -1 < x < 2\}$
36. $\{x \mid -2 < x < 1\}$
37. $\{x \mid -3 < x < 1\}$
38. $\{x \mid 1 \leq x \leq 2\}$
39. $\{x \mid -3 \leq x < 0\}$
40. $\{x \mid 2 \leq x < 4\}$
41. $\{x \mid -4 < x \leq 1\}$
42. $\{x \mid -1 < x \leq 5\}$

The inequalities below extend the work you have already done with inequalities to some slightly more complicated graphs. They are compound and continued inequalities together. In each case show the graph.

43. $\{x \mid x < -3 \text{ or } 2 < x < 4\}$
44. $\{x \mid -4 \leq x \leq -2 \text{ or } x \geq 3\}$
45. $\{x \mid -5 \leq x \text{ or } 0 \leq x \leq 3\}$
46. $\{x \mid -3 < x < 0 \text{ or } x > 5\}$
47. $\{x \mid -5 < x < -2 \text{ or } 2 < x < 5\}$
48. $\{x \mid -3 \leq x \leq -1 \text{ or } 1 \leq x \leq 3\}$

49. Suppose a person knows that if he invests $1,000 in a certain stock, he will make between $60 and $90 in dividends. Write an inequality that indicates how much he will make in dividends if he invests $3,000 in these stocks.

50. A spring 3 inches long will increase its length by 3 inches when a weight is hung from it. Write a continued inequality that shows all the lengths the spring stretches through when the weight is attached.

51. Write an inequality that gives all the numbers that are 5 or less units from 2 on the number line.

52. Write an inequality that gives all the numbers within 3 units of -2 on the number line.

53. Specify, by using a compound inequality, all the numbers that are more than 4 units from 0 on the number line.

54. Can you think of a single inequality that uses an absolute value symbol that will give the same numbers as the inequality you found in Problem 53?

In this section we will list all the things we know to be true of real numbers and the operation symbols listed in Section 1.1. Mathematics is a game we play with real numbers. The rules of the game are the properties of real numbers listed in this section. We play the game by taking real numbers and their properties and applying them to as many new situations as possible.

The list of properties given in this section is actually just an organized summary of the things we know from past experience to be true about numbers in general. For instance, we know that adding 3 and 7 gives the same answer as adding 7 and 3. The order of two numbers in an addition problem can be changed without changing the result. This fact about numbers and addition is called the *commutative property of addition*. We say addition is a commutative operation. Likewise, multiplication is a commutative operation.

**1.4
Properties of Real
Numbers**

We now state these properties formally. For all the properties listed in this section, a, b, and c represent real numbers.

Commutative Property of Addition
In symbols: $a + b = b + a$
In words: The *order* of the numbers in a sum does not affect the result.

Commutative Property of Multiplication
In symbols: $a \cdot b = b \cdot a$
In words: The *order* of the numbers in a product does not affect the result.

▼ **Example 1**

a.   The statement $3 + 7 = 7 + 3$ is an example of the commutative property of addition.
b.   The statement $3 \cdot x = x \cdot 3$ is an example of the commutative property of multiplication.
c.   Using the commutative property of addition, the expression $3 + x + 7$ can be simplified:

$$3 + x + 7 = 3 + 7 + x \qquad \text{Commutative property}$$
$$= 10 + x \qquad \text{Addition} \qquad ▲$$

The other two basic operations (subtraction and division) are not commutative. If we change the order in which we are subtracting or dividing two numbers we will change the result.

Another property of numbers you have used many times has to do with grouping. When adding $3 + 5 + 7$ we can add the 3 and 5 first and then the 7, or we can add the 5 and 7 first and then the 3. Mathematically it looks like this: $(3 + 5) + 7 = 3 + (5 + 7)$. Operations that behave in this manner are called *associative* operations. The answers will not change when we change the grouping. Here is the formal definition.

Associative Property of Addition
In symbols: $a + (b + c) = (a + b) + c$
In words: The *grouping* of the numbers in a sum does not affect the result.

Associative Property of Multiplication
In symbols: $a(bc) = (ab)c$
In words: The *grouping* of the numbers in a product does not affect the result.

The following examples illustrate how the associative properties can be used to simplify expressions that involve both numbers and variables.

▼ **Example 2**

a. $2 + (3 + y) = (2 + 3) + y$ Associative property
$= 5 + y$ Addition
b. $5(4x) = (5 \cdot 4)x$ Associative property
$= 20x$ Multiplication

c. $\dfrac{1}{4}(4a) = \left(\dfrac{1}{4} \cdot 4\right)a$ Associative property

$= 1a$ Multiplication

$= a$ ▲

Our next property involves both addition and multiplication. It is called the *distributive property* and is stated as follows.

Distributive Property

In symbols: $a(b + c) = ab + ac$
In words: Multiplication *distributes* over addition.

You will see as we progress through the book that the distributive property is used very frequently in algebra. To see that the distributive property works, compare the following.

$$3(4 + 5) \qquad 3(4) + 3(5)$$
$$3(9) \qquad\quad 12 + 15$$
$$27 \qquad\qquad 27$$

In both cases the result is 27. Since the results are the same, the original two expressions must be equal. Or, $3(4 + 5) = 3(4) + 3(5)$.

▼ **Example 3** Apply the distributive property to the following expressions, then simplify each one.

a. $3(7 + 1) = 3(7) + 3(1) = 21 + 3 = 24$
b. $4(x + 2) = 4(x) + 4(2) = 4x + 8$
c. $5(a + b) = 5a + 5b$ ▲

The distributive property can also be used to combine similar terms. (For now, a term is the product of a number with one or more variables. We will give a precise definition in Chapter 3.) Similar terms are terms with the same variable part. The terms $3x$ and $5x$ are similar, as are $2y$, $7y$, and $-3y$, because the variable parts are the same. To combine similar terms we use the distributive property in the reverse direction from that in Example 3. The following example illustrates.

▼   **Example 4**   Use the distributive property to combine similar terms.

a. $3x + 5x = (3 + 5)x$ Distributive property
              $= 8x$ Addition
b. $4a + 7a = (4 + 7)a$ Distributive property
              $= 11a$ Addition
c. $3y + y = (3 + 1)y$ Distributive property
              $= 4y$ Addition   ▲

The distributive property is also used to add fractions. For example, to add $\frac{3}{7}$ and $\frac{2}{7}$ we first write each as the product of a whole number and $\frac{1}{7}$. Then we apply the distributive property as we did in the example above.

$$\frac{3}{7} + \frac{2}{7} = 3 \cdot \frac{1}{7} + 2 \cdot \frac{1}{7}$$

$$= (3 + 2)\frac{1}{7} \qquad \text{Distributive property}$$

$$= 5 \cdot \frac{1}{7}$$

$$= \frac{5}{7}$$

To add fractions using the distributive property, each fraction must have the same denominator. Here are some further examples.

▼   **Example 5**   Add

a. $\dfrac{3}{9} + \dfrac{4}{9} + \dfrac{1}{9} = 3 \cdot \dfrac{1}{9} + 4 \cdot \dfrac{1}{9} + 1 \cdot \dfrac{1}{9}$

$$= (3 + 4 + 1)\frac{1}{9} \qquad \text{Distributive property}$$

$$= 8 \cdot \frac{1}{9}$$

$$= \frac{8}{9}$$

b. $\dfrac{4}{\sqrt{2}} + \dfrac{5}{\sqrt{2}} = 4 \cdot \dfrac{1}{\sqrt{2}} + 5 \cdot \dfrac{1}{\sqrt{2}}$

$\qquad = (4 + 5)\dfrac{1}{\sqrt{2}}$ Distributive property

$\qquad = 9 \cdot \dfrac{1}{\sqrt{2}}$

$\qquad = \dfrac{9}{\sqrt{2}}$

c. $\dfrac{4}{x} + \dfrac{2}{x} = 4 \cdot \dfrac{1}{x} + 2 \cdot \dfrac{1}{x}$

$\qquad = (4 + 2)\dfrac{1}{x}$ Distributive property

$\qquad = 6 \cdot \dfrac{1}{x}$

$\qquad = \dfrac{6}{x}$ ▲

We can use the commutative, associative, and distributive properties together to simplify expressions such as $3x + 4 + 5x + 8$. We begin by applying the commutative property to change the order of the terms and write:

$$3x + 5x + 4 + 8$$

Next we use the associative property to group similar terms together:

$$(3x + 5x) + (4 + 8)$$

Applying the distributive property to the first two terms we have

$$(3 + 5)x + (4 + 8)$$

Finally, we add 3 and 5, and 4 and 8 to get

$$8x + 12$$

Here are some additional examples.

▼ **Example 6** Simplify

a. $7x + 4 + 6x + 3 = (7x + 6x) + (4 + 3)$ Commutative and associative properties

$$= (7 + 6)x + (4 + 3) \qquad \text{Distributive property}$$
$$= 13x + 7 \qquad \text{Addition}$$

b. $8a + 4 + a + 6a = (8a + a + 6a) + 4 \qquad$ Commutative and associative properties

$$= (8 + 1 + 6)a + 4 \qquad \text{Distributive property}$$
$$= 15a + 4 \qquad \text{Addition} \qquad \blacktriangle$$

In actual practice you may not show all the steps we have shown in Example 6. We are showing them here so you can see that each manipulation we do in algebra can be justified by a property or definition.

The remaining properties of real numbers have to do with the numbers 0 and 1.

Additive Identity Property
There exists a unique number 0 such that:
In symbols: $a + 0 = a$ and $0 + a = a$
In words: Zero preserves identities under addition. (The identity of the number is unchanged after addition with 0.)

Multiplicative Identity Property
There exists a unique number 1 such that:
In symbols: $a(1) = a$ and $1(a) = a$
In words: The number 1 preserves identities under multiplication. (The identity of the number is unchanged after multiplication by 1.)

Additive Inverse Property
For each real number a, there exists a unique number $-a$ such that:

In symbols: $a + (-a) = 0$
In words: Opposites add to 0.

Multiplicative Inverse Property
For every real number a, except 0, there exists a unique real number $1/a$ such that:

In symbols : $a\left(\dfrac{1}{a}\right) = 1$

In words :   Reciprocals multiply to 1.

Of all the basic properties listed, the commutative, associative, and distributive properties are the ones we will use most often. They are important because they are used as justifications or reasons for many of the things we will do in the future.

The following example illustrates how we use the properties listed above. Each line contains an algebraic expression that has been changed in some way. The property that justifies the change is written to the right.

▼ **Example 7**

a. $5(x + 3) = 5x + 15$ Distributive property

b. $7(1) = 7$ 1 is the identity element for
                                             multiplication.

c. $11 + 5 = 5 + 11$ Commutative property of
                                             addition

d. $11 \cdot 5 = 5 \cdot 11$ Commutative property of
                                             multiplication

e. $4 + (-4) = 0$ Additive inverse property

f. $3 + (x + 1) = (3 + x) + 1$ Associative property of addition

g. $6(\frac{1}{6}) = 1$ Multiplicative inverse property

h. $3 + 9 = 9 + 3$ Commutative property of addition

i. $(2 + 5) + x = 5 + (2 + x)$ Commutative and associative
                                             properties

j. $(1 + y) + 3 = 3 + (1 + y)$ Commutative property of
                                             addition

k. $(5 + 0) + 2 = 5 + 2$ 0 is the identity element for
                                             addition.   ▲

Notice in part i of Example 7 that both the order and grouping have changed from the left to the right expression, so both the commutative and associative properties were used.

In part j the only change is in the order of the numbers in the expression, not in the grouping, so only the commutative property was used.

Certainly there are other properties of numbers not contained in the list. For example, we know that multiplying by 0 always gives us 0. This property of 0 is not listed with the other basic properties because it is

implied by the properties listed. That is, we can (but we won't) prove it is true using the properties already listed. Remember, we want to keep our list of assumptions as short as possible. We will call any additional properties *theorems*.

THEOREM 1.1 For any real number a,

$$a(0) = 0$$

As a final note on the properties of real numbers we should mention that although some of the properties are stated for only two or three real numbers, they hold for as many numbers as needed. For example, the distributive property holds for expressions like $3(x + y + z + 5 + 2)$. That is,

$$3(x + y + z + 5 + 2) = 3x + 3y + 3z + 15 + 6$$

It is not important how many numbers are contained in the sum, only that it is a sum. Multiplication, you see, distributes over addition, whether there are two numbers in the sum or two hundred.

Problem Set 1.4

Use the associative property to rewrite each of the following expressions, and then simplify the result.

1. $4 + (2 + x)$
2. $6 + (5 + 3x)$
3. $(a + 3) + 5$
4. $(4a + 5) + 7$
5. $5(3y)$
6. $7(4y)$
7. $\frac{1}{3}(3x)$
8. $\frac{1}{5}(5x)$
9. $4(\frac{1}{4}a)$
10. $7(\frac{1}{7}a)$
11. $\frac{2}{3}(\frac{3}{2}x)$
12. $\frac{4}{3}(\frac{3}{4}x)$

Apply the distributive property to each expression. Simplify when possible.

13. $3(x + 6)$
14. $5(x + 9)$
15. $2(6x + 4)$
16. $3(7x + 8)$
17. $5(3a + 2b)$
18. $7(2a + 3b)$
19. $4(7 + 3y)$
20. $8(6 + 2y)$
21. $(5x + 1)8$
22. $(6x + 1)7$
23. $6(3x + 2 + 4y)$
24. $2(5x + 1 + 3y)$

Use the distributive property to combine similar terms.

25. $5x + 8x$
26. $7x + 4x$
27. $8y + 2y + 6y$
28. $9y + 3y + 4y$
29. $6a + a + 2a$
30. $a + 3a + 4a$

Use the distributive property to add the following fractions.

31. $\dfrac{3}{7} + \dfrac{1}{7} + \dfrac{2}{7}$

32. $\dfrac{3}{8} + \dfrac{1}{8} + \dfrac{1}{8}$

33. $\dfrac{4}{\sqrt{3}} + \dfrac{5}{\sqrt{3}}$

34. $\dfrac{1}{\sqrt{5}} + \dfrac{8}{\sqrt{5}}$

35. $\dfrac{4}{x} + \dfrac{7}{x}$

36. $\dfrac{5}{y} + \dfrac{9}{y}$

37. $\dfrac{3}{a} + \dfrac{5}{a} + \dfrac{1}{a}$

38. $\dfrac{4}{x} + \dfrac{1}{x} + \dfrac{3}{x}$

Use the commutative, associative, and distributive properties to simplify the following.

39. $3x + 5 + 4x + 2$

40. $5x + 1 + 7x + 8$

41. $x + 3 + 4x + 9$

42. $5x + 2 + x + 10$

43. $5a + 7 + 8a + a$

44. $6a + 4 + a + 4a$

45. $3y + y + 5 + 2y + 1$

46. $4y + 2y + 3 + y + 7$

47. $x + 1 + x + 2 + x + 3$

48. $5 + x + 6 + x + 7 + x$

Each of the following problems has a mistake in it. Correct the right-hand side.

49. $5(2x + 4) = 10x + 4$

50. $7(x + 8) = 7x + 15$

51. $3x + 4x = 7(2x)$

52. $3x + 4x = 7x^2$

53. $\frac{3}{5} + \frac{1}{5} = \frac{4}{10}$

54. $\frac{5}{9} + \frac{2}{9} = \frac{7}{18}$

Identify the property of real numbers that justifies each of the following.

55. $3 + 2 = 2 + 3$

56. $3(ab) = (3a)b$

57. $5x = x5$

58. $2 + 0 = 2$

59. $4 + (-4) = 0$

60. $1(6) = 6$

61. $x + (y + 2) = (y + 2) + x$

62. $(a + 3) + 4 = a + (3 + 4)$

63. $4(5 \cdot 7) = 5(4 \cdot 7)$

64. $6(xy) = (xy)6$

65. $4 + (x + y) = (4 + y) + x$

66. $(r + 7) + s = (r + s) + 7$

67. $3(4x + 2) = 12x + 6$

68. $5(\frac{1}{5}) = 1$

Use the given property to complete each of the following.

69. $5y = ?$ Commutative property

70. $4 + 0 = ?$ Additive identity property

71. $3 + a = ?$ Commutative property

72. $5(x + y) = ?$ Distributive property

73. $7(1) = ?$ Multiplicative identity property

74. $6(\frac{1}{6}) = ?$ Multiplicative inverse property

75. $2 + (x + 6) = ?$ Associative property

76. $5(x \cdot 3) = ?$ Associative property

77. $(x + 2)y = ?$ Distributive property

78. $7a + 7b + 7c = ?$ Distributive property

79. $11 \cdot 1 = ?$ Multiplicative identity property

80. $-6 + 6 = ?$ Additive identity property

81. Show that the statement $5x - 5 = x$ is not correct by replacing x with 4 and simplifying both sides.

82. Show that the statement $8x - x = 8$ is not correct by replacing x with 5 and simplifying both sides.

83. Simplify the expressions $15 - (8 - 2)$ and $(15 - 8) - 2$ to show that subtraction is not an associative operation.

84. Simplify the expression $(48 \div 6) \div 2$ and the expression $48 \div (6 \div 2)$ to show that division is not an associative operation.

85. Suppose we defined a new operation with numbers this way: $a * b = ab + a$. (For example, $3 * 5 = 3 \cdot 5 + 3 = 15 + 3 = 18$.) Is the operation $*$ a commutative operation?

86. Is the operation defined by $a \nabla b = aa + bb$ a commutative operation?

1.5
Addition and
Subtraction of
Real Numbers

The purpose of this section is to review the rules for addition and subtraction of real numbers and the justification for those rules. The goal here is the ability to add and subtract positive and negative real numbers quickly and accurately, the latter being the more important.

Addition of Real Numbers

We can justify the rules for addition of real numbers geometrically by use of the real number line. Since real numbers can be thought of as having both a distance from 0 (absolute value) and a direction from 0 (positive or negative), we can visualize addition of two numbers as follows.

Consider the sum of -5 and 3.

$$-5 + 3$$

We can interpret this expression as meaning "start at the origin and move 5 units in the negative direction and then 3 units in the positive direction." With the aid of a number line we can visualize the process:

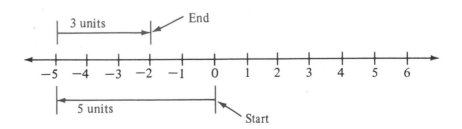

Since the process ends at -2, we say the sum of -5 and 3 is -2.

$$-5 + 3 = -2$$

We can use the real number line in this way to add any combination of positive and negative numbers.

The sum of -4 and -2, $-4 + (-2)$, can be interpreted as starting at the origin, moving 4 units in the negative direction, and then 2 more units in the negative direction:

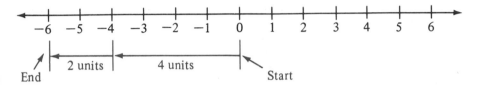

Since the process ends at -6, we say the sum of -4 and -2 is -6.

$$-4 + (-2) = -6$$

We can eliminate actually drawing a number line by simply visualizing it mentally. The following example gives the results of all possible sums of positive and negative 5 and 7.

▼ **Example 1**   Add all possible combinations of positive and negative 5 and 7.

$$5 + 7 = 12$$
$$-5 + 7 = 2$$
$$5 + (-7) = -2$$
$$-5 + (-7) = -12$$ ▲

Looking closely at the relationships in Example 1 (and trying other similar examples if necessary), we can arrive at the following rule for adding two real numbers.

RULE   To add two real numbers with

a.   the *same* sign: simply add absolute values and use the common sign. If both numbers are positive, the answer is positive. If both numbers are negative, the answer is negative.

b.   *different* signs: subtract the smaller absolute value from the larger. The answer will have the sign of the number with the larger absolute value.

Here are other examples of addition of real numbers:

▼ **Example 2**

a. $-2 + (-3) + (-4) = -5 + (-4)$
$= -9$

b. $-3 + 5 + (-7) = 2 + (-7)$
$= -5$

c. $-6 + (-3 + 5) + 4 = -6 + 2 + 4$
$= -4 + 4$
$= 0$ ▲

Subtraction of Real Numbers

In order to have as few rules as possible, we will not attempt to list new rules for the difference of two real numbers. We will define subtraction in terms of addition and apply the rule for addition.

DEFINITION (SUBTRACTION)   If a and b are any two real numbers, then the difference of a and b is

$$\underbrace{a - b} \quad = \quad \underbrace{a + (-b)}$$

To subtract b,   add the opposite of b.

We define the process of subtracting b from a to be equivalent to adding the opposite of b to a. In short, we say, "subtraction is addition of the opposite."
Here is how it works:

▼ **Example 3**   Subtract.

a. $5 - 3 = 5 + (-3) = 2$ Subtracting 3 is equivalent to adding -3.

b. $-7 - 6 = -7 + (-6) = -13$ Subtracting 6 is equivalent to adding -6.

c. $9 - (-2) = 9 + 2 = 11$ Subtracting -2 is equivalent to adding 2.

d. $-6 - (-5) = -6 + 5 = -1$ Subtracting -5 is equivalent to adding 5.   ▲

The following example involves combinations of sums and differences. Generally, the differences are first changed to appropriate sums; the additions are then performed left to right.

▼ **Example 4** Perform the indicated operations.

a. $9 - 5 + 2 = 9 + (-5) + 2$
$$= 4 + 2$$
$$= 6$$

b. $6 - (-3) + 2 = 6 + 3 + 2$
$$= 9 + 2$$
$$= 11$$

c. $-4 - 2 - (-5) = -4 + (-2) + 5$
$$= -6 + 5$$
$$= -1$$

d. $10 - |-3| + |-2| = 10 - 3 + 2$
$$= 10 + (-3) + 2$$
$$= 9$$

e. $-8 - (-3) - |-7| = -8 - (-3) - 7$
$$= -8 + 3 + (-7)$$
$$= -12$$ ▲

Since subtraction is defined in terms of addition, we can state the distributive property in terms of subtraction. That is, if a, b, and c are real numbers, then

$$a(b - c) = a(b) - a(c)$$

Here are some examples that use the distributive property and the rules for addition and subtraction of real numbers.

▼ **Example 5** Apply the distributive property.

a. $3(x - 4) = 3x - 12$
b. $2(y - 6) = 2y - 12$
c. $5(a - 3) = 5a - 15$
d. $3(x - 2 + y) = 3x - 6 + 3y$ ▲

We can also use the distributive property to combine similar terms as we did in Section 1.4.

▼ **Example 6** Combine similar terms.

a. $8x - 2x = (8 - 2)x$ Distributive property
$$= 6x$$ Subtraction
b. $-4x - 7x = (-4 - 7)x$ Distributive property
$$= -11x$$ Subtraction

c. $5y - y = (5 - 1)y$ Distributive property
    $= 4y$ Subtraction
    (Remember, $y = 1y$. Multiplying by 1 leaves y unchanged.)
d. $-9a + 7a = (-9 + 7)a$ Distributive property
    $= -2a$ Addition
e. $-7x + 5x - 8x = (-7 + 5 - 8)x$ Distributive property
    $= -10x$ Addition and subtraction
                                                                    ▲

We can extend the idea of combining similar terms as shown in Example 6 to some slightly more complicated expressions. Suppose we want to simplify the expression $7x + 4 - 3x$. We can first change the subtraction to addition of the opposite, and then change the order of the terms since addition is a commutative operation.

$7x + 4 - 3x = 7x + 4 + (-3x)$ Definition of subtraction
$\qquad\qquad = 7x + (-3x) + 4$ Commutative property
$\qquad\qquad = 7x - 3x + 4$ Definition of subtraction
$\qquad\qquad = 4x + 4$ Combine similar terms

In actual practice you will not show all the steps we have shown above. They are shown here simply so you can see that subtraction can be written in terms of addition, and then the order of the terms rearranged because addition is commutative. The point is, if you move the $3x$ term to another position, you have to take the negative (or subtraction) sign with it. Here are some examples that show only a few of the steps.

▼ **Example 7** Simplify by combining similar terms.

a. $8x - 4 - 5x + 9 = 8x - 5x - 4 + 9$
    $= 3x + 5$
b. $7x - x + 4 - 2x = 7x - x - 2x + 4$
    $= 4x + 4$
c. $x - 7 + 4 - 6x = x - 6x - 7 + 4$
    $= -5x - 3$ ▲

Problem Set 1.5   Find each of the following sums.

1. $6 + (-2)$ 2. $11 + (-5)$
3. $-6 + 2$ 4. $-11 + 5$
5. $-6 + (-2)$ 6. $-11 + (-5)$
7. $-3 + 5 + (-7)$ 8. $-1 + 4 + (-6)$

9. $3 + (-5) - (-7)$
10. $1 + (-4) + (-7)$
11. $-9 + (-3) + (-1)$
12. $-10 + (-5) + (-2)$
13. $-3 + (-8 + 1) + (-4)$
14. $-2 + (-9 + 2) + (-5)$

Find each of the following differences.

15. $7 - 3$
16. $6 - 9$
17. $-7 - 3$
18. $-6 - 9$
19. $-7 - (-3)$
20. $-6 - (-9)$
21. $7 - (-3)$
22. $6 - (-9)$
23. $-15 - 20$
24. $-11 - 15$
25. $12 - (-4)$
26. $5 - (-2)$
27. $-8 - (-11)$
28. $-4 - (-12)$

Perform the indicated operations.

29. $3 + (-2) - 6$
30. $8 + (-3) - 5$
31. $-4 - 3 + 8$
32. $-9 - 5 + 7$
33. $6 - (-2) + 11$
34. $8 - (-3) + 12$
35. $-8 - (-3) - 9$
36. $-1 - (-2) - 3$
37. $4 - (-5) - (-1)$
38. $7 - (-2) - (-6)$
39. $|-2| + |-3| - |-6|$
40. $|-5| + |-2| - |-7|$
41. $|-2| + |-3| - (-6)$
42. $|-5| + |-2| - (-7)$
43. $|-4| - (-2) - |-10|$
44. $|-2| - (-3) - |-12|$
45. $9 - (-2) - 5 - 6 + (-3)$
46. $11 - (-6) - 2 - 4 + (-3)$
47. $|-2| - |-3| - (-8) + |-(-1)|$
48. $|-7| - |-2| - (-4) + |-(-6)|$
49. $-14 - (-20) + |-26| - |-16|$
50. $-11 - (-14) + |-17| - |-20|$

Apply the distributive property.

51. $2(x - 4)$
52. $7(x - 8)$
53. $5(y - 3)$
54. $6(y - 1)$
55. $7(x - 4)$
56. $3(x - 7)$
57. $4(a - 2 + b)$
58. $5(a - 5 + b)$
59. $5(x - y - 4)$
60. $7(a - b - 2)$

Combine similar terms.

61. $5x - 2x$
62. $8x - 2x$
63. $-4x - 9x$
64. $-7x - 10x$
65. $5y - y$
66. $3y - y$
67. $9a - 8a$
68. $3a - 2a$
69. $5x - 3x - 8x$
70. $7x - 3x - 5x$
71. $-4a - 9a + 2a$
72. $-7a - 8a + a$
73. $-3x - x + 6x$
74. $-4x + x - 7x$

Simplify each expression.

75. $3x - 5 + 4x - 9$
76. $7x - 4 - 5x + 8$
77. $8x - 3x + 7 - 4x$
78. $7x - 4x + 3 - 2x$
79. $3a + 2 - a - 5a$
80. $5a + 5 - a - 6a$
81. $6y + 4 - 5y - y + 9$
82. $5y - 3 - 4y - y + 2$
83. $-5t - 3 + 8t + 2 - t$
84. $-9t - 5 + 7t + 3 - t$
85. Subtract 5 from -3.
86. Subtract -3 from 5.
87. Subtract $4x$ from $-3x$.
88. Subtract $-5x$ from $7x$.
89. What number do you subtract from 5 to get -8?
90. What number do you subtract from -3 to get 9?
91. Subtract $3a$ from the sum of $8a$ and a.
92. Subtract $-3a$ from the sum of $3a$ and $5a$.
93. Find five times the sum of $3x$ and -4.
94. Find six times the sum of $3x$ and -5.
95. ＊Add 7 to the difference of $3y$ and -1.
96. Subtract $4a$ from the sum of $-5a$ and 4.

1.6
Multiplication, Division, and Order of Operation for Real Numbers
Multiplication of Real Numbers

Multiplication with whole numbers is simply a shorthand way of writing repeated addition. That is, the product 3(2) can be interpreted as the sum of three 2s:

$$3(2) = 2 + 2 + 2$$

Although this definition of multiplication does not hold for other kinds of numbers, such as fractions, it does, however, give us ways of interpreting products of positive and negative numbers. For example, $3(-2)$ can be evaluated as follows:

$$3(-2) = -2 + (-2) + (-2)$$
$$= -6$$

We can evaluate the product $-3(2)$ in a similar manner if we first apply the commutative property of multiplication.

$$-3(2) = 2(-3) \qquad \text{Commutative property}$$
$$= -3 + (-3) \quad \text{Repeated Addition}$$
$$= -6$$

From these results it seems reasonable to say that the product of a positive and a negative is a negative number.

The last case we must consider is the product of two negative numbers. For example:

$$-3(-2)$$

To evaluate this product we will look at the expression $-3[2 + (-2)]$ in two different ways. First, since $2 + (-2) = 0$, we have

$$-3[2 + (-2)] = -3(0) = 0$$

So we know this expression is equal to 0. On the other hand, we can apply the distributive property to get

$$-3[2 + (-2)] = -3(2) + (-3)(-2)$$
$$= -6 + ?$$

Since we know the expression is equal to 0, it must be true that our ? is 6, since 6 is the only number we can add to -6 to get 0. Therefore, we have:

$$-3(-2) = 6$$

Here is a summary of what we have so far.

$$3(2) = 6$$
$$3(-2) = -6$$
$$-3(2) = -6$$
$$-3(-2) = 6$$

This discussion justifies writing the following rule for multiplication of real numbers.

RULE To multiply two real numbers, simply multiply their absolute values. The product is

a. *positive* if both numbers have the same sign; that is, both are $+$ or both are $-$; or

b. *negative* if the two numbers have opposite signs; that is, one $+$ and the other $-$.

The following example illustrates this rule.

▼ **Example 1**

a. $(7)(3) = 21$
b. $(7)(-3) = -21$
c. $(-7)(3) = -21$
d. $(-7)(-3) = 21$
e. $(4)(-3)(-2) = -12(-2)$
$$= 24$$
f. $(5)(-3 \cdot 2) = 5(-6)$
$$= -30$$

g. $-4(5x) = (-4 \cdot 5)x$ Associative property
      $= -20x$ Multiplication
h. $-2(7a) = (-2 \cdot 7)a$ Associative property
      $= -14a$ Multiplication
i. $-3(x + 4) = -3(x) + (-3)4$ Distributive property
          $= -3x - 12$ Multiplication
j. $-2(3a + 5) = -2(3a) + (-2)(5)$ Distributive property
          $= -6a - 10$ Multiplication   ▲

Division of Real   In order to have as few rules as possible in building our system of
Numbers   algebra, we will now define division for two real numbers in terms of
                     multiplication.

DEFINITION If a and b are any two real numbers, where $b \neq 0$, then

$$\frac{a}{b} = a \cdot \left(\frac{1}{b}\right)$$

Dividing a by b is equivalent to multiplying a by the reciprocal of b.
In short, we say, "division is multiplication by the reciprocal."
    Since division is defined in terms of multiplication, the same rules
hold for assigning the correct sign to a quotient as held for assigning the
correct sign to a product. That is, *the quotient of two numbers with like
signs is positive, while the quotient of two numbers with unlike signs is
negative.*

▼ **Example 2** Divide.

a. $\dfrac{6}{3} = 6 \cdot \left(\dfrac{1}{3}\right) = 2$

b. $\dfrac{6}{-3} = 6 \cdot \left(-\dfrac{1}{3}\right) = -2$

c. $\dfrac{-6}{3} = -6 \cdot \left(\dfrac{1}{3}\right) = -2$

d. $\dfrac{-6}{-3} = -6 \cdot \left(\dfrac{1}{-3}\right) = 2$

Notice these examples
indicate that if a and b are
positive real numbers then

$$\frac{-a}{b} = \frac{a}{-b} = -\frac{a}{b}$$

and

$$\frac{-a}{-b} = \frac{a}{b} \qquad ▲$$

    The second step in the above examples is written only to show that
each quotient can be written as a product. It is not actually necessary to
show this step when working problems.

We can also use the definition of division to review division with fractions. Since the definition of division was stated for real numbers in general, it applies to fractions as well as integers. That is, to divide by a fraction, we simply multiply by its reciprocal.

Review of Division with Fractions

▼ **Example 3** Divide $\frac{3}{5} \div \frac{2}{7}$.

Solution To divide by $\frac{2}{7}$, we multiply by its reciprocal, $\frac{7}{2}$.

$$\frac{3}{5} \div \frac{2}{7} = \frac{3}{5} \cdot \frac{7}{2}$$

$$= \frac{21}{10} \qquad \blacktriangle$$

The next examples of division with fractions include reducing to lowest terms. Recall that to reduce a fraction to lowest terms we divide the numerator and denominator by the largest number that divides both of them.

▼ **Example 4** Divide and reduce to lowest terms.

a. $\dfrac{3}{4} \div \dfrac{6}{11} = \dfrac{3}{4} \cdot \dfrac{11}{6}$ Definition of division

$\qquad = \dfrac{33}{24}$ Multiply numerators, multiply denominators

$\qquad = \dfrac{11}{8}$ Divide numerator and denominator by 3

b. $-7 \div \dfrac{1}{3} = -7 \cdot \dfrac{3}{1}$ Definition of division

$\qquad = -21$

c. $10 \div \dfrac{5}{6} = \dfrac{10}{1} \cdot \dfrac{6}{5}$ Definition of division

$\qquad = \dfrac{60}{5}$ Multiply numerators, multiply denominators

$\qquad = 12$ Divide

d. $-\dfrac{3}{8} \div 6 = -\dfrac{3}{8} \cdot \dfrac{1}{6}$ Definition of division

$\qquad = -\dfrac{3}{48}$ Multiply numerators, multiply denominators

$$= -\frac{1}{16} \qquad \text{Divide numerator and denominator by 3} \qquad \blacktriangle$$

Order of Operation

It is important when evaluating arithmetic expressions in mathematics that each expression have only one answer in reduced form. Consider the expression

$$3 \cdot 7 + 2$$

If we find the product of 3 and 7 first, then add 2, the answer is 23. On the other hand, if we first combine the 7 and 2, then multiply by 3, we have 27. The problem seems to have two distinct answers depending on whether we multiply first or add first. To avoid this situation we will decide that multiplication in a situation like this will always be done before addition. In this case, only the first answer, 23, is correct.

Here is the complete set of rules for evaluating expressions. It is intended to avoid the type of confusion found in the illustration above:

RULE (ORDER OF OPERATION) When evaluating a mathematical expression, we will perform the operations in the following order.

1. Perform operations inside the innermost parentheses first if possible.
2. Then do all multiplications and divisions left to right.
3. Perform all additions and subtractions left to right.

Here are some more complicated examples using combinations of the four basic operations:

▼   **Example 5** Simplify as much as possible:

a. $\dfrac{5(-3) - 10}{-4 - 1} = \dfrac{-15 - 10}{-4 - 1}$

$$= \frac{-25}{-5}$$

$$= 5$$

Notice that the division rule (the line used to separate the numerator from the denominator) is treated like parentheses. It serves to group the numbers on top separately from the numbers on the bottom.

b. $3 - 5(4 - 7) - (-3) = 3 - 5(-3) + 3$
$$= 3 + 15 + 3$$
$$= 21$$

c. $\dfrac{-5(-4) + 2(-3)}{2(-1) - 5} = \dfrac{20 - 6}{-2 - 5}$

$= \dfrac{14}{-7}$

$= -2$

d. $\dfrac{3(-2 + 7) - (-5)}{4 - 2(-3)} = \dfrac{3(5) + 5}{4 + 6}$

$= \dfrac{15 + 5}{10}$

$= \dfrac{20}{10}$

$= 2$ ▲

In the next examples we apply the rule for order of operation to expressions that contain parentheses and variables.

▼ **Example 6** Simplify $3(2y - 1) + y$.

Solution We begin by multiplying the 3 and $2y - 1$. Then we combine similar terms.

$3(2y - 1) + y = 6y - 3 + y$ Distributive property
$\qquad\qquad\quad = 7y - 3$ Combine similar terms   ▲

▼ **Example 7** Simplify $8 - 3(4x - 2) + 5x$

Solution First we distribute the -3 across the $4x - 2$. Then we combine similar terms.

$8 - 3(4x - 2) + 5x = 8 - 12x + 6 + 5x$
$\qquad\qquad\qquad\qquad = -7x + 14$ ▲

▼ **Example 8** Simplify $5(2a + 3) - (6a - 4)$

Solution We begin by applying the distributive property to remove the parentheses. The expression $-(6a - 4)$ can be thought of as $-1(6a - 4)$. Thinking of it in this way allows us to apply the distributive property.

$-1(6a - 4) = -1(6a) - (-1)(4) = -6a + 4$

Here is the complete problem:

$$5(2a + 3) - (6a - 4) = 10a + 15 - 6a + 4 \qquad \text{Distributive property}$$
$$= 4a + 19 \qquad \text{Combine similar terms} \quad \blacktriangle$$

Problem Set 1.6

Find the following products:

1. $3(-5)$
2. $-3(5)$
3. $-3(-5)$
4. $4(-6)$
5. $-8(3)$
6. $-7(-6)$
7. $-5(-4)$
8. $-4(0)$
9. $-2(-1)(-6)$
10. $-3(-2)(5)$
11. $2(-3)(4)$
12. $-2(3)(-4)$
13. $-1(-2)(-3)(4)$
14. $-3(-2)(1)(4)$
15. $-2(4)(-3)(1)$
16. $-5(6)(-3)(-2)$
17. $-2(5x)$
18. $-5(4x)$
19. $-7(3a)$
20. $-6(5a)$
21. $4(-8y)$
22. $6(-2y)$
23. $-3(-5x)$
24. $-2(-9x)$
25. $-5(x + 8)$
26. $-7(x + 4)$
27. $-2(4x + 3)$
28. $-6(2x + 1)$
29. $-6(2x - 5)$
30. $-7(3x - 2)$

Use the definition of division to write each division problem as a multiplication problem, then simplify.

31. $\dfrac{8}{-4}$
32. $\dfrac{-8}{4}$

33. $\dfrac{-8}{-4}$
34. $\dfrac{-12}{-4}$

35. $-\dfrac{3}{4} \div \dfrac{9}{8}$
36. $-\dfrac{2}{3} \div \dfrac{4}{9}$

37. $-8 \div \left(-\dfrac{1}{4}\right)$
38. $-12 \div \left(-\dfrac{2}{3}\right)$

39. $-40 \div \left(-\dfrac{3}{8}\right)$
40. $-30 \div \left(-\dfrac{5}{6}\right)$

41. $\dfrac{4}{9} \div (-8)$ **42.** $\dfrac{3}{7} \div (-6)$

43. $-\dfrac{7}{12} \div \left(-\dfrac{21}{48}\right)$ **44.** $\dfrac{-9}{10} \div \left(-\dfrac{27}{40}\right)$

Simplify each expression as much as possible.

45. $3(-4) - 2$ **46.** $-3(-4) - 2$
47. $5(-2) - (-3)$ **48.** $-8(-11) - (-1)$
49. $4(-3) - 6(-5)$ **50.** $-6(-3) - 5(-7)$
51. $-8(4) - (-6)(-2)$ **52.** $9(-1) - 4(-3)$
53. $2 - 4[3 - 5(-1)]$ **54.** $6 - 5[2 - 4(-8)]$
55. $(8 - 7)[4 - 7(-2)]$ **56.** $(6 - 9)[15 - 3(-4)]$

57. $\dfrac{6(-2) - 8}{-15 - (-10)}$ **58.** $\dfrac{8(-3) - 6}{-7 - (-2)}$

59. $\dfrac{3(-1) - 4(-2)}{8 - 5}$ **60.** $\dfrac{6(-4) - 5(-2)}{7 - 6}$

61. $8 - (-6)\left[\dfrac{2(-3) - 5(4)}{-8(6) - 4}\right]$ **62.** $-9 - 5\left[\dfrac{11(-1) - 9}{4(-3) - 2(5)}\right]$

63. $6 - (-3)\left[\dfrac{2 - 4(3 - 8)}{6 - 5(1 - 3)}\right]$ **64.** $8 - (-7)\left[\dfrac{6 - 1(6 - 10)}{4 - 3(5 - 7)}\right]$

Simplify each expression.

65. $3(5x + 4) - x$ **66.** $4(7x + 3) - x$
67. $3(2a - 4) - 7a$ **68.** $-2(3a - 2) - 7a$
69. $7 + 3(x + 2)$ **70.** $5 + 2(3x - 4)$
71. $6 - 7(m - 3)$ **72.** $3 - 5(2m - 5)$
73. $7 - 2(3x - 1) + 4x$ **74.** $8 - 5(2x - 3) + 4x$
75. $5(y + 2) - 4(y + 1)$ **76.** $6(y - 3) - 5(y + 2)$
77. $5(3y + 1) - (8y - 5)$ **78.** $4(6y + 3) - (6y - 6)$
79. $10 - 4(2x + 1) - (3x - 4)$
80. $7 - 2(3x + 5) - (2x - 3)$
81. Subtract -5 from the product of 12 and $-\frac{2}{3}$.
82. Subtract -3 from the product of -12 and $\frac{3}{4}$.
83. Add -5 to the quotient of -3 and $\frac{1}{2}$.
84. Add -7 to the quotient of 6 and $-\frac{1}{2}$.
85. Add $8x$ to the product of -2 and $3x$.
86. Add $7x$ to the product of -5 and $-2x$.

Examples: The margins of the chapter summaries will be used for brief examples of the topics being reviewed, whenever it is convenient.

Chapter 1   Summary and Review

The number(s) in brackets next to each heading indicates the section(s) in which that topic is discussed.

SYMBOLS [1.1]

$a = b$	a is equal to b
$a \neq b$	a is not equal to b
$a < b$	a is less than b
$a \leq b$	a is less than or equal to b
$a \geq b$	a is greater than or equal to b
$a > b$	a is greater than b
$a \ngtr b$	a is not greater than b
$a \nless b$	a is not less than b
$a + b$	the sum of a and b
$a - b$	the difference of a and b
$a \cdot b$	the product of a and b
a/b	the quotient of a and b

1. If $A = \{0, 1, 2\}$ and
  $B = \{2, 3\}$ then
  $A \cup B = \{0, 1, 2, 3\}$ and
  $A \cap B = \{2\}$

SETS [1.1]

A *set* is any well-defined collection of objects or things.

The *union* of two sets A and B, written $A \cup B$, is all the elements that are in A *or* are in B, *or* are in both A and B.

The *intersection* of two sets A and B, written $A \cap B$, is the set consisting of all elements common to both A *and* B.

Set A is a *subset* of set B, written $A \subset B$, if all elements in set A are also in set B.

SPECIAL SETS [1.2]

Counting numbers $= \{1, 2, 3, \ldots\}$
Whole numbers $= \{0, 1, 2, 3, \ldots\}$
Integers $= \{\ldots -3, -2, -1, 0, 1, 2, 3, \ldots\}$
Rational numbers $= \{\frac{a}{b} | a$ and b are integers, $b \neq 0\}$
Irrational numbers $= \{x | x$ is a nonrepeating, nonterminating decimal$\}$
Real numbers $= \{x | x$ is rational or x is irrational$\}$

OPPOSITES [1.2, 1.4]

Any two real numbers the same distance from 0 on the number line, but in opposite directions from 0, are called *opposites* or *additive inverses*. Opposites always add to 0.

2. The numbers 5 and -5 are opposites; their sum is 0.

$$5 + (-5) = 0$$

RECIPROCALS [1.2, 1.4]

Any two real numbers whose product is 1 are called *reciprocals*. Every real number has a reciprocal except 0.

3. The numbers 3 and $\frac{1}{3}$ are reciprocals; their product is 1.

$$3\left(\frac{1}{3}\right) = 1$$

ABSOLUTE VALUE [1.2]

The *absolute value* of a real number is its distance from 0 on the number line. If $|x|$ represents the absolute value of x, then

$$|x| = \begin{cases} x \text{ if } x \geq 0 \\ -x \text{ if } x < 0 \end{cases}$$

The absolute value of a real number is never negative.

4. $\quad |5| = 5$
$\quad\quad |-5| = 5$

PROPERTIES OF REAL NUMBERS [1.4]

	For Addition	*For Multiplication*
Commutative	$a+b=b+a$	$a \cdot b = b \cdot a$
Associative	$a+(b+c)=(a+b)+c$	$a \cdot (b \cdot c) = (a \cdot b) \cdot c$
Identity	$a+0=a$	$a \cdot 1 = a$
Inverse	$a+(-a)=0$	$a\left(\dfrac{1}{a}\right) = 1$
Distributive		$a(b+c)=ab+ac$

ADDITION [1.5]

To add two real numbers with

1. *the same sign:* simply add absolute values and use the common sign.

2. *different signs:* subtract the smaller absolute value from the larger absolute value. The answer has the same sign as the number with the larger absolute value.

5. $\quad\quad 5 + 3 = 8$
$\quad\quad 5 + (-3) = 2$
$\quad\quad -5 + 3 = -2$
$\quad\quad -5 + (-3) = -8$

6. $6 - 2 = 6 + (-2) = 4$
            $6 - (-2) = 6 + 2 = 8$

SUBTRACTION [1.5]

If a and b are real numbers,

$$a - b = a + (-b)$$

To subtract b, add the opposite of b.

7. $5(4) = 20$
             $5(-4) = -20$
             $-5(4) = -20$
             $-5(-4) = 20$

MULTIPLICATION [1.6]

To multiply two real numbers simply multiply their absolute values. Like signs give a positive answer. Unlike signs give a negative answer.

8. $\dfrac{12}{-3} = -4$

$\dfrac{-12}{-3} = 4$

DIVISION [1.6]

If a and b are real numbers and $b \neq 0$, then

$$\frac{a}{b} = a \cdot \left(\frac{1}{b}\right)$$

To divide by b, multiply by the reciprocal of b.

9. $6 - 3(4 + 1) = 6 - 3(5)$
                    $= 6 - 15$
                    $= -9$

ORDER OF OPERATION [1.6]

1. Do what is inside the parentheses first.
2. Then perform all multiplications and divisions left to right.
3. Finally, do all additions and subtractions left to right.

COMMON MISTAKES

1. Interpreting absolute value as changing the sign of the number inside the absolute value symbols. That is, $|-5| = +5, |+5| = -5$. To avoid this mistake, remember, absolute value is defined as a distance and distance is always measured in positive units.

2. Confusing $-(-5)$ with $-|-5|$. The first answer is $+5$, while the second answer is -5.

The numbers in brackets indicate the section to which the problems correspond.   **Chapter 1   Test**

Write each of the following in symbols. [1.1]

1. Twice the sum of $3x$ and $4y$.
2. The difference of $2a$ and $3b$ is less than their sum.

If $A = \{1, 2, 3, 4\}$, $B = \{2, 4, 6\}$, and $C = \{1, 3, 5\}$, find: [1.1]

3. $(A \cup B) \cap C$ **4.** $A \cap (B \cup C)$
5. $\{x \mid x \in B \text{ and } x \in C\}$

Give the opposite and reciprocal of each of the following. [1.2]

6. -3 **7.** $\frac{4}{3}$
8. $-\sqrt{5}$

Simplify each of the following. [1.2]

9. $-(-3)$ **10.** $|-4| + |-3|$
11. $-|-2|$

For the set $\{-5, -4.1, -3.75, -\frac{5}{6}, -\sqrt{2}, 0, \sqrt{3}, 1, 1.8, 4\}$, list all the elements belonging to the following sets. [1.2]

12. Integers   **13.** Rational numbers
14. Irrational numbers

Graph each of the following. [1.3]

15. $\{x \mid x > 2\}$ **16.** $\{x \mid x \le -1 \text{ or } x > 5\}$
17. $\{x \mid -2 \le x \le 4\}$

State the property or properties that justify each of the following. [1.4]

18. $4 + x = x + 4$ **19.** $5(1) = 5$
20. $3(x \cdot y) = (3y) \cdot x$ **21.** $(a + 1) + b = (a + b) + 1$

Simplify each of the following as much as possible. [1.5, 1.6, 1.7]

22. $5(-4) + 1$ **23.** $-4(-3) + 2$
24. $-3(5) - 4$ **25.** $12 \div \frac{2}{3} - 4$

26. $-5 - 15(\frac{7}{5})$ **27.** $\dfrac{6(-3) - 2}{-6 - 2}$

28. $\dfrac{-4(-1) - (-10)}{5 - (-2)}$ **29.** $-4\left[\dfrac{-3 - (-6)}{2(-4) - 4}\right]$

30. $3 - 2\left[\dfrac{8(-1) - 5}{-3(2) - 4}\right]$ **31.** $(6 - 5)[4 - 3(2 - 1)]$

32. $6(4x)$ **33.** $-5(2x)$
34. $5(x + 3)$ **35.** $-2(x + 9)$
36. $-3(2x + 4)$ **37.** $-4(3x + 2)$
38. $-2(3x - 5y + 4)$ **39.** $2x + 5x$

40. $-3x + 7x$ **41.** $-9x + 2x + 3x$

42. $-x + 7x - 4x$

43. $8 - 3(2x + 4)$

44. $5(2y - 3) - (6y - 5)$

45. $3 + 4(2x - 5) - 5x$

46. $2 + 5a + 3(2a - 4)$

47. Add $-\frac{2}{3}$ to the product of -2 and $\frac{5}{6}$.

48. Subtract $\frac{3}{4}$ from the product of -4 and $\frac{7}{16}$.

49. Subtract -4 from the quotient of -4 and $-\frac{1}{3}$.

50. Add -7 to the quotient of -10 and -5.

First-Degree Equations and Inequalities

To the student:

One of the best-known mathematical formulas is the formula $E = mc^2$ from Einstein's theory of relativity. Einstein viewed the universe as if everything in it were in one of two states, matter or energy. His theory states that matter and energy are constantly being transformed into one another. The amount of energy (E) that can be obtained from an object with mass m is given by the formula $E = mc^2$, where c is the speed of light. Now as far as we are concerned, the theory behind the formula is not important. What is important is that the formula $E = mc^2$ describes a certain characteristic of the universe. The universe has always had this characteristic. The formula simply gives us a way of stating it in symbols. Once this property of matter and energy has been stated in symbols we can apply any of our mathematical knowledge to it, if we need to. The idea behind all of this is that mathematics can be used to describe the world around us symbolically. Mathematics is the language of science.

In this chapter we will begin our work with equations. The information in this chapter is some of the most important information in the book. You will learn the basic steps used in solving equations. Probably the most useful tool in algebra is the ability to solve first-degree equations in one variable. The methods we develop to solve first-degree

equations in one variable will be used again and again throughout the rest of the book.

In this chapter we will also consider first-degree inequalities in one variable as well as provide a section on formulas and a section on word problems.

A large part of your success in this chapter depends on how well you mastered the concepts from Chapter 1. Here is a list of the more important concepts needed to begin this chapter:

1. You must know how to add, subtract, multiply, and divide positive and negative numbers.
2. You should be familiar with the commutative, associative, and distributive properties.
3. You should understand that opposites add to 0 and reciprocals multiply to 1.
4. You must know the definition of absolute value.

2.1
First-Degree
Equations

In this section we will solve some first-degree equations. A first-degree equation is any equation that can be put in the form

$$ax + b = c$$

where a, b, and c are constants.

Some examples of first-degree equations are

$$5x + 3 = 2 \qquad 2x = 7 \qquad 2x + 5 = 0$$

Each is a first-degree equation because it can be put in the form $ax + b = c$. In the first equation above, $5x$, 3, and 2 are called *terms* of the equation. $5x$ is a variable term; 3 and 2 are constant terms.

DEFINITION   The *solution set* for an equation is the set of all numbers which, when used in place of the variable, make the equation a true statement.

▼   **Example 1** The solution set for $2x - 3 = 9$ is 6, since replacing x with 6 makes the equation a true statement.

$$
\begin{array}{ll}
\text{If} & x = 6 \\
\text{then} & 2x - 3 = 9 \\
\text{becomes} & 2(6) - 3 = 9 \\
& 12 - 3 = 9 \\
& 9 = 9 \leftarrow \text{a true statement}
\end{array}
$$

▲

DEFINITION   Two or more equations with the same solution set are called *equivalent equations*.

▼   **Example 2**   The equations $2x - 5 = 9$, $x - 1 = 6$, and $x = 7$ are all equivalent equations since the solution set for each is $\{7\}$.   ▲

In addition to the properties from Chapter 1, we need two new properties—one for addition or subtraction and one for multiplication or division—to assist us in solving first-degree equations.

The first property states that adding the same quantity to both sides of an equation preserves equality. Or, more importantly, adding the same amount to both sides of an equation *never changes* the solution set. This property is called the *addition property of equality* and is stated in symbols as follows.

Addition Property of Equality

For any three algebraic expressions A, B, and C,

$$\text{If} \qquad A = B$$
$$\text{then } A + C = B + C$$

In words: Adding the same quantity to both sides of an equation will not change the solution.

Our second new property is called the multiplication property of equality and is stated like this.

Multiplication Property of Equality

For any three algebraic expressions A, B, and C, where $C \neq 0$,

$$\text{If} \quad A = B$$
$$\text{then } AC = BC$$

In words: Multiplying both sides of an equation by the same nonzero quantity will not change the solution.

Since subtraction is defined in terms of addition (subtraction is addition of the opposite) and division is defined in terms of multiplication (division by a number gives the same result as multiplication by its reciprocal), we do not need to introduce separate properties for subtraction and division. The solution set for an equation will never be

changed by subtracting the same amount from both sides or by dividing both sides by the same nonzero quantity.

The following examples illustrate how we use the properties from Chapter 1 along with the addition property of equality and the multiplication property of equality to solve first-degree equations.

▼   **Example 3**   Solve for x: $2x - 3 = 9$.

Solution   We begin by using the addition property of equality to add $+3$, the opposite of -3, to both sides of the equation:

$$2x - 3 + 3 = 9 + 3$$
$$2x = 12$$

To get x alone on the left side, we use the multiplication property of equality and multiply both sides by $\frac{1}{2}$, the reciprocal of 2.

$$\tfrac{1}{2}(2x) = \tfrac{1}{2}(12)$$
$$x = 6$$

Since the addition and multiplication properties of equality always produce equations equivalent to the original equations, our last equation, $x = 6$, is equivalent to our first equation, $2x - 3 = 9$. The solution set is, therefore, $\{6\}$.   ▲

▼   **Example 4**   Solve $\frac{3}{4}x + 5 = -4$.

Solution   We begin by adding -5 to both sides of the equation. Once this has been done, we multiply both sides by the reciprocal of $\frac{3}{4}$, which is $\frac{4}{3}$.

$$\frac{3}{4}x + 5 = -4$$

$$\frac{3}{4}x + 5 + (-5) = -4 + (-5) \qquad \text{Add } -5 \text{ to both sides}$$

$$\frac{3}{4}x = -9$$

$$\frac{4}{3}\left(\frac{3}{4}x\right) = \frac{4}{3}(-9) \qquad\qquad \text{Multiply both sides by } \tfrac{4}{3}$$

$$x = -12 \qquad\qquad\qquad \tfrac{4}{3}(-9) = \tfrac{4}{3}(\tfrac{-9}{1}) = \tfrac{-36}{3}$$
$$= -12 \qquad ▲$$

Our next example involves solving an equation that has variable terms on both sides of the equal sign.

▼ **Example 5**   Find the solution set for $3a - 5 = -6a + 1$.

Solution   To solve for a we must isolate it on one side of the equation. Let's decide to isolate a on the left side. To do this we must remove the $-6a$ from the right side. We accomplish this by adding $6a$ to both sides of the equation.

$$3a - 5 = -6a + 1$$
$$3a + 6a - 5 = -6a + 6a + 1 \qquad \text{Add } 6a \text{ to both sides}$$
$$9a - 5 = 1$$
$$9a - 5 + 5 = 1 + 5 \qquad \text{Add 5 to both sides}$$
$$9a = 6$$
$$\frac{1}{9}(9a) = \frac{1}{9}(6) \qquad \text{Multiply both sides by } \tfrac{1}{9}$$
$$a = \frac{2}{3} \qquad\qquad \tfrac{1}{9}(6) = \tfrac{6}{9} = \tfrac{2}{3} \qquad ▲$$

We can check our solution in Example 5 by replacing a in the original equation with $\tfrac{2}{3}$.

When $\qquad\qquad\qquad a = \dfrac{2}{3}$

the equation $\qquad 3a - 5 = -6a + 1$

becomes $\qquad 3\left(\dfrac{2}{3}\right) - 5 = -6\left(\dfrac{2}{3}\right) + 1$

$$2 - 5 = -4 + 1$$
$$-3 = -3 \leftarrow \text{a true statement.}$$

▼ **Example 6**   Solve for x: $7x - x + 4 - 2x = 4$

Solution   We begin by simplifying the left side as much as possible.
$$7x - x + 4 - 2x = 4$$
$$4x + 4 = 4 \qquad\qquad \text{Simplify the left side}$$
$$4x + 4 - 4 = 4 - 4 \qquad \text{Subtract 4 from both sides}$$
$$\qquad\qquad\qquad\qquad\qquad \text{(or, add } -4 \text{ to both sides)}$$
$$4x = 0$$

$$\frac{1}{4}(4x) = \frac{1}{4}(0) \qquad \text{Multiply both sides by } \tfrac{1}{4}$$

$$x = 0 \qquad\qquad\qquad \blacktriangle$$

As the equations become more complicated, it is sometimes helpful to use the following steps as a guide in solving the equations:

Step 1.   Use the distributive property to separate terms.

Step 2.   Use the commutative and associative properties to simplify both sides as much as possible.

Step 3.   Use the addition property of equality to get all the terms containing the variable (variable terms) on one side and all other terms (constant terms) on the other side.

Step 4.   Use the multiplication property of equality to get x alone on one side of the equal sign.

Step 5.   Check your results in the original equation, if necessary.

Let's see how our steps apply by solving another equation

▼ **Example 7**   Solve $3(2y - 1) + y = 5y + 3$.

Solution   We begin by using the distributive property to separate terms.

Step 1
$$3(2y - 1) + y = 5y + 3$$
$$6y - 3 \;+ y = 5y + 3 \qquad\qquad \text{Distributive property.}$$

Step 2 $7y - 3 = 5y + 3 \qquad\qquad 6y + y = 7y$

Step 3
$$7y + (-5y) - 3 = 5y + (-5y) + 3 \qquad \text{Add } -5y \text{ to both sides.}$$
$$2y - 3 = 3$$
$$2y - 3 + 3 = 3 + 3 \qquad\qquad \text{Add } +3 \text{ to both sides.}$$
$$2y = 6$$

Step 4
$$\tfrac{1}{2}(2y) = \tfrac{1}{2}(6) \qquad\qquad \text{Multiply by } \tfrac{1}{2}.$$
$$y = 3$$

The solution set is $\{3\}$.

We should mention here that after step 2 has been completed, there are, at most, four terms left—two variable terms and two constant terms. After step 3, there are two terms left—one variable and one constant term.   ▲

▼ **Example 8** Solve the equation $8 - 3(4x - 2) + 5x = 35$.

Solution We must begin by distributing the -3 across the quantity $4x - 2$. (It would be a mistake to subtract 3 from 8 first, since the rule for order of operation indicates we are to do multiplication before subtraction.)

$Step\ 1$ $\begin{cases} 8 - 3(4x - 2) + 5x = 35 \\ \qquad \downarrow \quad \downarrow \\ 8 - 12x + 6 + 5x = 35 \end{cases}$ $\qquad$ Distributive property

$Step\ 2$ $\qquad\qquad -7x + 14 = 35$

$Step\ 3$ $\begin{cases} -7x + 14 - 14 = 35 - 14 \\ \\ \qquad\qquad -7x = 21 \end{cases}$ $\qquad$ Subtract 14 from both sides.

$Step\ 4$ $\begin{cases} -\dfrac{1}{7}(-7x) = \left(-\dfrac{1}{7}\right)21 \\ \\ \qquad\qquad x = -3 \end{cases}$ Multiply both sides by $-\frac{1}{7}$

$\qquad\qquad\qquad\qquad\qquad\qquad\qquad\qquad\qquad\qquad\qquad\qquad$ ▲

Solve each of the following equations: $\qquad\qquad\qquad$ *Problem Set 2.1*

1. $x - 5 = 3$
2. $x + 2 = 7$
3. $2x - 4 = 6$
4. $3x - 5 = 4$
5. $4a - 1 = 7$
6. $3a - 5 = 10$
7. $3 - y = 10$
8. $5 - 2y = 11$
9. $-3 - 4x = 15$
10. $-8 - 5x = -6$
11. $2x + 5 = -3$
12. $9x + 6 = -12$
13. $-3y + 1 = 5$
14. $-2y + 8 = 3$
15. $-5x - 4 = 16$
16. $-6x - 5 = 11$
17. $3 - 4a = -11$
18. $8 - 2a = -13$
19. $9 + 5a = -2$
20. $3 + 7a = -7$
21. $\frac{2}{3}x = 8$
22. $\frac{3}{2}x = 9$
23. $-\frac{3}{5}a + 2 = 8$
24. $-\frac{5}{3}a + 3 = 23$
25. $6 + \frac{2}{7}y = 8$
26. $4 + \frac{3}{7}y = 1$
27. $9 - \frac{3}{4}t = 12$
28. $3 - \frac{2}{3}t = 1$

Simplify each side of the following equations, then find the solution set:

29. $2x - 5 = 3x + 2$
30. $5x - 1 = 4x + 3$
31. $-3a + 2 = -2a - 1$
32. $-4a - 8 = -3a + 7$
33. $2x - 3 - x = 3x + 5$
34. $3x - 5 - 2x = 2x - 3$
35. $5y - 2 + 4y = 2y + 12$
36. $7y - 3 + 2y = 7y - 9$
37. $11x - 5 + 4x - 2 = 8x$
38. $2x + 7 - 3x + 4 = -2x$

39. $3(x - 2) = 12$ **40.** $4(x + 1) = 12$
41. $2(3k - 5) = k$ **42.** $3(4k - 1) = 9k$
43. $-5(2x + 1) + 5 = 3x$ **44.** $-3(5x + 7) - 4 = -10x$
45. $5(y + 2) - 4(y + 1) = 3$ **46.** $6(y - 3) - 5(y + 2) = 8$
47. $6 - 7(m - 3) = -1$ **48.** $3 - 5(2m - 5) = -2$
49. $4(a - 3) + 5 = 7(3a - 1)$ **50.** $6(a - 4) + 6 = 2(5a + 2)$
51. $7 + 3(x + 2) = 4(x - 1)$ **52.** $5 + 2(3x - 4) = 3(2x - 1)$
53. $7 - 2(3x - 1) + 4x = 5$ **54.** $8 - 5(2x - 3) + 4x = 20$
55. $10 - 4(2x + 1) - (3x - 4) = -9x + 4 - 4x$
56. $7 - 2(3x + 5) - (2x - 3) = -5x + 3 - 2x$
57. The equations you have solved so far have had exactly one solution. Because of the absolute value symbols, the equation $|x + 2| = 5$ has two solutions. One of the solutions is $x = -7$. Without showing any work, what do you think is the other solution?
58. One solution to the equation $|x - 3| = 2$ is $x = 5$. Without showing any work, what is the other solution?
59. Is $x = -3$ a solution to $2 - 4x = -x + 17$?
60. Is $x = -3$ a solution to $2 - 4x = -x + 11$?

The solution set for the equations below is either all real numbers or the empty set (no real numbers are solutions). Solve each equation using the methods developed in Section 2.1, and see if you can tell from the results what the solution set is in each case.

61. $3x - 6 = 3(x + 4)$ **62.** $7x - 14 = 7(x - 2)$
63. $4y + 2 - 3y + 5 = 3 + y + 4$
64. $7y + 5 - 2y - 3 = 6 + 5y - 4$
65. $2(4t - 1) + 3 = 5t + 4 + 3t$
66. $5(2t - 1) + 1 = 2t - 4 + 8t$

Review Problems   From here on, each problem set will end with a series of review problems. In mathematics it is very important to review. The more you review, the better you will understand the topics we cover and the longer you will remember them. Also, there are times when material that seemed confusing earlier will be less confusing the second time around.

The problems below review material we covered in Section 1.3. Reviewing these problems will help you with the next section.

Graph each inequality.

67. $\{x \mid x > -5\}$ **68.** $\{x \mid x \leq 4\}$
69. $\{x \mid x \leq -2 \text{ or } x > 5\}$ **70.** $\{x \mid x < 3 \text{ or } x \geq 5\}$
71. $\{x \mid x > -4 \text{ and } x < 0\}$ **72.** $\{x \mid x \geq 0 \text{ and } x \leq 2\}$
73. $\{x \mid 1 \leq x \leq 4\}$ **74.** $\{x \mid -4 < x < -2\}$

A first-degree inequality is any inequality that can be put in the following form:

$$ax + b < c \qquad (a, b, \text{ and } c \text{ constants, } a \neq 0)$$

where the inequality symbol ($<$) can be replaced with any of the other three inequality symbols ($\leq$, $>$, or $\geq$).

Some examples of first-degree inequalities are

$$3x - 2 \geq 7 \qquad -5y < 25 \qquad 3(x - 4) > 2x$$

Each of these is a first-degree inequality because it can be put in the form $ax + b < c$. They may not be in that form to begin with, but each can be put in the correct form.

Solving first-degree inequalities is similar to solving first-degree equations. We need to develop two new properties to solve inequalities. Here is our first property.

Addition Property for Inequalities

For any algebraic expressions A, B, and C,

$$\text{If} \qquad A < B$$
$$\text{then } A + C < B + C$$

In words: Adding the same quantity to both sides of an inequality will not change the solution set.

Before we state the multiplication property for inequalities, we will take a look at what happens to an inequality statement when we multiply both sides by a positive number and what happens when we multiply by a negative number.

We begin by writing three true inequality statements:

$$3 < 5 \qquad -3 < 5 \qquad -5 < -3$$

We multiply both sides of each inequality by a positive number—say, 4.

$$4(3) < 4(5) \qquad 4(-3) < 4(5) \qquad 4(-5) < 4(-3)$$
$$12 < 20 \qquad\quad -12 < 20 \qquad\quad -20 < -12$$

Notice in each case that the resulting inequality symbol points in the same direction as the original inequality symbol. Multiplying both sides of an inequality by a positive number preserves the *sense* of the inequality.

Let's take the same three original inequalities and multiply both sides by -4.

$$3 < 5 \qquad\qquad -3 < 5 \qquad\qquad -5 < -3$$
$$\downarrow \qquad\qquad\qquad \downarrow \qquad\qquad\qquad\quad \downarrow$$
$$-4(3) > -4(5) \qquad -4(-3) > -4(5) \qquad -4(-5) > -4(-3)$$
$$-12 > -20 \qquad\qquad 12 > -20 \qquad\qquad\quad 20 > 12$$

Notice in this case that the resulting inequality symbol always points in the opposite direction from the original one. Multiplying both sides of an inequality by a negative number *reverses* the sense of the inequality. Keeping this in mind, we will now state the multiplication property for inequalities.

Multiplication Property for Inequalities

Let A, B, and C represent algebraic expressions.

If $A < B$
then $AC < BC$ if C is positive $(C > 0)$
or $AC > BC$ if C is negative $(C < 0)$

In words: Multiplying both sides of an inequality by a positive number always produces an equivalent inequality. Multiplying both sides of an inequality by a negative number reverses the sense of the inequality.

The multiplication property for inequalities states that we can multiply both sides of an inequality by any nonzero number we choose. If that number happens to be *negative,* we must also *reverse* the direction of the inequality.

We have stated both properties using the "less than" ($<$) symbol. The properties also hold for the other three incquality symbols.

Since subtraction is defined as addition of the opposite and division as multiplication by the reciprocal, our two new properties hold for both subtraction and division. (Note too that multiplication or division of both sides of an inequality by a negative number *always* reverses the direction of the inequality.)

We will follow the same basic steps in solving inequalities as we did with equations. With inequalities, we will graph the solution set as an extra step.

▼ **Example 1** Solve $3x - 5 \leq 7$.

Solution

$$3x - 5 \leq 7$$
$$3x - 5 + 5 \leq 7 + 5 \qquad \text{Add } +5 \text{ to both sides.}$$
$$3x \leq 12$$
$$\tfrac{1}{3}(3x) \leq \tfrac{1}{3}(12) \qquad \text{Multiply by } \tfrac{1}{3}.$$
$$x \leq 4$$

The solution set is $\{x \mid x \leq 4\}$, the graph of which is

▲

▼ **Example 2** Find the solution set for $-2y - 3 < 7$.

Solution

$$-2y - 3 < 7$$
$$-2y < 10 \qquad \text{Add } +3 \text{ to both sides.}$$
$$\downarrow$$
$$-\tfrac{1}{2}(-2y) > -\tfrac{1}{2}(10) \qquad \text{Multiply by } -\tfrac{1}{2} \text{ and reverse the}$$
$$y > -5 \qquad\qquad \text{direction of the inequality symbol.}$$

The solution set is $\{y \mid y > -5\}$, the graph of which is

▲

▼ **Example 3** Solve $3(2x - 4) - 7x \leq -3x$.

Solution We begin by using the distributive property to separate terms. Next, simplify both sides:

$$3(2x - 4) - 7x \leq -3x$$
$$6x - 12 - 7x \leq -3x$$
$$-x - 12 \leq -3x \qquad\qquad 6x - 7x = (6 - 7)x = -x$$
$$-12 \leq -2x \qquad\qquad \text{Add } x \text{ to both sides.}$$
$$\downarrow$$
$$-\tfrac{1}{2}(-12) \geq -\tfrac{1}{2}(-2x) \qquad \text{Multiply both sides by } -\tfrac{1}{2} \text{ and}$$
$$6 \geq x \qquad\qquad\qquad \text{reverse the direction of the}$$
$$\text{inequality symbol.}$$

The solution set is $\{x \mid x \le 6\}$, and the graph is

   Notice in Examples 2 and 3 that each time we multiplied both sides of the inequality by a negative number, we also reversed the direction of the inequality. If we fail to reverse the inequality in this situation, the graph of our solution set will be on the wrong side of the end point.

▼   **Example 4**   Solve the continued inequality $-2 \le 5k - 7 \le 13$.

   Solution   We can extend our properties for addition and multiplication to cover this situation. If we add a number to the middle expression, we must add the same number to the outside expressions. If we multiply the center expression by a number, we must do the same to the outside expressions, remembering to reverse the direction of the inequality symbols if we multiply by a negative number.

$$
\begin{aligned}
-2 &\le & 5k - 7 & & \le 13 & \\
-2 + 7 &\le & 5k - 7 + 7 & & \le 13 + 7 & \qquad \text{Add 7 to each expression.} \\
5 &\le & 5k & & \le 20 & \\
\tfrac{1}{5}(5) &\le & \tfrac{1}{5}(5k) & & \le \tfrac{1}{5}(20) & \qquad \text{Multiply each expression by } \tfrac{1}{5}. \\
1 &\le & k & & \le 4 &
\end{aligned}
$$

The graph of the solution set is

▼   **Example 5**   Solve the compound inequality $3x + 1 \le -5$ or $2x - 3 > 7$.

   Solution   We solve each half of the compound inequality separately, then graph the solution set:

$$3x + 1 \le -5 \quad \text{or} \quad 2x - 3 > 7$$

$$3x \leq -6 \quad \text{or} \quad 2x > 10$$
$$x \leq -2 \quad \text{or} \quad x > 5$$

$$\overset{-2}{\qquad} \quad \overset{5}{\qquad}$$

Solve each of the following inequalities and graph each solution set: *Problem Set 2.2*

1. $2x \leq 3$ 2. $5x \geq -15$
3. $-5x \leq 25$ 4. $-7x \geq 35$
5. $-6a < -12$ 6. $-3a \leq -18$
7. $2x + 1 \geq 3$ 8. $5x - 4 < 16$
9. $-3x + 1 > 10$ 10. $-2x - 5 \leq 15$
11. $6 - m \leq 7$ 12. $5 - m > -2$
13. $-3 - 4x \leq 9$ 14. $-2 - 5x < 18$
15. $3 - 2y \geq -4$ 16. $5 - 3y < -2$

Simplify each side first, then solve the following inequalities:

17. $2(3y + 1) \leq -10$ 18. $3(2y - 4) > 0$
19. $2(x - 5) \leq -3x$ 20. $5(2x - 1) < 9x$
21. $-(a + 1) - 4a \leq 2a - 8$ 22. $-(a - 2) - 5a \leq 3a + 7$
23. $4x - 5 + 3x - 2 \leq 2x + 3$ 24. $3x + 7 - 2x + 1 \leq -2x - 4$
25. $2t - 3(5 - t) < 0$ 26. $3t - 4(2t - 5) < 0$
27. $5 - 7(2a + 3) \geq -2$ 28. $3 - 4(3a - 1) > 1$
29. $-3(x + 5) \leq -2(x - 1)$ 30. $-4(2x + 1) \leq -3(x + 2)$
31. $5(y + 3) + 4 < 6y - 1 - 5y$ 32. $4(y - 1) + 2 \geq 3y + 8 - 2y$
33. $3(x + 1) - 2(x - 2) \leq 0$ 34. $5(x + 4) - 3(x + 7) \leq 5$

Solve the following continued inequalities. Be sure to graph the solution sets.

35. $2 \leq m - 5 \leq 7$ 36. $-3 < m + 1 \leq 5$
37. $-8 < -5x + 2 < 3$ 38. $-4 \leq -3x - 1 \leq 5$
39. $-6 < 2a + 2 < 6$ 40. $-6 < 5a - 4 < 6$
41. $5 \leq -3a - 7 \leq 11$ 42. $1 \leq -4a + 1 \leq 3$

Graph the solution sets for the following compound inequalities:

43. $x + 5 \leq -2 \quad \text{or} \quad 2x + 1 > 7$
44. $3x + 2 < -3 \quad \text{or} \quad x - 5 \geq 3$
45. $5y + 1 \leq -4 \quad \text{or} \quad 2y - 4 \geq 6$
46. $7y - 5 \leq 2 \quad \text{or} \quad 3y - 6 \geq 3$
47. $2x + 5 > 3x - 1 \quad \text{or} \quad x - 4 < 2x + 6$
48. $3x - 1 < 2x + 4 \quad \text{or} \quad 5x - 2 > 3x + 4$
49. $3(a + 1) < 2(a - 5) \quad \text{or} \quad 2(a - 1) \geq a + 2$
50. $5(a + 3) < 2(a - 4) \quad \text{or} \quad 4(a - 6) \geq a + 3$

51. Suppose that a number lies somewhere between 2 and 5 on the number line. (If we let x represent this number, then an inequality that describes this situation is $2 < x < 5$.) Between what two numbers does the opposite of the number lie? Write an inequality that shows where the opposite of the number lies.

52. A number lies between -2 and 3 on the number line. Write an inequality that shows where the opposite of the number lies.

53. Is -5 one of the numbers in the solution set for the inequality $|x| < 2$?

54. Is -3 one of the numbers in the solution set for the inequality $|x| > 2$?

55. Checking the solution set for an inequality is not as easy as checking the solution to an equation. You can, however, check to see that the end point for your solution set is correct. (If $x < 5$ is the solution set for an inequality, then 5 is the end point for the solution set.) Suppose you are taking a test and your solution set for the inequality $-3x + 2 > 2x + 12$ is $x < -2$. Check the end point by replacing x in the original inequality with 5 and observing whether both sides simplify to the same number.

56. The solution set for $-7(x - 2) \geq -2x + 1$ is $x \leq 3$. Check the end point of the solution set in the original inequality.

Review Problems   The problems below review some of the material we covered in Sections 1.2 and 1.5.

Simplify each expression.

57. $|-4|$ **58.** $\left|-\frac{3}{4}\right|$

59. $-\left|-\frac{1}{2}\right|$ **60.** $-\left(-\frac{1}{2}\right)$

61. $|7| - |-3|$ **62.** $|8| - |-2|$

63. $|-2| - |-3| - (-2)$ **64.** $|-4| - (-3) - |-2|$

65. Give a definition for the absolute value of x that involves the number line. (This is the geometric definition.)

66. Give a definition of the absolute value of x that does not involve the number line. (This is the algebraic definition.)

**2.3
Equations with
Absolute Value**

In Chapter 1 we defined the absolute value of x, $|x|$, to be the distance between x and 0 on the number line. The absolute value of a number measures its distance from 0.

In this section we will solve some first-degree equations involving absolute value. In order to solve these equations, we must rely heavily on the definition of absolute value.

▼   **Example 1**   Solve for x: $|x| = 5$.

Solution   Using the definition of absolute value, we can read the equation as, "The distance between x and 0 on the number line is 5." If x is 5 units from 0, then x can be 5 or -5.

$$\text{If } |x| = 5, \qquad \text{then } x = 5 \quad \text{or} \quad x = -5$$

The solution set is $\{5, -5\}$.   ▲

In general, then, we can see that any equation of the form $|a| = b$ is equivalent to the equations $a = b$ or $a = -b$.

▼   **Example 2**   Solve $|2a - 1| = 7$.

Solution   We can read this equation as "$2a - 1$ is 7 units from 0 on the number line." The quantity $2a - 1$ must be equal to 7 or -7.

$$|2a - 1| = 7$$
$$2a - 1 = 7 \quad \text{or} \quad 2a - 1 = -7$$

We have transformed our absolute value equation into two first-degree equations that do not involve absolute value. We can solve each equation using the method in Section 2.1:

$$
\begin{array}{llll}
2a - 1 = 7 & \text{or} & 2a - 1 = -7 & \\
\quad 2a = 8 & \text{or} & \quad 2a = -6 & \text{Add } +1 \text{ to both sides.} \\
\quad\;\; a = 4 & \text{or} & \quad\;\; a = -3 & \text{Multiply by } \tfrac{1}{2}.
\end{array}
$$

Our solution set is $\{4, -3\}$.   ▲

To check our solutions, we put them into the original absolute value equation:

When	$a = 4$	When	$a = -3$
the equation	$|2a - 1| = 7$	the equation	$|2a - 1| = 7$
becomes	$|2(4) - 1| = 7$	becomes	$|2(-3) - 1| = 7$
	$|7| = 7$		$|-7| = 7$
	$7 = 7$		$7 = 7$

▼   **Example 3**   Solve $|3y - 5| = 2$.

Solution   Proceeding as in Example 1, we have

$$|3y - 5| = 2$$

$$3y - 5 = 2 \quad \text{or} \quad 3y - 5 = -2$$

$$3y = 7 \quad \text{or} \quad\quad 3y = 3 \qquad \text{Add } +5 \text{ to both sides.}$$

$$y = \tfrac{7}{3} \quad \text{or} \quad\quad y = 1 \qquad \text{Multiply by } \tfrac{1}{3}.$$

The solution set is $\{\tfrac{7}{3}, 1\}$. ▲

▼　**Example 4**　Solve $|5x - 3| + 5 = 12$.

Solution　In order to use the definition of absolute value to solve this equation, we must isolate the absolute value on the left side of the equal sign. We can delete the $+5$ from the left side by adding -5 to both sides of the equation.

$$|5x - 3| + 5 + (-5) = 12 + (-5)$$
$$|5x - 3| = 7$$

Now that the equation is in the correct form, we can see that $5x - 3$ is 7 or -7. We continue as in Examples 2 and 3:

$$5x - 3 = 7 \quad \text{or} \quad 5x - 3 = -7$$
$$5x = 10 \quad \text{or} \quad\quad 5x = -4 \qquad \text{Add } +3 \text{ to both sides.}$$
$$x = 2 \quad \text{or} \quad\quad x = -\tfrac{4}{5} \qquad \text{Multiply by } \tfrac{1}{5}.$$

The solution set is $\{2, -\tfrac{4}{5}\}$. ▲

Notice in each of the last three examples that one of the solutions comes out negative. This does not contradict our statement in Chapter 1 that absolute value is always a positive quantity, or zero. The quantity inside the absolute value symbols can be negative. When we take its absolute value, the *result* is always positive.

▼　**Example 5**　Solve $|3a - 6| = -4$.

Solution　The solution set is ∅ because the left side cannot be negative and the right side is negative. No matter what we try to substitute for the variable a, the quantity $|3a - 6|$ will always be positive, or zero. It can never be -4. ▲

As a last case in absolute value equations, let's consider the situation where we have two quantities whose absolute values are equal.

Consider the statement $|x| = |y|$. What can we say about x and y? We know they are equal in absolute value. By the definition of absolute

value, they are the same distance from 0 on the number line. They must be equal to each other or opposites of each other. In symbols we write

$$|x| = |y| \quad \Leftrightarrow \quad x = y \quad \text{or} \quad x = -y$$

$$\underset{\substack{\uparrow \\ \text{Equal in} \\ \text{absolute value}}}{} \qquad \underset{\substack{\uparrow \\ \text{Equals}}}{} \quad \text{or} \quad \underset{\substack{\uparrow \\ \text{Opposites}}}{}$$

▼ **Example 6** Solve $|3a + 2| = |2a + 3|$.

Solution The quantities $(3a + 2)$ and $(2a + 3)$ have equal absolute values. They are, therefore, the same distance from 0 on the number line. They must be equals or opposites.

$$|3a + 2| = |2a + 3|$$

Equals	*Opposites*
$3a + 2 = 2a + 3$ or	$3a + 2 = -(2a + 3)$
$a + 2 = 3$	$3a + 2 = -2a - 3$
$a = 1$	$5a + 2 = -3$
	$5a = -5$
	$a = -1$

The solution set is $\{1, -1\}$.

It makes no difference in the outcome of the problem if we take the opposite of the first or second expression. It is very important, once we have decided which one to take the opposite of, that we take the opposite of both its terms and not just the first term. That is, the opposite of $2a + 3$ is $-(2a + 3)$, which we can think of as $-1(2a + 3)$. Distributing the -1 across *both* terms, we have

$$-1(2a + 3) = -1(2a) + (-1)(3)$$
$$= -2a - 3 \qquad \blacktriangle$$

▼ **Example 7** Solve $|x - 5| = |x - 7|$.

Solution As was the case in Example 6, the quantities $x - 5$ and $x - 7$ must be equal or they must be opposites, because their absolute values are equal.

Equals	*Opposites*
$x - 5 = x - 7$ or	$x - 5 = -(x - 7)$
$-5 = -7$	$x - 5 = -x + 7$
no solution here	$2x - 5 = 7$
	$2x = 12$
	$x = 6$

Since the first equation leads to a false statement, we do not get any solutions from it. (If either of the two equations was to reduce to a true statement, it would mean all real numbers would satisfy the original equation.) In this case our only solution is $x = 6$.   ▲

Problem Set 2.3

Use the definition of absolute value to solve each of the following problems:

1. $\|x\| = 4$		**2.** $\|x\| = 7$
3. $\|a\| = 2$		**4.** $\|a\| = 5$
5. $\|x\| = -3$		**6.** $\|x\| = -4$
7. $\|a\| + 2 = 3$		**8.** $\|a\| - 5 = 2$
9. $\|y\| + 4 = 3$		**10.** $\|y\| + 3 = 1$
11. $\|x\| - 2 = 4$		**12.** $\|x\| - 5 = 3$
13. $\|x - 2\| = 5$		**14.** $\|x + 1\| = 2$
15. $\|a - 4\| = 1$		**16.** $\|a + 2\| = 7$
17. $\|3 - x\| = 1$		**18.** $\|4 - x\| = 2$
19. $\|2x + 1\| = -3$		**20.** $\|2x - 5\| = -5$
21. $\|3a + 1\| = 5$		**22.** $\|2x - 3\| - 4 = 3$
23. $\|3x + 4\| + 1 = 7$		**24.** $\|5x - 3\| - 4 = 3$
25. $\|2y - 3\| + 4 = 3$		**26.** $\|7y - 8\| + 9 = 1$
27. $\|1 + 2(x - 1)\| = 7$		**28.** $\|3 + 4(x - 2)\| = 5$
29. $\|2(k + 4) - 3\| = 1$		**30.** $\|3(k - 2) + 1\| = 4$

Solve the following equations by the method used in Example 6:

31. $\|3a + 1\| = \|2a - 4\|$		**32.** $\|5a + 2\| = \|4a + 7\|$
33. $\|6x - 2\| = \|3x + 1\|$		**34.** $\|x - 5\| = \|2x + 1\|$
35. $\|y - 2\| = \|y + 3\|$		**36.** $\|y - 5\| = \|y - 4\|$
37. $\|3x - 1\| = \|3x + 1\|$		**38.** $\|5x - 8\| = \|5x + 8\|$
39. $\|3 - m\| = \|m + 4\|$		**40.** $\|5 - m\| = \|m + 8\|$
41. $\|3 - x\| = \|4 + 5x\|$		**42.** $\|7 - x\| = \|8 - 2x\|$
43. $\|x - 3\| = \|3 - x\|$		**44.** $\|x - 5\| = \|5 - x\|$
45. $\|2x - 3\| = \|3 - 2x\|$		**46.** $\|4x - 1\| = \|1 - 4x\|$

47. Each of the equations in Problems 43 through 46 has the form $\|a - b\| = \|b - a\|$. The solution set for each of these equations is all real numbers. This means that the statement $\|a - b\| = \|b - a\|$ must be true no matter what numbers a and b are. The statement itself is a property of absolute value. Show that the statement is true when $a = 4$ and $b = -7$, as well as when $a = -5$ and $b = -8$.

48. Show that the statement $\|ab\| = \|a\|\|b\|$ is true when $a = 3$ and $b = -6$, and when $a = -8$ and $b = -2$.

49. Name all the numbers in the set $\{-4, -3, -2, -1, 0, 1, 2, 3, 4\}$ that are solutions to the equation $\|x + 2\| = x + 2$.

50. Name all the numbers in the set $\{-4, -3, -2, -1, 0, 1, 2, 3, 4\}$ that are solutions to the equation $|x - 2| = x - 2$.

51. The equation $|a| = a$ is only true if $a \geq 0$. We can use this fact to actually solve equations like the one in Problem 49. The only way the equation $|x + 2| = x + 2$ can be true is if $x + 2 \geq 0$. Adding -2 to both sides of this last inequality we see that $x \geq -2$. Use the same kind of reasoning to solve the equation $|x - 2| = x - 2$.

52. Solve the equation $|x + 3| = x + 3$.

Review Problems The problems below review material we covered in Sections 1.3 and 2.2. Reviewing these problems will help you with the next section.

Graph each inequality.

53. $x < -2$ or $x > 8$ **54.** $1 < x < 4$

55. $-\frac{11}{3} \leq x \leq -1$ **56.** $x \leq -\frac{3}{2}$ or $x \geq 3$

Solve each inequality

57. $4t - 3 \leq -9$ **58.** $-3 < 2a - 5 < 3$

59. $-3x > 15$ **60.** $-2x \leq 10$

In this section we will again apply the definition of absolute value to solve inequalities involving absolute value. Again, the absolute value of x, which is $|x|$, represents the distance that x is from 0 on the number line. We will begin by considering three absolute value expressions and their English translations:

Expression	*In Words*
$|x| = 7$	x is exactly 7 units from 0 on the number line.
$|a| < 5$	a is less than 5 units from 0 on the number line.
$|y| \geq 4$	y is greater than or equal to 4 units from 0 on the number line.

2.4
Inequalities Involving Absolute Value

Once we have translated the expression into words, we can use the translation to graph the original equation or inequality. The graph is then used to write a final equation or inequality that does not involve absolute value.

Original Expression	*Graph*	*Final Expression*
$|x| = 7$	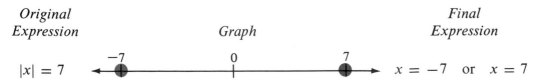	$x = -7$ or $x = 7$

Original Expression	Graph	Final Expression
$\lvert a \rvert < 5$	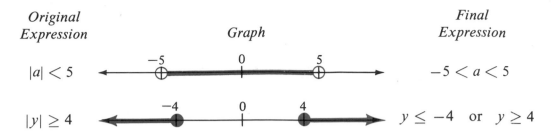	$-5 < a < 5$
$\lvert y \rvert \geq 4$		$y \leq -4$ or $y \geq 4$

Although we will not always write out the English translation of an absolute value inequality, it is important that we understand the translation. Our second expression, $\lvert a \rvert < 5$, means a is within 5 units of 0 on the number line. The graph of this relationship is

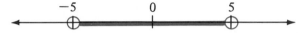

which can be written with the following continued inequality:

$$-5 < a < 5$$

We can follow this same kind of reasoning to solve more complicated absolute value inequalities.

▼   **Example 1**   Graph the solution set $\lvert 2x - 5 \rvert < 3$.

Solution   The absolute value of $2x - 5$ is the distance that $2x - 5$ is from 0 on the number line. We can translate the inequality as, "$2x - 5$ is less than 3 units from 0 on the number line." That is, $2x - 5$ must appear between -3 and 3 on the number line.

A picture of this relationship is

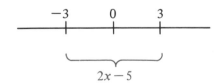

Using the picture, we can write an inequality without absolute value that describes the situation:

$$-3 < 2x - 5 < 3$$

Next, we solve the continued inequality by first adding $+5$ to all three members and then multiplying all three by $\frac{1}{2}$:

$$-3 < 2x - 5 < 3$$
$$2 < \quad 2x \quad < 8 \qquad \text{Add } +5 \text{ to all three members.}$$
$$1 < \quad x \quad < 4 \qquad \text{Multiply each member by } \tfrac{1}{2}.$$

The graph of the solution set is

We can see from the solution that in order for the absolute value of $2x - 5$ to be within 3 units of 0 on the number line, x must be between 1 and 4.   ▲

▼   **Example 2**   Solve and graph $|3a + 7| \leq 4$.

Solution   We can read the inequality as, "The distance between $3a + 7$ and 0 is less than or equal to 4." Or, "$3a + 7$ is within 4 units of 0 on the number line." This relationship can be written without absolute value as

$$-4 \leq 3a + 7 \leq 4$$

Solving as usual, we have

$$-4 \leq 3a + 7 \leq 4$$
$$-11 \leq \quad 3a \quad \leq -3 \qquad \text{Add } -7 \text{ to all three members.}$$
$$\frac{-11}{3} \leq \quad a \quad \leq -1 \qquad \text{Multiply each by } \tfrac{1}{3}.$$

We can see from Examples 1 and 2 that in order to solve an inequality involving absolute value, we must be able to write an equivalent expression that does not involve absolute value.

▼   **Example 3**   Solve $|x - 3| > 5$.

Solution   We interpret the absolute value inequality to mean that $x - 3$ is more than 5 units from 0 on the number line. The quantity $x - 3$ must be either above $+5$ or below -5. Here is a picture of the relationship:

An inequality without absolute value that also describes this situation is

$$x - 3 < -5 \quad \text{or} \quad x - 3 > 5$$

Adding $+3$ to both sides of each inequality we have

$$x < -2 \quad \text{or} \quad x > 8$$

the graph of which is

From our results we see that in order for the absolute value of $x - 3$ to be more than 5 units from 0, x must be below -2 or above $+8$. ▲

▼ **Example 4**   Graph the solution set $|4t - 3| \geq 9$.

Solution   The quantity $4t - 3$ is greater than or equal to 9 units from 0. It must be either above $+9$ or below -9.

$$
\begin{array}{llll}
4t - 3 \leq -9 & \text{or} & 4t - 3 \geq 9 & \\
4t \leq -6 & \text{or} & 4t \geq 12 & \text{Add } +3. \\
t \leq -\frac{6}{4} & \text{or} & t \geq \frac{12}{4} & \text{Multiply by } \frac{1}{4}. \\
t \leq -\frac{3}{2} & \text{or} & t \geq 3 &
\end{array}
$$

▲

Since absolute value always results in a nonnegative quantity, we sometimes come across special solution sets when a negative number appears on the right side of an absolute value inequality.

▼ **Example 5**   Solve $|7y - 1| < -2$.

Solution   The *left* side is never negative because it is an absolute value. The *right* side is negative. We have a positive quantity less than a negative quantity, which is impossible. The solution set is the empty set, $\varnothing$. There is no real number to substitute for y to make the above inequality a true statement. ▲

▼ **Example 6** Solve $|6x + 2| > -5$.

Solution This is the opposite case from that in Example 5. No matter what real number we use for x on the *left* side, the result will always be positive, or zero. The *right* side is negative. We have a positive quantity greater than a negative quantity. Every real number we choose for x gives us a true statement. The solution set is the set of all real numbers.   ▲

Solve each of the following inequalities using the definition of absolute value. Graph the solution set in each case.   *Problem Set 2.4*

1. $|x| < 3$ 2. $|x| \leq 7$
3. $|x| \geq 2$ 4. $|x| > 4$
5. $|a| < 1$ 6. $|a| \leq 4$
7. $|x| + 2 < 5$ 8. $|x| - 3 < -1$
9. $|t| - 3 > 4$ 10. $|t| + 5 > 8$
11. $|y| < -5$ 12. $|y| > -3$
13. $|x| \geq -2$ 14. $|x| \leq -4$
15. $|x - 3| < 7$ 16. $|x + 4| < 2$
17. $|a + 5| \geq 4$ 18. $|a - 6| \geq 3$
19. $|a - 1| < -3$ 20. $|a + 2| \geq -5$
21. $|2x - 3| \leq 5$ 22. $|2x + 7| \leq 3$
23. $|2x - 4| < 6$ 24. $|2x + 6| < 2$
25. $|3y + 9| \geq 6$ 26. $|5y - 1| \geq 4$
27. $|6x - 7| > 3$ 28. $|4x + 2| > 5$
29. $|2k + 3| \geq 7$ 30. $|2k - 5| \geq 3$
31. $|x - 3| + 2 < 6$ (*Hint:* Isolate $x - 3$ on the left side)
32. $|x + 4| - 3 < -1$ 33. $|2a + 1| + 4 \geq 7$
34. $|2a - 6| - 1 \geq 2$ 35. $|3x + 5| - 8 < 5$
36. $|6x - 1| - 4 \leq 2$ 37. $|3(x - 5)| \leq -8$
38. $|4(2x - 1)| \geq -1$ 39. $|3(2a + 1)| \leq 6$
40. $|2(4a - 5)| \leq 8$

Solve each inequality below and graph the solution set. Keep in mind that if you multiply or divide both sides of an inequality by a negative number you must reverse the sense of the inequality.

41. $|5 - x| > 3$ 42. $|7 - x| > 2$
43. $|3 - \frac{2}{3}x| \geq 5$ 44. $|3 - \frac{3}{4}x| \geq 9$
45. $|2 - \frac{1}{2}x| < 1$ 46. $|3 - \frac{1}{3}x| < 1$
47. Write the continued inequality $-4 \leq x \leq 4$ as a single inequality involving absolute value.

48. Write the continued inequality $-8 \leq x \leq 8$ as a single inequality involving absolute value.

49. Write the continued inequality $-3 \leq x \leq 7$ as a single inequality involving absolute value. (Look for a number to add to all three parts of the continued inequality that will make the two outside numbers opposites.)

50. Write the continued inequality $-1 \leq x \leq 7$ as a single inequality involving absolute value.

Review Problems The problems below review some of the material we covered in Section 1.4.

Identify the property (or properties) that justifies each of the following statements.

51. $ax = xa$ **52.** $5(\frac{1}{5}) = 1$

53. $3 + (x + y) = (3 + x) + y$ **54.** $3 + (x + y) = (x + y) + 3$

55. $3 + (x + y) = (3 + y) + x$ **56.** $7(3x - 5) = 21x - 35$

57. $4(xy) = 4(yx)$ **58.** $4(xy) = (4y)x$

2.5
Formulas

A formula in mathematics is an equation which contains more than one variable. There are probably some formulas which are already familiar to you—for example, the formula for the area (A) of a rectangle with length l and width w:

$$A = lw$$ $w = \text{width}$
$$l = \text{length}$$

There are some others with which you are probably not as familiar—for example, the formula for the surface area (s) of a closed cylinder with a given height h and radius r:

$$s = 2\pi r^2 + 2\pi rh$$ $h = \text{height}$
$$r = \text{radius}$$

Formulas are very common in the application of algebra to other

disciplines. They are found in chemistry, physics, biology, and business, among others.

There are generally two main types of problems associated with formulas. We can solve for one of the variables in a formula if we are given numerical replacements for the other variables. Or we can solve for one of the variables in a formula without being given replacements for the other variables.

To illustrate the first case, let's consider a fairly simple example involving the formula for the area of a triangle.

▼ **Example 1** Using the formula $A = \frac{1}{2}bh$, find h if $b = 6$ in. (inches) and $A = 15$ sq in. (square inches).

Solution We have been given the formula for the area (A) of a triangle with base b and height h:

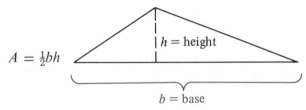

$A = \frac{1}{2}bh$ $h = \text{height}$

$b = \text{base}$

When $A = 15$ and $b = 6$, we have

$$15 = \frac{1}{2}(6)h$$
$$15 = 3h$$
$$5 = h$$

The height is $h = 5$ in.   ▲

Let's look at a slightly more difficult example.

▼ **Example 2** Given the formula $A = \frac{1}{2}(b + B)h$, find B when $A = 20$ sq ft (square feet), $b = 3$ ft (feet), and $h = 4$ ft.

Solution The formula is for the area of a trapezoid with bases b and B and height h:

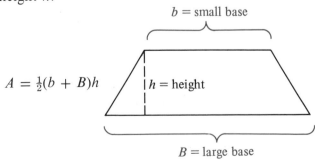

$b = \text{small base}$

$A = \frac{1}{2}(b + B)h$ $h = \text{height}$

$B = \text{large base}$

When $A = 20$, $b = 3$, and $h = 4$, we have

$$20 = \tfrac{1}{2}(3 + B)4$$

Multiplication is commutative, so we can multiply the $\tfrac{1}{2}$ and the 4:

$$20 = 2(3 + B)$$
$$10 = 3 + B \qquad \text{Multiply both sides by } \tfrac{1}{2}.$$
$$7 = B \qquad \text{Add } -3 \text{ to both sides.}$$

The larger base is $B = 7$ ft.   ▲

We will now consider examples where we are given a formula and are asked to solve for one of the variables without having replacements for the others.

▼   **Example 3**   Given the formula $P = 2w + 2l$, solve for w.

Solution   The formula represents the relationship between the perimeter P (the distance around the outside), the length l, and the width w of a rectangle.

To solve for w we must isolate it on one side of the equation. We can accomplish this if we delete the $2l$ term and the coefficient 2 from the right side of the equation.

To begin, we add $-2l$ to both sides:

$$P + (-2l) = 2w + 2l + (-2l)$$
$$P - 2l = 2w$$

To delete the 2 from the right side, we can multiply both sides by $\tfrac{1}{2}$:

$$\tfrac{1}{2}(P - 2l) = \tfrac{1}{2}(2w)$$
$$\frac{P - 2l}{2} = w$$

We know we are finished because w appears alone on the right side of the equal sign and does not appear on the left side.

The two formulas

$$P = 2l + 2w \quad \text{and} \quad w = \frac{P - 2l}{2}$$

give the relationship between P, l, and w. They look different, but they both say the same thing about P, l, and w. The first formula gives P in terms of l and w, and the second formula gives w in terms of P and l.   ▲

▼ **Example 4** Solve the formula $s = 2\pi rh + \pi r^2$ for h.

Solution To isolate h, we first add $-\pi r^2$ to both sides and then multiply both sides by $1/2\pi r$:

$$s = 2\pi rh + \pi r^2$$

$$s + (-\pi r^2) = 2\pi rh + \pi r^2 + (-\pi r^2) \qquad \text{Add } -\pi r^2$$
$$\text{to both sides.}$$

$$s - \pi r^2 = 2\pi rh$$

$$\frac{1}{2\pi r}(s - \pi r^2) = \frac{1}{2\pi r}(2\pi rh) \qquad \text{Multiply by } \frac{1}{2\pi r}.$$

$$\frac{s - \pi r^2}{2\pi r} = h$$

$$\text{or} \qquad h = \frac{s - \pi r^2}{2\pi r} \qquad\qquad\qquad\qquad ▲$$

Solve each of the following formulas for the variable that does not have a numerical replacement:

Problem Set 2.5

1. $A = lw$; $A = 30$, $l = 5$
2. $A = lw$; $A = 16$, $w = 2$
3. $A = \frac{1}{2}bh$; $A = 2$, $b = 2$
4. $A = \frac{1}{2}bh$; $A = 10$, $b = 10$
5. $I = prt$; $I = 200$, $p = 1{,}000$, $t = .05$
6. $I = prt$; $I = 100$, $p = 2{,}000$, $t = 2$
7. $P = 2l + 2w$; $P = 16$, $l = 3$
8. $P = 2l + 2w$; $P = 40$, $w = 5$
9. $A = 2\pi r^2 + 2\pi rh$; $r = 1$, $\pi = 3.14$, $h = 2$
10. $A = \frac{1}{2}(b + B)h$; $A = 7$, $b = 2$, $B = 3$
11. $A = \frac{1}{2}(b + B)h$; $A = 9$, $b = 3$, $h = 6$
12. $A = \frac{1}{2}(b + B)h$; $A = 15$, $B = 2$, $h = 8$
13. $y = mx + b$; $y = 18$, $b = 2$, $m = 2$
14. $y = mx + b$; $y = 30$, $m = 2$, $x = 5$
15. $s = \frac{1}{2}(a + b + c)$; $s = 12$, $a = 1$, $b = 2$
16. $s = \frac{1}{2}(a + b + c)$; $s = 15$, $a = 8$, $c = 4$

Solve each of the following formulas for the indicated variables:

17. $A = lw$ for l
18. $A = \frac{1}{2}bh$ for b
19. $I = prt$ for t
20. $I = prt$ for r

21. $E = mc^2$ for m
22. $C = 2\pi r$ for r
23. $PV = nRT$ for T
24. $PV = nRT$ for R
25. $c^2 = a^2 + b^2$ for b^2
26. $c^2 = a^2 + b^2$ for a^2
27. $y = mx + b$ for b
28. $y = mx + b$ for x
29. $A = \frac{1}{2}(b + B)h$ for h
30. $A = \frac{1}{2}(b + B)h$ for b
31. $s = \frac{1}{2}(a + b + c)$ for c
32. $s = \frac{1}{2}(a + b + c)$ for b
33. $d = vt + \frac{1}{2}at^2$ for a
34. $d = vt - \frac{1}{2}at^2$ for a
35. $A = P + Prt$ for r
36. $A = P + Prt$ for t
37. $C = \frac{5}{9}(F - 32)$ for F
38. $F = \frac{9}{5}C + 32$ for C
39. $A = a + (n - 1)d$ for d
40. $A = a + (n - 1)d$ for n
41. $9x - 3y = 6$ for y
42. $9x + 3y = 15$ for y

43. $z = \dfrac{x - \mu}{s}$ for x

44. $z = \dfrac{x - \mu}{s}$ for μ

45. Solve the inequality $-2.5 < \dfrac{x - \mu}{s} < 2.5$ for x.

46. Solve the absolute value inequality $\left| \dfrac{x - \mu}{s} \right| < 1.96$ for x.

47. The formula $F = \frac{9}{5}C + 32$ gives the relationship between the Celsius and Fahrenheit temperature scales. If the temperature range on a certain day is 86° to 104° Fahrenheit (that is, $86 \le F \le 104$), what is the temperature range in degrees Celsius?

48. If the temperature in degrees Fahrenheit is between 68° and 95°, what is the corresponding temperature range in degrees Celsius?

Review Problems   The problems below review some of the material we covered in Section 1.1. Reviewing these problems will help you with the next section.

Translate each of the following into symbols.

49. Twice the sum of x and 3.
50. Twice the sum of x and 3 is 16.
51. Five times the difference of x and 3.
52. Five times the difference of x and 3 is 10.
53. The sum of $3x$ and 2 is equal to the difference of x and 4.
54. The sum of x and $x + 2$ is 12 more than their difference.

2.6
Word Problems

There are a number of word problems whose solutions depend on solving first-degree equations in one variable. We will begin our study of word problems by considering some simple problems stated in words.

Admittedly, the problems in this section are a bit contrived. That is, the problems themselves are not the kind of problems you would find in

those fields of study that require a good background in mathematics, like chemistry or physics. (Although you may see these problems on entrance exams or aptitude tests.) As we progress through the book we will solve word problems of a more realistic nature. In the meantime, the problems in this section will allow you to practice the procedures used in setting up and solving word problems.

Here are some general steps we will follow in solving word problems:

Step 1. Let x represent the quantity asked for in the problem.
Step 2. Write expressions, using the variable x, that represent any other unknown quantities in the problem.
Step 3. Write an equation, in x, that describes the situation.
Step 4. Solve the equation found in step 3.
Step 5. Check the solution in the original words of the problem.

Step 3 is usually the most difficult step. Step 3 is really what word problems are all about—translating a problem stated in words into an algebraic equation.

As an aid in simplifying step 3, we will look at a number of phrases written in English and the equivalent mathematical expression.

English Phrase	*Algebraic Expression*
The sum of a and b	$a + b$
The difference of a and b	$a - b$
The product of a and b	$a \cdot b$
The quotient of a and b	a/b

The word *sum* always indicates addition. The word *difference* always implies subtraction. *Product* indicates multiplication, and *quotient* means division.

Let's continue our list:

English Phrase	*Algebraic Expression*
4 more than x	$4 + x$
Twice the sum of a and 5	$2(a + 5)$
The sum of twice a and 5	$2a + 5$
8 decreased by y	$8 - y$
3 less than m	$m - 3$
7 times the difference of x and 2	$7(x - 2)$

We could extend the list even more. For every English sentence involving a relationship with numbers, there is an associated mathematical expression.

▼   **Example 1**   Twice the sum of a number and 3 is 16. Find the number.

Solution

Step 1.   Let $x = $ the number asked for.
Step 2.   Twice the sum of x and 3 $= 2(x + 3)$.
Step 3.   An equation that describes the situation is

$$2(x + 3) = 16$$

(The word *is* always translates to $=$.)
Step 4.   Solving the equation, we have

$$2(x + 3) = 16$$
$$2x + 6 = 16$$
$$2x = 10$$
$$x = 5$$

Step 5.   Checking $x = 5$ in the original problem, we see that twice the sum of 5 and 3 is twice 8 or 16.   ▲

▼   **Example 2**   The sum of two consecutive even integers is 12 more than their positive difference. Find the two integers.

Solution

Step 1.   Let $x = $ the smaller integer.
Step 2.   The next consecutive even integer after x is $x + 2$. The sum of the two integers is $x + (x + 2)$, or $2x + 2$. Their positive difference is $(x + 2) - x$, or simply 2.
Step 3.   An equation that describes the situation is

$$2x + 2 = 2 + 12$$
$$2x = 12$$
$$x = 6$$

The two integers are $x = 6$ and $x + 2 = 8$.
Step 4.   Checking the original problem, we see that the sum of 6 and 8 is 14. Their difference is 2, and 14 is 12 more than 2. The answers check.   ▲

▼   **Example 3**   A rectangle is 3 times as long as it is wide. The
perimeter is 40 feet. Find the dimensions.

Solution

Step 1.   Let x = the width of the rectangle.
Step 2.   The length is 3 times the width, or $3x$.

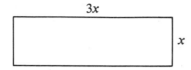

Step 3.   The perimeter of a rectangle is twice the width plus
twice the length:
$$2x + 2(3x) = 40$$

Step 4.   Solve: $8x = 40$
                                      $x = 5$

The width is 5 feet. The length is 15 feet.
Step 5.   Since the sum of twice 5 plus twice 15 is 40, the
solutions check in the original problem.   ▲

In all problems that involve geometric figures (rectangles, triangles,
squares, etc.), it is helpful to draw the figure and label the dimensions.

▼   **Example 4**   Diane is 4 years older than JoAnn. In 6 years the sum
of their ages will be 68. What are their ages now?

Solution

Step 1.   Let x = JoAnn's age now.
Step 2.   With age problems like this, it is usually helpful to use
a table like the following:

	Now	In 6 years
Diane		
JoAnn		

There is one column for each of the different times

given ("Now" and "In 6 years") and a row for each of the people mentioned (Diane and JoAnn).

To fill in the table, we use JoAnn's age as x. In 6 years she will be $x + 6$ years old. Diane is 4 years older than JoAnn or $x + 4$ years old. In 6 years she will be $x + 10$ years old.

	Now	In 6 years
Diane	$x + 4$	$x + 10$
JoAnn	x	$x + 6$

Step 3.   In 6 years the sum of their ages will be 68.

$$(x + 6) + (x + 10) = 68$$

Step 4.   Solve: $2x + 16 = 68$
$$2x = 52$$
$$x = 26$$

JoAnn is 26 and Diane is 30.

Step 5.   In 6 years JoAnn will be 32 and Diane will be 36. The sum of their ages will then be $32 + 36 = 68$. The solutions check in the original problem.   ▲

Problem Set 2.6

Number Problems

Solve each of the following word problems. Be sure to show the equation used in each case.

1. A number increased by 2 is 5 less than twice the number. Find the number.
2. Twice a number decreased by 3 is equal to the number. Find the number.
3. Three times the sum of a number and 4 is 3. Find the number.
4. Five times the difference of a number and 3 is 10. Find the number.
5. Twice the sum of 2 times a number and 1 is the same as 3 times the difference of the number and 5. Find the number.
6. The sum of 3 times a number and 2 is the same as the difference of the number and 4. Find the number.
7. If 5 times a number is increased by 2, the result is 8 more than 3 times the number. Find the number.
8. If 6 times a number is decreased by 5, the result is 7 more than 4 times the number decreased by 5. Find the number.

9. The sum of two consecutive even integers is 16 more than their positive difference. Find the two integers.

10. The sum of two consecutive odd integers is 14 more than their positive difference. Find the two integers.

11. The sum of two consecutive integers is 1 less than 3 times the smaller. Find the two integers.

12. The sum of two consecutive integers is 5 less than 3 times the larger. Find the integers.

13. If twice the smaller of two consecutive integers is added to the larger, the result is 7. Find the smaller one.

14. If twice the larger of two consecutive integers is added to the smaller, the result is 23. Find the smaller one.

15. If the larger of two consecutive odd integers is subtracted from twice the smaller, the result is 5. Find the two integers.

16. If the smaller of two consecutive even integers is subtracted from twice the larger, the result is 12. Find the two integers.

17. A rectangle is twice as long as it is wide. The perimeter is 60 feet. Find the dimensions.

18. The length of a rectangle is 5 times the width. The perimeter is 48 inches. Find the dimensions.

19. A square has a perimeter of 28 feet. Find the length of the side.

20. A square has a perimeter of 36 centimeters. Find the length of the side.

21. A triangle has a perimeter of 23 inches. The medium side is 3 more than the smallest side, and the longest side is twice the shortest side. Find the shortest side.

22. The longest side of a triangle is 3 times the shortest side. While the medium side is twice the shortest side, the perimeter is 18 meters. Find the dimensions.

23. The length of a rectangle is 3 less than twice the width. The perimeter is 18 meters. Find the width.

24. The length of a rectangle is one more than twice the width. The perimeter is 20 feet. Find the dimensions.

25. Patrick is 4 years older than Amy. In 10 years the sum of their ages will be 36. How old are they now?

26. Mr. Lloyd is 2 years older than Mrs. Lloyd. Five years ago the sum of their ages was 44. How old are they now?

27. John is twice as old as Bill. In 5 years John will be $1\frac{1}{2}$ times as old as Bill. How old is each boy now?

28. Bob is 6 years older than his little brother Ron. In 4 years he will be twice as old as his little brother. How old are they now?

29. Jane is 3 times as old as Kate. In 5 years Jane's age will be 2 less than twice Kate's. How old are the girls now?

30. Carol is 2 years older than Mike. In 4 years her age will be 8 less than twice his. How old are they now?

Consecutive Integer Problems

Geometry Problems

Age Problems

Review Problems The problems below review some of the material we covered in Section 1.6.

Simplify each expression as much as possible.

31. $4(-3) - 5(-6)$ **32.** $-4(-5) - 7(-2)$

33. $-18 \div (-\frac{3}{4})$ **34.** $-14 \div (\frac{7}{10})$

35. $4 - (-2)\left[\dfrac{3(-4) - 6}{2 - 2(4)}\right]$ **36.** $7 - 2\left[\dfrac{6 - 3(5 - 9)}{-3 - 3}\right]$

37. $8 - 3(4x - 5) - (3x + 2)$ **38.** $6 - 4(5x - 7) - (2x + 3)$

39. Subtract -4 from the product of 15 and $-\frac{2}{5}$.

40. Subtract -3 from the product of -20 and $-\frac{3}{5}$.

Examples

1. We can solve $x + 3 = 5$ by adding -3 to both sides.

$$x + 3 + (-3) = 5 + (-3)$$
$$x = 2$$

2. We can solve $3x = 12$ by multiplying both sides by $\frac{1}{3}$.

$$3x = 12$$
$$\tfrac{1}{3}(3x) = \tfrac{1}{3}(12)$$
$$x = 4$$

3. Solve $3(2x - 1) = 9$.

$$3(2x - 1) = 9$$
$$6x - 3 = 9$$
$$6x - 3 + 3 = 9 + 3$$
$$6x = 12$$
$$x = 2$$

Chapter 2 Summary and Review

ADDITION PROPERTY OF EQUALITY [2.1]

For algebraic expressions A, B, and C,

$$\text{if} \qquad A = B$$
$$\text{then} \qquad A + C = B + C$$

This property states that we can add the same quantity to both sides of an equation without changing the solution set.

MULTIPLICATION PROPERTY OF EQUALITY [2.1]

For algebraic expressions A, B, and C,

$$\text{if} \qquad A = B$$
$$\text{then} \qquad AC = BC, \quad C \neq 0$$

Multiplying both sides of an equation by the same nonzero quantity never changes the solution set.

SOLVING A FIRST-DEGREE EQUATION IN ONE VARIABLE [2.1]

Step 1. Use the distributive property to separate terms.

Step 2. Simplify the left and right sides of the equation separately whenever possible.

Step 3. Use the addition property of equality to write all terms containing the variable on one side of

Step 4. the equal sign and all remaining terms on the other.

Step 4. Use the multiplication property of equality in order to get x alone on one side of the equal sign.

Step 5. Check your solution in the original equation, if necessary.

ADDITION PROPERTY FOR INEQUALITIES [2.2]

For expressions A, B, and C,

$$\text{if} \quad A < B$$
$$\text{then} \quad A + C < B + C$$

Adding the same quantity to both sides of an inequality never changes the solution set.

4. Adding 5 to both sides of the inequality $x - 5 < -2$ gives

$$x - 5 + 5 < -2 + 5$$
$$x < 3$$

MULTIPLICATION PROPERTY FOR INEQUALITIES [2.2]

For expressions A, B, and C,

$$\text{if} \quad A < B$$
$$\text{then} \quad AC < BC \quad \text{if} \quad C > 0$$
$$\text{or} \quad AC > BC \quad \text{if} \quad C < 0$$

We can multiply both sides of an inequality by the same nonzero number without changing the solution set as long as each time we multiply by a negative number we also reverse the direction of the inequality symbol.

5. Multiplying both sides of $-2x \geq 6$ by $-\frac{1}{2}$ gives

$$-2x \geq 6$$
$$-\tfrac{1}{2}(-2x) \leq -\tfrac{1}{2}(6)$$
$$x \leq -3$$

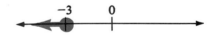

SOLVING A FIRST-DEGREE INEQUALITY IN ONE VARIABLE [2.2]

Follow the same first four steps used to solve a first-degree equation—using, of course, the addition and multiplication properties for inequalities. As a fifth step, graph the solution set.

6. $2(3x - 5) > 14$
$6x - 10 > 14$
$6x > 24$
$x > 4$

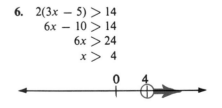

ABSOLUTE VALUE EQUATIONS AND INEQUALITIES [2.3, 2.4]

Use the definition of absolute value to write an equivalent equation or inequality that does not involve absolute value.

7. The key to solving absolute value equations and inequalities is to write equivalent equations or inequalities that do not contain absolute value symbols.

$$|x| = 5 \quad \text{is equivalent to} \quad x = 5 \quad \text{or} \quad x = -5$$
$$|x| < 5 \quad \text{is equivalent to} \quad -5 < x < 5$$
$$|x| > 5 \quad \text{is equivalent to} \quad x < -5 \quad \text{or} \quad x > 5$$

8. Solve for w.

$$P = 2l + 2w$$
$$P - 2l = 2w$$
$$\frac{P - 2l}{2} = w$$

FORMULAS [2.5]

A formula in algebra is an equation involving more than one variable. To solve a formula for one of its variables, simply isolate that variable on one side of the equation.

9. If the perimeter of a rectangle is 32 inches and the length is 3 times the width, we can find the dimensions by letting x be the width and $3x$ the length.

$$x + x + 3x + 3x = 32$$
$$8x = 32$$
$$x = 4$$

Width is 4 inches, length is 12 inches.

SOLVING A PROBLEM STATED IN WORDS [2.6]

Step 1. Let x represent the quantity asked for.
Step 2. If possible, write all other unknown quantities in terms of x.
Step 3. Write an equation, using x, that describes the situation.
Step 4. Solve the equation in step 3.
Step 5. Check the solution from step 4 with the original words of the problem.

COMMON MISTAKE

A very common mistake in solving inequalities is to forget to reverse the direction of the inequality symbol when multiplying both sides by a negative number. When this mistake occurs, the graph of the solution set is always to the wrong side of the end point.

Chapter 2 Test

Solve the following equations using the steps outlined in Section 2.1:

1. $x - 5 = 7$ **2.** $3y = -4$
3. $3a - 6 = 11$ **4.** $5(x - 1) - 2(2x + 3) = 5x - 4$

Solve the following inequalities by the method developed in Section 2.2:

5. $-5t \le 30$ **6.** $8 - 2x \ge 6$
7. $4x - 5 < 2x + 7$ **8.** $3(2y + 4) \ge 5(y - 8)$

Solve the following equations and inequalities using the definition of absolute value:

9. $|x - 4| = 2$ **10.** $|2a + 7| = 5$

11. $|6x - 1| > 7$ **12.** $|3x - 5| - 4 \leq 3$
13. Solve the formula $A = 2l + 2w$ for w.
14. Solve the formula $A = \frac{1}{2}h(b + B)$ for B.
15. Find two consecutive even integers whose sum is 18. (Be sure to write an equation that describes the situation.)
16. Patrick is 4 years older than Amy. In 5 years he will be twice as old as she is now. Find their ages now.

3

Exponents and Polynomials

To the student:

There are many expressions and equations in mathematics that involve exponents. For example, in genetics, the phenotypic variance P^2 of a given trait is the sum of the genetic variance (G^2), the environmental variance (E^2), and the variance due to the interaction between them (I^2), or $P^2 = G^2 + E^2 + I^2$. The 2's in the equation are exponents. Before we get to the point where we can work with equations like this, we must review the properties of exponents and practice working with expressions that involve exponents.

In addition to being able to add, subtract, multiply, and divide real numbers, your success in this chapter depends on your ability to understand and apply the distributive property. The fact that multiplication distributes over addition allows us to add and multiply polynomials. Understanding addition and multiplication of polynomials allows us to understand subtraction and factoring of polynomials. Actually, we could probably rename this chapter "Exponents and Applications of the Distributive Property." Understanding the distributive property is the key to understanding this chapter.

In Chapter 1 we showed multiplication by positive integers as repeated addition. That is, $3 \cdot 4 = 4 + 4 + 4$. In this section we will define positive integer exponents to be shorthand notation for repeated multiplication.

Consider the expression 3^4, which reads "3 to the fourth power" or just "3 to the fourth." The 3 is called the *base* and the 4 is called the *exponent*. The exponent gives us the number of times the base is used as a factor in the expansion of the expression. For example,

$$3^4 = \underbrace{3 \cdot 3 \cdot 3 \cdot 3}_{\text{4 factors}} = 81$$

The expression 3^4 is said to be in *exponential form,* while $3 \cdot 3 \cdot 3 \cdot 3$ is said to be in *expanded form.*

Here are some more examples of expressions with integer exponents and their expansions.

▼ **Example 1** Expand and multiply, if possible.

a. $(-4)^3 = (-4)(-4)(-4) = -64$ Base -4, exponent 3

b. $(-5)^2 = (-5)(-5) = 25$ Base -5, exponent 2

c. $-5^2 = -5 \cdot 5 = -25$ Base 5, exponent 2

d. $(\frac{2}{3})^4 = \frac{2}{3} \cdot \frac{2}{3} \cdot \frac{2}{3} \cdot \frac{2}{3} = \frac{16}{81}$ Base $\frac{2}{3}$, exponent 4

e. $(x + y)^2 = (x + y)(x + y)$ Base $(x + y)$, exponent 2

f. $x^6 = x \cdot x \cdot x \cdot x \cdot x \cdot x$ Base x, exponent 6   ▲

In this section we will be concerned with the simplification of those products that involve more than one base and more than one exponent. Actually, almost all of the expressions can be simplified by applying the definition of positive integer exponents; that is, by writing all expressions in expanded form and then simplifying by using ordinary arithmetic. This process, however, is usually very time-consuming. We can shorten the process considerably by making some generalizations about exponents and applying the generalizations whenever it is convenient.

▼ **Example 2** Write the product $x^3 \cdot x^4$ with a single exponent.

Solution $x^3 \cdot x^4 = (x \cdot x \cdot x)(x \cdot x \cdot x \cdot x)$
                  $= (x \cdot x \cdot x \cdot x \cdot x \cdot x \cdot x)$
                  $= x^7$ *Notice:* $3 + 4 = 7$ ▲

We can generalize this result into the first property of exponents.

PROPERTY 1   If a is a real number, and r and s are positive integers, then

$$a^r \cdot a^s = a^{r+s}$$

The product of two expressions with the same base is equivalent to the base raised to the sum of the exponents from the original two expressions.

Expressions of the form $(a^r)^s$ occur frequently with exponents. Here is an example:

▼   **Example 3**   Write $(5^3)^2$ with a single exponent.

   Solution $(5^3)^2 = 5^3 \cdot 5^3$
                      $= 5^6$ *Notice:* $3 \cdot 2 = 6$ ▲

Generalizing this result we have a second property of exponents:

PROPERTY 2   If a is a real number and r and s are positive integers, then

$$(a^r)^s = a^{r \cdot s}$$

An expression with an exponent, raised to another power, is the same as the base from the original expression raised to the product of the powers.

A third property of exponents arises when we have the product of two or more numbers raised to an integer power. For example:

▼   **Example 4**   Expand $(3x)^4$ and then multiply.

   Solution $(3x)^4 = (3x)(3x)(3x)(3x)$
                   $= (3 \cdot 3 \cdot 3 \cdot 3)(x \cdot x \cdot x \cdot x)$
                   $= 3^4 \cdot x^4$ *Notice:* The exponent 4 dis-
                                          tributes over the product $3x$.
                   $= 81x^4$ ▲

PROPERTY 3   If a and b are any two real numbers, and r is a positive integer, then

$$(ab)^r = a^r \cdot b^r$$

Here are some examples that use combinations of the first three properties of exponents to simplify expressions involving exponents:

▼ **Example 5** Simplify each expression using the properties of exponents.

a. $(-3x^2)(5x^4) = -3(5)(x^2 \cdot x^4)$ Commutative, associative
                                                                  properties
$\qquad\qquad\qquad = -15x^6$ Property 1 for exponents

b. $(-2x^2)^3(4x^5) = (-2)^3(x^2)^3(4x^5)$ Property 3
$\qquad\qquad\quad = -8x^6 \cdot (4x^5)$ Property 2
$\qquad\qquad\quad = (-8 \cdot 4)(x^6 \cdot x^5)$ Commutative and associative
$\qquad\qquad\quad = -32x^{11}$ Property 1

c. $(x^2)^4(x^2y^3)^2(y^4)^3 = x^8 \cdot x^4 \cdot y^6 \cdot y^{12}$ Properties 2 and 3
$\qquad\qquad\qquad = x^{12}y^{18}$ Property 1   ▲

The last property of exponents for this section deals with negative integer exponents. We will state this property as a definition.

DEFINITION (PROPERTY 4) If a is any nonzero real number and r is a positive integer, then

$$a^{-r} = \frac{1}{a^r}$$

This is the definition for negative exponents. It says negative exponents indicate reciprocals.

The following example illustrates the definition for negative exponents:

▼ **Example 6** Write with positive exponents, then simplify.

a. $5^{-2} = \dfrac{1}{5^2} = \dfrac{1}{25}$

b. $(-2)^{-3} = \dfrac{1}{(-2)^3} = \dfrac{1}{-8} = -\dfrac{1}{8}$

c. $\left(\dfrac{3}{4}\right)^{-2} = \dfrac{1}{(\frac{3}{4})^2} = \dfrac{1}{\frac{9}{16}} = \dfrac{16}{9}$ ▲

As you can see, negative exponents give us another way of writing fractions. For example, the fraction $\frac{1}{6}$ can be written as 6^{-1}. Likewise, the fraction $\frac{1}{9}$ can be written as 3^{-2}.

Here is an example that involves negative exponents and reviews adding fractions with different denominators.

▼   **Example 7**   Simplify $6^{-1} + 3^{-2}$.

Solution   Applying the definition of negative exponents we have

$$6^{-1} + 3^{-2} = \frac{1}{6^1} + \frac{1}{3^2}$$

$$= \frac{1}{6} + \frac{1}{9}$$

The least common denominator for these two fractions is 18. (It is the smallest number that is exactly divisible by both 6 and 9.) We multiply $\frac{1}{6}$ by $\frac{3}{3}$ and $\frac{1}{9}$ by $\frac{2}{2}$ to produce two fractions with the same value as the original two fractions, but both of which have the least common denominator.

$$\frac{1}{6} + \frac{1}{9} = \frac{1}{6} \cdot \frac{3}{3} + \frac{1}{9} \cdot \frac{2}{2}$$

$$= \frac{3}{18} + \frac{2}{18}$$

$$= \frac{5}{18} \qquad\qquad ▲$$

The next example shows how we simplify expressions that contain both positive and negative exponents.

▼   **Example 8**   Simplify and write your answers with positive exponents only. (Assume all variables are nonzero.)

a. $(2x^{-3})^4 = 2^4(x^{-3})^4$ Property 3

$\qquad\quad = 16x^{-12}$ Property 2

$\qquad\quad = 16 \cdot \dfrac{1}{x^{12}}$ Property 4

$\qquad\quad = \dfrac{16}{x^{12}}$ Multiplication of fractions

b. $(5y^{-4})^2(2y^5) = 25y^{-8}(2y^5)$ Properties 2 and 3

$\qquad\qquad\quad = (25 \cdot 2)(y^{-8}y^5)$ Commutative and associative

$\qquad\qquad\quad = 50x^{-3}$ Property 1

$$= 50 \cdot \frac{1}{x^3} \qquad \text{Property 4}$$

$$= \frac{50}{x^3} \qquad \text{Multiplication of fractions} \qquad \blacktriangle$$

The last topic we will cover in this section is scientific notation. Scientific notation is a way in which to write very large or very small numbers in a more manageable form. Here is the definition.

DEFINITION   A number is written in *scientific notation* if it is written as the product of a number between 1 and 10 with an integer power of 10. A number written in scientific notation has the form

$$n \times 10^r$$

where $1 \leq n < 10$ and $r =$ an integer.

The table below lists some numbers and their forms in scientific notation. Each number on the left is equal to the number on the right-hand side of the equal sign.

Number Written the Long Way		*Number Written Again in Scientific Notation*
376,000	=	3.76×10^5
49,500	=	4.95×10^4
3,200	=	3.2×10^3
591	=	5.91×10^2
46	=	4.6×10^1
8	=	8×10^0
0.47	=	4.7×10^{-1}
0.093	=	9.3×10^{-2}
0.00688	=	6.88×10^{-3}
0.0002	=	2×10^{-4}
0.000098	=	9.8×10^{-5}

Notice that in each case, when the number is written in scientific notation, the decimal point in the first number is placed so that the number is between 1 and 10. The exponent on 10 in the second number keeps track of the number of places we moved the decimal point in the original number to get a number between 1 and 10:

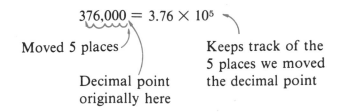

$$376{,}000 = 3.76 \times 10^5$$

Moved 5 places

Decimal point
originally here

Keeps track of the
5 places we moved
the decimal point

If a number written in expanded form is greater than or equal to 10, then when the number is written in scientific notation the exponent on 10 will be positive. A number that is less than 10 will have a negative exponent when written in scientific notation.

Problem Set 3.1

Evaluate each of the following.

1. 4^2 2. $(-4)^2$ 3. -4^2 4. $-(-4)^2$
5. -3^3 6. $(-3)^3$ 7. 2^5 8. 2^4
9. $(\frac{1}{2})^3$ 10. $(\frac{3}{4})^2$ 11. $(-\frac{5}{6})^2$ 12. $(-\frac{7}{8})^2$

Use the properties of exponents to simplify each of the following as much as possible.

13. $x^5 \cdot x^4$ 14. $x^6 \cdot x^3$ 15. $(2^3)^2$ 16. $(3^2)^2$
17. $(-2x^2)^3$ 18. $(-3x^4)^3$ 19. $-3a^2(2a^4)$ 20. $5a^7(-4a^6)$
21. $6x^2(-3x^4)(2x^5)$ 22. $(5x^3)(-7x^4)(-2x^6)$
23. $(-3n)^4(2n^3)^2(-n^6)^4$ 24. $(5n^6)^2(-2n^3)^2(-3n^7)^2$

Write each of the following with positive exponents. Then simplify as much as possible.

25. 3^{-2} 26. $(-5)^{-2}$ 27. $(-2)^{-5}$ 28. 2^{-5}
29. $(-3)^{-2}$ 30. $(-7)^{-2}$ 31. $(\frac{3}{4})^{-2}$ 32. $(\frac{2}{5})^{-2}$
33. $3^{-1} + 2^{-2}$ 34. $4^{-1} + 2^{-3}$
35. $3^{-2} - 2^{-3}$ 36. $2^{-3} - 4^{-2}$
37. $(\frac{1}{3})^{-2} + (\frac{1}{2})^{-3}$ 38. $(\frac{1}{2})^{-2} + (\frac{1}{3})^{-3}$
39. $(\frac{2}{3})^{-2} - (\frac{2}{5})^{-2}$ 40. $(\frac{3}{2})^{-2} - (\frac{3}{4})^{-2}$

Simplify each expression. Write all answers with positive exponents only. (Assume all variables are nonzero.)

41. $x^{-4}x^7$ 42. $x^{-3}x^8$
43. $(a^2b^{-5})^3$ 44. $(a^4b^{-3})^3$
45. $(2x^{-3})^3(6x^4)$ 46. $(4x^{-4})^3(2x^8)$
47. $(5y^4)^{-3}(2y^{-2})^3$ 48. $(3y^5)^{-2}(2y^{-4})^3$
49. $x^{m+2} \cdot x^{-2m} \cdot x^{m-5}$ 50. $x^{m-4} \cdot x^{m+9} \cdot x^{-2m}$
51. $(y^m)^2(y^{-3m})(y^{m+3})$ 52. $(y^m)^{-4}(y^{3m})(y^{m-6})$

Write each number in scientific notation.

53. 378,000   54. 3,780,000
55. 4,900   56. 490
57. 0.00037   58. 0.000037
59. 0.00495   60. 0.0495
61. 0.562   62. 0.0562

Write each number in expanded form.

63. 5.34×10^3 64. 5.34×10^2
65. 7.8×10^6 66. 7.8×10^4
67. 3.44×10^{-3} 68. 3.44×10^{-5}
69. 4.9×10^{-1} 70. 4.9×10^{-2}

Property 4 for exponents states that $a^{-r} = \dfrac{1}{a^r}$. We could also have stated it this

way $\dfrac{1}{a^{-r}} = a^r$ because

$$\frac{1}{a^{-r}} = \frac{1}{\dfrac{1}{a^r}} = 1 \cdot \frac{a^r}{1} = a^r.$$

Use this idea to simplify the following.

71. $\dfrac{1}{2^{-3}}$ 72. $\dfrac{1}{3^{-2}}$

73. $\dfrac{1}{5^{-2}}$ 74. $\dfrac{1}{3^{-4}}$

75. Which of these two expressions is larger, $(2^2)^3$ or 2^{2^3}?
76. Which of these two expressions is larger, $(3^4)^0$ or 3^{4^0}?
77. Let $d = 174$, $v = 10$, and $t = 3$ in the formula $d = vt + \frac{1}{2}gt^2$, and then solve for g.
78. Let $d = 36$, $v = 60$, and $t = 3$ in the formula $d = vt - \frac{1}{2}gt^2$, and then solve for g.
79. The statement $(a + b)^{-1} = a^{-1} + b^{-1}$ is false for all pairs of real numbers a and b. Show that it is false when $a = 2$ and $b = 4$.
80. Is the statement $(a + b)^2 = a^2 + b^2$ true in general? Try it with $a = 2$ and $b = 3$ and see.
81. The mass of the earth is approximately 5.98×10^{24} kilograms. If this number were written in expanded form, how many zeros would it contain?
82. The mass of a single hydrogen atom is approximately 1.67×10^{-27} kilograms. If this number were written in expanded form, how many digits would there be to the right of the decimal point?

Review Problems   The following problems review material we covered in Section 1.5.

Simplify each expression.

83. $6 - (-8)$ **84.** $-6 - (-8)$
85. $8 - (-6)$ **86.** $-8 - (-6)$
87. $-4 - (-3)$ **88.** $4 - (-3)$
89. $-4 - (-9)$ **90.** $4 - (-9)$

**3.2
Properties of
Exponents II**

We begin this section by stating two properties of exponents that show how exponents affect division. The first property shows that exponents distribute over quotients. This is property 5 in our list of properties.

PROPERTY 5 If a and b are any two real numbers with $b \neq 0$, and r is a positive integer, then

$$\left(\frac{a}{b}\right)^r = \frac{a^r}{b^r}$$

Proof

$$\left(\frac{a}{b}\right)^r = \underbrace{\left(\frac{a}{b}\right)\left(\frac{a}{b}\right)\left(\frac{a}{b}\right) \cdots \left(\frac{a}{b}\right)}_{r \text{ factors}}$$

$$= \frac{a \cdot a \cdot a \cdots a}{b \cdot b \cdot b \cdots b} \begin{array}{l} \leftarrow r \text{ factors} \\ \leftarrow r \text{ factors} \end{array}$$

$$= \frac{a^r}{b^r}$$

Since we will be working with quotients in this section, let's assume all our variables are represent nonzero numbers.

▼ **Example 1** Simplify each expression and write all answers with positive exponents only.

a. $\left(\dfrac{x^2}{y^3}\right)^4 = \dfrac{(x^2)^4}{(y^3)^4}$ Property 5

$\qquad = \dfrac{x^8}{y^{12}}$ Property 2

b. $\left(\dfrac{2x^{-2}}{y^{-1}}\right)^3 = \dfrac{(2x^{-2})^3}{(y^{-1})^3}$ Property 5

$\qquad = \dfrac{8x^{-6}}{y^{-3}}$ Properties 2 and 3

$$= \frac{\dfrac{8}{x^6}}{\dfrac{1}{y^3}} \qquad \text{Definition of negative} \\ \text{exponents. (Property 4)}$$

$$= \frac{8}{x^6} \cdot \frac{y^3}{1} \qquad \text{Division of fractions}$$

$$= \frac{8y^3}{x^6} \qquad\qquad\qquad\qquad\qquad \blacktriangle$$

We will now consider some examples of the quotient of two expressions involving exponents in which the bases are the same. Since multiplication with the same base resulted in addition of exponents, and since division is to multiplication as subtraction is to addition, it seems reasonable to expect division with the same base to result in subtraction of exponents.

Let's begin with an example in which the exponent in the numerator is larger than the exponent in the denominator:

$$\frac{6^5}{6^3} = \frac{6 \cdot 6 \cdot 6 \cdot 6 \cdot 6}{6 \cdot 6 \cdot 6}$$

Dividing out the 6's common to the numerator and denominator, we have

$$\frac{6^5}{6^3} = \frac{\not6 \cdot \not6 \cdot \not6 \cdot 6 \cdot 6}{\not6 \cdot \not6 \cdot \not6}$$
$$= 6 \cdot 6$$
$$= 6^2 \qquad \textit{Notice: } 5 - 3 = 2$$

If we were simply to subtract the exponent in the denominator from the exponent in the numerator, we would obtain the correct result, which is 6^2.

Let's consider an example where the exponent in the denominator is larger than the exponent in the numerator:

$$\frac{5^3}{5^7} = \frac{\not5 \cdot \not5 \cdot \not5}{\not5 \cdot \not5 \cdot \not5 \cdot 5 \cdot 5 \cdot 5 \cdot 5}$$
$$= \frac{1}{5 \cdot 5 \cdot 5 \cdot 5}$$
$$= \frac{1}{5^4}$$
$$= 5^{-4} \qquad \textit{Notice: } 3 - 7 = -4$$

Here again, simply subtracting the exponent in the denominator from the exponent in the numerator would give the correct result.

We summarize this discussion with property 6 for exponents.

PROPERTY 6   If a is any nonzero real number, and r and s are any two integers, then

$$\frac{a^r}{a^s} = a^{r-s}$$

Notice we have specified r and s to be any integers instead of only positive integers as was the case in our previous properties. Our definition of negative exponents is such that the properties of exponents hold for all integer exponents, whether positive integers or negative.

▼   **Example 2**   Apply Property 6 to each expression and then simplify the result. All answers that contain exponents should contain positive exponents only.

a. $\dfrac{2^8}{2^3} = 2^{8-3} = 2^5 = 32$

b. $\dfrac{x^2}{x^{18}} = x^{2-18} = x^{-16} = \dfrac{1}{x^{16}}$

c. $\dfrac{a^6}{a^{-8}} = a^{6-(-8)} = a^{14}$

d. $\dfrac{m^{-5}}{m^{-7}} = m^{-5-(-7)} = m^2$ ▲

In the next example we use more than one property of exponents to simplify an expression.

▼   **Example 3**   Simplify $\left(\dfrac{x^{-3}}{x^5}\right)^{-2}$

Solution   Let's begin by applying Property 6 to simplify inside the parentheses.

$$\left(\frac{x^{-3}}{x^5}\right)^{-2} = (x^{-3-5})^{-2} \qquad \text{Property 6}$$

$$= (x^{-8})^{-2}$$

$$= x^{16} \qquad\qquad \text{Property 2}$$

Notice that we could have obtained the same result by applying Property 5 first.

$$\left(\frac{x^{-3}}{x^5}\right)^{-2} = \frac{(x^{-3})^{-2}}{(x^5)^{-2}} \qquad \text{Property 5}$$

$$= \frac{x^6}{x^{-10}} \qquad \text{Property 2}$$

$$= x^{6-(-10)} \qquad \text{Property 6}$$

$$= x^{16} \qquad \qquad \blacktriangle$$

Let's complete our list of properties by looking at how the numbers 0 and 1 behave when used as exponents.

We can use the original definition for exponents when the number 1 is used as an exponent.

$$a^1 = \underbrace{a}_{\text{1 factor}}$$

For 0 as an exponent, consider the expression $3^4/3^4$. Since $3^4 = 81$, we have

$$\frac{3^4}{3^4} = \frac{81}{81} = 1$$

On the other hand, since we have the quotient of two expressions with the same base, we can subtract exponents.

$$\frac{3^4}{3^4} = 3^{4-4} = 3^0$$

Hence, 3^0 must be the same as 1.

Summarizing these results, we have our last property for exponents:

PROPERTY 7 If a is any real number, then

$$\text{and} \quad \begin{array}{l} a^1 = a \\ a^0 = 1 \quad (\text{as long as } a \neq 0) \end{array}$$

▼ **Example 4** Simplify.

a. $(2x^2y^4)^0 = 1$
b. $(2x^2y^4)^1 = 2x^2y^4$ ▲

Here are two examples that use many of the properties of exponents. There are a number of different ways to proceed on problems like these. You should use the method that works best for you.

▼ **Example 5** Simplify.

a. $\dfrac{(2x^2)^4(-3x^{-2})^2}{4x^{-6}(9x^{-3})} = \dfrac{16x^8 \cdot 9x^{-4}}{4x^{-6} \cdot 9x^{-3}}$

$\qquad\qquad = \dfrac{4x^4}{x^{-9}} \qquad$ Property 1

$\qquad\qquad = 4x^{13} \qquad$ Property 6

b. $\left(\dfrac{2x^0y^4z^{-3}}{8x^5y^{-3}z^{-2}}\right)^{-2} = \left(\dfrac{x^{-5} \cdot y^7 \cdot z^{-1}}{4}\right)^{-2} \qquad$ Property 6

$\qquad\qquad = \dfrac{x^{10}y^{-14}z^2}{4^{-2}} \qquad$ Properties 2 and 3

$\qquad\qquad = \dfrac{x^{10}\left(\dfrac{1}{y^{14}}\right)z^2}{\frac{1}{16}} \qquad$ Property 4

$\qquad\qquad = \dfrac{16x^{10}z^2}{y^{14}}$ ▲

We can use our properties of exponents to do arithmetic with numbers written in scientific notation. Here is an example.

▼ **Example 6** Simplify each expression and write all answers in scientific notation.

a. $(2 \times 10^8)(3 \times 10^{-3}) = (2)(3) \times (10^8)(10^{-3})$

$\qquad\qquad = 6 \times 10^5$

b. $\dfrac{4.8 \times 10^9}{2.4 \times 10^{-3}} = \dfrac{4.8}{2.4} \times \dfrac{10^9}{10^{-3}}$

$\qquad\qquad = 2 \times 10^{12}$

c. $\dfrac{(6.8 \times 10^5)(3.9 \times 10^{-7})}{7.8 \times 10^{-4}} = \dfrac{(6.8)(3.9)}{7.8} \times \dfrac{(10^5)(10^{-7})}{10^{-4}}$

$\qquad\qquad = 3.4 \times 10^2$ ▲

Use the properties of exponents to simplify each expression. All answers should   *Problem Set 3.2*
contain positive exponents only. Assume all variables are nonzero.

1. $\left(\dfrac{x^3}{y^2}\right)^2$ **2.** $\left(\dfrac{x^5}{y^2}\right)^3$ **3.** $\left(\dfrac{2a^{-2}}{b^{-1}}\right)^2$ **4.** $\left(\dfrac{2a^{-3}}{b^{-2}}\right)^3$

5. $\dfrac{3^4}{3^6}$ **6.** $\dfrac{3^6}{3^4}$ **7.** $\dfrac{2^{-2}}{2^{-5}}$ **8.** $\dfrac{2^{-5}}{2^{-2}}$

9. $\dfrac{x^{-1}}{x^9}$ **10.** $\dfrac{x^{-3}}{x^5}$ **11.** $\dfrac{2a^{-5}}{a^{-6}}$ **12.** $\dfrac{a^{-3}}{3a^{-4}}$

13. $\dfrac{3^5 \cdot 3^2}{3^6 \cdot 3^5}$ **14.** $\dfrac{2^7 \cdot 2^3}{2^6 \cdot 2^8}$ **15.** $\left(\dfrac{x^{-2}}{x^4}\right)^{-1}$ **16.** $\left(\dfrac{x^3}{x^{-2}}\right)^{-2}$

17. $\dfrac{a^{-4}b^5}{a^{-3}b^{-2}}$ **18.** $\dfrac{a^5 b^{-3}}{a^{-2}b^{-1}}$

19. $\left(\dfrac{x^{-5}y^2}{x^{-3}y^5}\right)^{-2}$ **20.** $\left(\dfrac{x^{-8}y^{-3}}{x^{-5}y^6}\right)^{-1}$

21. $\dfrac{12m^{-6}n^0}{3m^{-4}n^{-5}}$ **22.** $\dfrac{18m^8n^{-4}}{6m^0n^{-7}}$

23. $\dfrac{(2x^2)^3(3x^4)^{-1}}{8x^{-3}}$ **24.** $\dfrac{(3x)^4(2x^2)^{-3}}{9x^{-2}}$

25. $\left(\dfrac{2x^{-3}y^0}{4x^6y^{-5}}\right)^{-2}$ **26.** $\left(\dfrac{2x^6y^4z^0}{8x^{-3}y^0z^{-5}}\right)^{-1}$

27. $\left(\dfrac{ab^{-3}c^{-2}}{a^{-3}b^0c^{-5}}\right)^{-1}$ **28.** $\left(\dfrac{a^3b^2c^1}{a^{-1}b^{-2}c^{-3}}\right)^{-2}$

29. $\left(\dfrac{x^{-3}y^2}{x^4y^{-5}}\right)^{-2}\left(\dfrac{x^{-4}y}{x^0y^2}\right)$ **30.** $\left(\dfrac{x^{-1}y^4}{x^{-5}y^0}\right)^{-1}\left(\dfrac{x^3y^{-1}}{xy^{-3}}\right)$

31. $\dfrac{x^{n+2}}{x^{n-3}}$ **32.** $\dfrac{x^{n-3}}{x^{n-7}}$

33. $\dfrac{a^{3m}a^{m+1}}{a^{4m}}$ **34.** $\dfrac{a^{2m}a^{m-5}}{a^{3m-7}}$

35. $\dfrac{(y^r)^{-2}}{y^{-2r}}$ **36.** $\dfrac{(y^r)^2}{y^{2r-1}}$

Use the properties of exponents to simplify each of the following expressions.
Write all answers in scientific notation.

37. $(4 \times 10^{10})(2 \times 10^{-6})$ **38.** $(3 \times 10^{-12})(3 \times 10^4)$

39. $(4.5 \times 10^6)(2 \times 10^4)$ **40.** $(4.3 \times 10^8)(2 \times 10^5)$

41. $\dfrac{8 \times 10^{14}}{4 \times 10^5}$ **42.** $\dfrac{6 \times 10^8}{2 \times 10^3}$

43. $\dfrac{6.8 \times 10^6}{3.4 \times 10^8}$ **44.** $\dfrac{9.6 \times 10^{11}}{4.8 \times 10^{15}}$

45. $\dfrac{(5 \times 10^6)(4 \times 10^{-8})}{8 \times 10^4}$

46. $\dfrac{(6 \times 10^{-7})(3 \times 10^9)}{5 \times 10^6}$

47. $\dfrac{(2.4 \times 10^{-3})(3.6 \times 10^{-7})}{(4.8 \times 10^6)(1 \times 10^{-9})}$

48. $\dfrac{(7.5 \times 10^{-6})(1.5 \times 10^9)}{(1.8 \times 10^4)(2.5 \times 10^{-2})}$

Convert each number to scientific notation and then simplify. Write all answers in scientific notation.

49. (2,000,000)(0.0000249)

50. (30,000)(0.000192)

51. $\dfrac{69,800}{0.000349}$

52. $\dfrac{0.000545}{1,090,000}$

53. $\dfrac{(40,000)(0.0007)}{0.0014}$

54. $\dfrac{(800,000)(0.00002)}{4,000}$

55. Write the number 278×10^5 in scientific notation.

56. Write the number 278×10^{-5} in scientific notation.

For each expression that follows, find a value of x that makes it a true statement.

57. $4^3 \cdot 4^x = 4^9$

58. $(4^3)^x = 4^9$

59. $\dfrac{3^x}{3^5} = 3^4$

60. $\dfrac{3^x}{3^5} = 3^{-4}$

61. $2^x \cdot 2^3 = \dfrac{1}{16}$

62. $(2^x)^3 = \dfrac{1}{8}$

63. A light-year, the distance light travels in one year, is approximately 5.9×10^{12} miles. The Andromeda Galaxy is approximately 1.7×10^6 light years from our galaxy. Find the distance in miles between our galaxy and the Andromeda Galaxy.

64. The distance from the earth to the sun is approximately 9.3×10^7 miles. If light travels 1.12×10^7 miles in one minute, how many minutes does it take the light from the sun to reach the earth?

Review Problems   The problems below review some of the material we covered in Section 1.6.

Simplify each expression.

65. $6 + 2(x + 3)$

66. $8 + 3(x + 4)$

67. $3(2a + 1) - 6a$
69. $3(4y - 2) - (6y - 3)$
71. $3 - 7(x - 5) + 3x$

68. $5(3a + 3) - 8a$
70. $2(5y - 6) - (3y + 4)$
72. $4 - 9(x - 3) + 5x$

We begin this section with the definition of the basic unit, a term, around which polynomials are defined. Once we have listed all the terminology associated with polynomials, we will show how the distributive property is used to find sums and differences of polynomials.

**3.3
Polynomials, Sums,
and Differences**

DEFINITION   A *term* or *monomial* is a constant or the product of a constant and one or more variables raised to whole-number exponents.

*Polynomials in
General*

The following are monomials or terms:

$$-16, \qquad 3x^2y, \qquad -\tfrac{2}{5}a^3b^2c, \qquad xy^2z$$

The numerical part of each monomial is called the *numerical coefficient*, or just *coefficient* for short. For the above terms the coefficients are -16, 3, $-\tfrac{2}{5}$, and 1. Notice that the coefficient for xy^2z is understood to be 1.

DEFINITION   A *polynomial* is any finite sum of terms. Since subtraction can be written in terms of addition, finite differences are also included in this definition.

The following are polynomials:

$$2x^2 - 6x + 3, \qquad -5x^2y + 2xy^2, \qquad 4a - 5b + 6c + 7d$$

Polynomials can be classified further according to the number of terms present. If a polynomial consists of two terms, it is said to be a *binomial*. If it has three terms, it is called a *trinomial*. And as stated above, a polynomial with only one term is said to be a *monomial*.

DEFINITION   The *degree* of a polynomial with one variable is the highest power to which the variable is raised in any one term.

▼   **Example 1**

a. $6x^2 + 2x - 1$ A trinomial of degree 2
b. $5x - 3$ A binomial of degree 1
c. $7x^6 - 5x^3 + 2x - 4$ A polynomial of degree 6

d. $-7x^4$ A monomial of degree 4
e.   15   A monomial of degree 0   ▲

Polynomials in one variable are usually written in decreasing powers of the variable. When this is the case, the coefficient of the first term is called the *leading coefficient*. In part a of Example 1 above, the leading coefficient is 6. In part b it is 5. The leading coefficient in part c is 7.

Addition and Subtraction of Polynomials

Combining polynomials to find sums and differences is a very simple process. For the most part it involves applying the distributive property to combine similar terms.

DEFINITION   Two or more terms which differ only in their numerical coefficients are called *similar* or *like* terms. Since similar terms differ only in their coefficients, they have identical variable parts—that is, the same variables raised to the same power. For example, $3x^2$ and $-5x^2$ are similar terms. So are $15x^2y^3z$, $-27x^2y^3z$ and $\frac{3}{4}x^2y^3z$.

We can use the distributive property to combine the similar terms $6x^2$ and $9x^2$ as follows:

$$6x^2 + 9x^2 = (6 + 9)x^2 \qquad \text{Distributive property.}$$
$$= 15x^2 \qquad \text{The sum of 6 and 9 is 15.}$$

The distributive property can also be used to combine more than two like terms.

▼   **Example 2**   Combine $7x^2y + 4x^2y - 10x^2y + 2x^2y$.

Solution

$$7x^2y + 4x^2y - 10x^2y + 2x^2y = (7 + 4 - 10 + 2)x^2y \qquad \text{Distributive property}$$
$$= 3x^2y \qquad \text{Addition} \quad ▲$$

To add two polynomials, we simply apply the commutative and associative properties to group similar terms and then use the distributive property as we have in the example above.

▼   **Example 3**   Add $5x^2 - 4x + 2$ and $3x^2 + 9x - 6$.

Solution

$$(5x^2 - 4x + 2) + (3x^2 + 9x - 6)$$

$$= (5x^2 + 3x^2) + (-4x + 9x) + (2 - 6) \qquad \text{Commutative and associative properties}$$

$$= (5 + 3)x^2 + (-4 + 9)x + (2 - 6) \qquad \text{Distributive property}$$

$$= 8x^2 + 5x + (-4)$$
$$= 8x^2 + 5x - 4 \qquad\qquad\qquad\qquad\qquad \blacktriangle$$

In actual practice it is not necessary to show all the steps shown in Example 3. It is important to understand that addition of polynomials is equivalent to combining similar terms. We add similar terms by combining coefficients.

Sometimes it is convenient to add polynomials vertically in columns.

▼   **Example 4**   Find the sum of $-8x^3 + 7x^2 - 6x + 5$ and $10x^3 + 3x^2 - 2x - 6$.

Solution   We can add the two polynomials using the method of Example 3, or we can arrange similar terms in columns and add vertically. Using the column method, we have:

$$\begin{array}{r}
-8x^3 + 7x^2 - 6x + 5 \\
10x^3 + 3x^2 - 2x - 6 \\
\hline
2x^3 + 10x^2 - 8x - 1
\end{array} \qquad \blacktriangle$$

It is important to notice that no matter which method is used to combine two polynomials, the variable part of the combined terms never changes. It is sometimes tempting to say $7x^2 + 3x^2 = 10x^4$, which is incorrect. Remember, we combine similar terms by using the distributive property: $7x^2 + 3x^2 = (7 + 3)x^2 = 10x^2$. The variable part common to each term is unchanged.

To find the difference of two polynomials, we need to use the fact that the opposite of a sum is the sum of the opposites. That is,

$$-(a + b) = -a + (-b)$$

One way to remember this is to observe that $-(a + b)$ is equivalent to $-1(a + b) = (-1)a + (-1)b = -a + (-b)$.

If there is a negative sign directly preceding the parentheses surrounding a polynomial, we may remove the parentheses and preceding negative sign by changing the sign of each term within the parentheses. For example:

$$-(3x + 4) = -3x + (-4) = -3x - 4$$
$$-(5x^2 - 6x + 9) = -5x^2 + 6x - 9$$
$$-(-x^2 + 7x - 3) = x^2 - 7x + 3$$

To find the difference of two or more polynomials, we simply apply this principle and proceed as we did when finding sums.

▼   **Example 5**

$(9x^2 - 3x + 5) - (4x^2 + 2x - 3)$

$= 9x^2 - 3x + 5 + (-4x^2) + (-2x) + 3$ The opposite of a sum is the sum of the opposites.

$= (9x^2 - 4x^2) + (-3x - 2x) + (5 + 3)$ Commutative and associative properties

$= 5x^2 - 5x + 8$ Combine similar terms.   ▲

▼   **Example 6**

$(2x^3 + 5x^2 + 3) - (4x^2 - 2x - 7) - (6x^3 - 3x + 1)$
$= (2x^3 + 5x^2 + 3) + (-4x^2 + 2x + 7) + (-6x^3 + 3x - 1)$
$= 2x^3 + 5x^2 + 3 - 4x^2 + 2x + 7 - 6x^3 + 3x - 1$
$= (2x^3 - 6x^3) + (5x^2 - 4x^2) + (2x + 3x) + (3 + 7 - 1)$
$= -4x^3 + x^2 + 5x + 9$ ▲

▼   **Example 7**   Subtract $4x^2 - 9x + 1$ from $-3x^2 + 5x - 2$.

Solution

$(-3x^2 + 5x - 2) - (4x^2 - 9x + 1)$
$= (-3x^2 + 5x - 2) + (-4x^2 + 9x - 1)$
$= -3x^2 + 5x - 2 - 4x^2 + 9x - 1$
$= (-3x^2 - 4x^2) + (5x + 9x) + (-2 - 1)$
$= -7x^2 + 14x - 3$ ▲

When one set of grouping symbols is contained within another, it is best to begin the process of simplification with the innermost grouping symbol and work out from there.

▼ **Example 8**   Simplify $4x - 3[2 - (3x + 4)]$

   Solution   Removing the innermost parentheses first, we have

$$\begin{aligned} 4x - 3[2 - (3x + 4)] &= 4x - 3(2 - 3x - 4) \\ &= 4x - 3(-3x - 2) \\ &= 4x + 9x + 6 \\ &= 13x + 6 \end{aligned}$$ ▲

▼ **Example 9**   Simplify $(2x + 3) - [(3x + 1) - (x - 7)]$

   Solution

$$\begin{aligned} (2x + 3) - [(3x + 1) - (x - 7)] &= (2x + 3) - (3x + 1 - x + 7) \\ &= (2x + 3) - (2x + 8) \\ &= -5 \end{aligned}$$ ▲

In the example that follows we will find the value of a polynomial for a given value of the variable.

▼ **Example 10**   Find the value of $5x^3 - 3x^2 + 4x - 5$ when x is 2.

   Solution   We begin by substituting 2 for x in the original polynomial.

When $x = 2$
the polynomial $5x^3 - 3x^2 + 4x - 5$
becomes $5 \cdot 2^3 - 3 \cdot 2^2 + 4 \cdot 2 - 5$

When we simplify a numerical term that contains an exponent, like $5 \cdot 2^3$, we evaluate the power first, and then multiply; $5 \cdot 2^3 = 5 \cdot 8 = 40$.

$$\begin{aligned} 5 \cdot 2^3 - 3 \cdot 2^2 + 4 \cdot 2 - 5 &= 5 \cdot 8 - 3 \cdot 4 + 4 \cdot 2 - 5 \\ &= 40 - 12 + 8 - 5 \\ &= 31 \end{aligned}$$ ▲

Problem Set 3.3

Identify those of the following that are monomials, binomials, or trinomials. Give the degree of each and name the leading coefficient.

1. $5x^2 - 3x + 2$
2. $2x^2 + 4x - 1$
3. $3x - 5$
4. $5y + 3$
5. $8a^2 + 3a - 5$
6. $9a^2 - 8a - 4$
7. $4x^3 - 6x^2 + 5x - 3$
8. $9x^4 + 4x^3 - 2x^2 + x$
9. $-\frac{3}{4}$
10. -16
11. $4x - 5 + 6x^3$
12. $9x + 2 + 3x^3$

Simplify each of the following by combining similar terms:

13. $(4x + 2) + (3x - 1)$
14. $(8x - 5) + (-5x + 4)$
15. $2x^2 - 3x + 10x - 15$
16. $6x^2 - 4x - 15x + 10$
17. $12a^2 + 8ab - 15ab - 10b^2$
18. $28a^2 - 8ab + 7ab - 2b^2$
19. $(5x^2 - 6x + 1) - (4x^2 + 7x - 2)$
20. $(11x^2 - 8x) - (4x^2 - 2x - 7)$
21. $(6x^2 - 4x - 2) - (3x^2 + 7x) + (4x - 1)$
22. $(8x^2 - 6x) - (3x^2 + 2x + 1) - (6x^2 + 3)$
23. $(y^3 - 2y^2 - 3y + 4) - (2y^3 - y^2 + y - 3)$
24. $(8y^3 - 3y^2 + 7y + 2) - (-4y^3 + 6y^2 - 5y - 8)$
25. $(5x^3 - 4x^2) - (3x + 4) + (5x^2 - 7) - (3x^3 + 6)$
26. $(x^3 - x) - (x^2 + x) + (x^3 - 1) - (-3x + 2)$
27. $(8x^2 - 2xy + y^2) - (7x^2 - 4xy - 9y^2)$
28. $(2x^2 - 5xy + y^2) + (-3x^2 + 4xy - 5)$
29. $(3a^3 + 2a^2b + ab^2 - b^3) - (6a^3 - 4a^2b + 6ab^2 - b^3)$
30. $(a^3 - 3a^2b + 3ab^2 - b^3) - (a^3 + 3a^2b + 3ab^2 + b^3)$
31. Subtract $2x^2 - 4x$ from $2x^2 - 7x$
32. Subtract $-3x + 6$ from $-3x + 9$
33. Find the sum of $x^2 - 6xy + y^2$ and $2x^2 - 6xy - y^2$.
34. Find the sum of $9x^3 - 6x^2 + 2$ and $3x^2 - 5x + 4$.
35. Subtract $-8x^5 - 4x^3 + 6$ from $9x^5 - 4x^3 - 6$.
36. Subtract $4x^4 - 3x^3 - 2x^2$ from $2x^4 + 3x^3 + 4x^2$.
37. Find   the   sum   of $11a^2 + 3ab + 2b^2$, $9a^2 - 2ab + b^2$,   and $-6a^2 - 3ab + 5b^2$.
38. Find the sum of $a^2 - ab - b^2$, $a^2 + ab - b^2$, and $a^2 + 2ab + b^2$.

Simplify each of the following. Begin by working on the innermost parentheses first.

39. $-[2 - (4 - x)]$
40. $-[-3 - (x - 6)]$
41. $-5[-(x - 3) - (x + 2)]$
42. $-6[(2x - 5) - 3(8x - 2)]$

43. $4x - 5[3 - (x - 4)]$

44. $x - 7[3x - (2 - x)]$

45. $-(3x - 4y) - [(4x + 2y) - (3x + 7y)]$

46. $(8x - y) - [-(2x + y) - (-3x - 6y)]$

47. $4a - \{3a + 2[a - 5(a + 1) + 4]\}$

48. $6a - \{-2a - 6[2a + 3(a - 1) - 6]\}$

49. Find the value of $2x^2 - 3x - 4$ when x is 2.

50. Find the value of $4x^2 + 3x - 2$ when x is -1.

51. Find the value of $x^2 - 6x + 5$ when x is 0.

52. Find the value of $4x^2 - 4x + 4$ when x is 0.

53. Find the value of $x^3 - x^2 + x - 1$ when x is -2.

54. Find the value of $x^3 + x^2 + x + 1$ when x is -2.

55. Let $a = 3$ in each of the following expressions and then simplify each one.

$$(a + 4)^2 \qquad a^2 + 16 \qquad a^2 + 8a + 16$$

56. Let $a = 2$ in each of the following expressions and then simplify each one.

$$(2a - 3)^2 \qquad 4a^2 - 9 \qquad 4a^2 - 12a + 9$$

57. If an object is thrown straight up into the air with a velocity of 128 feet/second, then its height h above the ground t seconds later is given by the formula

$$h = -16t^2 + 128t.$$

Find the height after 3 seconds, and after 5 seconds.

58. The formula for the height of an object that has been thrown straight up with a velocity of 64 feet/second is

$$h = -16t^2 + 64t.$$

Find the height after 1 second and after 3 seconds.

59. There is a surprising relationship between the set of odd whole numbers $\{1, 3, 5, 7, 9, \ldots\}$ and the set of whole number squares $\{1, 4, 9, 16, 25, \ldots\}$. To see this relationship, add the first two odd numbers, then the first three odd numbers, and then the first four odd numbers. See if you can write this relationship in words.

60. If x is a positive integer, then the formula

$$S = \frac{x^2 + x}{2}$$

gives the sum S of all the integers from 1 to x. Use this formula to find the sum of the first 10 positive integers.

Review Problems The problems below review material we covered in Section 3.1. Reviewing these problems will help you with the next section.

Simplify each expression.

61. $4x^3(5x^2)$

62. $4x^3(-3x)$

63. $2a^2b(ab^2)$

64. $2a^2b(-6a^2b)$

65. $2x(3x^2)$ **66.** $-3y(4y^2)$
67. $2x(-xy)$ **68.** $-3y(-xy)$

3.4
Multiplication of Polynomials

The distributive property is the key to multiplying polynomials. The simplest type of multiplication occurs when we multiply a polynomial by a monomial.

▼ **Example 1** Find the product of $4x^3$ and $5x^2 - 3x + 1$.

Solution

$4x^3(5x^2 - 3x + 1)$
$$= 4x^3(5x^2) + 4x^3(-3x) + 4x^3(1) \qquad \text{Distributive property}$$
$$= 20x^5 - 12x^4 + 4x^3$$

Notice we multiply coefficients and add exponents.   ▲

▼ **Example 2** Multiply $-2a^2b(4a^3 - 6a^2b + ab^2 - 3b^3)$.

Solution

$-2a^2b(4a^3 - 6a^2b + ab^2 - 3b^3)$
$$= (-2a^2b)(4a^3) + (-2a^2b)(-6a^2b) + (-2a^2b)(ab^2) + (-2a^2b)(-3b^3)$$
$$= -8a^5b + 12a^4b^2 - 2a^3b^3 + 6a^2b^4 \qquad\qquad ▲$$

The distributive property can also be applied to multiply a polynomial by a polynomial. Let's consider the case where both polynomials have two terms:

▼ **Example 3** Multiply $2x - 3$ and $x + 5$.

Solution

$(2x - 3)(x + 5)$
$$= (2x - 3)x + (2x - 3)5 \qquad\qquad\quad \text{Distributive property}$$
$$= 2x(x) + (-3)x + 2x(5) + (-3)5 \qquad \text{Distributive property}$$
$$= 2x^2 - 3x + 10x - 15$$
$$= 2x^2 + 7x - 15 \qquad\qquad\qquad\qquad \text{Combine like terms.}$$

Notice the third line in this example. It consists of all possible products of terms in the first binomial and those of the second

binomial. We can generalize this into a rule for multiplying two polynomials. ▲

RULE To multiply two polynomials, multiply each term in the first polynomial by each term in the second polynomial.

Multiplying polynomials can be accomplished by a method that looks very similar to long multiplication with whole numbers. We line up the polynomials vertically and then apply our rule for multiplication of polynomials. Here's how it looks using the same two binomials used in the last example:

$$
\begin{array}{r}
2x - 3 \\
x + 5 \\
\hline
10x - 15 \\
2x^2 - 3x \\
\hline
2x^2 + 7x - 15
\end{array}
$$

Multiply $+5$ times $2x - 3$.
Multiply x times $2x - 3$.
Add in columns.

The vertical method of multiplying polynomials does not directly show the use of the distributive property. It is, however, very useful since it always gives the correct result and is easy to remember.

▼ **Example 4** Multiply $(2x - 3y)$ and $(3x^2 - xy + 4y^2)$ vertically.

Solution

$$
\begin{array}{r}
3x^2 - xy + 4y^2 \\
2x - 3y \\
\hline
-9x^2y + 3xy^2 - 12y^3 \\
6x^3 - 2x^2y + 8xy^2 \\
\hline
6x^3 - 11x^2y + 11xy^2 - 12y^3
\end{array}
$$

Multiply $(3x^2 - xy + 4y^2)$ by $-3y$.
Multiply $(3x^2 - xy + 4y^2)$ by $2x$.
Add similar terms. ▲

▼ **Example 5** Multiply $(x - 3)(2x + 1)(3x - 4)$.

Solution By the associative property we can first find the product of any two of the binomials and then multiply the result by the other binomial.

We begin by multiplying the first two binomials:

$$
\begin{array}{r}
2x + 1 \\
x - 3 \\
\hline
-6x - 3 \\
2x^2 + x \\
\hline
2x^2 - 5x - 3
\end{array}
$$

We now multiply this product by the last binomial, $3x - 4$:

$$
\begin{array}{r}
2x^2 - 5x - 3 \\
3x - 4 \\
\hline
-8x^2 + 20x + 12 \\
6x^3 - 15x^2 - 9x \\
\hline
6x^3 - 23x^2 + 11x + 12
\end{array}
$$

The product is

$$(x - 3)(2x + 1)(3x - 4) = 6x^3 - 23x^2 + 11x + 12 \qquad \blacktriangle$$

The product of two binomials occurs very frequently in algebra. Since this type of product is so common, we have a special method of multiplication that applies only to products of binomials.

Consider the product of $(2x - 5)$ and $(3x - 2)$. Distributing $(3x - 2)$ over $2x$ and -5 we have

$$
\begin{aligned}
(2x - 5)(3x - 2) &= (2x)(3x - 2) + (-5)(3x - 2) \\
&= (2x)(3x) + (2x)(-2) + (-5)(3x) + (-5)(-2) \\
&= 6x^2 - 4x - 15x + 10 \\
&= 6x^2 - 19x + 10
\end{aligned}
$$

Looking closely at the second and third lines we notice the following relationships:

1. $6x^2$ comes from multiplying the *first* terms in each binomial:

    $$(2x - 5)(3x - 2) \qquad 2x(3x) = 6x^2 \qquad \textit{First } \text{terms}$$

2. $-4x$ comes from multiplying the *outside* terms in the product:

    $$(2x - 5)(3x - 2) \qquad 2x(-2) = -4x \qquad \textit{Outside } \text{terms}$$

3. $-15x$ comes from multiplying the *inside* terms in the product:

    $$(2x - 5)(3x - 2) \qquad -5(3x) = -15x \qquad \textit{Inside } \text{terms}$$

4. 10 comes from multiplying the *last* two terms in the product:

    $$(2x - 5)(3x - 2) \qquad -5(-2) = 10 \qquad \textit{Last } \text{terms}$$

Once we know where the terms in the answer come from, we can reduce the number of steps used in finding the product:

$$
\begin{array}{ccccccc}
(2x - 5)(3x - 2) = & 6x^2 & - & 4x & - & 15x & + & 10 \\
& \text{First} & & \text{Outside} & & \text{Inside} & & \text{Last}
\end{array}
$$

$$= 6x^2 - 19x + 10$$

This method is called the FOIL Method—*F*irst-*O*utside-*I*nside-*L*ast. The FOIL method does not show the properties used in multiplying two binomials. It is simply a way of finding products of binomials quickly. Remember, the FOIL method only applies to products of two binomials. The vertical method applies to all products of polynomials with two or more terms.

▼ **Example 6** Multiply $(4a - 5b)(3a + 2b)$.

Solution

$$(4a - 5b)(3a + 2b) = 12a^2 + 8ab - 15ab - 10b^2$$

$$\qquad\qquad\qquad\quad\ \ \updownarrow\qquad\updownarrow\qquad\ \updownarrow\qquad\ \ \updownarrow$$

$$\qquad\qquad\qquad\qquad\ \ \text{F}\qquad\text{O}\qquad\text{I}\qquad\ \text{L}$$

$$\qquad\qquad\quad\ = 12a^2 - 7ab - 10b^2 \qquad\qquad ▲$$

▼ **Example 7** Find $(4x - 6)^2$.

Solution

$$(4x - 6)^2 = (4x - 6)(4x - 6)$$

$$\qquad\quad\ = 16x^2 - 24x - 24x + 36$$

$$\qquad\qquad\quad\ \text{F}\qquad\text{O}\qquad\text{I}\qquad\text{L}$$

$$\qquad\quad\ = 16x^2 - 48x + 36 \qquad\qquad ▲$$

The last example is the square of a binomial. This type of product occurs frequently enough in algebra that we have a special formula for it.

Here are the formulas for binomial squares:

$$(a + b)^2 = (a + b)(a + b) = a^2 + ab + ab + b^2 = a^2 + 2ab + b^2$$

$$(a - b)^2 = (a - b)(a - b) = a^2 - ab - ab + b^2 = a^2 - 2ab + b^2$$

Observing the results in both cases we have the following rule:

RULE The square of a binomial is the sum of the square of the first term, twice the product of the two terms, and the square of the last term. Or:

$$(a + b)^2 = \quad a^2 \quad + \quad 2ab \quad + \quad b^2$$

	Square of first term	Twice the product of the two terms	Square of last term

$$(a - b)^2 = \quad a^2 \quad - \quad 2ab \quad + \quad b^2$$

▼   Example 8

a. $(x + y)^2 =$ x^2 $+$ $2xy$ $+$ y^2 $= x^2 + 2xy + y^2$
b. $(x + 7)^2 =$ x^2 $+$ $2(x)(7)$ $+$ 7^2 $= x^2 + 14x + 49$
c. $(3x - 5)^2 = (3x)^2 + 2(3x)(-5) + (-5)^2 = 9x^2 - 30x + 25$
d. $(4x - 2y)^2 = (4x)^2 + 2(4x)(-2y) + (-2y)^2 = 16x^2 - 16xy + 4y^2$

       First term   Twice their   Last term   Answer
        squared   product   squared   ▲

Note   From the above rule and examples it should be obvious that $(a + b)^2 \neq a^2 + b^2$. That is, the square of a sum is not the same as the sum of the squares.

Another frequently occurring kind of product is found when multiplying two binomials which differ only in the sign between their terms.

▼   **Example 9**   Multiply $(3x - 5)$ and $(3x + 5)$.

Solution

$$(3x - 5)(3x + 5) = 9x^2 + 15x - 15x - 25 \qquad \text{Two middle terms}$$
$$\text{add to 0.}$$

$$= 9x^2 - 25 \qquad\qquad\qquad ▲$$

The outside and inside products in Example 9 are opposites and therefore add to 0.
Here it is in general:

$$(a - b)(a + b) = a^2 + ab - ab + b^2 \qquad \text{Two middle terms}$$
$$\text{add to 0.}$$

$$= a^2 - b^2$$

RULE   To multiply two binomials which differ only in the sign between their two terms, simply subtract the square of the second term from the square of the first term:

$$(a - b)(a + b) = a^2 - b^2$$

The expression $a^2 - b^2$ is called the *difference of two squares*.

Once we memorize and understand this rule, we can multiply binomials of this form with a minimum of work.

▼   Example 10

a. $(x - 5)(x + 5) = x^2 - 25$
b. $(2a - 3)(2a + 3) = 4a^2 - 9$
c. $(x^2 + 4)(x^2 - 4) = x^4 - 16$
d. $(3x + 7y)(3x - 7y) = 9x^2 - 49y^2$
e. $(x^3 - 2a)(x^3 + 2a) = x^6 - 4a^2$ ▲

When working through the problems in the problem set, pay par-
ticular attention to products of binomials. We want to be aware of the
relationship between the terms in the answers and the terms in the
original binomials. In the next section we will do factoring. Since
factoring is actually the reverse of multiplication, the more familiar we
are with products, the easier factoring will be.

Multiply the following by applying the distributive property: *Problem Set 3.4*

1. $2x(6x^2 - 5x + 4)$ 2. $-3x(5x^2 - 6x - 4)$
3. $-3a^2(a^3 - 6a^2 + 7)$ 4. $4a^3(3a^2 - a + 1)$
5. $2a^2b(a^3 - ab + b^3)$ 6. $-5a^2b^2(8a^2 - 2ab + b^2)$
7. $-4x^2y^3(7x^2 - 3xy + 6y^2)$ 8. $-3x^3y^2(6x^2 - 3xy + 4y^2)$
9. $3r^3s^2(r^3 - 2r^2s + 3rs^2 + s^3)$
10. $-5r^2s^3(2r^3 + 3r^2s - 4rs^2 + 5s^3)$

Multiply the following vertically:

11. $(x - 5)(x + 3)$ 12. $(x + 4)(x + 6)$
13. $(2x - 3)(3x - 5)$ 14. $(3x + 4)(2x - 5)$
15. $(x + 3)(x^2 + 6x + 5)$ 16. $(x - 2)(x^2 - 5x + 7)$
17. $(3a + 5)(2a^3 - 3a^2 + a)$ 18. $(2a - 3)(3a^2 - 5a + 1)$
19. $(a - b)(a^2 + ab + b^2)$ 20. $(a + b)(a^2 - ab + b^2)$
21. $(2x + y)(4x^2 - 2xy + y^2)$ 22. $(x - 3y)(x^2 + 3xy + 9y^2)$
23. $(2a - 3b)(a^2 + ab + b^2)$ 24. $(5a - 2b)(a^2 - ab - b^2)$
25. $(3x - 4y)(6x^2 + 3xy + 4y^2)$ 26. $(2x - 6y)(7x^2 - 6xy + 3y^2)$
27. $2x^2(x - 5)(3x - 7)$ 28. $-5x^3(3x - 2)(x + 4)$
29. $(x - 2)(2x + 3)(3x - 4)$ 30. $(x + 5)(2x - 6)(x - 3)$

Multiply the following using the FOIL method:

31. $(x - 2)(x + 3)$ 32. $(x + 2)(x - 3)$
33. $(x - 2)(x - 3)$ 34. $(x + 2)(x + 3)$
35. $(2a + 3)(3a + 2)$ 36. $(5a - 4)(2a + 1)$
37. $(3x - 5)(2x + 4)$ 38. $(x - 7)(3x + 6)$
39. $(5x - 4)(x - 5)$ 40. $(7x - 5)(3x + 1)$
41. $(4a + 1)(5a + 1)$ 42. $(3a - 1)(2a - 1)$

43. $(5x - 6y)(4x + 3y)$ 44. $(6x - 5y)(2x - 3y)$
45. $(2x - 3y)(4x - 5y)$ 46. $(6x - 2y)(3x + y)$
47. $(4a + b)(7a - 2b)$ 48. $(3a - 4b)(6a + b)$

Find the following special products:

49. $(x + 2)^2$ 50. $(x - 5)^2$
51. $(2a - 3)^2$ 52. $(3a + 2)^2$
53. $(5x + 2y)^2$ 54. $(3x - 4y)^2$
55. $(x - 4)(x + 4)$ 56. $(x + 5)(x - 5)$
57. $(2a + 3b)(2a - 3b)$ 58. $(6a - 1)(6a + 1)$
59. $(3r + 7s)(3r - 7s)$ 60. $(5r - 2s)(5r + 2s)$
61. $(5x - 4y)(5x + 4y)$ 62. $(4x - 5y)(4x + 5y)$

Find the following products:

63. $(x - 2)^3$ 64. $(x + 4)^3$
65. $3(x - 1)(x - 2)(x - 3)$ 66. $2(x + 1)(x + 2)(x + 3)$
67. $(x^N + 3)(x^N - 2)$ 68. $(x^N + 4)(x^N - 1)$
69. $(x^{2N} - 3)(x^{2N} + 3)$ 70. $(x^{3N} + 4)(x^{3N} - 4)$
71. $(x + 3)(y + 5)$ 72. $(x - 4)(y + 6)$
73. $(b^2 + 8)(a^2 + 1)$ 74. $(b^2 + 1)(a^4 - 5)$
75. $(x - 2)(3y^2 + 4)$ 76. $(x - 4)(2y^3 + 1)$
77. Multiply $(x + y - 4)(x + y + 5)$ by first writing it like this:

$$[(x + y) - 4][(x + y) + 5]$$

and then applying the FOIL method.

78. Multiply $(x - 5 - y)(x - 5 + y)$ by first writing it like this:

$$[(x - 5) - y][(x - 5) + y].$$

79. Expand and multiply $(x + y + z)^2$ by first grouping the first two terms within the parentheses together. Like this:

$$[(x + y) + z]^2$$

80. Repeat Problem 79, but this time group the last two terms together. Like this:

$$[x + (y + z)]^2$$

81. The flower color (either red, pink, or white) of a certain species of sweet pea plant is due to a single gene. If p is the proportion of the dominant form of the gene in a population, and q is the proportion of the recessive form of this gene in the same population, then the proportion of plants in the next generation that have pink flowers is given by the middle term in the expansion of $(p + q)^2$. If $p = \frac{1}{4}$ and $q = \frac{3}{4}$, find the proportion of the next generation that will have pink flowers.

82. Repeat Problem 81 with $p = \frac{1}{2}$ and $q = \frac{1}{2}$.

83. Let $a = 2$ and $b = 3$ and evaluate each of the following expressions.

$$a^4 - b^4 \qquad (a - b)^4 \qquad (a^2 + b^2)(a + b)(a - b)$$

84. Let $a = 2$ and $b = 3$ and evaluate each of the following expressions.

$$a^3 + b^3 \qquad (a + b)^3 \qquad a^3 + 3a^2b + 3ab^2 + b^3$$

Review Problems   The problems that follow review material we covered in Sections 2.1 and 2.3.

Solve each equation.

85. $7x - 4 = 3x + 12$ **86.** $9x - 10 = 3x + 8$

87. $5 - 2(3a + 1) = -7$ **88.** $7 - 2(4a - 3) = -11$

89. $|4y| + 3 = 5$ **90.** $|6y| - 3 = 9$

91. $|5x + 10| = 15$ **92.** $|4x - 8| = 12$

In this section we will consider two basic types of factoring: factoring out the greatest common factor, and factoring by grouping. Both types of factoring rely heavily on the distributive property.

    In general, factoring is the reverse of multiplication. The following diagram illustrates the relationship between factoring and multiplication:

Multiplication

Factors $3 \cdot 7 = 21$ Product

Factoring

    Reading from left to right we say the product of 3 and 7 is 21. Reading in the other direction, from right to left, we say 21 factors into 3 times 7. Or, 3 and 7 are factors of 21.

DEFINITION   The *greatest common factor* for a polynomial is the largest monomial that divides (is a factor of) each term of the polynomial.

    The greatest common factor for the polynomial $25x^5 + 20x^4 - 30x^3$ is $5x^3$ since it is the largest monomial that is a factor of each term. We can apply the distributive property and write

$$25x^5 + 20x^4 - 30x^3 = 5x^3(5x^2) + 5x^3(4x) + 5x^3(-6)$$
$$= 5x^3(5x^2 + 4x - 6)$$

    The last line is written in factored form.

    Once we recognize the greatest common factor for a polynomial, we apply the distributive property and factor it from each term. We rewrite the original polynomial as the product of its greatest common factor and

the polynomial that remains after the greatest common factor has been factored from each term.

▼   **Example 1**   Factor the greatest common factor from

$$16a^5b^4 - 24a^2b^5 - 8a^3b^3$$

Solution   The largest monomial that divides each term is $8a^2b^3$. We write each term of the original polynomial in terms of $8a^2b^3$ and apply the distributive property to write the polynomial in factored form:

$16a^5b^4 - 24a^2b^5 - 8a^3b^3$

$$= 8a^2b^3(2a^3b) + 8a^2b^3(-3b^2) + 8a^2b^3(-a)$$
$$= 8a^2b^3(2a^3b - 3b^2 - a) \qquad\qquad ▲$$

Note   The term *largest monomial* as used here refers to the monomial with the largest integer exponents whose coefficient has the greatest absolute value. We could have factored the polynomial in Example 1 correctly by taking out $-8a^2b^3$. We usually keep the coefficient of the greatest common factor positive. However, it is not incorrect (and is sometimes useful) to have the coefficient negative.

Here are some further examples of factoring out the greatest common factor:

▼   **Example 2**

a. $16x^3y^4z^5 + 10x^4y^3z^2 - 12x^2y^3z^4$
$$= 2x^2y^3z^2(8xyz^3) + 2x^2y^3z^2(5x^2) - 2x^2y^3z^2(6z^2)$$
$$= 2x^2y^3z^2(8xyz^3 + 5x^2 - 6z^2)$$

b. $5(a + b)^3 - 10(a + b)^2 + 5(a + b)$
$$= 5(a + b)(a + b)^2 - 5(a + b)2(a + b) + 5(a + b)(1)$$
$$= 5(a + b)[(a + b)^2 - 2(a + b) + 1] \qquad\qquad ▲$$

The second step in the two parts of the above example is not necessary. It is shown here simply to emphasize use of the distributive property.

Factoring by Grouping   Many polynomials have no greatest common factor other than the number 1. Some of these can be factored using the distributive property if those terms with a common factor are grouped together.

For example, the polynomial $5x + 5y + x^2 + xy$ can be factored by

noticing that the first two terms have a 5 in common, whereas the last two have an x in common.

Applying the distributive property we have

$$5x + 5y + x^2 + xy = 5(x + y) + x(x + y)$$

This last expression can be thought of as having two terms, $5(x + y)$ and $x(x + y)$, each of which has a common factor $(x + y)$. We apply the distributive property again to factor $(x + y)$ from each term:

$$5(x + y) + x(x + y)$$
$$= (5 + x)(x + y)$$

▼ **Example 3** Factor $a^2b^2 + b^2 + 8a^2 + 8$.

Solution The first two terms have b^2 in common; the last two have 8 in common.

$$a^2b^2 + b^2 + 8a^2 + 8 = b^2(a^2 + 1) + 8(a^2 + 1)$$
$$= (b^2 + 8)(a^2 + 1) \qquad ▲$$

▼ **Example 4** Factor $15 - 5y^4 - 3x^3 + x^3y^4$.

Solution Let's try factoring a 5 from the first two terms and an x^3 from the last two terms:

$$15 - 5y^4 - 3x^3 + x^3y^4 = 5(3 - y^4) + x^3(-3 + y^4)$$

Now, $3 - y^4$ and $-3 + y^4$ are not equal and we cannot factor further. Notice, however, that we can factor $-x^3$ instead of x^3 from the last two terms and obtain the desired result:

$$15 - 5y^4 - 3x^3 + x^3y^4 = 5(3 - y^4) - x^3(3 - y^4)$$
$$= (5 - x^3)(3 - y^4) \qquad ▲$$

The type of factoring shown in these last two examples is called factoring by grouping. With some practice it becomes fairly mechanical.

Factor the greatest common factor from each of the following. (The answers in the back of the book all show greatest common factors whose coefficients are positive.)

Problem Set 3.5

1. $10x^3 - 15x^2$ 2. $12x^5 + 18x^7$
3. $9y^6 + 18y^3$ 4. $24y^4 - 8y^2$

5. $9a^2b - 6ab^2$
6. $30a^3b^4 + 20a^4b^3$
7. $21xy^4 + 7x^2y^2$
8. $14x^6y^3 - 6x^2y^4$
9. $3a^2 - 21a + 30$
10. $3a^2 - 3a - 6$
11. $4x^3 - 16x^2 - 20x$
12. $2x^3 - 14x^2 + 20x$
13. $10x^4y^2 + 20x^3y^3 - 30x^2y^4$
14. $6x^4y^2 + 18x^3y^3 - 24x^2y^4$
15. $-x^2y + xy^2 - x^2y^2$
16. $-x^3y^2 - x^2y^3 - x^2y^2$
17. $4x^3y^2z - 8x^2y^2z^2 + 6xy^2z^3$
18. $7x^4y^3z^2 - 21x^2y^2z^2 - 14x^2y^3z^4$
19. $20a^2b^2c^2 - 30ab^2c + 25a^2bc^2$
20. $8a^3bc^5 - 48a^2b^4c + 16ab^3c^5$
21. $5x(a - 2b) - 3y(a - 2b)$
22. $3a(x - y) - 7b(x - y)$
23. $3x^2(x + y)^2 - 6y^2(x + y)^2$
24. $10x^3(2x - 3y) - 15x^2(2x - 3y)$
25. $2x^2(x + 5) + 7x(x + 5) + 6(x + 5)$
26. $2x^2(x + 2) + 13x(x + 2) + 15(x + 2)$

Factor each of the following by grouping:

27. $3xy + 3y + 2ax + 2a$
28. $5xy^2 + 5y^2 + 3ax + 3a$
29. $x^2y + x + 3xy + 3$
30. $x^3y^3 + 2x^3 + 5x^2y^3 + 10x^2$
31. $a + b + 5ax + 5bx$
32. $a + b + 7ax + 7bx$
33. $3x - 2y + 6x - 4y$
34. $4x - 5y + 12x - 15y$
35. $3xy^2 - 6y^2 + 4x - 8$
36. $8x^2y - 4x^2 + 6y - 3$
37. $2xy^3 - 8y^3 + x - 4$
38. $8x^2y^4 - 4y^4 + 14x^2 - 7$
39. $x^2 - ax - bx + ab$
40. $ax - x^2 - bx + ab$
41. $ab + 5a - b - 5$
42. $x^2 - xy - ax + ay$
43. $a^4b^2 + a^4 - 5b^2 - 5$
44. $2a^2 - bc^2 - a^2b + 2c^2$
45. $4x^2y + 5y - 8x^2 - 10$
46. $x^3y^2 + 8y^2b^2 - x^3 - 8b^2$

47. The greatest common factor of the binomial $3x - 9$ is 3. The greatest common factor of the binomial $6x - 2$ is 2. What is the greatest common factor of their product, $(3x - 9)(6x - 2)$, when it has been multiplied out?

48. The greatest common factors of the binomials $5x - 10$ and $2x + 4$ are 5 and 2 respectively. What is the greatest common factor of their product, $(5x - 10)(2x + 4)$, when it has been multiplied out?

49. If P dollars are placed in a savings account in which the rate of interest r is compounded yearly, then at the end of 1 year the amount of money in the account can be written as $P + Pr$. At the end of two years the amount of money in the account is

$$P + Pr + (P + Pr)r.$$

Use factoring by grouping to show that this last expression can be written as $P(1 + r)^2$

50. At the end of three years, the amount of money in the savings account in Problem 49 will be

$$P(1 + r)^2 + P(1 + r)^2r.$$

Use factoring by grouping to show that this last expression can be written as $P(1 + r)^3$.

Review Problems   The problems below review material we covered in Section 3.4. Reviewing these problems will help you with the next section.

Multiply using the FOIL method.

51. $(x + 2)(x + 3)$ 52. $(x - 2)(x - 3)$
53. $(x + 2)(x - 3)$ 54. $(x - 2)(x + 3)$
55. $(x - 6)(x - 1)$ 56. $(x + 6)(x + 1)$
57. $(x - 6)(x + 1)$ 58. $(x + 6)(x - 1)$

Factoring trinomials is probably the most common type of factoring found in algebra. We begin this section by considering trinomials which have a leading coefficient of 1. The remainder of the section is concerned with trinomials with leading coefficients other than 1. The more familiar we are with multiplication of binomials, the easier factoring trinomials will be.

3.6
Factoring Trinomials

In Section 3.4 we multiplied binomials:

$$(x - 2)(x + 3) = x^2 + x - 6$$
$$(x + 5)(x + 2) = x^2 + 7x + 10$$

In each case the product of two binomials is a trinomial. The first term in the resulting trinomial is obtained by multiplying the first term in each binomial. The middle term comes from adding the product of the two inside terms and the two outside terms. The last term is the product of the last term in each binomial.

In general,

$$(x + a)(x + b) = x^2 + ax + bx + ab$$
$$= x^2 + (a + b)x + ab$$

Writing this as a factoring problem we have

$$x^2 + (a + b)x + ab = (x + a)(x + b)$$

To factor a trinomial with a leading coefficient of 1, we simply find the two numbers a and b whose sum is the coefficient of the middle term and whose product is the constant term.

▼   **Example 1**   Factor $x^2 + 5x + 6$.

Solution   The leading coefficient is 1. We need two numbers whose sum is 5 and whose product is 6. The numbers are 2 and 3. (6 and 1 do not work because their sum is not 5.)

$$x^2 + 5x + 6 = (x + 2)(x + 3)$$

To check our work we simply multiply:

$$(x + 2)(x + 3) = x^2 + 3x + 2x + 6$$
$$= x^2 + 5x + 6 \qquad \blacktriangle$$

▼ **Example 2** Factor $x^2 + 2x - 15$.
Solution Again the leading coefficient is 1. We need two integers whose product is -15 and whose sum is $+2$. The integers are $+5$ and -3.

$$x^2 + 2x - 15 = (x + 5)(x - 3) \qquad \blacktriangle$$

If a trinomial is factorable, then its factors are unique. For instance, in the example above we found factors of $x + 5$ and $x - 3$. These are the only two factors for $x^2 + 2x - 15$. There is no other pair of binomials whose product is $x^2 + 2x - 15$.

▼ **Example 3** Factor $x^2 - xy - 12y^2$.

Solution We need two numbers whose product is $-12y^2$ and whose sum is $-y$. The numbers are $-4y$ and $3y$:

$$x^2 - xy - 12y^2 = (x - 4y)(x + 3y)$$

Checking this result gives

$$(x - 4y)(x + 3y) = x^2 + 3xy - 4xy - 12y^2$$
$$= x^2 - xy - 12y^2 \qquad \blacktriangle$$

▼ **Example 4** Factor $x^2 - 8x + 6$.

Solution Since there is no pair of integers whose product is 6 and whose sum is 8, the trinomial $x^2 - 8x + 6$ is not factorable. We say it is a *prime polynomial*. $\qquad \blacktriangle$

▼ **Example 5** Factor $3x^4 - 15x^3y - 18x^2y^2$.

Solution The leading coefficient is not 1. However, each term is divisible by $3x^2$. Factoring this out to begin with we have

$$3x^4 - 15x^3y - 18x^2y^2 = 3x^2(x^2 - 5xy - 6y^2)$$

Factoring the resulting trinomial as in the examples above gives

$$3x^2(x^2 - 5xy - 6y^2) = 3x^2(x - 6y)(x + y) \qquad \blacktriangle$$

Note As a general rule, it is best to factor out the greatest common factor first.

We want to turn our attention now to trinomials with leading coefficients other than 1 and with no greatest common factor other than 1.

Suppose we want to factor $3x^2 - x - 2$. The factors will be a pair of binomials. The product of the first terms will be $3x^2$ and the product of the last terms will be -2. We can list all the possible factors along with their products as follows.

Possible factors	First term	Middle term	Last term
$(x + 2)(3x - 1)$	$3x^2$	$+5x$	-2
$(x - 2)(3x + 1)$	$3x^2$	$-5x$	-2
$(x + 1)(3x - 2)$	$3x^2$	$+x$	-2
$(x - 1)(3x + 2)$	$3x^2$	$-x$	-2

From the last line we see that the factors of $3x^2 - x - 2$ are $(x - 1)(3x + 2)$. That is,

$$3x^2 - x - 2 = (x - 1)(3x + 2)$$

Unlike the first examples in this section, there is no straightforward way of factoring trinomials with leading coefficients other than 1, when the greatest common factor is 1. We must use trial and error or list all the possible factors. In either case the idea is this: look only at pairs of binomials whose products give the correct first and last terms, then look for combinations that will give the correct middle term.

▼ **Example 6** Factor $2x^2 + 13xy + 15y^2$.

Solution Listing all possible factors the product of whose first terms is $2x^2$ and the product of whose last terms is $+15y^2$ yields

Possible factors	Middle term of product
$(2x - 5y)(x - 3y)$	$-11xy$
$(2x - 3y)(x - 5y)$	$-13xy$
$(2x + 5y)(x + 3y)$	$+11xy$
$(2x + 3y)(x + 5y)$	$+13xy$

The last line has the correct middle term:

$$2x^2 + 13xy + 15y^2 = (2x + 3y)(x + 5y)$$

Actually, we did not need to check the first two pairs of possible factors in the above list. All the signs in the trinomial $2x^2 + 13xy + 15y^2$ are positive. The binomial factors must then be of the form $(ax + b)(cx + d)$, where a, b, c, and d are all positive.

▲

There are other ways to reduce the number of possible factors to consider. For example, if we were to factor the trinomial $2x^2 - 11x + 12$, we would not have to consider the pair of possible factors $(2x - 4)(x - 3)$. If the original trinomial has no greatest common factor other than 1, then neither of its binomial factors will either. The trinomial $2x^2 - 11x + 12$ has a greatest common factor of 1, but the possible factor $2x - 4$ has a greatest common factor of 2: $2x - 4 = 2(x - 2)$. Therefore, we do not need to consider $2x - 4$ as a possible factor.

▼   **Example 7**   Factor $18x^3y + 3x^2y^2 - 36xy^3$.

Solution   First factor out the greatest common factor $3xy$. Then factor the remaining trinomial:

$$18x^3y + 3x^2y^2 - 36xy^3 = 3xy(6x^2 + xy - 12y^2)$$
$$= 3xy(3x - 4y)(2x + 3y) \qquad ▲$$

▼   **Example 8**   Factor $12x^4 + 17x^2 + 6$.

Solution   This is a trinomial in x^2:

$$12x^4 + 17x^2 + 6 = (4x^2 + 3)(3x^2 + 2)$$

We could have made the substitution $y = x^2$ to begin with, in order to simplify the trinomial, and then factor as

$$12y^2 + 17y + 6 = (4y + 3)(3y + 2)$$

Again using $y = x^2$, we have:

$$(4y + 3)(3y + 2) = (4x^2 + 3)(3x^2 + 2) \qquad ▲$$

Factor each of the following trinomials: *Problem Set 3.6*

1. $x^2 + 7x + 12$ 2. $x^2 - 7x + 12$
3. $x^2 - x - 12$ 4. $x^2 + x - 12$
5. $y^2 + y - 6$ 6. $y^2 - y - 6$
7. $x^2 - 6x - 16$ 8. $x^2 + 2x - 3$
9. $x^2 + 8x + 12$ 10. $x^2 - 2x - 15$
11. $3a^2 - 21a + 30$ 12. $3a^2 - 3a - 6$
13. $4x^3 - 16x^2 - 20x$ 14. $2x^3 - 14x^2 + 20x$
15. $x^2 + 3xy + 2y^2$ 16. $x^2 - 5xy - 24y^2$
17. $a^2 + 3ab - 18b^2$ 18. $a^2 - 8ab - 9b^2$
19. $x^2 - 2xa - 48a^2$ 20. $x^2 + 13xa + 48a^2$
21. $x^2 - 12xb + 36b^2$ 22. $x^2 + 10xb + 25b^2$
23. $3x^2 - 6xy - 9y^2$ 24. $5x^2 + 45xy + 20y^2$
25. $2a^5 + 4a^4b + 4a^3b^2$ 26. $3a^4 - 18a^3b + 27a^2b^2$
27. $10x^4y^2 + 20x^3y^3 - 30x^2y^4$
28. $6x^4y^2 + 18x^3y^3 - 24x^2y^4$

Factor each of the following.

29. $2x^2 + 7x - 15$ 30. $2x^2 - 7x - 15$
31. $2x^2 + x - 15$ 32. $2x^2 - x - 15$
33. $2x^2 - 13x + 15$ 34. $2x^2 + 13x + 15$
35. $2x^2 - 11x + 15$ 36. $2x^2 + 11x + 15$
37. $2x^2 + 7x + 15$ 38. $2x^2 + x + 15$
39. $6a^2 + 7a + 2$ 40. $6a^2 - 7a + 2$
41. $4y^2 - y - 3$ 42. $6y^2 + 5y - 6$
43. $6x^2 - x - 2$ 44. $3x^2 + 2x - 5$
45. $4r^2 - 12r + 9$ 46. $4r^2 + 20r + 25$
47. $4x^2 - 11xy - 3y^2$ 48. $3x^2 + 19xy - 14y^2$
49. $10x^2 - 3xa - 18a^2$ 50. $9x^2 + 9xa - 10a^2$
51. $18a^2 + 3ab - 28b^2$ 52. $6a^2 - 7ab - 5b^2$
53. $8x^2 + 8x - 6$ 54. $35x^2 - 60x - 20$
55. $9y^4 + 9y^3 - 10y^2$ 56. $4y^5 + 7y^4 - 2y^3$
57. $12a^4 - 2a^3 - 24a^2$ 58. $20a^4 + 65a^3 - 60a^2$
59. $8x^4y^2 - 2x^3y^3 - 6x^2y^4$ 60. $8x^4y^2 - 47x^3y^3 - 6x^2y^4$
61. $3x^4 + 10x^2 + 3$ 62. $6x^4 - x^2 - 7$
63. $20a^4 + 37a^2 + 15$ 64. $20a^4 + 13a^2 - 15$
65. $12r^4 + 3r^2 - 9$ 66. $30r^4 - 4r^2 - 2$

Factor each of the following by first factoring out the greatest common factor,
and then factoring the trinomial that remains.

67. $2x^2(x + 5) + 7x(x + 5) + 6(x + 5)$
68. $2x^2(x + 2) + 13x(x + 2) + 15(x + 2)$
69. $x^2(2x + 3) + 7x(2x + 3) + 10(2x + 3)$
70. $2x^2(x + 1) + 7x(x + 1) + 6(x + 1)$

71. What polynomial, when factored, gives $(3x + 5y)(3x - 5y)$?

72. What polynomial, when factored, gives $(7x + 2y)(7x - 2y)$?

73. One factor of the trinomial $a^2 + 260a + 2500$ is $a + 10$. What is the other factor?

74. One factor of the trinomial $a^2 - 75a - 2500$ is $a + 25$. What is the other factor?

Review Problems The problems below review some of the material we covered in Section 3.4. Reviewing these problems will help you with the next section.

Multiply.

75. $(x + 3)^2$ **76.** $(x - 3)^2$

77. $(2x - 5)^2$ **78.** $(2x + 5)^2$

79. $(x + 2)(x^2 - 2x + 4)$ **80.** $(x - 2)(x^2 + 2x + 4)$

**3.7
Special Factoring**

In this section we will end our study of factoring by considering some special formulas. Some of the formulas are familiar, others are not. In any case, the formulas will work best if they are memorized.

*Perfect Square
Trinomials*

We previously listed some special products found in multiplying polynomials. Two of the formulas looked like this:

$$(a + b)^2 = a^2 + 2ab + b^2$$
$$(a - b)^2 = a^2 - 2ab + b^2$$

If we exchange the left and right sides of each formula we have two special formulas for factoring:

$$a^2 + 2ab + b^2 = (a + b)^2$$
$$a^2 - 2ab + b^2 = (a - b)^2$$

The left side of each formula is called a *perfect square trinomial*. The right sides are binomial squares. Perfect square trinomials can always be factored using the usual methods for factoring trinomials. However, if we notice that the first and last terms of a trinomial are perfect squares, it is wise to see if the trinomial factors as a binomial square before attempting to factor by the usual method.

▼ **Example 1** Factor $x^2 - 6x + 9$.

Solution Since the first and last terms are perfect squares, we attempt to factor according to the formulas above:

$$x^2 - 6x + 9 = (x - 3)^2$$

If we expand $(x - 3)^2$, we have $x^2 - 6x + 9$, indicating we have factored correctly. ▲

▼ **Example 2** Factor $8x^2 - 24xy + 18y^2$.

Solution We begin by factoring the greatest common factor 2 from each term. We then proceed as in Example 1:

$$8x^2 - 24xy + 18y^2 = 2(4x^2 - 12xy + 9y^2)$$
$$= 2(2x - 3y)^2 \qquad ▲$$

Recall the formula that results in the difference of two squares: $(a - b)(a + b) = a^2 - b^2$. Writing this as a factoring formula we have

The Difference of Two Squares

$$a^2 - b^2 = (a - b)(a + b)$$

▼ **Example 3** Factor $16x^4 - 81y^4$.

Solution The first and last terms are perfect squares. We factor according to the formula above:

$$16x^4 - 81y^4 = (4x^2)^2 - (9y^2)^2$$
$$= (4x^2 - 9y^2)(4x^2 + 9y^2)$$

Notice that the first factor is also the difference of two squares. Factoring completely we have

$$16x^4 - 81y^4 = (2x - 3y)(2x + 3y)(4x^2 + 9y^2) \qquad ▲$$

Note The sum of two squares never factors into the product of two binomials. That is, if we were to attempt to factor $(4x^2 + 9y^2)$ in the last example, we would be unable to find two binomials (or any other polynomials) whose product was $4x^2 + 9y^2$. The factors do not exist as polynomials.

Here is another example of the difference of two squares:

▼ **Example 4** Factor $(x - 3)^2 - 25$

Solution This example has the form $a^2 - b^2$ where a is $x - 3$ and b is 5. We factor it according to the formula for the difference of two squares.

$$(x - 3)^2 - 25 = (x - 3)^2 - 5^2 \qquad \text{Write 25 as } 5^2$$
$$= [(x - 3) - 5][(x - 3) + 5] \qquad \text{Factor}$$
$$= (x - 8)(x + 2) \qquad \text{Simplify}$$

Notice in this example we could have expanded $(x - 3)^2$, subtracted 25, and then factored to obtain the same result. Like this

$$
\begin{aligned}
(x - 3)^2 - 25 &= x^2 - 6x + 9 - 25 &&\text{Expand } (x - 3)^2. \\
&= x^2 - 6x - 16 &&\text{Simplify.} \\
&= (x - 8)(x + 2) &&\text{Factor.} \qquad \blacktriangle
\end{aligned}
$$

▼   **Example 5**   Factor $x^2 - 10x + 25 - y^2$

Solution   Notice the first three terms form a perfect square trinomial. That is, $x^2 - 10x + 25 = (x - 5)^2$. If we replace the first three terms by $(x - 5)^2$ the expression that results has the form $a^2 - b^2$. We can then factor as we did in Example 4.

$$
\begin{aligned}
&x^2 - 10x + 25 - y^2 \\
&= (x^2 - 10x + 25) - y^2 &&\text{Group first 3 terms together.} \\
&= (x - 5)^2 - y^2 &&\text{This has the form } a^2 - b^2. \\
&= [(x - 5) - y][(x - 5) + y] &&\text{Factor according to the formula} \\
& &&\quad a^2 - b^2 = (a - b)(a + b). \\
&= (x - 5 - y)(x - 5 + y) &&\text{Simplify.}
\end{aligned}
$$

We could check this result by multiplying the two factors together. (You may want to do that to convince yourself that we have the correct result.)   ▲

The Sum and Difference of Two Cubes

Here are the formulas for factoring the sum and difference of two cubes:

$$
\begin{aligned}
a^3 + b^3 &= (a + b)(a^2 - ab + b^2) \\
a^3 - b^3 &= (a - b)(a^2 + ab + b^2)
\end{aligned}
$$

Since these formulas are unfamiliar, it is important that we verify them.

▼   **Example 6**   Verify the two formulas given above.

Solution   We verify the formulas by multiplying the right sides and comparing the results with the left sides:

$$
\begin{array}{r}
a^2 - ab + b^2 \\
a + b \\
\hline
a^2b - ab^2 + b^3 \\
a^3 - a^2b + ab^2 \\
\hline
a^3 + b^3
\end{array}
$$

The first formula is correct.

$$
\begin{array}{r}
a^2 + ab + b^2 \\
a - b \\
\hline
- a^2b - ab^2 - b^3 \\
a^3 + a^2b + ab^2 \\
\hline
a^3 \qquad\qquad - b^3
\end{array}
$$

The second formula is correct.   ▲

Here are some examples using the formulas for factoring the sum and difference of two cubes:

▼ **Example 7** Factor $x^3 - 8$.

Solution Since the two terms are perfect cubes, we write them as such and apply the formula

$$
\begin{aligned}
x^3 - 8 &= x^3 - 2^3 \\
&= (x - 2)(x^2 + 2x + 4)
\end{aligned}
$$ ▲

Note The second factor does not factor further. If you have the idea that it does factor, you should try it to convince yourself that it does not.

▼ **Example 8** Factor $27x^3 + 125y^3$.

Solution Writing both terms as perfect cubes, we have

$$
\begin{aligned}
27x^3 + 125y^3 &= (3x)^3 + (5y)^3 \\
&= (3x + 5y)(9x^2 - 15xy + 25y^2)
\end{aligned}
$$ ▲

▼ **Example 9** Factor $x^6 - y^6$.

Solution We have a choice of how we want to write the two terms to begin with. We can write the expression as the difference of two squares, $(x^3)^2 - (y^3)^2$, or as the difference of two cubes, $(x^2)^3 - (y^2)^3$. It is better to use the difference of two squares if we have a choice:

$$
\begin{aligned}
x^6 - y^6 &= (x^3)^2 - (y^3)^2 \\
&= (x^3 - y^3)(x^3 + y^3) \\
&= (x - y)(x^2 + xy + y^2)(x + y)(x^2 - xy + y^2)
\end{aligned}
$$

Try this example again writing the first line as the difference of two cubes instead of the difference of two squares. It will become apparent why it is better to use the difference of two squares. ▲

Problem Set 3.7

Factor each perfect square trinomial:

1. $x^2 - 6x + 9$
2. $x^2 + 10x + 25$
3. $a^2 - 12a + 36$
4. $36 - 12a + a^2$
5. $9x^2 + 24x + 16$
6. $4x^2 - 12xy + 9y^2$
7. $16a^2 + 40ab + 25b^2$
8. $25a^2 - 40ab + 16b^2$
9. $16x^2 - 48x + 36$
10. $36x^2 + 48x + 16$
11. $75a^3 + 30a^2 + 3a$
12. $45a^4 - 30a^3 + 5a^2$

Factor each as the difference of two squares. Be sure to factor completely.

13. $x^2 - 9$
14. $x^2 - 16$
15. $4a^2 - 1$
16. $25a^2 - 1$
17. $9x^2 - 16y^2$
18. $25x^2 - 49y^2$
19. $x^4 - 81$
20. $x^4 - 16$
21. $16a^4 - 81$
22. $81a^4 - 16b^4$
23. $x^6 - y^6$
24. $x^6 - 1$
25. $a^6 - 64$
26. $64a^6 - 1$
27. $5x^2 - 125$
28. $16x^2 - 36$
29. $3a^4 - 48$
30. $7a^4 - 7$
31. $(x - 2)^2 - 9$
32. $(x + 2)^2 - 9$
33. $(y - 4)^2 - 16$
34. $(y - 4)^2 - 16$
35. $25 - (a + 3)^2$
36. $49 - (a - 1)^2$
37. $x^2 - 10x + 25 - y^2$
38. $x^2 - 6x + 9 - y^2$
39. $a^2 + 8a + 16 - b^2$
40. $a^2 + 12a + 36 - b^2$
41. $x^2 + 2xy + y^2 - a^2$
42. $a^2 + 2ab + b^2 - y^2$

Factor each of the following as the sum or difference of two cubes.

43. $x^3 - y^3$
44. $x^3 + y^3$
45. $a^3 + 8$
46. $a^3 - 8$
47. $y^3 - 1$
48. $y^3 + 1$
49. $r^3 - 125$
50. $r^3 + 125$
51. $8x^3 - 27y^3$
52. $27x^3 - 8y^3$
53. $3x^3 - 81$
54. $4x^3 + 32$
55. $(x - 4)^3 + 8$
56. $(x + 4)^3 - 8$
57. $(a + 1)^3 - b^3$
58. $(a - 1)^3 + b^3$
59. Factor $4(x + 2)^2 - 24(x + 2) + 36$ by replacing $(x + 2)$ with a, factoring the result, and then replacing a with $x + 2$ and simplifying.
60. Simplify $4(x + 2)^2 - 24(x + 2) + 36$ and then factor the result. (The factors should be the same as those found in Problem 59.)
61. Find two values of b that will make $9x^2 + bx + 25$ a perfect square trinomial.
62. Find a value of c that will make $49x^2 - 42x + c$ a perfect square trinomial.

Review Problems   The problems that follow review the material we covered in Section 2.2.

Solve each inequality.

63. $-5x \leq 35$ **64.** $-3x > -12$
65. $9 - 2a > -3$ **66.** $7 - 4a \leq -9$
67. $3x + 5 < 7x - 3$ **68.** $2x + 7 \geq 6x - 9$
69. $4(3y + 7) - 5 \geq -1$ **70.** $3(5y + 7) - 4 < 2$

Chapter 3   Summary and Review

Examples

PROPERTIES OF EXPONENTS [3.1, 3.2]

1. These expressions illustrate the properties of exponents.

If a and b represent real numbers and r and s represent integers, then

1. $a^r \cdot a^s = a^{r+s}$

2. $(a^r)^s = a^{r \cdot s}$

3. $(ab)^r = a^r \cdot b^r$

4. $a^{-r} = \dfrac{1}{a^r}$ $\qquad (a \neq 0)$

5. $\left(\dfrac{a}{b}\right)^r = \dfrac{a^r}{b^r}$ $\qquad (b \neq 0)$

6. $\dfrac{a^r}{a^s} = a^{r-s}$ $\qquad (a \neq 0)$

7. $a^1 = a$
$\quad\;\; a^0 = 1$ $\qquad (a \neq 0)$

a. $x^2 \cdot x^3 = x^{2+3} = x^5$

b. $(x^2)^3 = x^{2\cdot 3} = x^6$

c. $(3x)^2 = 3^2 \cdot x^2 = 9x^2$

d. $2^{-3} = \dfrac{1}{2^3} = \dfrac{1}{8}$

e. $\left(\dfrac{x}{5}\right)^2 = \dfrac{x^2}{5^2} = \dfrac{x^2}{25}$

f. $\dfrac{x^7}{x^5} = x^{7-5} = x^2$

g. $3^1 = 3$
$\quad\;\; 3^0 = 1$

SCIENTIFIC NOTATION [3.1]

2. $49{,}800{,}000 = 4.98 \times 10^7$
$\quad\;\; 0.00462 = 4.62 \times 10^{-3}$

A number is written in scientific notation when it is written as the product of a number between 1 and 10 with an integer power of 10. That is, when it has the form

$$n \times 10^r$$

where $1 \leq n < 10$ and $r =$ an integer.

3. $(3x^2 + 2x - 5) + (4x^2 - 7x + 2)$
  $= 7x^2 - 5x - 3$

ADDITION OF POLYNOMIALS [3.3]

To add two polynomials simply combine the coefficients of similar terms.

4. $-(2x^2 - 8x - 9)$
  $= -2x^2 + 8x + 9$

NEGATIVE SIGN PRECEDING PARENTHESES [3.3]

If there is a negative sign directly preceding the parentheses surrounding a polynomial, we may remove the parentheses and preceding negative sign by changing the sign of each term within the parentheses.

5. $(3x - 5)(x + 2)$
  $= 3x^2 + 6x - 5x - 10$
  $= 3x^2 + x - 10$

MULTIPLICATION OF POLYNOMIALS [3.4]

To multiply two polynomials, multiply each term in the first by each term in the second.

6. The following are examples of the three special products.
$$(x + 3)^2 = x^2 + 6x + 9$$
$$(x - 5)^2 = x^2 - 10x + 25$$
$$(x + 7)(x - 7) = x^2 - 49$$

SPECIAL PRODUCTS [3.4]

$$(a + b)^2 = a^2 + 2ab + b^2$$
$$(a - b)^2 = a^2 - 2ab + b^2$$
$$(a + b)(a - b) = a^2 - b^2$$

7. The greatest common factor of $10x^5 - 15x^4 + 30x^3$ is $5x^3$. Factoring it out of each term we have
$$5x^3(2x^2 - 3x + 6)$$

GREATEST COMMON FACTOR [3.5]

The greatest common factor of a polynomial is the largest monomial (the monomial with the largest coefficient and highest exponent) that divides each term of the polynomial. The first step in factoring a polynomial is to factor the greatest common factor (if it is other than 1) out of each term.

8. $x^2 + 5x + 6 = (x + 2)(x + 3)$
  $x^2 - 5x + 6 = (x - 2)(x - 3)$
  $x^2 + x - 6 = (x - 2)(x + 3)$
  $x^2 - x - 6 = (x + 2)(x - 3)$

FACTORING TRINOMIALS [3.6]

We factor a trinomial by writing it as the product of two binomials. (This refers to trinomials whose greatest common factor is 1.) Each factorable trinomial has a unique set of factors. Finding the factors is sometimes a matter of trial and error.

SPECIAL FACTORING [3.7]

$a^2 + 2ab + b^2 = (a + b)^2$
$a^2 - 2ab + b^2 = (a - b)^2$
$a^2 - b^2 = (a - b)(a + b)$ Difference of two squares
$a^3 - b^3 = (a - b)(a^2 + ab + b^2)$ Difference of two cubes
$a^3 + b^3 = (a + b)(a^2 - ab + b^2)$ Sum of two cubes

9. Here are some binomials that have been factored this way.

$x^2 + 6x + 9 = (x + 3)^2$
$x^2 - 6x + 9 = (x - 3)^2$
$x^2 - 9 = (x - 3)(x + 3)$
$x^3 - 27 = (x - 3)(x^2 + 3x + 9)$
$x^3 + 27 = (x + 3)(x^2 - 3x + 9)$

COMMON MISTAKES

1. Confusing the expressions $(-5)^2$ and -5^2. The base in the expression $(-5)^2$ is -5. The base in the expression -5^2 is just 5.

$$(-5)^2 = (-5)(-5) = 25$$
$$-5^2 = -5 \cdot 5 = -25$$

2. When subtracting one polynomial from another it is common to forget to add the opposite of each term in the second polynomial. For example:

$$(6x - 5) - (3x + 4) = 6x - 5 - 3x + 4 \quad \text{Mistake}$$
$$= 3x - 1$$

This mistake occurs if the negative sign outside the second set of parentheses is not distributed over all terms inside the parentheses. To avoid this mistake, remember: The opposite of a sum is the sum of the opposite, or,

$$-(3x + 4) = -3x + (-4)$$

3. Interpreting the square of a sum to be the sum of the squares. That is,

$$(x + y)^2 = x^2 + y^2 \quad \text{Mistake}$$

This can easily be shown as false by trying a couple of numbers for x and y. If $x = 4$ and $y = 3$, we have

$$(4 + 3)^2 = 4^2 + 3^2$$
$$7^2 = 16 + 9$$
$$49 = 25$$

There has obviously been a mistake. The correct formula for $(a + b)^2$ is

$$(a + b)^2 = a^2 + 2ab + b^2$$

Chapter 3
Test

Simplify. (Assume all variables are non-negative.)

1. $x^4 \cdot x^7 \cdot x^{-3}$

2. 2^{-5}

3. $(\frac{3}{4})^{-2}$

4. $(2x^2y)^3(2x^3y^4)^2$

5. $\dfrac{a^{-5}}{a^{-7}}$

6. $\dfrac{x^{n+1}}{x^{n-5}}$

7. $\dfrac{(2ab^3)^{-2}(a^4b^{-3})}{(a^{-4}b^3)^4(2a^{-2}b^2)^{-3}}$

8. $(3x^3 - 4x^2 - 6) - (x^2 + 8x - 2)$

9. $3 - 4[2x - 3(x + 6)]$

Write each number in scientific notation.

10. 6,530,000

11. 0.00087

Perform the indicated operations and write your answers in scientific notation.

12. $(2.9 \times 10^{12})(3 \times 10^{-5})$

13. $\dfrac{(6 \times 10^{-4})(4 \times 10^9)}{8 \times 10^{-3}}$

Multiply.

14. $(3y - 7)(2y + 5)$

15. $(2x - 5)(x^2 + 4x - 3)$

16. $(4a - 3b)^2$

17. $(6y - 1)(6y + 1)$

18. $2x(x - 3)(2x + 5)$

Factor completely.

19. $x^2 + x - 12$

20. $12x^2 + 26x - 10$

21. $16a^4 - 81y^4$

22. $7ax^2 - 14ay - b^2x^2 + 2b^2y$

23. $x^3 + 27$

24. $4a^5b - 24a^4b^2 - 64a^3b^3$

25. $x^2 - 10x + 25 - b^2$

Rational Expressions

To the student:

This chapter is mostly concerned with simplifying a certain kind of algebraic expression. The expressions are called rational expressions because they are to algebra what rational numbers are to arithmetic. Most of the work we will do with rational expressions parallels the work you have done in previous math classes with fractions.

Once we have learned to add, subtract, multiply, and divide rational expressions, we will turn our attention to equations involving rational expressions. Equations of this type are used to describe a number of concepts in science, medicine, and other fields. For example, in electronics, it is a well-known fact that if R is the equivalent resistance of two resistors, R_1 and R_2, connected in parallel, then

$$\frac{1}{R} = \frac{1}{R_1} + \frac{1}{R_2}$$

This formula involves three fractions or rational expressions. If two of the resistances are known, we can solve for the third using the methods we will develop in this chapter.

The single most important tool needed for success in this chapter is factoring. Almost every problem encountered in this chapter involves factoring at one point or another. You may be able to understand all the theory and steps involved in solving the problems, but unless you can factor the polynomials in the problems, you will be unable to work any of them. Essentially, this chapter is a review of the properties of fractions and an exercise in factoring.

4.1
Basic Properties and
Reducing to Lowest
Terms

Recall from Chapter 1 that a *rational number* is any number that can be expressed as the ratio of two integers:

$$\text{Rational numbers} = \left\{ \frac{a}{b} \,\middle|\, a \text{ and } b \text{ are integers, } b \neq 0 \right\}$$

A rational expression is defined similarly as any expression that can be written as the ratio of two polynomials:

$$\text{Rational expressions} = \left\{ \frac{P}{Q} \,\middle|\, P \text{ and } Q \text{ are polynomials, } Q \neq 0 \right\}$$

Some examples of rational expressions are

$$\frac{2x - 3}{x + 5} \qquad \frac{x^2 - 5x - 6}{x^2 - 1} \qquad \frac{a - b}{b - a}$$

Note A polynomial can be considered as a rational expression since it can be thought of as the ratio of itself to 1. (You see, any real number is a polynomial of degree 0. That is, $5 = 5x^0$, $1 = 1x^0$, $2 = 2x^0$, etc.)

Basic Properties

The basic properties associated with rational expressions are equivalent to the properties of fractions. It is important when working with fractions that we are able to change the form of a fraction (as when reducing to lowest terms or writing with a common denominator) without changing the value of the fraction.

The two procedures that may change the form of a fraction but will never change its value are multiplying the numerator and denominator by the same nonzero number and dividing the numerator and denominator by the same nonzero number. That is, if a, b, and c are real numbers with $b \neq 0$ and $c \neq 0$, then

$$\frac{a}{b} = \frac{ac}{bc} \quad \text{and} \quad \frac{a}{b} = \frac{a/c}{b/c}$$

For rational expressions, multiplying the numerator and denominator by the same nonzero expression may change the form of the rational expression, but it will always produce an expression equivalent to the original one. The same is true when dividing the numerator and denominator by the same nonzero quantity.

If P, Q, and K are polynomials with $Q \neq 0$ and $K \neq 0$, then

$$\frac{P}{Q} = \frac{PK}{QK} \quad \text{and} \quad \frac{P}{Q} = \frac{P/K}{Q/K}$$

The two statements above are equivalent since division is defined as multiplication by the reciprocal. We choose to state them separately for clarity.

The fraction $\frac{6}{8}$ can be written in lowest terms as $\frac{3}{4}$. The process is shown below:

Reducing to Lowest Terms

$$\frac{6}{8} = \frac{3 \cdot 2}{4 \cdot 2} = \frac{3}{4}$$

Reducing $\frac{6}{8}$ to $\frac{3}{4}$ involves dividing the numerator and denominator by 2, the factor they have in common. Before dividing out the common factor 2, we must notice that the common factor *is* 2! (This may not be obvious since we are very familiar with the numbers 6 and 8 and therefore do not have to put much thought into finding what number divides both of them.)

We reduce rational expressions to lowest terms by first factoring the numerator and denominator and then dividing both numerator and denominator by any factors they have in common.

▼ **Example 1** Reduce $\dfrac{x^2 - 9}{x - 3}$ to lowest terms.

Solution Factoring, we have

$$\frac{x^2 - 9}{x - 3} = \frac{(x + 3)(x - 3)}{x - 3}$$

The numerator and denominator have the factor $x - 3$ in common.

Dividing the numerator and denominator by $x - 3$, we have

$$\frac{(x + 3)(x - 3)}{x - 3} = \frac{x + 3}{1} = x + 3 \qquad \blacktriangle$$

Note The lines drawn through the $(x - 3)$ in the numerator and denominator indicate that we have divided through by $(x - 3)$. As the problems become more involved these lines will help keep track of which factors have been divided out and which have not. The lines do not indicate some new property of rational expressions. The only way to reduce a rational expression to lowest terms is to divide the numerator and denominator by the factors they have in common.

Here are some other examples of reducing rational expressions to lowest terms:

▼ **Example 2** Reduce to lowest terms.

a. $\dfrac{y^2 - 5y - 6}{y^2 - 1} = \dfrac{(y - 6)(y + 1)}{(y - 1)(y + 1)}$ Factor numerator and denominator.

$= \dfrac{y - 6}{y - 1}$ Divide out common factor $(y + 1)$.

b. $\dfrac{2a^3 - 16}{4a^2 - 12a + 8} = \dfrac{2(a^3 - 8)}{4(a^2 - 3a + 2)}$

$= \dfrac{2(a - 2)(a^2 + 2a + 4)}{4(a - 2)(a - 1)}$ Factor numerator and denominator.

$= \dfrac{a^2 + 2a + 4}{2(a - 1)}$ Divide out common factor $2(a - 2)$.

c. $\dfrac{x^2 - 3x + ax - 3a}{x^2 - ax - 3x + 3a} = \dfrac{x(x - 3) + a(x - 3)}{x(x - a) - 3(x - a)}$

$= \dfrac{(x - 3)(x + a)}{(x - 3)(x - a)}$ Factor numerator and denominator.

$= \dfrac{x + a}{x - a}$ Divide out common factor $(x - 3)$. ▲

The answer to part c in Example 2 is $(x + a)/(x - a)$. The problem cannot be reduced further. It is a fairly common mistake to attempt to divide out an x or an a in this last expression. Remember, we can divide out only the factors common to the numerator and denominator of a

rational expression. For the last expression in Example 2, part c, neither the numerator nor the denominator can be factored further; x is not a factor of the numerator or the denominator and neither is a. The expression is in lowest terms.

The next example involves what we may call a trick. It is a way of changing one of the factors that may not be obvious the first time we see it. This is very useful in some situations.

▼ **Example 3** Reduce to lowest terms $\dfrac{x^2 - 25}{5 - x}$.

Solution We begin by factoring the numerator:

$$\frac{x^2 - 25}{5 - x} = \frac{(x - 5)(x + 5)}{5 - x}$$

The factors $(x - 5)$ and $(5 - x)$ are similar but are not exactly the same. We can reverse the order of either by factoring -1 from them. That is:

$$5 - x = -1(-5 + x) = -1(x - 5).$$

$$\frac{(x - 5)(x + 5)}{5 - x} = \frac{(x - 5)(x + 5)}{-1(x - 5)}$$

$$= \frac{x + 5}{-1}$$

$$= -(x + 5) \qquad \blacktriangle$$

Sometimes we can apply the trick before we actually factor the polynomials in our rational expression.

▼ **Example 4** Reduce to lowest terms $\dfrac{x^2 - 6xy + 9y^2}{9y^2 - x^2}$.

Solution We begin by factoring -1 from the denominator to reverse the order of the terms in $9y^2 - x^2$:

$$\frac{x^2 - 6xy + 9y^2}{9y^2 - x^2} = \frac{x^2 - 6xy + 9y^2}{-1(x^2 - 9y^2)}$$

$$= \frac{(x - 3y)(x - 3y)}{-1(x - 3y)(x + 3y)}$$

$$= -\frac{x - 3y}{x + 3y} \qquad \blacktriangle$$

Problem Set 4.1

Reduce each fraction to lowest terms:

1. $\dfrac{8}{24}$

2. $\dfrac{13}{39}$

3. $-\dfrac{12}{36}$

4. $-\dfrac{45}{60}$

5. $\dfrac{9x^3}{3x}$

6. $\dfrac{14x^5}{7x^2}$

7. $\dfrac{2a^2b^3}{4a^2}$

8. $\dfrac{3a^3b^2}{6b^2}$

9. $-\dfrac{24x^3y^5}{16x^4y^2}$

10. $-\dfrac{36x^6y^8}{24x^3y^9}$

11. $\dfrac{144a^2b^3c^4}{56a^4b^3c^2}$

12. $\dfrac{108a^5b^2c^5}{27a^2b^5c^2}$

Reduce each rational expression to lowest terms:

13. $\dfrac{x^2 - 16}{6x + 24}$

14. $\dfrac{5x + 25}{x^2 - 25}$

15. $\dfrac{12x - 9y}{3x^2 + 3xy}$

16. $\dfrac{x^3 - xy^2}{4x + 4y}$

17. $\dfrac{a^4 - 81}{a - 3}$

18. $\dfrac{a + 4}{a^2 - 16}$

19. $\dfrac{y^2 - y - 12}{y - 4}$

20. $\dfrac{y^2 + 7y + 10}{y + 5}$

21. $\dfrac{a^2 - 4a - 12}{a^2 + 8a + 12}$

22. $\dfrac{a^2 - 7a + 12}{a^2 - 9a + 20}$

23. $\dfrac{4y^2 - 9}{2y^2 - y - 3}$

24. $\dfrac{9y^2 - 1}{3y^2 - 10y + 3}$

25. $\dfrac{x^2 + x - 6}{x^2 + 2x - 3}$

26. $\dfrac{x^2 + 10x + 25}{x^2 - 25}$

27. $\dfrac{(x - 3)^2(x + 2)}{(x + 2)^2(x - 3)}$

28. $\dfrac{(x - 4)^3(x + 3)}{(x + 3)^2(x - 4)}$

29. $\dfrac{a^3 + b^3}{a^2 - b^2}$

30. $\dfrac{a^2 - b^2}{a^3 - b^3}$

31. $\dfrac{6x^2 + 7xy - 3y^2}{6x^2 + xy - y^2}$

32. $\dfrac{4x^2 - y^2}{4x^2 - 8xy - 5y^2}$

33. $\dfrac{ax + 2x + 3a + 6}{ay + 2y - 4a - 8}$

34. $\dfrac{ax - x - 5a + 5}{ax + x - 5a - 5}$

35. $\dfrac{x^2 + bx - 3x - 3b}{x^2 - 2bx - 3x + 6b}$

36. $\dfrac{x^2 - 3ax - 2x + 6a}{x^2 - 3ax + 2x - 6a}$

Refer to Examples 3 and 4 in this section and reduce the following to lowest terms:

37. $\dfrac{x - 4}{4 - x}$

38. $\dfrac{6 - x}{x - 6}$

39. $\dfrac{y^2 - 36}{6 - y}$

40. $\dfrac{1 - y}{y^2 - 1}$

41. $\dfrac{1 - 9a^2}{9a^2 - 6a + 1}$

42. $\dfrac{1 - a^2}{a^2 - 2a + 1}$

43. $\dfrac{6 - 5x - x^2}{x^2 + 5x - 6}$

44. $\dfrac{x^2 - 5x + 6}{6 - x - x^2}$

Explain the mistake made in each of the following problems:

45. $\dfrac{\overset{x}{\cancel{x^2}} - \overset{3}{\cancel{9}}}{\cancel{x} - \cancel{3}} = x - 3$

46. $\dfrac{\cancel{x^2} - 6\cancel{x} + \overset{3}{\cancel{9}}}{\cancel{x^2} + \cancel{x} - \underset{2}{\cancel{6}}} = \dfrac{3}{2}$

47. $\dfrac{\cancel{x} + y}{\cancel{x}} = y$

48. $\dfrac{x + \cancel{3}}{\cancel{3}} = x$

49. Replace x with 3 in the expression $\dfrac{x^3 - 1}{x - 1}$, and then simplify. The result should be the same as what you would get if you replaced x with 3 in the expression $x^2 + x + 1$.

50. Replace x with 7 in the expression $\dfrac{x - 4}{4 - x}$ and simplify. Now, replace x with 10 and simplify. The result in both cases should be the same. Can you think of a number to replace x with, that will not give the same result?

Review Problems   The problems below review material we covered in Section 3.3. Reviewing these problems will help you with the next section.

51. Subtract $x^2 + 2x + 1$ from $4x^2 - 5x + 5$.
52. Subtract $3x^2 - 5x + 2$ from $7x^2 + 6x + 4$.
53. Subtract $10x - 20$ from $10x - 11$.
54. Subtract $-6x - 18$ from $-6x + 5$.
55. Subtract $4x^3 - 8x^2$ fom $4x^3$.
56. Subtract $2x^2 + 6x$ from $2x^2$.

We begin this section by considering division of a polynomial by a monomial. This is the simplest kind of polynomial division. The rest of the section is devoted to division of a polynomial by a polynomial.

4.2
Division of
Polynomials

Dividing a Polynomial
by a Monomial

To divide a polynomial by a monomial we use the definition of division and apply the distributive property. The following example illustrates the procedure.

▼   **Example 1** Divide $10x^5 - 15x^4 + 20x^3$ by $5x^2$.

Solution

$$\frac{10x^5 - 15x^4 + 20x^3}{5x^2}$$

$$= (10x^5 - 15x^4 + 20x^3) \cdot \frac{1}{5x^2}$$ Dividing by $5x^2$ is the same as multiplying by $1/5x^2$.

$$= 10x^5 \cdot \frac{1}{5x^2} - 15x^4 \cdot \frac{1}{5x^2} + 20x^3 \cdot \frac{1}{5x^2}$$ Distributive property

$$= \frac{10x^5}{5x^2} - \frac{15x^4}{5x^2} + \frac{20x^3}{5x^2}$$ Multiplying by $1/5x^2$ is the same as dividing by $5x^2$.

$$= 2x^3 - 3x^2 + 4x$$ Divide coefficients, subtract exponents.

Notice that division of a polynomial by a monomial is accomplished by dividing each term of the polynomial by the monomial. The first two steps are usually not shown in a problem like this. They are part of Example 1 to justify distributing $5x^2$ under all three terms of the polynomial $10x^5 - 15x^4 + 20x^3$.   ▲

Here are some more examples of this kind of division:

▼   **Example 2**

a. $$\frac{8x^3y^5 - 16x^2y^2 + 4x^4y^3}{-2x^2y} = \frac{8x^3y^5}{-2x^2y} + \frac{-16x^2y^2}{-2x^2y} + \frac{4x^4y^3}{-2x^2y}$$

$$= -4xy^4 + 8y - 2x^2y^2$$

b. $$\frac{10a^4b^2 + 8ab^3 - 12a^3b + 6ab}{4a^2b^2} = \frac{10a^4b^2}{4a^2b^2} + \frac{8ab^3}{4a^2b^2} - \frac{12a^3b}{4a^2b^2} + \frac{6ab}{4a^2b^2}$$

$$= \frac{5a^2}{2} + \frac{2b}{a} - \frac{3a}{b} + \frac{3}{2ab}$$ ▲

Notice in part b of Example 2 that the result is not a polynomial because of the last three terms. If we were to write each as a product, some of the variables would have negative exponents. For example, the second term would be

$$\frac{2b}{a} = 2a^{-1}b$$

The divisor in each of the examples above was a monomial. We now want to turn our attention to division of polynomials in which the divisor has two or more terms.

Dividing a Polynomial by a Polynomial

▼   **Example 3**   Divide $\dfrac{x^2 - 6xy - 7y^2}{x + y}$.

Solution   In this case we can factor the numerator and perform our division by simply dividing out common factors, just like we did in the previous section.

$$\frac{x^2 - 6xy - 7y^2}{x + y} = \frac{(x + y)(x - 7y)}{x + y}$$
$$= x - 7y \qquad ▲$$

For the type of division shown in Example 3, the denominator must be a factor of the numerator. When the denominator is not a factor of the numerator, or in the case where we can't factor the numerator, the method used in Example 3 won't work. We need to develop a new method for these cases. Since this new method is very similar to long division with whole numbers, we will review it here.

▼   **Example 4**   Divide $25\overline{)4628}$.

Solution

$$
\begin{array}{r}
1 \\
25\overline{)4628} \\
\underline{25} \\
21
\end{array}
$$

⟵ Estimate: 25 into 46
⟵ Multiply: $1 \times 25 = 25$
⟵ Subtract: $46 - 25 = 21$

$$
\begin{array}{r}
1 \\
25\overline{)4628} \\
\underline{25\downarrow} \\
212
\end{array}
$$

⟵ Bring down the 2.

These are the four basic steps in long division: estimate, multiply, subtract, and bring down the next term. To complete the problem we simply perform the same four steps again.

$$
\begin{array}{r}
18 \quad\leftarrow\ 8\ \text{is the estimate.} \\
25\overline{)4628} \\
25 \\
\overline{212} \\
200\downarrow \quad\leftarrow\ \text{Multiply to get 200.} \\
\overline{128} \quad\leftarrow\ \text{Subtract to get 12, then} \\
\text{bring down the 8.}
\end{array}
$$

One more time:

$$
\begin{array}{r}
185 \quad\leftarrow\ 5\ \text{is the estimate.} \\
25\overline{)4628} \\
25 \\
\overline{212} \\
200\downarrow \\
\overline{128} \\
125 \quad\leftarrow\ \text{Multiply to get 125.} \\
\overline{3} \quad\leftarrow\ \text{Subtract to get 3.}
\end{array}
$$

Since 3 is less than 25, we have our answer:

$$
\frac{4628}{25} = 185 + \frac{3}{25}
$$

To check our answer, we multiply 185 by 25, then add 3 to the result:

$$
25(185) + 3 = 4625 + 3 = 4628 \qquad\qquad \blacktriangle
$$

Note You may realize when looking over this last example that you don't have a very good idea why you proceed as you do with the steps in long division. What you do know is the process always works. We are going to approach the explanation for long division with two polynomials with this in mind. That is, we won't always be sure why the steps we use are important, only that they always produce the correct result.

Long division with polynomials is very similar to long division with whole numbers. Both use the same four basic steps: estimate, multiply, subtract, and bring down the next term. We use long division with polynomials when the denominator has two or more terms and is not a factor of the numerator. Here is an example:

▼   **Example 5**   Divide $\dfrac{2x^2 - 7x + 9}{x - 2}$.

Solution

$$
\begin{array}{r}
2x \\
x - 2 \overline{)\ 2x^2 - 7x + 9}
\end{array}
$$
$\longleftarrow$ Estimate: $2x^2 \div x = 2x$.

$$
\begin{array}{r}
2x \\
x - 2 \overline{)\ 2x^2 - 7x + 9} \\
-\quad + \\
\cancel{2x^2}\ \cancel{4x}
\end{array}
$$
$\longleftarrow$ Multiply: $2x(x - 2) = 2x^2 - 4x$
$\qquad\qquad\quad - 3x$ $\longleftarrow$ Subtract: $(2x^2 - 7x) - (2x^2 - 4x) = -3x$

$$
\begin{array}{r}
2x \\
x - 2 \overline{)\ 2x^2 - 7x + 9} \\
-\quad + \\
\cancel{2x^2}\ \cancel{4x} \\
- 3x + 9
\end{array}
$$
$\longleftarrow$ Bring down the 9.

Notice we change the signs on $2x^2 - 4x$ and add in the subtraction step. Subtracting a polynomial is equivalent to adding its opposite.

We repeat the four steps again:

$$
\begin{array}{r}
2x\ -3 \\
x - 2 \overline{)\ 2x^2 - 7x + 9} \\
-\quad + \\
\cancel{2x^2}\ \cancel{4x} \\
- 3x + 9 \\
+\quad - \\
\cancel{3x}\ \cancel{6} \\
3
\end{array}
$$

$\longleftarrow$ -3 is the estimate: $-3x \div x = -3$

$\longleftarrow$ Multiply: $-3(x - 2) = -3x + 6$
$\longleftarrow$ Subtract: $(-3x + 9) - (-3x + 6) = 3$

Since we have no other term to bring down, we have our answer:

$$
\frac{2x^2 - 7x + 9}{x - 2} = 2x - 3 + \frac{3}{x - 2}
$$

To check we multiply $(2x - 3)(x - 2)$ to get $2x^2 - 7x + 6$; then, adding the remainder 3 to this result, we have $2x^2 - 7x + 9$.   ▲

In setting up a long division problem involving two polynomials there are two things to remember: (1) both polynomials should be in decreasing powers of the variable, and (2) neither should skip any powers from the highest power down to the constant term. If there are any missing terms, they can be filled in using a coefficient of 0.

▼   **Example 6**   Divide $2x - 4 \overline{) 4x^3 - 6x - 11}$.

Solution   Since the first polynomial is missing a term in x^2, we can fill it in with $0x^2$:

$$4x^3 - 6x - 11 = 4x^3 + 0x^2 - 6x - 11$$

Adding $0x^2$ does not change our original problem.

$$
\begin{array}{r}
2x^2 + 4x\ + 5 \\
2x - 4 \overline{)\ 4x^3 + 0x^2 -\ \ 6x - 11} \\
\end{array}
$$

Notice: adding the $0x^2$ term gives us a column in which to write $+8x^2$.

$$
\begin{array}{r}
2x^2 + 4x\ + 5 \\
2x - 4 \overline{)\ 4x^3 + 0x^2 -\ \ 6x - 11} \\
\underline{\cancel{-}\ \cancel{4x^3}\ \cancel{+}\ \cancel{8x^2}} \\
+ 8x^2 -\ \ 6x \\
\underline{\cancel{-}\ \cancel{8x^2}\ \cancel{+}\ \cancel{16x}} \\
+ 10x - 11 \\
\underline{\cancel{-}\ \cancel{10x}\ \cancel{+}\ \cancel{20}} \\
+9
\end{array}
$$

$$\frac{4x^3 - 6x - 11}{2x - 4} = 2x^2 + 4x + 5 + \frac{9}{2x - 4}$$ ▲

For our final example in this section, let's do Example 3 again, but this time use long division.

▼   **Example 7**   Divide $\dfrac{x^2 - 6xy - 7y^2}{x + y}$.

Solution

$$
\begin{array}{r}
x\ - 7y \\
x + y \overline{)\ x^2 - 6xy - 7y^2} \\
\underline{\cancel{-}\ \cancel{x^2}\ \cancel{-}\ \cancel{xy}} \\
- 7xy - 7y^2 \\
\underline{\cancel{+}\ \cancel{7xy}\ \cancel{+}\ \cancel{7y^2}} \\
0
\end{array}
$$

In this case the remainder is 0 and we have

$$\frac{x^2 - 6xy - 7y^2}{x + y} = x - 7y$$ ▲

As you can see, the result is the same as we obtained previously. If we have a choice of methods, the method used in Example 3 will usually give us our results with fewer steps.

Find the following quotients:

1. $\dfrac{4x^3 - 8x^2 + 6x}{2x}$

2. $\dfrac{6x^3 + 12x^2 - 9x}{3x}$

3. $\dfrac{10x^4 + 15x^3 - 20x^2}{-5x^2}$

4. $\dfrac{12x^5 - 18x^4 - 6x^3}{6x^3}$

5. $\dfrac{8y^5 + 10y^3 - 6y}{4y^3}$

6. $\dfrac{6y^4 - 3y^3 + 18y^2}{9y^2}$

7. $\dfrac{5x^3 - 8x^2 - 6x}{-2x^2}$

8. $\dfrac{-9x^5 + 10x^3 - 12x}{-6x^4}$

9. $\dfrac{28a^3b^5 + 42a^4b^3}{7a^2b^2}$

10. $\dfrac{a^2b + ab^2}{ab}$

11. $\dfrac{10x^3y^2 - 20x^2y^3 - 30x^3y^3}{-10x^2y}$

12. $\dfrac{9x^4y^4 + 18x^3y^4 - 27x^2y^4}{-9xy^3}$

Divide by factoring numerators and then dividing out common factors.

13. $\dfrac{x^2 - x - 6}{x - 3}$

14. $\dfrac{x^2 - x - 6}{x + 2}$

15. $\dfrac{2a^2 - 3a - 9}{2a + 3}$

16. $\dfrac{2a^2 + 3a - 9}{2a - 3}$

17. $\dfrac{5x^2 - 14xy - 24y^2}{x - 4y}$

18. $\dfrac{5x^2 - 26xy - 24y^2}{5x + 4y}$

19. $\dfrac{x^3 - y^3}{x - y}$

20. $\dfrac{x^3 + 8}{x + 2}$

21. $\dfrac{y^4 - 16}{y - 2}$

22. $\dfrac{y^4 - 81}{y - 3}$

Divide using the long division method.

23. $\dfrac{x^2 - 5x - 7}{x + 2}$

24. $\dfrac{x^2 + 4x - 8}{x - 3}$

25. $\dfrac{6x^2 + 7x - 18}{3x - 4}$

26. $\dfrac{8x^2 - 26x - 9}{2x - 7}$

27. $\dfrac{2x^3 - 3x^2 - 4x + 5}{x + 1}$

28. $\dfrac{3x^3 - 5x^2 + 2x - 1}{x - 2}$

29. $\dfrac{2y^3 - 9y^2 - 17y + 39}{2y - 3}$

30. $\dfrac{3y^3 - 19y^2 + 17y + 4}{3y - 4}$

31. $\dfrac{2x^3 - 9x^2 + 11x - 6}{2x^2 - 3x + 2}$ **32.** $\dfrac{6x^3 + 7x^2 - x + 3}{3x^2 - x + 1}$

33. $\dfrac{6y^3 - 8y + 5}{2y - 4}$ **34.** $\dfrac{9y^3 - 6y^2 + 8}{3y - 3}$

35. $\dfrac{a^4 - 2a + 5}{a - 2}$ **36.** $\dfrac{a^4 + a^3 - 1}{a + 2}$

37. $\dfrac{y^4 - 16}{y - 2}$ **38.** $\dfrac{y^4 - 81}{y - 3}$

39. $\dfrac{x^4 + x^3 - 3x^2 - x + 2}{x^2 + 3x + 2}$ **40.** $\dfrac{2x^4 + x^3 + 4x - 3}{2x^2 - x + 3}$

41. Problems 21 and 37 are the same problem. Are the two answers you obtained equivalent?

42. Problems 22 and 38 are the same problem. Are the two answers you obtained equivalent?

43. Find the value of the polynomial $x^2 - 5x - 7$ when x is -2. Compare it with the remainder in Problem 23.

44. Find the value of the polynomial $x^2 + 4x - 8$ when x is 3. Compare it with the remainder in Problem 24.

Review Problems The problems below review material we covered in Section 1.6. Reviewing these problems will help you get started in the next section.

Divide.

45. $\frac{3}{5} \div \frac{2}{7}$ **46.** $\frac{2}{7} \div \frac{3}{5}$

47. $\frac{3}{4} \div \frac{6}{11}$ **48.** $\frac{6}{8} \div \frac{3}{5}$

49. $\frac{4}{9} \div 8$ **50.** $\frac{3}{7} \div 6$

51. $8 \div \frac{1}{4}$ **52.** $12 \div \frac{2}{3}$

**4.3
Multiplication and
Division of Rational
Expressions**

In Section 4.1 we found the process of reducing rational expressions to lowest terms to be the same process used in reducing fractions to lowest terms. The similarity also holds for the process of multiplication or division of rational expressions.

Let's review multiplication and division with fractions.

▼ **Example 1** Multiply $\frac{6}{7} \times \frac{14}{18}$.

Solution

$$\frac{6}{7} \times \frac{14}{18} = \frac{6(14)}{7(18)} \qquad \text{Multiply numerators and denominators.}$$

$$= \frac{\cancel{2}\cdot\cancel{3}(2\cdot\cancel{1})}{\cancel{1}(\cancel{2}\cdot\cancel{3}\cdot 3)} \qquad \text{Factor.}$$

$$= \frac{2}{3} \qquad\qquad \begin{array}{l}\text{Divide out common}\\ \text{factors.}\end{array} \qquad \blacktriangle$$

▼ **Example 2**   Divide $\frac{6}{8} \div \frac{3}{5}$.

Solution

$$\frac{6}{8} \div \frac{3}{5} = \frac{6}{8} \times \frac{5}{3} \qquad \begin{array}{l}\text{Write division in terms}\\ \text{of multiplication.}\end{array}$$

$$= \frac{6(5)}{8(3)} \qquad\qquad \begin{array}{l}\text{Multiply numerators and}\\ \text{denominators.}\end{array}$$

$$= \frac{\cancel{2}\cdot\cancel{3}(5)}{\cancel{2}\cdot 2\cdot 2(\cancel{3})} \qquad \text{Factor.}$$

$$= \frac{5}{4} \qquad\qquad \text{Divide out common factors.} \qquad \blacktriangle$$

The product of two rational expressions is the product of their numerators over the product of their denominators.

▼ **Example 3**   Multiply $\dfrac{x-3}{x^2-4} \cdot \dfrac{x+2}{x^2-6x+9}$.

Solution   We begin by multiplying numerators and denominators. We then factor all polynomials and divide out factors common to the numerator and denominator:

$$\frac{x-3}{x^2-4} \cdot \frac{x+2}{x^2-6x+9}$$

$$= \frac{(x-3)(x+2)}{(x^2-4)(x^2-6x+9)} \qquad \text{Multiply.}$$

$$= \frac{(\cancel{x-3})(\cancel{x+2})}{(\cancel{x+2})(x-2)(\cancel{x-3})(x-3)} \qquad \text{Factor.}$$

$$= \frac{1}{(x-2)(x-3)} \qquad \begin{array}{l}\text{Divide out}\\ \text{common factors.}\end{array} \qquad \blacktriangle$$

The first two steps can be combined to save time. We can perform the multiplication and factoring steps together.

▼   **Example 4**   Multiply $\dfrac{2y^2 - 4y}{2y^2 - 2} \cdot \dfrac{y^2 - 2y - 3}{y^2 - 5y + 6}$.

Solution

$$\frac{2y^2 - 4y}{2y^2 - 2} \cdot \frac{y^2 - 2y - 3}{y^2 - 5y + 6} = \frac{2y(y-2)(y-3)(y+1)}{2(y+1)(y-1)(y-3)(y-2)}$$

$$= \frac{y}{y-1} \qquad\qquad ▲$$

Notice in both of the above examples that we did not actually multiply the polynomials as we did in Chapter 3. It would be senseless to do that since we would then have to factor each of the resulting products to reduce them to lowest terms.

The quotient of two rational expressions is the product of the first and the reciprocal of the second. That is, we find the quotient of two rational expressions the same way we find the quotient of two fractions.

▼   **Example 5**   Divide $\dfrac{x^2 - y^2}{x^2 - 2xy + y^2} \div \dfrac{x^3 + y^3}{x^3 - x^2y}$.

Solution   We begin by writing the problem as the product of the first and the reciprocal of the second and then proceed as in the previous two examples:

$$\frac{x^2 - y^2}{x^2 - 2xy + y^2} \div \frac{x^3 + y^3}{x^3 - x^2y}$$

$$= \frac{x^2 - y^2}{x^2 - 2xy + y^2} \cdot \frac{x^3 - x^2y}{x^3 + y^3} \qquad \begin{array}{l}\text{Multiply by the} \\ \text{reciprocal of the} \\ \text{divisor.}\end{array}$$

$$= \frac{(x-y)(x+y)(x^2)(x-y)}{(x-y)(x-y)(x+y)(x^2 - xy + y^2)} \qquad \begin{array}{l}\text{Factor and} \\ \text{multiply.}\end{array}$$

$$= \frac{x^2}{x^2 - xy + y^2} \qquad\qquad \begin{array}{l}\text{Divide out common} \\ \text{factors.} \qquad ▲\end{array}$$

Here are some more examples of multiplication and division with rational expressions:

▼   **Example 6**   Perform the indicated operations.

a. $\dfrac{a^2 - 8a + 15}{a + 4} \cdot \dfrac{a + 2}{a^2 - 5a + 6} \div \dfrac{a^2 - 3a - 10}{a^2 + 2a - 8}$

$= \dfrac{(a^2 - 8a + 15)(a + 2)(a^2 + 2a - 8)}{(a + 4)(a^2 - 5a + 6)(a^2 - 3a - 10)}$ Change division to multiplication by using the reciprocal.

$= \dfrac{(a - 5)(a - 3)(a + 2)(a + 4)(a - 2)}{(a + 4)(a - 3)(a - 2)(a - 5)(a + 2)}$ Factor.

$= 1$ Divide out common factors.

b. $\dfrac{xa + xb + ya + yb}{xa - xb - ya + yb} \cdot \dfrac{xa + xb - ya - yb}{xa - xb + ya - yb}$

$= \dfrac{x(a + b) + y(a + b)}{x(a - b) - y(a - b)} \cdot \dfrac{x(a + b) - y(a + b)}{x(a - b) + y(a - b)}$ Factor by grouping.

$= \dfrac{(x + y)(a + b)(x - y)(a + b)}{(x - y)(a - b)(x + y)(a - b)}$

$= \dfrac{(a + b)^2}{(a - b)^2}$ ▲

▼   **Example 7**   Multiply $(4x^2 - 36) \cdot \dfrac{12}{4x + 12}$.

Solution   We can think of $4x^2 - 36$ as having a denominator of 1. Thinking of it in this way allows us to proceed as we did in the previous examples.

$(4x^2 - 36) \cdot \dfrac{12}{4x + 12}$

$= \dfrac{4x^2 - 36}{1} \cdot \dfrac{12}{4x + 12}$ Write $4x^2 - 36$ with denominator 1

$= \dfrac{4(x - 3)(x + 3)12}{4(x + 3)}$ Factor

$= 12(x - 3)$ Divide out common factors   ▲

Perform the indicated operations involving fractions: *Problem Set 4.3*

1. $\frac{2}{9} \cdot \frac{3}{4}$ 2. $\frac{5}{6} \cdot \frac{7}{8}$

3. $\frac{3}{4} \div \frac{1}{3}$ 4. $\frac{3}{8} \div \frac{5}{4}$

5. $\frac{3}{7} \cdot \frac{14}{24} \div \frac{1}{2}$

6. $\frac{6}{5} \cdot \frac{10}{36} \div \frac{3}{4}$

7. $\frac{10x^2}{5y^2} \cdot \frac{15y^3}{2x^4}$

8. $\frac{8x^3}{7y^4} \cdot \frac{14y^6}{16x^2}$

9. $\frac{11a^2b}{5ab^2} \div \frac{22a^3b^2}{10ab^4}$

10. $\frac{8ab^3}{9a^2b} \div \frac{16a^2b^2}{18ab^3}$

11. $\frac{6x^2}{5y^3} \cdot \frac{11z^2}{2x^2} \div \frac{33z^5}{10y^8}$

12. $\frac{4x^3}{7y^2} \cdot \frac{6z^5}{5x^6} \div \frac{24z^2}{35x^6}$

Perform the indicated operations. Be sure to write all answers in lowest terms.

13. $\frac{x^2 - 9}{x^2 - 4} \cdot \frac{x - 2}{x - 3}$

14. $\frac{x^2 - 16}{x^2 - 25} \cdot \frac{x - 5}{x - 4}$

15. $\frac{y^2 - 1}{y + 2} \cdot \frac{y^2 + 5y + 6}{y^2 + 2y - 3}$

16. $\frac{y - 1}{y^2 - y - 6} \cdot \frac{y^2 + 5y + 6}{y^2 - 1}$

17. $\frac{3x - 12}{x^2 - 4} \cdot \frac{x^2 + 6x + 8}{x - 4}$

18. $\frac{x^2 + 5x + 1}{4x - 4} \cdot \frac{x - 1}{x^2 + 5x + 1}$

19. $\frac{a^2 - 5a + 6}{a^2 - 2a - 3} \div \frac{a - 5}{a^2 + 3a + 2}$

20. $\frac{a^2 + 7a + 12}{a - 5} \div \frac{a^2 + 9a + 18}{a^2 - 7a + 10}$

21. $\frac{2x^2 - 5x - 12}{4x^2 + 8x + 3} \div \frac{x^2 - 16}{2x^2 + 7x + 3}$

22. $\frac{x^2 - 2x + 1}{3x^2 + 7x - 20} \div \frac{x^2 + 3x - 4}{3x^2 - 2x - 5}$

23. $\frac{x^2 + 5x + 6}{x + 1} \cdot \frac{x^2 - 1}{x^2 + 7x + 10} \div \frac{x^2 + 2x - 3}{x + 5}$

24. $\frac{2x^2 - x - 1}{x^2 - 2x - 15} \cdot \frac{x - 5}{4x^2 - 1} \div \frac{x - 1}{x^2 - 9}$

25. $\frac{x^3 - 1}{x^4 - 1} \cdot \frac{x^2 - 1}{x^2 + x + 1}$

26. $\frac{x^3 - 8}{x^4 - 16} \cdot \frac{x^2 + 4}{x^2 + 2x + 4}$

27. $\frac{a^2 - 16}{a^2 - 8a + 16} \cdot \frac{a^2 - 9a + 20}{a^2 - 7a + 12} \div \frac{a^2 - 25}{a^2 - 6a + 9}$

28. $\frac{a^2 - 6a + 9}{a^2 - 4} \cdot \frac{a^2 - 5a + 6}{(a - 3)^2} \div \frac{a^2 - 9}{a^2 - a - 6}$

29. $\frac{xy - 2x + 3y - 6}{xy + 2x - 4y - 8} \cdot \frac{xy + x - 4y - 4}{xy - x + 3y - 3}$

30. $\frac{ax + bx + 2a + 2b}{ax - 3a + bx - 3b} \cdot \frac{ax - bx - 3a + 3b}{ax - bx - 2a + 2b}$

31. $\dfrac{xy - y + 4x - 4}{xy - 3y + 4x - 12} \div \dfrac{xy + 2x + y + 2}{xy - 3y + 2x - 6}$

32. $\dfrac{xb - 2b + 3x - 6}{xb + 3b + 3x + 9} \cdot \dfrac{xb - 2b - 2x + 4}{xb + 3b - 2x - 6}$

Use the method shown in Example 7 to find the following products.

33. $(3x - 6) \cdot \dfrac{x}{x - 2}$ **34.** $(4x + 8) \cdot \dfrac{x}{x + 2}$

35. $(x^2 - 25) \cdot \dfrac{2}{x - 5}$ **36.** $(x^2 - 49) \cdot \dfrac{5}{x + 7}$

37. $(x^2 - 3x + 2) \cdot \dfrac{3}{3x - 3}$

38. $(x^2 - 3x + 2) \cdot \dfrac{-1}{x - 2}$

39. $(y - 3)(y - 4)(y + 3) \cdot \dfrac{-1}{y^2 - 9}$

40. $(y + 1)(y + 4)(y - 1) \cdot \dfrac{3}{y^2 - 1}$

41. $a(a + 5)(a - 5) \cdot \dfrac{a + 1}{a^2 + 5a}$

42. $a(a + 3)(a - 3) \cdot \dfrac{a - 1}{a^2 - 3a}$

Review Problems   The problems that follow review material we covered in Section 2.4.

Solve each inequality.

43. $|x - 3| > 5$ **44.** $|x - 5| > 2$
45. $|2x - 5| < 3$ **46.** $|3x - 6| < 12$
47. $|2x + 1| \le 3$ **48.** $|4x - 3| \le 9$

We begin this section by considering two examples of addition with fractions. Since subtraction is defined as addition of the opposite, we will not need to show a separate example for subtraction.

To add two fractions with the same denominator, we simply add numerators and use the common denominator.

4.4
Addition and
Subtraction of
Rational Expressions

▼   **Example 1** Add $\frac{4}{9} + \frac{2}{9}$.

Solution   Since both fractions have the same denominator, we simply combine numerators and reduce to lowest terms:

$$\frac{4}{9} + \frac{2}{9} = \frac{4+2}{9}$$

$$= \frac{6}{9}$$

$$= \frac{2}{3} \qquad\qquad \blacktriangle$$

The main step in adding fractions is actually another application of the distributive property, even if we don't show it. To show the use of the distributive property we must first observe that $\frac{4}{9} = 4(\frac{1}{9})$ and $\frac{2}{9} = 2(\frac{1}{9})$, since division by 9 is equivalent to multiplication by $\frac{1}{9}$. Here is the solution to Example 1 again showing the use of the distributive property:

$$\frac{4}{9} + \frac{2}{9} = 4\left(\frac{1}{9}\right) + 2\left(\frac{1}{9}\right)$$

$$= (4 + 2)\left(\frac{1}{9}\right) \qquad \text{Distributive property}$$

$$= 6\left(\frac{1}{9}\right)$$

$$= \frac{6}{9}$$

$$= \frac{2}{3}$$

Adding fractions is always done by using the distributive property, even if it is not shown. The distributive property is the reason we begin all addition (or subtraction) problems involving fractions by making sure all fractions have the same denominator.

DEFINITION   The *least common denominator* (LCD) for a set of denominators is the smallest quantity divisible by each of the denominators.

The first step in adding two fractions is to find a common denominator. Once we have a common denominator we write each fraction again (without changing the value of either one) as a fraction with the common denominator. The procedure then follows that of Example 1 from this section.

Example 2 is a step-by-step solution to an addition problem involving fractions with unlike denominators. The purpose behind spending so much time here working with fractions is that addition of rational expressions follows the same procedure. If we understand addition of fractions, addition of rational expressions will follow naturally.

▼   **Example 2**   Add $\frac{3}{14} + \frac{7}{30}$.

Solution

Step 1.   Find the least common denominator.
To do this we first factor both denominators into prime factors.

$$\text{Factor 14:} \quad 14 = 2 \cdot 7$$
$$\text{Factor 30:} \quad 30 = 2 \cdot 3 \cdot 5$$

Since the LCD must be divisible by 14, it must have factors of $2 \cdot 7$. It must also be divisible by 30 and therefore have factors of $2 \cdot 3 \cdot 5$. We do not need to repeat the 2 that appears in both the factors of 14 and those of 30. Therefore,

$$\text{LCD} = 2 \cdot 3 \cdot 5 \cdot 7 = 210$$

Step 2.   Change to equivalent fractions.
Since we want each fraction to have a denominator of 210 and at the same time keep its original value, we multiply each by 1 in the appropriate form.

Change $\frac{3}{14}$ to a fraction with denominator 210:

$$\frac{3}{14} \cdot \frac{15}{15} = \frac{45}{210}$$

Change $\frac{7}{30}$ to a fraction with denominator 210:

$$\frac{7}{30} \cdot \frac{7}{7} = \frac{49}{210}$$

Step 3.   Add numerators of equivalent fractions found in step 2:

$$\frac{45}{210} + \frac{49}{210} = \frac{94}{210}$$

Step 4.   Reduce to lowest terms if necessary:

$$\frac{94}{210} = \frac{47}{105} \qquad\qquad\qquad ▲$$

The main idea in adding fractions is to write each fraction again with the LCD for a denominator. In doing so, we must be sure not to change the value of either of the original fractions.

When adding rational expressions we follow the same steps used in Example 2 to add fractions.

▼ **Example 3**   Add $\dfrac{x}{x^2 - 1} + \dfrac{1}{x^2 - 1}$.

Solution   Since the denominators are the same, we simply add numerators:

$$\frac{x}{x^2 - 1} + \frac{1}{x^2 - 1} = \frac{x + 1}{x^2 - 1} \qquad \text{Add numerators.}$$

$$= \frac{\cancel{x + 1}}{(x - 1)(\cancel{x + 1})} \qquad \text{Factor denominator.}$$

$$= \frac{1}{x - 1} \qquad \begin{array}{l}\text{Divide out common} \\ \text{factor } x + 1.\end{array} \quad ▲$$

▼ **Example 4**   Add $\dfrac{5}{x^2 - 2x - 3} + \dfrac{4}{x^2 - 9}$.

Solution   In this case the denominators are not the same, so we follow the steps from Example 2 of this section.

Step 1.   Factor each denominator and build the LCD from the factors:

$$\left.\begin{array}{l} x^2 - 2x - 3 = (x - 3)(x + 1) \\ x^2 - 9 \qquad\; = (x - 3)(x + 3) \end{array}\right\} \text{LCD} = (x - 3)(x + 3)(x + 1)$$

Step 2.   Change each rational expression to an equivalent expression that has the LCD for a denominator:

$$\frac{5}{x^2 - 2x - 3} = \frac{5}{(x - 3)(x + 1)} \cdot \frac{(x + 3)}{(x + 3)} = \frac{5x + 15}{(x - 3)(x + 3)(x + 1)}$$

$$\frac{4}{x^2 - 9} = \frac{4}{(x - 3)(x + 3)} \cdot \frac{(x + 1)}{(x + 1)} = \frac{4x + 4}{(x - 3)(x + 3)(x + 1)}$$

Step 3.   Add numerators of the rational expressions found in step 2:

$$\frac{5x + 15}{(x - 3)(x + 3)(x + 1)} + \frac{4x + 4}{(x - 3)(x + 3)(x + 1)}$$

$$= \frac{9x + 19}{(x - 3)(x + 3)(x + 1)}$$

The numerator and denominator of the resulting expression do not have any factors in common. It is in lowest terms. It is best to leave the denominator in factored form rather than multiply it out.   ▲

▼ **Example 5**   Subtract $\dfrac{2x}{x^2 + 7x + 10} - \dfrac{3}{5x + 10}$.

Solution   Factoring the denominators, we have:

$$\frac{2x}{x^2 + 7x + 10} - \frac{3}{5x + 10} = \frac{2x}{(x + 2)(x + 5)} + \frac{-3}{5(x + 2)}$$

The LCD is $5(x + 2)(x + 5)$. Completing the problem, we have

$$= \frac{5}{5} \cdot \frac{2x}{(x + 2)(x + 5)} + \frac{-3}{5(x + 2)} \cdot \frac{(x + 5)}{(x + 5)}$$

$$= \frac{10x}{5(x + 2)(x + 5)} + \frac{-3x - 15}{5(x + 2)(x + 5)}$$

$$= \frac{7x - 15}{5(x + 2)(x + 5)}$$

Notice that in the first step we wrote subtraction as addition of the opposite. There seems to be less chance for error when this is done.   ▲

▼ **Example 6**   Add $\dfrac{x^2}{x - 7} + \dfrac{6x + 7}{7 - x}$.

Solution   Recall Section 4.1 where we were able to reverse the terms in a factor such as $7 - x$ by factoring -1 from each term. In a problem like this, the same result can be obtained by multiplying the numerator and denominator by -1:

$$\frac{x^2}{x - 7} + \frac{6x + 7}{7 - x} \cdot \frac{-1}{-1} = \frac{x^2}{x - 7} + \frac{-6x - 7}{x - 7}$$

$$= \frac{x^2 - 6x - 7}{x - 7} \qquad \text{Add numerators.}$$

$$= \frac{(x - 7)(x + 1)}{(x - 7)} \qquad \text{Factor numerator.}$$

$$= x + 1 \qquad \text{Divide out } (x - 7).$$

▲

Problem Set 4.4

Combine the following fractions:

1. $\frac{3}{4} + \frac{1}{2}$ 2. $\frac{5}{6} + \frac{1}{3}$

3. $\frac{2}{5} - \frac{1}{15}$ 4. $\frac{5}{8} - \frac{1}{4}$

5. $\frac{5}{6} + \frac{7}{8}$ 6. $\frac{3}{4} + \frac{2}{3}$

7. $\frac{9}{48} - \frac{3}{54}$ 8. $\frac{6}{28} - \frac{5}{42}$

9. $\frac{3}{4} - \frac{1}{8} + \frac{2}{3}$ 10. $\frac{1}{3} - \frac{5}{6} + \frac{5}{12}$

Combine the following rational expressions. Reduce all answers to lowest terms.

11. $\frac{x}{x + 3} + \frac{3}{x + 3}$ 12. $\frac{5x}{5x + 2} + \frac{2}{5x + 2}$

13. $\frac{4}{y - 4} - \frac{y}{y - 4}$ 14. $\frac{8}{y + 8} + \frac{y}{y + 8}$

15. $\frac{x}{x^2 - y^2} - \frac{y}{x^2 - y^2}$ 16. $\frac{x}{x^2 - y^2} + \frac{y}{x^2 - y^2}$

17. $\frac{5}{3x + 6} + \frac{2}{x + 2}$ 18. $\frac{7}{5x - 10} - \frac{3}{x - 2}$

19. $\frac{1}{a} + \frac{2}{a^2} + \frac{3}{a^3}$ 20. $\frac{3}{a} - \frac{2}{a^2} + \frac{1}{a^3}$

21. $\frac{4}{2x^2} - \frac{5}{3x}$ 22. $\frac{6}{4x^3} - \frac{2}{x^2}$

23. $\frac{5}{x - 1} + \frac{x}{x^2 - 1}$ 24. $\frac{1}{x + 3} + \frac{x}{x^2 - 9}$

25. $\frac{x}{x^2 - 5x + 6} - \frac{3}{3 - x}$ 26. $\frac{x}{x^2 + 4x + 4} - \frac{2}{2 + x}$

27. $\frac{3}{a^2 - 5a + 6} - \frac{2}{a^2 - a - 2}$ 28. $\frac{1}{a^2 + a - 2} + \frac{4}{a^2 - a - 6}$

29. $\frac{8}{y^2 - 16} - \frac{7}{y^2 - y - 12}$

30. $\frac{6}{y^2 - 9} - \frac{5}{y^2 - y - 6}$

31. $\frac{4a}{a^2 + 6a + 5} - \frac{3a}{a^2 + 5a + 4}$

32. $\frac{3a}{a^2 + 7a + 10} - \frac{2a}{a^2 + 6a + 8}$

33. $x + \dfrac{1}{3 - x} + \dfrac{1}{x^2 - 9}$

34. $x - \dfrac{1}{2 + x} - \dfrac{2}{x^2 - 4}$

35. $\dfrac{3}{x^2 - x - 6} + \dfrac{2}{x^2 - 4} + \dfrac{1}{x^2 - 5x + 6}$

36. $\dfrac{1}{x^2 - 25} + \dfrac{1}{x^2 - 4x - 5} + \dfrac{1}{x^2 + 6x + 5}$

37. $\dfrac{xy}{x^3 - y^3} + \dfrac{1}{x - y}$

38. $\dfrac{1}{x - 2} - \dfrac{2x}{x^3 - 8}$

39. $\dfrac{x}{x + 2} + \dfrac{1}{2x + 4} - \dfrac{3}{x^2 + 2x}$

40. $\dfrac{x}{x + 3} + \dfrac{7}{3x + 9} - \dfrac{2}{x^2 + 3x}$

41. $\dfrac{1}{x} + \dfrac{x}{2x + 4} - \dfrac{2}{x^2 + 2x}$

42. $\dfrac{1}{x} + \dfrac{x}{3x + 9} - \dfrac{3}{x^2 + 3x}$

43. The formula $P = \dfrac{1}{a} + \dfrac{1}{b}$ is used by optometrists to help determine how strong to make the lenses for a pair of eyeglasses. If a is 10 and b is .2, find the corresponding value of P.

44. Show that the formula in Problem 43 can be written $P = \dfrac{a + b}{ab}$, and then let $a = 10$ and $b = .2$ in this new form of the formula to find P.

45. Simplify the expression below by first subtracting inside each set of parentheses and then simplifying the result.
$$(1 - \tfrac{1}{2})(1 - \tfrac{1}{3})(1 - \tfrac{1}{4})(1 - \tfrac{1}{5})$$

46. Simplify the expression below by adding inside each set of parentheses first and then simplifying the result.
$$(1 + \tfrac{1}{2})(1 + \tfrac{1}{3})(1 + \tfrac{1}{4})(1 + \tfrac{1}{5})$$

47. Show that the expressions $(x + y)^{-1}$ and $x^{-1} + y^{-1}$ are not equal when $x = 3$ and $y = 4$.

48. Show that the expressions $(x + y)^{-1}$ and $x^{-1} + y^{-1}$ are not equal. (Begin by writing each with positive exponents only.)

Review Problems   The problems below review material we covered in Section 4.3.

Multiply or divide as indicated.

49. $xy \cdot \dfrac{1}{x}$ **50.** $xy \cdot \dfrac{1}{y}$

51. $\dfrac{2x - 1}{x^2 - x} \div \dfrac{1}{x^2 - x}$ **52.** $\dfrac{x + y}{2x^2} \div \dfrac{x + y}{4x}$

53. $\dfrac{x + 1}{x^2 - 4} \div \dfrac{x^2 - 1}{x + 2}$ **54.** $\dfrac{x - 2}{x^2 - 9} \div \dfrac{x^2 - 4}{x + 3}$

4.5
Complex Fractions

The quotient of two fractions or two rational expressions is called a *complex fraction.* This section is concerned with the simplification of complex fractions. There are no new properties. The problems in this section can be worked using the methods already developed in the chapter.

▼   **Example 1**   Simplify $\dfrac{3/4}{5/8}$.

Solution   There are generally two methods that can be used to simplify complex fractions.

Method 1.   Instead of dividing by 5/8 we can multiply by 8/5:

$$\frac{3/4}{5/8} = \frac{3}{4} \times \frac{8}{5} = \frac{24}{20} = \frac{6}{5}$$

Method 2.   We can multiply the numerator and denominator of the complex fraction by the LCD for both of the fractions, which in this case is 8.

$$\frac{3/4}{5/8} = \frac{\frac{3}{4} \cdot 8}{\frac{5}{8} \cdot 8} = \frac{6}{5} \qquad\qquad \blacktriangle$$

Here are some examples of complex fractions involving rational expressions. Most can be solved using either of the two methods shown in Example 1.

▼   **Example 2** Simplify $\dfrac{\dfrac{1}{x} + \dfrac{1}{y}}{\dfrac{1}{x} - \dfrac{1}{y}}$.

Solution This problem is most easily solved using method 2. We begin by multiplying both the numerator and denominator by the quantity xy, which is the LCD for all the fractions:

$$\frac{\dfrac{1}{x} + \dfrac{1}{y}}{\dfrac{1}{x} - \dfrac{1}{y}} = \frac{\left(\dfrac{1}{x} + \dfrac{1}{y}\right) \cdot xy}{\left(\dfrac{1}{x} - \dfrac{1}{y}\right) \cdot xy}$$

$$= \frac{\dfrac{1}{x}(xy) + \dfrac{1}{y}(xy)}{\dfrac{1}{x}(xy) - \dfrac{1}{y}(xy)}$$

Apply the distributive property to distribute xy over both terms in the numerator and denominator.

$$= \frac{y + x}{y - x}$$ ▲

▼   **Example 3** Simplify $\dfrac{\dfrac{x - 2}{x^2 - 9}}{\dfrac{x^2 - 4}{x + 3}}$.

Solution Applying method 1, we have

$$\frac{\dfrac{x - 2}{x^2 - 9}}{\dfrac{x^2 - 4}{x + 3}} = \frac{x - 2}{x^2 - 9} \cdot \frac{x + 3}{x^2 - 4}$$

$$= \frac{(x - 2)(x + 3)}{(x + 3)(x - 3)(x + 2)(x - 2)}$$

$$= \frac{1}{(x - 3)(x + 2)}$$ ▲

▼   **Example 4** Simplify

$$\frac{1 - \dfrac{4}{x^2}}{1 - \dfrac{1}{x} - \dfrac{6}{x^2}}$$

Solution   The simplest way to simplify this complex fraction is to multiply the numerator and denominator by the LCD, x^2:

$$\frac{1 - \dfrac{4}{x^2}}{1 - \dfrac{1}{x} - \dfrac{6}{x^2}} = \frac{x^2\left(1 - \dfrac{4}{x^2}\right)}{x^2\left(1 - \dfrac{1}{x} - \dfrac{6}{x^2}\right)}$$ Multiply numerator and denominator by x^2.

$$= \frac{x^2 \cdot 1 - x^2 \cdot \dfrac{4}{x^2}}{x^2 \cdot 1 - x^2 \cdot \dfrac{1}{x} - x^2 \cdot \dfrac{6}{x^2}}$$ Distributive property

$$= \frac{x^2 - 4}{x^2 - x - 6}$$ Simplify.

$$= \frac{(x - 2)(x + 2)}{(x - 3)(x + 2)}$$ Factor.

$$= \frac{x - 2}{x - 3}$$ Reduce.   ▲

▼   **Example 5**   Simplify $x - \dfrac{3}{x + \frac{1}{3}}$.

Solution   We begin by simplifying the denominator:

$$x + \frac{1}{3} = \frac{3x}{3} + \frac{1}{3} = \frac{3x + 1}{3}$$

Here is the complete solution:

$$x - \frac{3}{x + \dfrac{1}{3}} = x - \frac{3}{\dfrac{3x + 1}{3}}$$

$$= x - 3 \cdot \frac{3}{3x + 1}$$

$$= x - \frac{9}{3x + 1}$$

$$= \frac{x}{1} \cdot \frac{3x + 1}{3x + 1} - \frac{9}{3x + 1}$$

$$= \frac{3x^2 + x - 9}{3x + 1}$$ ▲

Simplify each of the following as much as possible: *Problem Set 4.5*

1. $\dfrac{3/4}{2/3}$

2. $\dfrac{5/9}{7/12}$

3. $\dfrac{\frac{1}{3}-\frac{1}{4}}{\frac{1}{2}+\frac{1}{8}}$

4. $\dfrac{\frac{1}{6}-\frac{1}{3}}{\frac{1}{4}-\frac{1}{8}}$

5. $\dfrac{3+\frac{2}{5}}{1-\frac{3}{7}}$

6. $\dfrac{2+\frac{5}{6}}{1-\frac{7}{8}}$

7. $\dfrac{\dfrac{1}{x}}{1+\dfrac{1}{x}}$

8. $\dfrac{1-\dfrac{1}{x}}{\dfrac{1}{x}}$

9. $\dfrac{1+\dfrac{1}{a}}{1-\dfrac{1}{a}}$

10. $\dfrac{1-\dfrac{2}{a}}{1-\dfrac{3}{a}}$

11. $\dfrac{\dfrac{1}{x}-\dfrac{1}{y}}{\dfrac{1}{x}+\dfrac{1}{y}}$

12. $\dfrac{\dfrac{1}{x}+\dfrac{2}{y}}{\dfrac{2}{x}+\dfrac{1}{y}}$

13. $\dfrac{\dfrac{x-5}{x^2-4}}{\dfrac{x^2-25}{x+2}}$

14. $\dfrac{\dfrac{3x+1}{x^2-49}}{\dfrac{9x^2-1}{x-7}}$

15. $\dfrac{\dfrac{4a}{2a^3+2}}{\dfrac{8a}{4a+4}}$

16. $\dfrac{\dfrac{2a}{3a^3-3}}{\dfrac{4a}{6a-6}}$

17. $\dfrac{1-\dfrac{9}{x^2}}{1-\dfrac{1}{x}-\dfrac{6}{x^2}}$

18. $\dfrac{4-\dfrac{1}{x^2}}{4+\dfrac{4}{x}+\dfrac{1}{x^2}}$

19. $\dfrac{2+\dfrac{5}{a}-\dfrac{3}{a^2}}{2-\dfrac{5}{a}+\dfrac{2}{a^2}}$

20. $\dfrac{3+\dfrac{5}{a}-\dfrac{2}{a^2}}{3-\dfrac{10}{a}+\dfrac{3}{a^2}}$

21. $\dfrac{1-\dfrac{1}{a+1}}{1+\dfrac{1}{a-1}}$

22. $\dfrac{\dfrac{1}{a-1}+1}{\dfrac{1}{a+1}-1}$

23. $\dfrac{\dfrac{y+1}{y-1}+\dfrac{y-1}{y+1}}{\dfrac{y+1}{y-1}+\dfrac{y-1}{y+1}}$

24. $\dfrac{\dfrac{y-1}{y+1}-\dfrac{y+1}{y-1}}{\dfrac{y-1}{y+1}+\dfrac{y+1}{y-1}}$

25. $1-\dfrac{x}{1-\dfrac{1}{x}}$

26. $x-\dfrac{1}{x-\dfrac{1}{2}}$

27. $1+\dfrac{1}{1+\dfrac{1}{1+1}}$

28. $1-\dfrac{1}{1-\dfrac{1}{1-\frac{1}{2}}}$

29. $\dfrac{1-\dfrac{1}{x-\frac{1}{2}}}{1+\dfrac{1}{x+\frac{1}{2}}}$

30. $\dfrac{2+\dfrac{1}{x+\frac{1}{3}}}{2-\dfrac{1}{x-\frac{1}{3}}}$

31. The formula $f=\dfrac{ab}{a+b}$ is used in optics to find the focal length of a lens. Show that the formula $f=(a^{-1}+b^{-1})^{-1}$ is equivalent to the formula above by rewriting it without the negative exponents, and then simplifying the results.

32. Show that the expression $(a^{-1}-b^{-1})^{-1}$ can be simplified to $\dfrac{ab}{b-a}$ by first writing it without the negative exponents, and then simplifying the result.

33. Show that the expression $\dfrac{1-x^{-1}}{1+x^{-1}}$ can be written as $\dfrac{x-1}{x+1}$.

34. Show that the formula $(1+x^{-1})^{-1}$ can be written as $\dfrac{x}{x+1}$.

Review Problems   The problems that follow review material we covered in Section 2.1. Reviewing these problems will help you with the next section.

Solve each equation.

35. $3x+60=15$

36. $3x-18=4$

37. $3(a-4)=48$

38. $2(a+5)=28$

39. $3(y-3)=2(y-2)$

40. $5(y+2)=4(y+1)$

41. $10-2(x+3)=x+1$

42. $15-3(x-1)=x-2$

4.6

Equations Involving Rational Expressions

The first step in solving an equation that contains one or more rational expressions is to find the LCD for all denominators in the equation. We then multiply both sides of the equation by the LCD to clear the equation of all fractions. That is, after we have multiplied through by the LCD, each term in the resulting equation will have a denominator of 1.

▼ **Example 1** Solve $\dfrac{x}{2} - 3 = \dfrac{2}{3}$.

Solution The LCD for 2 and 3 is 6. Multiplying both sides by 6, we have

$$6\left(\frac{x}{2} - 3\right) = 6\left(\frac{2}{3}\right)$$

$$6\left(\frac{x}{2}\right) - 6(3) = \overset{2}{\cancel{6}}\left(\frac{2}{\cancel{3}}\right)$$

$$3x - 18 = 4$$

$$3x = 22$$

$$x = \frac{22}{3} \qquad\qquad ▲$$

Multiplying both sides of an equation by the LCD clears the equation of fractions because the LCD has the property that all the denominators divide it evenly. That is, the LCD contains all factors of each denominator.

▼ **Example 2** Solve $\dfrac{6}{a - 4} = \dfrac{3}{8}$.

Solution The LCD for $a - 4$ and 8 is $8(a - 4)$. Multiplying both sides by this quantity yields

$$8(a - 4) \cdot \frac{6}{a - 4} = 8(a - 4) \cdot \frac{3}{8}$$

$$48 = (a - 4) \cdot 3$$

$$48 = 3a - 12$$

$$60 = 3a$$

$$20 = a$$

The solution set is {20}, which checks in the original equation. ▲

When we multiply both sides of an equation by an expression containing the variable, we must be sure to check our solutions. The multiplication property of equality does not allow multiplication by 0. If the expression we multiply by contains the variable, then it has the possibility of being 0. In the last example we multiplied both sides by $8(a - 4)$. This gives a restriction $a \neq 4$ for any solution we come up with.

▼ **Example 3**   Solve $\dfrac{x}{x-2} + \dfrac{2}{3} = \dfrac{2}{x-2}$.

Solution   The LCD is $3(x-2)$. We are assuming $x \neq 2$ when we multiply both sides of the equation by $3(x-2)$:

$$3(x-2) \cdot \left[\dfrac{x}{x-2} + \dfrac{2}{3} \right] = 3(x-2) \cdot \dfrac{2}{x-2}$$

$$3x + (x-2) \cdot 2 = 3 \cdot 2$$
$$3x + 2x - 4 = 6$$
$$5x - 4 = 6$$
$$5x = 10$$
$$x = 2$$

The only possible solution is $x = 2$. Checking this value back in the original equation gives

$$\dfrac{2}{2-2} + \dfrac{2}{3} = \dfrac{2}{2-2}$$

$$\dfrac{2}{0} + \dfrac{2}{3} = \dfrac{2}{0}$$

The first and last terms are undefined. The proposed solution, $x = 2$, does not check in the original equation. In the process of solving the equation, we multiplied both sides by $3(x-2)$, solved for x, and got $x = 2$. When $x = 2$, the quantity $3(x-2)$ is 0, which means we multiplied both sides of our original equation by 0. Multiplying both sides by 0 does not produce an equation equivalent to the original one.

The solution set is $\emptyset$. There is no real number x such that

$$\dfrac{x}{x-2} + \dfrac{2}{3} = \dfrac{2}{x-2}$$ ▲

When the proposed solution to an equation is not actually a solution, it is called an *extraneous* solution. In the last example, $x = 2$ is an extraneous solution.

▼ **Example 4**   Solve $\dfrac{5}{x^2 - 3x + 2} - \dfrac{1}{x-2} = \dfrac{1}{3x-3}$.

Solution   Writing the equation again with the denominators in factored form, we have

$$\frac{5}{(x-2)(x-1)} - \frac{1}{x-2} = \frac{1}{3(x-1)}$$

The LCD is $3(x-2)(x-1)$. Multiplying through by the LCD, we have

$$3(x-2)(x-1)\frac{5}{(x-2)(x-1)} - 3(x-2)(x-1)\cdot\frac{1}{(x-2)}$$

$$= 3(x-2)(x-1)\cdot\frac{1}{3(x-1)}$$

$$\begin{aligned}
3\cdot 5 - 3(x-1)\cdot 1 &= (x-2)\cdot 1 \\
15 - 3x + 3 &= x - 2 \\
-3x + 18 &= x - 2 \\
-4x + 18 &= -2 \\
-4x &= -20 \\
x &= 5
\end{aligned}$$

Checking the proposed solution $x = 5$ in the original equation gives

$$\frac{5}{(5-2)(5-1)} - \frac{1}{5-2} = \frac{1}{3(5-1)}$$

$$\frac{5}{12} - \frac{1}{3} = \frac{1}{12}$$

$$\frac{5}{12} - \frac{4}{12} = \frac{1}{12}$$

$$\frac{1}{12} = \frac{1}{12}$$

The proposed solution $x = 5$ checks. The solution set is $\{5\}$. ▲

Note   We can check the proposed solution in any of the equations obtained before multiplying through by the LCD. We cannot check the proposed solution in an equation obtained after multiplying through by the LCD since, if we have multiplied by 0, the resulting equations will not be equivalent to the original one. Checking solutions is required whenever we have multiplied both sides of the equation by an expression containing the variable.

Problem Set 4.6   Solve each of the following equations:

1. $\dfrac{x}{5} + 4 = \dfrac{5}{3}$

2. $\dfrac{x}{5} = \dfrac{x}{2} - 9$

3. $\dfrac{a}{3} + 2 = \dfrac{4}{5}$

4. $\dfrac{a}{4} + \dfrac{1}{2} = \dfrac{2}{3}$

5. $\dfrac{y}{2} + \dfrac{y}{4} + \dfrac{y}{6} = 3$

6. $\dfrac{y}{3} - \dfrac{y}{6} + \dfrac{y}{2} = 1$

7. $\dfrac{5}{2x} = \dfrac{1}{x} + \dfrac{3}{4}$

8. $\dfrac{1}{2a} = \dfrac{2}{a} - \dfrac{3}{8}$

9. $\dfrac{1}{x} = \dfrac{1}{3} - \dfrac{2}{3x}$

10. $\dfrac{5}{2x} = \dfrac{2}{x} - \dfrac{1}{12}$

11. $\dfrac{2x}{x-3} + 2 = \dfrac{2}{x-3}$

12. $\dfrac{2}{x+5} = \dfrac{2}{5} - \dfrac{x}{x+5}$

13. $\dfrac{x+2}{x+1} = \dfrac{1}{x+1} + 2$

14. $\dfrac{x+6}{x+3} = \dfrac{3}{x+3} + 2$

15. $\dfrac{3}{a-2} = \dfrac{2}{a-3}$

16. $\dfrac{5}{a+1} = \dfrac{4}{a+2}$

17. $\dfrac{1}{x-1} - \dfrac{1}{x+1} = \dfrac{3x}{x^2-1}$

18. $\dfrac{5}{x-1} + \dfrac{2}{x-1} = \dfrac{4}{x+1}$

19. $\dfrac{2}{x-3} + \dfrac{x}{x^2-9} = \dfrac{4}{x+3}$

20. $\dfrac{2}{x+5} + \dfrac{3}{x+4} = \dfrac{2x}{x^2+9x+20}$

21. $\dfrac{3}{2} - \dfrac{1}{x-4} = \dfrac{-2}{2x-8}$

22. $\dfrac{2}{x} - \dfrac{1}{x+1} = \dfrac{-2}{5x+5}$

23. $\dfrac{3}{y-4} - \dfrac{2}{y+1} = \dfrac{5}{y^2-3y-4}$

24. $\dfrac{1}{y+2} - \dfrac{2}{y-3} = \dfrac{-2y}{y^2-y-6}$

25. $\dfrac{2}{1+a} = \dfrac{3}{1-a} + \dfrac{5}{a}$

26. $\dfrac{1}{a+3} - \dfrac{a}{a^2-9} = \dfrac{2}{3-a}$

27. $\dfrac{3}{2x-6} - \dfrac{x+1}{4x-12} = 4$

28. $\dfrac{2x-3}{5x+10} + \dfrac{3x-2}{4x+8} = 1$

29. $\dfrac{4}{2x-6} - \dfrac{12}{4x+12} = \dfrac{12}{x^2-9}$

30. $\dfrac{1}{x+2} + \dfrac{1}{x-2} = \dfrac{4}{x^2-4}$

31. $\dfrac{2}{y^2-7y+12} - \dfrac{1}{y^2-9} = \dfrac{4}{y^2-y-12}$

32. $\dfrac{1}{y^2 + 5y + 4} + \dfrac{3}{y^2 - 1} = \dfrac{-1}{y^2 + 3y - 4}$

33. The following diagram shows a section of an electronic circuit with a 3-ohm resistor and a 5-ohm resistor connected in parallel.

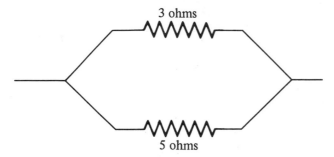

3 ohms

5 ohms

If R is the resistance equivalent to the two resistors connected in parallel, then the relationship between them is given by

$$\frac{1}{R} = \frac{1}{3} + \frac{1}{5}$$

Solve for R.

34. If a convex lens has a focal length of 15 cm, the image of an object 20 cm from the lens will appear d centimeters from the lens, where d is given by the equation.

$$\frac{1}{d} = \frac{1}{15} - \frac{1}{20}.$$

Solve this equation for d.

35. Solve the equation $6x^{-1} + 4 = 7$ by multiplying both sides by x. (Remember, $x^{-1} \cdot x = x^{-1} \cdot x^1 = x^0 = 1$.)

36. Solve the equation $3x^{-1} - 5 = 2x^{-1} - 3$ by multiplying both sides by x.

Review Problems The problems below review material we covered in Section 2.6. Reviewing these problems will get you ready for the next section. In each case, be sure to show the equation used.

37. Twice the sum of a number and 3 is 16. Find the number.
38. The sum of two consecutive odd integers is 48. Find the two integers.
39. The length of a rectangle is 3 less than twice the width. The perimeter is 42 meters. Find the length and width.
40. Kelley is 3 years older than his sister Lisa. In 4 years the sum of their ages will be 35. How old are they now?

4.7
Word Problems

The procedure used to solve the word problems in this section is the same procedure used in the past to solve word problems. Here, however, translating the problems from words into symbols will result in equations that involve rational expressions.

▼   **Example 1** One number is twice another. The sum of their reciprocals is 2. Find the numbers.

Solution   Let x = the smaller number. The larger number is $2x$. Their reciprocals are $1/x$ and $1/2x$. The equation that describes the situation is

$$\frac{1}{x} + \frac{1}{2x} = 2$$

Multiplying both sides by the LCD $2x$, we have

$$2x \cdot \frac{1}{x} + 2x \cdot \frac{1}{2x} = 2x(2)$$

$$2 + 1 = 4x$$

$$3 = 4x$$

$$x = \tfrac{3}{4}$$

The smaller number is $\tfrac{3}{4}$. The larger is $2(\tfrac{3}{4}) = \tfrac{6}{4} = \tfrac{3}{2}$. Adding their reciprocals, we have

$$\tfrac{4}{3} + \tfrac{2}{3} = \tfrac{6}{3} = 2$$

The sum of the reciprocals of $\tfrac{3}{4}$ and $\tfrac{3}{2}$ is 2.   ▲

▼   **Example 2** The speed of a boat in still water is 20 mi/hr. It takes the same amount of time for the boat to travel 3 miles downstream (with the current) as it does to travel 2 miles upstream (against the current). Find the speed of the current.

Solution   The following table will be helpful in finding the equation necessary to solve this problem.

	d (distance)	r (rate)	t (time)
Upstream			
Downstream			

If we let $x =$ the speed of the current, the speed (rate) of the boat upstream is $(20 - x)$ since it is traveling against the current. The rate downstream is $(20 + x)$ since the boat is then traveling with the current. The distance traveled upstream is 2 miles, while the distance traveled downstream is 3 miles. Putting the information given here into the table, we have

	d	r	t
Upstream	2	$20 - x$	
Downstream	3	$20 + x$	

To fill in the last two spaces in the table we must use the relationship $d = r \cdot t$. Since we know the spaces to be filled in are in the time column, we solve the equation $d = r \cdot t$ for t and get

$$t = \frac{d}{r}$$

The completed table then is

	d	r	t
Upstream	2	$20 - x$	$\dfrac{2}{20 - x}$
Downstream	3	$20 + x$	$\dfrac{3}{20 + x}$

Reading the problem again, we find that the time moving upstream is equal to the time moving downstream, or

$$\frac{2}{20 - x} = \frac{3}{20 + x}$$

Multiplying both sides by the LCD $(20 - x)(20 + x)$ gives

$$(20 + x) \cdot 2 = 3(20 - x)$$
$$40 + 2x = 60 - 3x$$
$$5x = 20$$
$$x = 4$$

The speed of the current is 4 mi/hr.   ▲

▼   **Example 3** John can do a certain job in 3 hours, while it takes Bob 5 hours to do the same job. How long will it take them, working together, to get the job done?

Solution In order to solve a problem like this we must assume that each person works at a constant rate. That is, they do the same amount of work in the first hour as they do in the last hour.

Solving a problem like this seems to be easier if we think in terms of how much work is done by each person in 1 hour.

If it takes John 3 hours to do the whole job, then in 1 hour he must do 1/3 of the job.

If we let $x =$ the amount of time it takes to complete the job working together, then in 1 hour they must do $1/x$ of the job. Here is the equation that describes the situation:

In 1 hour

$$\begin{bmatrix} \text{Amount of work} \\ \text{done by John} \end{bmatrix} + \begin{bmatrix} \text{Amount of work} \\ \text{done by Bob} \end{bmatrix} = \begin{bmatrix} \text{Total amount} \\ \text{of work done} \end{bmatrix}$$

$$\frac{1}{3} \qquad + \qquad \frac{1}{5} \qquad = \qquad \frac{1}{x}$$

Multiplying through by the LCD $15x$, we have

$$15x \cdot \frac{1}{3} + 15x \cdot \frac{1}{5} = 15x \cdot \frac{1}{x}$$

$$5x + 3x = 15$$

$$8x = 15$$

$$x = \frac{15}{8}$$

It takes them $\frac{15}{8}$ hours to do the job when they work together. ▲

▼   **Example 4** An inlet pipe can fill a pool in 10 hours, while an outlet pipe can empty it in 12 hours. If the pool is empty and both pipes are open, how long will it take to fill the pool?

Solution This problem is very similar to the problem in Example 3. It is helpful to think in terms of how much work is done by each pipe in 1 hour.

Let $x =$ the time it takes to fill the pool with both pipes open.

If the inlet pipe can fill the pool in 10 hours, then in 1 hour it is 1/10 full.

If the outlet pipe empties the pool in 12 hours, then in 1 hour it is 1/12 empty.

If the pool can be filled in x hours with both pipes open, then in 1 hour it is $1/x$ full when both pipes are open.

Here is the equation:

In 1 hour

$$\begin{bmatrix} \text{Amount full by} \\ \text{inlet pipe} \end{bmatrix} - \begin{bmatrix} \text{Amount empty by} \\ \text{outlet pipe} \end{bmatrix} = \begin{bmatrix} \text{Fraction of pool} \\ \text{filled by both} \end{bmatrix}$$

$$\frac{1}{10} \quad - \quad \frac{1}{12} \quad = \quad \frac{1}{x}$$

Multiplying through by $60x$, we have

$$60x \cdot \frac{1}{10} - 60x \cdot \frac{1}{12} = 60x \cdot \frac{1}{x}$$

$$6x - 5x = 60$$

$$x = 60$$

It takes 60 hours to fill the pool if both the inlet pipe and the outlet pipe are open.   ▲

Solve each of the following word problems. Be sure to show the equation in each case.

Problem Set 4.7

1. One number is 3 times another. The sum of their reciprocals is $\frac{20}{3}$. Find the numbers.

2. One number is 3 times another. The sum of their reciprocals is $\frac{4}{9}$. Find the numbers.

3. If a certain number is added to the numerator and denominator of $\frac{7}{9}$, the result is $\frac{5}{6}$. Find the number.

4. Find the number you would add to both the numerator and denominator of $\frac{8}{11}$ so the result would be $\frac{6}{7}$.

5. The speed of a boat in still water is 5 mi/hr. If the boat travels 3 miles downstream in the same amount of time it takes to travel 1.5 miles upstream, what is the speed of the current?

6. A boat, which moves at 18 mi/hr in still water, travels 14 miles downstream in the same amount of time it takes to travel 10 miles upstream. Find the speed of the current.

7. Train A has a speed 15 mi/hr greater than that of train B. If train A travels 150 miles in the same time train B travels 120 miles, what are the speeds of the two trains?

8. A train travels 30 mi/hr faster than a car. If the train covers 120 miles in the same time the car covers 80 miles, what is the speed of each of them?

9. If Sam can do a certain job in 3 days, while it takes Fred 6 days to do the same job, how long will it take them, working together, to complete the job?

10. Tim can finish a certain job in 10 hours. It takes his wife JoAnn only 8 hours to do the same job. If they work together, how long will it take them to complete the job?

11. Two people working together can complete a job in 6 hours. If one of them works twice as fast as the other, how long would it take the faster person, working alone, to do the job?

12. If two people working together can do a job in 3 hours, how long will it take the slower person to do the same job if one of them is 3 times as fast as the other?

13. A water tank can be filled by an inlet pipe in 8 hours. It takes twice that long for the outlet pipe to empty the tank. How long will it take to fill the tank if both pipes are open?

14. A sink can be filled from the faucet in 5 minutes. It takes only 3 minutes to empty the sink when the drain is open. If the sink is full and both the faucet and the drain are open, how long will it take to empty the sink?

15. It takes 10 hours to fill a pool with the inlet pipe. It can be emptied in 15 hours with the outlet pipe. If the pool is half-full to begin with, how long will it take to fill it from there if both pipes are open?

16. A sink is $\frac{1}{4}$ full when both the faucet and the drain are opened. The faucet alone can fill the sink in 6 minutes, while it takes 8 minutes to empty it with the drain. How long will it take to fill the remaining $\frac{3}{4}$ of the sink?

Review Problems   The problems below review material we covered in Section 2.5.

17. Solve $A = 2l + 2w$ for l.
18. Solve $A = P + Prt$ for t.
19. Solve $y = mx + b$ for m.
20. Solve $y = 3x - 4$ for x.
21. Solve $A = a + (n - 1)d$ for n.
22. Solve $A = \frac{1}{2}(b + B)h$ for b.

Chapter 4 Summary and Review

RATIONAL NUMBERS AND EXPRESSIONS [4.1]

A *rational number* is any number that can be expressed as the ratio of two integers:

$$\text{Rational numbers} = \left\{ \frac{a}{b} \;\middle|\; a \text{ and } b \text{ are integers, } b \neq 0 \right\}$$

A *rational expression* is any quantity that can be expressed as the ratio of two polynomials:

$$\text{Rational expressions} = \left\{ \frac{P}{Q} \;\middle|\; P \text{ and } Q \text{ are polynomials, } Q \neq 0 \right\}$$

1. $\frac{3}{4}$ is a rational number.

$\dfrac{x-3}{x^2-9}$ is a rational expression.

PROPERTIES OF RATIONAL EXPRESSIONS [4.1]

If P, Q, and K are polynomials with $Q \neq 0$ and $K \neq 0$, then

$$\frac{P}{Q} = \frac{PK}{QK} \quad \text{and} \quad \frac{P}{Q} = \frac{P/K}{Q/K}$$

which is to say that multiplying or dividing the numerator and denominator of a rational expression by the same nonzero quantity always produces an equivalent rational expression.

REDUCING TO LOWEST TERMS [4.1]

To reduce a rational expression to lowest terms we first factor the numerator and denominator and then divide the numerator and denominator by any factors they have in common.

2. $\dfrac{x-3}{x^2-9} = \dfrac{x-3}{(x-3)(x+3)}$

$= \dfrac{1}{x+3}$

DIVIDING A POLYNOMIAL BY A MONOMIAL [4.2]

To divide a polynomial by a monomial, divide each term of the polynomial by the monomial.

3. $\dfrac{15x^3 - 20x^2 + 10x}{5x}$

$= 3x^2 - 4x + 2$

4.
$$x - 3\overline{)x^2 - 5x + 8}$$
with work shown:
$$x - 2$$
$$\begin{array}{r} x - 2 \\ x-3\overline{)x^2 - 5x + 8} \\ \underline{-\ +} \\ \not{x^2} \not{\ 3x} \ \downarrow \\ -2x + 8 \\ \underline{+\quad -} \\ \not{\ 2x} \not{\ 6} \\ \hline 2 \end{array}$$

5.
$$\frac{x + 1}{x^2 - 4} \cdot \frac{x + 2}{3x + 3}$$

$$= \frac{(x + 1)(x + 2)}{(x - 2)(x + 2)(3)(x + 1)}$$

$$= \frac{1}{3(x - 2)}$$

6. the LCD for $\dfrac{2}{x - 3}$ and $\dfrac{3}{5}$ is $5(x - 3)$.

7.
$$\frac{2}{x - 3} + \frac{3}{5}$$

$$= \frac{2}{x - 3} \cdot \frac{5}{5} + \frac{3}{5} \cdot \frac{x - 3}{x - 3}$$

$$= \frac{3x + 1}{5(x - 3)}$$

8.
$$\frac{\dfrac{1}{x} + \dfrac{1}{y}}{\dfrac{1}{x} - \dfrac{1}{y}} = \frac{xy\left(\dfrac{1}{x} + \dfrac{1}{y}\right)}{xy\left(\dfrac{1}{x} - \dfrac{1}{y}\right)}$$

$$= \frac{y + x}{y - x}$$

LONG DIVISION WITH POLYNOMIALS [4.2]

If division with polynomials cannot be accomplished by dividing out factors common to the numerator and denominator, then we use a process similar to long division with whole numbers. The steps in the process are: estimate, multiply, subtract, and bring down the next term.

MULTIPLICATION AND DIVISION [4.3]

If P, Q, R, and S represent polynomials, then

$$\frac{P}{Q} \cdot \frac{R}{S} = \frac{PR}{QS} \qquad (Q \neq 0 \text{ and } S \neq 0)$$

$$\frac{P}{Q} \div \frac{R}{S} = \frac{P}{Q} \cdot \frac{S}{R} = \frac{PS}{QR} \qquad (Q \neq 0,\ S \neq 0,\ R \neq 0)$$

LEAST COMMON DENOMINATOR [4.4]

The *least common denominator,* LCD, for a set of denominators is the smallest quantity divisible by each of the denominators.

ADDITION AND SUBTRACTION [4.4]

If P, Q, and R represent polynomials, $R \neq 0$, then

$$\frac{P}{R} + \frac{Q}{R} = \frac{P + Q}{R} \quad \text{and} \quad \frac{P}{R} - \frac{Q}{R} = \frac{P - Q}{R}$$

When adding or subtracting rational expressions with different denominators, we must find the LCD for all denominators and change each rational expression to an equivalent expression that has the LCD.

COMPLEX FRACTIONS [4.5]

A rational expression that contains, in its numerator or denominator, other rational expressions is called a complex fraction. One method of simplifying a complex fraction is to multiply the numerator and denominator by the LCD for all denominators.

EQUATIONS INVOLVING RATIONAL EXPRESSIONS [4.6]

To solve an equation involving rational expressions we first find the LCD for all denominators appearing on either side of the equation. We then multiply both sides by the LCD to clear the equation of all fractions and solve as usual.

9. Solve $\frac{x}{2} + 3 = \frac{1}{3}$

$$6(\tfrac{x}{2}) + 6 \cdot 3 = 6 \cdot \tfrac{1}{3}$$

$$3x + 18 = 2$$

$$x = -\frac{16}{3}$$

COMMON MISTAKES

1. Attempting to divide the numerator and denominator of a rational expression by a quantity that is not a factor of both. Like this:

$$\overset{3}{\cancel{x^2}} - \overset{}{\cancel{9x}} - \overset{2}{\cancel{20}} \over \underset{1}{\cancel{x^2}} - \underset{}{\cancel{3x}} - \underset{1}{\cancel{10}}} \quad \text{Mistake}$$

This makes no sense at all. The numerator and denominator must be factored completely before any factors they have in common can be recognized:

$$\frac{x^2 - 9x + 20}{x^2 - 3x - 10} = \frac{(x - 5)(x - 4)}{(x - 5)(x + 2)}$$

$$= \frac{x - 4}{x + 2}$$

2. Forgetting to check solutions to equations involving rational expressions. When we multiply both sides of an equation by a quantity containing the variable, we must be sure to check for extraneous solutions (see Section 4.6).

Reduce to lowest terms:

1. $\dfrac{x^2 - y^2}{x - y}$

2. $\dfrac{2x^2 - 5x + 3}{2x^2 - x - 3}$

Divide:

3. $\dfrac{24x^3y + 12x^2y^2 - 16xy^3}{4xy}$

4. $\dfrac{2x^3 - 9x^2 + 10}{2x - 1}$

Perform the indicated operations:

5. $\dfrac{a^2 - 16}{5a - 15} \cdot \dfrac{10(a - 3)^2}{a^2 - 7a + 12}$

6. $\dfrac{a^4 - 81}{a^2 + 9} \div \dfrac{a^2 - 8a + 15}{4a - 20}$

7. $\dfrac{x^3 - 8}{2x^2 - 9x + 10} \div \dfrac{x^2 + 2x + 4}{2x^2 + x - 15}$

8. $\frac{4}{21} + \frac{6}{35}$

Chapter 4 Test

9. $\frac{3}{4} - \frac{1}{2} + \frac{5}{8}$

10. $\dfrac{a}{a^2 - 9} + \dfrac{3}{a^2 - 9}$

11. $\dfrac{1}{x} + \dfrac{2}{x - 3}$

12. $\dfrac{4x}{x^2 + 6x + 5} - \dfrac{3x}{x^2 + 5x + 4}$

Simplify each complex fraction:

13. $\dfrac{\dfrac{3}{8}}{\dfrac{6}{40}}$

14. $\dfrac{3 - \dfrac{1}{a + 3}}{3 + \dfrac{1}{a + 3}}$

15. $\dfrac{1 - \dfrac{9}{x^2}}{1 + \dfrac{1}{x} - \dfrac{6}{x^2}}$

Solve each of the following equations:

16. $\dfrac{1}{x} + 3 = \dfrac{4}{3}$

17. $\dfrac{x}{x - 3} + 3 = \dfrac{3}{x - 3}$

18. $\dfrac{y + 3}{2y} + \dfrac{5}{y - 1} = \dfrac{1}{2}$

Solve the following word problems. Be sure to show the equation in each case.

19. What number must be subtracted from the denominator of $\frac{10}{23}$ to make the result $\frac{1}{3}$?

20. An inlet pipe can fill a pool in 10 hours while an outlet pipe can empty it in 15 hours. If the pool is half-full and both pipes are left open, how long will it take to fill the pool the rest of the way?

5

Rational Exponents and Roots

To the student:

This chapter is concerned with fractional exponents, roots, and complex numbers. As we will see, expressions involving fractional exponents are actually just radical expressions. That is, fractional exponents are used to denote square roots, cube roots, and so forth. We use fractional exponents as an alternative to radical notation (radical notation involves the use of the symbol $\sqrt{}$).

Many of the formulas that describe the characteristics of objects in the universe involve roots. The formula for the length of the diagonal of a square involves a square root. The length of time it takes a pendulum (as on a grandfather clock) to swing through one complete cycle depends on the square root of the length of the pendulum. The formulas that describe the changes in length, mass, and time for objects traveling at velocities close to the speed of light also contain roots.

We begin the chapter with some simple roots and the correlation between fractional exponents and roots. We will then list the properties associated with radicals and use these properties to write some radical expressions in simplified form. Combinations of radical expressions and equations involving radicals are considered next. The chapter concludes with the definition for complex numbers and some applications of this definition. The work we have done previously with exponents and

polynomials will be very useful in understanding the concepts developed here. Radical expressions and complex numbers behave like polynomials. As was the case in the preceding chapter, the distributive property is used extensively as justification for many of the properties developed in this chapter.

5.1
Rational Exponents

In Chapter 3 we developed notation (exponents) to give us the square, cube, or any power of a number. For instance, if we wanted the square of 3, we wrote $3^2 = 9$. If we wanted the cube of 3, we wrote $3^3 = 27$. In this section we will develop notation that will take us in the reverse direction, that is, from the square of a number, say 25, back to the original number, 5. There are two kinds of notation that allow us to do this. One is radical notation (square roots, cube roots, etc.), and the other involves fractional exponents. This section is concerned with the definitions associated with the two types of notation.

The number 49 has two square roots, 7 and -7. The positive square root (principal root) of 49 is written $\sqrt{49}$, while the negative square root of 49 is written $-\sqrt{49}$. All positive real numbers have two square roots, one positive and the other negative. The positive square root of 11 is $\sqrt{11}$. The negative square root of 11 is written $-\sqrt{11}$. Here is a definition for square roots:

DEFINITION   If x is a positive real number, then the expression $\sqrt{x}$ is called the *positive square root* of x and is such that

$$(\sqrt{x})^2 = x$$

In words: $\sqrt{x}$ is the positive number we square to get x.

The negative square root of x, $-\sqrt{x}$, is defined in a similar manner.

▼   **Example 1**   The positive square root of 64 is 8 because 8 is the positive number with the property $8^2 = 64$. The negative square root of 64 is -8 since -8 is the negative number whose square is 64. We can summarize both of these facts by saying

$$\sqrt{64} = 8 \quad \text{and} \quad -\sqrt{64} = -8 \qquad \blacktriangle$$

Note It is a common mistake to assume that an expression like $\sqrt{25}$ indicates both square roots, $+5$ and -5. The expression $\sqrt{25}$ indicates only the positive square root of 25, which is 5. If we want the negative square root, we must use a negative sign: $-\sqrt{25} = -5$.

The higher roots, cube roots, fourth roots, and so on, are defined by definitions similar to that of square roots.

DEFINITION   If x is a real number, and n is a positive integer, then

The positive square root of x, $\sqrt{x}$, is such that $(\sqrt{x})^2 = x$ (x positive)

The cube root of x, $\sqrt[3]{x}$, is such that $(\sqrt[3]{x})^3 = x$

The positive fourth root of x, $\sqrt[4]{x}$, is such that $(\sqrt[4]{x})^4 = x$ (x positive)

The fifth root of x, $\sqrt[5]{x}$, is such that $(\sqrt[5]{x})^5 = x$

$$\vdots \qquad \vdots \qquad \vdots$$

The nth root of x, $\sqrt[n]{x}$, is such that $(\sqrt[n]{x})^n = x$ (x positive if n is even)

Note   We have restricted the even roots in this definition to positive numbers. Even roots of negative numbers exist, but are not represented by real numbers. That is, $\sqrt{-4}$ is not a real number since there is no real number whose square is -4. We will have to wait until the last two sections of this chapter to see how to deal with even roots of negative numbers.

Here is a table of the most common roots used in this book. Any of the roots that are unfamiliar should be memorized.

Square roots		Cube roots	Fourth roots
$\sqrt{0} = 0$	$\sqrt{49} = 7$	$\sqrt[3]{0} = 0$	$\sqrt[4]{0} = 0$
$\sqrt{1} = 1$	$\sqrt{64} = 8$	$\sqrt[3]{1} = 1$	$\sqrt[4]{1} = 1$
$\sqrt{4} = 2$	$\sqrt{81} = 9$	$\sqrt[3]{8} = 2$	$\sqrt[4]{16} = 2$
$\sqrt{9} = 3$	$\sqrt{100} = 10$	$\sqrt[3]{27} = 3$	$\sqrt[4]{81} = 3$
$\sqrt{16} = 4$	$\sqrt{121} = 11$	$\sqrt[3]{64} = 4$	
$\sqrt{25} = 5$	$\sqrt{144} = 12$	$\sqrt[3]{125} = 5$	
$\sqrt{36} = 6$	$\sqrt{169} = 13$		

We will now develop a second kind of notation involving exponents that will allow us to designate square roots, cube roots, and so on in another way.

Consider the equation $x = 8^{1/3}$. Although we have not encountered fractional exponents before, let's assume that all the properties of exponents hold in this case. Cubing both sides of the equation, we have

$$x^3 = (8^{1/3})^3$$
$$x^3 = 8^{(1/3)(3)}$$
$$x^3 = 8^1$$
$$x^3 = 8$$

The last line tells us that x is the number whose cube is 8. It must be true, then, that x is the cube root of 8, $x = \sqrt[3]{8}$. Since we started with $x = 8^{1/3}$, it follows that

$$8^{1/3} = \sqrt[3]{8}$$

It seems reasonable, then, to define fractional exponents as indicating roots.

DEFINITION   If x is a real number and n is a positive integer, then

$$x^{1/n} = \sqrt[n]{x} \qquad (x \geq 0 \text{ when } n \text{ is even})$$

In words: The quantity $x^{1/n}$ is the nth root of x.

With this definition we have a way of representing roots with exponents. Here are some examples.

▼   **Example 2**

a. $8^{1/3} = \sqrt[3]{8} = 2$

b. $36^{1/2} = \sqrt{36} = 6$

c. $-25^{1/2} = -\sqrt{25} = -5$

d. $(-25)^{1/2} = \sqrt{-25}$,   which is not a real number

e. $\left(\frac{4}{9}\right)^{1/2} = \sqrt{\frac{4}{9}} = \frac{2}{3}$

f. $16^{1/2} \cdot 27^{1/3} = \sqrt{16}\sqrt[3]{27} = 4 \cdot 3 = 12$

g. $125^{1/3} + 81^{1/4} = \sqrt[3]{125} + \sqrt[4]{81} = 5 + 3 = 8$ ▲

Once we become familiar with fractional exponents, the second step in problems like the above will not be necessary.

The properties of exponents developed in Chapter 3 apply to integer exponents only. We will now extend these properties to include rational exponents also. We do so without proof.

PROPERTIES OF EXPONENTS   If a and b are real numbers and r and s are rational numbers, and a and b are positive whenever r or s indicate even roots, then

1. $a^r \cdot a^s = a^{r+s}$

2. $(a^r)^s = a^{rs}$

3. $(ab)^r = a^r b^r$

4. $a^{-r} = \dfrac{1}{a^r}$ $(a \neq 0)$

5. $\left(\dfrac{a}{b}\right)^r = \dfrac{a^r}{b^r}$ $(b \neq 0)$

6. $\dfrac{a^r}{a^s} = a^{r-s}$ $(a \neq 0)$

We can use the properties of exponents to prove the following theorem:

THEOREM 5.1 If a is a positive real number, m is an integer, and n is a positive integer, then

$$a^{m/n} = (a^{1/n})^m = (a^m)^{1/n}$$

Proof

$$
\begin{array}{l|l}
a^{m/n} = a^{m(1/n)} & a^{m/n} = a^{(1/n)(m)} \\
\quad\quad = (a^m)^{1/n} & \quad\quad = (a^{1/n})^m
\end{array}
$$

Here are some examples that use the definition of fractional exponents, the properties of exponents, and Theorem 5.1 to simplify expressions:

▼ **Example 3** Simplify as much as possible:

a. $8^{2/3} = (8^{1/3})^2$ Theorem 5.1
    $\quad\quad = 2^2$ Definition of fractional exponents
    $\quad\quad = 4$ Square of 2 is 4.

b. $25^{3/2} = (25^{1/2})^3$ Theorem 5.1
    $\quad\quad\;\, = 5^3$ Definition of fractional exponents
    $\quad\quad\;\, = 125$ The cube of 5 is 125.

c. $9^{-3/2} = (9^{1/2})^{-3}$ Theorem 5.1
    $\quad\quad\;\; = 3^{-3}$ Definition of fractional exponents

    $\quad\quad\;\; = \dfrac{1}{3^3}$ Property 4 for exponents

    $\quad\quad\;\; = \dfrac{1}{27}$ The cube of 3 is 27.

d. $\left(\dfrac{27}{8}\right)^{-4/3} = \left[\left(\dfrac{27}{8}\right)^{1/3}\right]^{-4}$ Theorem 5.1

$\qquad\qquad = \left(\dfrac{3}{2}\right)^{-4}$ Definition of fractional exponents

$\qquad\qquad = \left(\dfrac{2}{3}\right)^{4}$ Property 4 for exponents

$\qquad\qquad = \dfrac{16}{81}$ $\left(\dfrac{2}{3}\right)^{4} = \dfrac{16}{81}$ ▲

The following examples show the application of the properties of exponents to rational exponents:

▼ **Example 4** Assume $x, y,$ and z all represent positive quantities and simplify as much as possible:

a. $x^{1/3} \cdot x^{5/6} = x^{1/3+5/6}$
$\qquad\qquad = x^{2/6+5/6}$
$\qquad\qquad = x^{7/6}$

b. $(y^{2/3})^{3/4} = y^{(2/3)(3/4)}$
$\qquad\qquad = y^{1/2}$

c. $(x^{1/3}y^{2/5}z^{3/7})^4 = x^{4/3}y^{8/5}z^{12/7}$

d. $\dfrac{x^{1/3}}{x^{1/4}} = x^{1/3-1/4}$

$\qquad\qquad = x^{4/12-3/12}$
$\qquad\qquad = x^{1/12}$

e. $\dfrac{x^{4/5}y^{5/6}z^{3}}{x^{1/5}yz^{2/3}} = x^{3/5}y^{-1/6}z^{7/3}$

$\qquad\qquad = \dfrac{x^{3/5}z^{7/3}}{y^{1/6}}$

f. $\dfrac{(x^{3}y^{1/2})^{2/3}}{x^{1/6}y^{1/6}} = \dfrac{x^{2}y^{1/3}}{x^{1/6}y^{1/6}}$

$\qquad\qquad = x^{11/6}y^{1/6}$ ▲

In the next example we show how we use our properties of exponents to simplify some expressions involving roots.

▼ .**Example 5** Rewrite each expression with a fractional exponent instead of a root, and then simplify. Assume all variables represent positive numbers.

a. $\sqrt{16x^8} = (16x^8)^{1/2}$ Write with fractional exponent
$$= 16^{1/2}(x^8)^{1/2} \quad \text{Property 3}$$
$$= 2x^4 \quad\quad\quad\quad (x^8)^{1/2} = x^{8/2} = x^4$$

b. $\sqrt[3]{8a^6} = (8a^6)^{1/3}$ Write with fractional exponent
$$= 8^{1/3}(a^6)^{1/3} \quad \text{Property 3}$$
$$= 2a^2 \quad\quad\quad\quad (a^6)^{1/3} = a^{6/3} = a^2$$

c. $\sqrt[3]{a^{15}b^9} = (a^{15}b^9)^{1/3}$ Write with fractional exponent
$$= (a^{15})^{1/3}(b^9)^{1/3} \quad \text{Property 3}$$
$$= a^5b^3$$
▲

Use the definition of fractional exponents to write each of the following with the appropriate root, then simplify: *Problem Set 5.1*

1. $36^{1/2}$ 2. $49^{1/2}$

3. $-9^{1/2}$ 4. $-16^{1/2}$

5. $8^{1/3}$ 6. $-8^{1/3}$

7. $(-8)^{1/3}$ 8. $-27^{1/3}$

9. $32^{1/5}$ 10. $81^{1/4}$

11. $\left(\frac{81}{25}\right)^{1/2}$ 12. $\left(\frac{9}{16}\right)^{1/2}$

13. $\left(\frac{64}{125}\right)^{1/3}$ 14. $\left(\frac{8}{27}\right)^{1/3}$

Use Theorem 5.1 to simplify each of the following as much as possible:

15. $27^{2/3}$ 16. $8^{4/3}$

17. $25^{3/2}$ 18. $9^{3/2}$

19. $16^{3/4}$ 20. $81^{3/4}$

21. $27^{-1/3}$ 22. $9^{-1/2}$

23. $81^{-3/4}$ 24. $4^{-3/2}$

25. $\left(\frac{25}{36}\right)^{-1/2}$ 26. $\left(\frac{16}{49}\right)^{-1/2}$

27. $\left(\frac{81}{16}\right)^{-3/4}$ 28. $\left(\frac{27}{8}\right)^{-2/3}$

29. $16^{1/2} + 27^{1/3}$ 30. $25^{1/2} + 100^{1/2}$

31. $8^{-2/3} + 4^{-1/2}$ 32. $49^{-1/2} + 25^{-1/2}$

Use the properties of exponents to simplify each of the following as much as possible. Assume all bases are positive.

33. $x^{3/5} \cdot x^{1/5}$ 34. $x^{3/4} \cdot x^{5/4}$

35. $(a^{3/4})^{4/3}$ 36. $(a^{2/3})^{3/4}$

37. $\dfrac{x^{1/5}}{x^{3/5}}$ 38. $\dfrac{x^{2/7}}{x^{5/7}}$

39. $(a^{3/4} \cdot b^{1/3})^2$ 40. $(a^{5/6} \cdot b^{2/5})^3$

41. $x^{2/3} \cdot x^{2/5}$ 42. $x^{1/3} \cdot x^{3/4}$

43. $\dfrac{x^{5/6}}{x^{2/3}}$ 44. $\dfrac{x^{7/8}}{x^{8/7}}$

45. $(x^{3/5}y^{5/6}z^{1/3})^{3/5}$

46. $(x^{3/4}y^{1/8}z^{5/6})^{4/5}$

47. $\dfrac{x^{3/4}y^{2/3}}{x^{1/4}y^{1/3}}$

48. $\dfrac{x^{5/6}y^{3/5}}{x^{1/6}y^{2/5}}$

49. $\dfrac{a^{3/4}b^2}{a^{7/8}b^{1/4}}$

50. $\dfrac{a^{1/3}b^4}{a^{3/5}b^{1/3}}$

51. $\dfrac{(y^{2/3})^{3/4}}{(y^{1/3})^{3/5}}$

52. $\dfrac{(y^{5/4})^{2/5}}{(y^{1/4})^{4/3}}$

Rewrite each expression using a fractional exponent instead of a root, and then simplify. Assume all variables represent positive numbers.

53. $\sqrt{25x^4}$

54. $\sqrt{16x^6}$

55. $\sqrt{36a^8}$

56. $\sqrt{49a^{10}}$

57. $\sqrt[3]{x^6}$

58. $\sqrt[3]{x^9}$

59. $\sqrt[3]{27a^{12}}$

60. $\sqrt[3]{8a^{15}}$

61. $\sqrt[3]{x^3y^6}$

62. $\sqrt[3]{x^6y^3}$

63. $\sqrt[5]{32x^{10}y^5}$

64. $\sqrt[5]{32x^5y^{10}}$

65. Show that the expression $(a^{1/2} + b^{1/2})^2$ is not equal to $a + b$ by replacing a with 9 and b with 4 in both expressions and then simplifying each.

66. Show that the statement $(a^2 + b^2)^{1/2} = a + b$ is not, in general, true by replacing a with 3 and b with 4 and then simplifying both sides.

67. Use the formula $(a + b)(a - b) = a^2 - b^2$ to multiply $(x^{1/2} + y^{1/2})(x^{1/2} - y^{1/2})$.

68. Use the formula $(a + b)(a - b) = a^2 - b^2$ to multiply $(x^{3/2} + y^{3/2})(x^{3/2} - y^{3/2})$.

69. You may have noticed, if you have been using a calculator to find roots, that you can find the fourth root of a number by pressing the square root button twice. Written in symbols, this fact looks like this:

$$\sqrt{\sqrt{a}} = \sqrt[4]{a} \qquad (a \geq 0).$$

Show that this statement is true by rewriting each side with exponents instead of radical notation, and then simplifying the left side.

70. Show that the statement below is true by rewriting each side with exponents instead of radical notation, and then simplifying the left side.

$$\sqrt[3]{\sqrt{a}} = \sqrt[6]{a} \qquad (a \geq 0)$$

71. The maximum speed (v) that an automobile can travel around a curve of radius r without skidding is given by the equation

$$v = \left(\frac{5r}{2}\right)^{1/2}$$

Where v is in mi/hr and r is measured in feet. What is the maximum speed a car can travel around a curve with a radius of 250 feet without skidding?

72. The equation $L = \left(1 - \dfrac{v^2}{c^2}\right)^{1/2}$ gives the relativistic length of a 1-foot ruler traveling with velocity v. Find L if $\dfrac{v}{c} = \dfrac{3}{5}$.

Review Problems   The problems below review material we covered in Section 3.1.

Simplify each expression. Write your answer with positive exponents only.

73. $x^5 \cdot x^4 \cdot x^{-7}$ **74.** $x^{10} \cdot x^{-3} \cdot x^{-4}$

75. $(27a^6c^3)(2b^2c)$ **76.** $(8a^3b^3)(5a^2b)$

77. $(6x^2)(-3x^4)(2x^5)$ **78.** $(5x^3)(-7x^4)(-2x^6)$

79. $(5y^4)^{-3}(2y^{-2})^3$ **80.** $(3y^5)^{-2}(2y^{-4})^3$

Any expression containing a radical is called a *radical expression*. In the expression $\sqrt[3]{8}$, the ^{3} is called the *index*, the $\sqrt{}$ is the *radical sign*, and 8 is called the *radicand*. The index of a radical must be a positive integer greater than 1. If no index is written, it is assumed to be 2.

There are two properties of radicals that we will prove using the definition of fractional exponents and the properties of exponents. For these two properties we will assume a and b are nonnegative real numbers whenever n is an even number.

5.2
Simplified Form for Radicals

PROPERTY 1

$$\sqrt[n]{ab} = \sqrt[n]{a}\sqrt[n]{b}$$

In words: The nth root of a product is the product of the nth roots.

Proof

$\sqrt[n]{ab} = (ab)^{1/n}$ Definition of fractional exponents.

$\phantom{\sqrt[n]{ab}} = a^{1/n}b^{1/n}$ Exponents distribute over products.

$\phantom{\sqrt[n]{ab}} = \sqrt[n]{a}\sqrt[n]{b}$ Definition of fractional exponents.

PROPERTY 2

$$\sqrt[n]{\frac{a}{b}} = \frac{\sqrt[n]{a}}{\sqrt[n]{b}} \qquad (b \neq 0)$$

In words: The nth root of a quotient is the quotient of the nth roots.

Proof $\sqrt[n]{\dfrac{a}{b}} = \left(\dfrac{a}{b}\right)^{1/n}$ Definition of fractional
exponents.

$\qquad\qquad = \dfrac{a^{1/n}}{b^{1/n}}$ Exponents distribute
over quotients.

$\qquad\qquad = \dfrac{\sqrt[n]{a}}{\sqrt[n]{b}}$ Definition of fractional
exponents.

Note   There is no property for radicals that says the *n*th root of a sum is the sum
of the *n*th roots. That is,

$$\sqrt[n]{a + b} \neq \sqrt[n]{a} + \sqrt[n]{b}$$

The two properties of radicals allow us to change the form and simplify radical expressions without changing their value.

DEFINITION   A radical expression is in *simplified form* if:

1. none of the factors of the radicand (the quantity under the
   radical sign) can be written as powers greater than or equal to
   the index—that is, no perfect squares can be factors of the
   quantity under a square root sign, no perfect cubes as factors of
   what is under a cube root sign, and so forth;
2. there are no fractions under the radical sign; and
3. there are no radicals in the denominator.

Writing a radical expression in simplified form does not always result
in a simpler-looking expression. Simplified form for radicals is a way of
writing radicals so they are easiest to work with.

Satisfying the first condition for simplified form actually amounts to
taking as much out from under the radical sign as possible. The following examples illustrate the first condition for simplified form:

▼   **Example 1**   Write $\sqrt{50}$ in simplified form.

Solution   The largest perfect square that divides 50 is 25. We write
50 as $25 \cdot 2$ and apply Property 1 for radicals.

$$\sqrt{50} = \sqrt{25 \cdot 2} \qquad 50 = 25 \cdot 2$$
$$= \sqrt{25}\sqrt{2} \qquad \text{Property 1}$$
$$= 5\sqrt{2} \qquad \sqrt{25} = 5$$

We have taken as much as possible out from under the radical sign—in this case, factoring 25 from 50 and then writing $\sqrt{25}$ as 5.

▲

▼ **Example 2** Write in simplified form: $\sqrt{48x^4y^3}$, where $x, y \geq 0$.

Solution The largest perfect square that is a factor of the radicand is $16x^4y^2$. Applying Property 1 again, we have

$$
\begin{aligned}
\sqrt{48x^4y^3} &= \sqrt{16x^4y^2 \cdot 3y} \\
&= \sqrt{16x^4y^2}\sqrt{3y} \\
&= 4x^2y\sqrt{3y}
\end{aligned}
$$

▲

▼ **Example 3** Write $\sqrt[3]{40a^5b^4}$ in simplified form.

Solution We now want to factor the largest perfect cube from the radicand. We write $40a^5b^4$ as $8a^3b^3 \cdot 5a^2b$ and proceed as we did in Examples 1 and 2 above.

$$
\begin{aligned}
\sqrt[3]{40a^5b^4} &= \sqrt[3]{8a^3b^3 \cdot 5a^2b} \\
&= \sqrt[3]{8a^3b^3}\sqrt[3]{5a^2b} \\
&= 2ab\sqrt[3]{5a^2b}
\end{aligned}
$$

▲

Here are some further examples concerning the first condition for simplified form:

▼ **Example 4** Write each expression in simplified form. (Assume $x, y \geq 0$.)

a.
$$
\begin{aligned}
\sqrt{12x^7y^6} &= \sqrt{4x^6y^6 \cdot 3x} \\
&= \sqrt{4x^6y^6}\sqrt{3x} \\
&= 2x^3y^3\sqrt{3x}
\end{aligned}
$$

b.
$$
\begin{aligned}
\sqrt[3]{54a^6b^2c^4} &= \sqrt[3]{27a^6c^3 \cdot 2b^2c} \\
&= \sqrt[3]{27a^6c^3}\sqrt[3]{2b^2c} \\
&= 3a^2c\sqrt[3]{2b^2c}
\end{aligned}
$$

c.
$$
\begin{aligned}
\sqrt[5]{32x^{10}y^7} &= \sqrt[5]{32x^{10}y^5 \cdot y^2} \\
&= \sqrt[5]{32x^{10}y^5}\sqrt[5]{y^2} \\
&= 2x^2y\sqrt[5]{y^2}
\end{aligned}
$$

▲

The second property of radicals is used to simplify a radical that contains a fraction.

▼ **Example 5**   Simplify $\sqrt{\dfrac{3}{4}}$.

Solution   Applying Property 2 for radicals, we have

$$\sqrt{\dfrac{3}{4}} = \dfrac{\sqrt{3}}{\sqrt{4}} \qquad \text{Property 2}$$

$$= \dfrac{\sqrt{3}}{2} \qquad \sqrt{4} = 2$$

The last expression is in simplified form because it satisfies all three conditions for simplified form.   ▲

▼ **Example 6**   Write $\sqrt{\dfrac{5}{6}}$ in simplified form.

Solution   Proceeding as in Example 5 above, we have

$$\sqrt{\dfrac{5}{6}} = \dfrac{\sqrt{5}}{\sqrt{6}}$$

The resulting expression satisfies the second condition for simplified form since neither radical contains a fraction. It does, however, violate condition 3 since it has a radical in the denominator. Getting rid of the radical in the denominator is called *rationalizing the denominator* and is accomplished, in this case, by multiplying the numerator and denominator by $\sqrt{6}$:

$$\dfrac{\sqrt{5}}{\sqrt{6}} = \dfrac{\sqrt{5}}{\sqrt{6}} \cdot \dfrac{\sqrt{6}}{\sqrt{6}}$$

$$= \dfrac{\sqrt{30}}{\sqrt{6^2}}$$

$$= \dfrac{\sqrt{30}}{6} \qquad\qquad ▲$$

The idea behind rationalizing the denominator is to produce a perfect square under the square root sign in the denominator. This is accomplished by multiplying both the numerator and denominator by the appropriate radical.

▼ **Example 7** Rationalize the denominator. (Assume $x, y > 0$.)

a. $\dfrac{4}{\sqrt{3}} = \dfrac{4}{\sqrt{3}} \cdot \dfrac{\sqrt{3}}{\sqrt{3}}$

$= \dfrac{4\sqrt{3}}{\sqrt{3^2}}$

$= \dfrac{4\sqrt{3}}{3}$

b. $\dfrac{2\sqrt{3x}}{\sqrt{5y}} = \dfrac{2\sqrt{3x}}{\sqrt{5y}} \cdot \dfrac{\sqrt{5y}}{\sqrt{5y}}$

$= \dfrac{2\sqrt{15xy}}{\sqrt{(5y)^2}}$

$= \dfrac{2\sqrt{15xy}}{5y}$ ▲

When the denominator involves a cube root, we must multiply by a radical that will produce a perfect cube under the cube root sign in the denominator.

▼ **Example 8** Rationalize the denominator in $\dfrac{7}{\sqrt[3]{4}}$.

Solution Since $4 = 2^2$, we can multiply both numerator and denominator by $\sqrt[3]{2}$ and obtain $\sqrt[3]{2^3}$ in the denominator.

$$\dfrac{7}{\sqrt[3]{4}} = \dfrac{7}{\sqrt[3]{2^2}}$$

$$= \dfrac{7}{\sqrt[3]{2^2}} \cdot \dfrac{\sqrt[3]{2}}{\sqrt[3]{2}}$$

$$= \dfrac{7\sqrt[3]{2}}{\sqrt[3]{2^3}}$$

$$= \dfrac{7\sqrt[3]{2}}{2}$$ ▲

▼ **Example 9** Rationalize the denominator in $\dfrac{5\sqrt[4]{2}}{\sqrt[4]{3x^3}}$. (Assume $x > 0$.)

Solution We need a perfect fourth power under the radical sign in the denominator. Multiplying numerator and denominator by $\sqrt[4]{3^3x}$ will give us what we want:

$$\frac{5\sqrt[4]{2}}{\sqrt[4]{3x^3}} = \frac{5\sqrt[4]{2}}{\sqrt[4]{3x^3}} \cdot \frac{\sqrt[4]{3^3x}}{\sqrt[4]{3^3x}}$$

$$= \frac{5\sqrt[4]{54x}}{\sqrt[4]{3^4x^4}}$$

$$= \frac{5\sqrt[4]{54x}}{3x} \qquad\qquad \blacktriangle$$

As a last example we consider a radical expression that requires the use of both properties to meet the three conditions for simplified form.

▼ **Example 10** Simplify $\sqrt{\dfrac{12x^5y^3}{5z}}$. (Assume x, y, $z > 0$.)

Solution We use Property 2 to write the numerator and denominator as two separate radicals:

$$\sqrt{\frac{12x^5y^3}{5z}} = \frac{\sqrt{12x^5y^3}}{\sqrt{5z}}$$

Simplifying the numerator, we have

$$\frac{\sqrt{12x^5y^3}}{\sqrt{5z}} = \frac{\sqrt{4x^4y^2}\sqrt{3xy}}{\sqrt{5z}}$$

$$= \frac{2x^2y\sqrt{3xy}}{\sqrt{5z}}$$

To rationalize the denominator we multiply the numerator and denominator by $\sqrt{5z}$:

$$\frac{2x^2y\sqrt{3xy}}{\sqrt{5z}} \cdot \frac{\sqrt{5z}}{\sqrt{5z}} = \frac{2x^2y\sqrt{15xyz}}{\sqrt{(5z)^2}}$$

$$= \frac{2x^2y\sqrt{15xyz}}{5z} \qquad\qquad \blacktriangle$$

Use Property 1 for radicals to write each of the following expressions in simplified form. (Assume all variables are positive throughout problem set.)   ***Problem Set 5.2***

1. $\sqrt{8}$ 2. $\sqrt{32}$

3. $\sqrt{18}$ 4. $\sqrt{98}$

5. $\sqrt{75}$ 6. $\sqrt{12}$

7. $\sqrt{288}$ 8. $\sqrt{128}$

9. $\sqrt{80}$ 10. $\sqrt{200}$

11. $\sqrt{48}$ 12. $\sqrt{27}$

13. $\sqrt{45}$ 14. $\sqrt{20}$

15. $\sqrt[3]{54}$ 16. $\sqrt[3]{24}$

17. $\sqrt[3]{128}$ 18. $\sqrt[3]{162}$

19. $\sqrt[5]{64}$ 20. $\sqrt[4]{48}$

21. $\sqrt{54}$ 22. $\sqrt{63}$

23. $\sqrt[3]{40}$ 24. $\sqrt[3]{48}$

25. $\sqrt{99}$ 26. $\sqrt{44}$

27. $\sqrt{18x^3}$ 28. $\sqrt{27x^5}$

29. $\sqrt{32y^7}$ 30. $\sqrt{20y^3}$

31. $\sqrt[3]{40x^4y^7}$ 32. $\sqrt[3]{128x^6y^2}$

33. $\sqrt{48a^2b^3c^4}$ 34. $\sqrt{72a^4b^3c^2}$

35. $\sqrt[3]{48a^2b^3c^4}$ 36. $\sqrt[3]{72a^4b^3c^2}$

37. $\sqrt[5]{64x^8y^{12}}$ 38. $\sqrt[4]{32x^9y^{10}}$

39. $\sqrt{12x^5y^2z^3}$ 40. $\sqrt{24x^8y^3z^5}$

Rationalize the denominator in each of the following expressions:

41. $\dfrac{2}{\sqrt{3}}$ 42. $\dfrac{3}{\sqrt{2}}$

43. $\dfrac{5}{\sqrt{6}}$ 44. $\dfrac{7}{\sqrt{5}}$

45. $\sqrt{\dfrac{1}{2}}$ 46. $\sqrt{\dfrac{1}{3}}$

47. $\sqrt{\dfrac{1}{5}}$ 48. $\sqrt{\dfrac{1}{6}}$

49. $\dfrac{4}{\sqrt[3]{2}}$ 50. $\dfrac{5}{\sqrt[3]{3}}$

51. $\dfrac{2}{\sqrt[3]{9}}$ 52. $\dfrac{3}{\sqrt[3]{4}}$

53. $\sqrt{\dfrac{3}{2x}}$ 54. $\sqrt{\dfrac{5}{3x}}$

55. $\sqrt{\dfrac{8}{y}}$

56. $\sqrt{\dfrac{27}{y}}$

57. $\sqrt[3]{\dfrac{4x}{3y}}$

58. $\sqrt[3]{\dfrac{7x}{6y}}$

59. $\sqrt[3]{\dfrac{2x}{9y}}$

60. $\sqrt[3]{\dfrac{5x}{4y}}$

61. $\sqrt[4]{\dfrac{1}{8x}}$

62. $\sqrt[4]{\dfrac{8}{9x}}$

Write each of the following in simplified form:

63. $\sqrt{\dfrac{27x^3}{5y}}$

64. $\sqrt{\dfrac{12x^5}{7y}}$

65. $\sqrt{\dfrac{75x^3y^2}{2z}}$

66. $\sqrt{\dfrac{50x^2y^3}{3z}}$

67. $\sqrt[3]{\dfrac{16a^4b^3}{9c}}$

68. $\sqrt[3]{\dfrac{54a^5b^4}{25c^2}}$

69. $\sqrt[3]{\dfrac{8x^3y^6}{9z}}$

70. $\sqrt[3]{\dfrac{27x^6y^3}{2z^2}}$

71. Suppose $x + 3$ is nonnegative and simplify $\sqrt{x^2 + 6x + 9}$ by first writing $x^2 + 6x + 9$ as $(x + 3)^2$.

72. Assume $x - 5$ is nonnegative and simplify $\sqrt{x^2 - 10x + 25}$.

73. Show that the statement $\sqrt{a + b} = \sqrt{a} + \sqrt{b}$ is not true by replacing a with 9 and b with 16 and simplifying both sides.

74. Find a pair of values for a and b that will make the statement $\sqrt{a + b} = \sqrt{a} + \sqrt{b}$ true.

75. Simplify each of the following expressions:
a. $\sqrt{5^2}$ **b.** $\sqrt{3^2}$ **c.** $\sqrt{(-5)^2}$ **d.** $\sqrt{(-3)^2}$

76. Notice from Problem 75 that the statement $\sqrt{x^2} = x$ is only true when $x > 0$. When x is a negative number $\sqrt{x^2} = -x$. (Like in part **c** above $\sqrt{(-5)^2} = 5$ which is $-(-5)$.) We can summarize the results from Problem 75 like this:

$$\sqrt{x^2} = \begin{cases} x & \text{if } x \geq 0 \\ -x & \text{if } x < 0 \end{cases}$$

Now, what other notation have we used in the past for this same situation?

77. The distance (d) between opposite corners of a rectangular room with length l and width w is given by

$$d = \sqrt{l^2 + w^2}$$

How far is it between opposite corners of a living room that measures 10 by 15 feet?

78.   The radius r of a sphere with volume V can be found by using the formula

$$r = \sqrt[3]{\frac{3V}{4\pi}}$$

Find the radius of a sphere with volume 9 cubic feet. Write your answer in simplified form. (Use 22/7 for π.)

Review Problems   The problems below review material we covered in Section 3.3. Reviewing these problems will help you with the next section.

Combine similar terms.

79. $3x^2 + 4x^2$ 80. $7x^2 + 5x^2$
81. $5a^3 - 4a^3 + 6a^3$ 82. $7a^4 - 2a^4 + 3a^4$
83. $(2x^2 - 5x + 3) - (x^2 - 3x + 7)$
84. $(6x^2 - 3x - 4) - (2x^2 - 3x + 5)$
85.   Subtract $2y - 5$ from $5y - 3$.
86.   Subtract $8y^2 + 2$ from $3y^2 - 7$.

In Chapter 3 we found that we could add only similar terms when combining polynomials. The same idea applies to addition and subtraction of radical expressions.

5.3
Addition and Subtraction of Radical Expressions

DEFINITION   Two radicals are said to be *similar radicals* if they have the same index and the same radicand.

The expressions $5\sqrt[3]{7}$ and $-8\sqrt[3]{7}$ are similar since the index is 3 in both cases and the radicands are 7. The expressions $3\sqrt[4]{5}$ and $7\sqrt[3]{5}$ are not similar since they have different indices, while the expressions $2\sqrt[5]{8}$ and $3\sqrt[5]{9}$ are not similar because the radicands are not the same.

We add and subtract radical expressions in the same way we add and subtract polynomials—by combining similar terms under the distributive property.

▼   **Example 1**   Combine $5\sqrt{3} - 4\sqrt{3} + 6\sqrt{3}$.

Solution   All three radicals are similar. We apply the distributive property to get

$$5\sqrt{3} - 4\sqrt{3} + 6\sqrt{3} = (5 - 4 + 6)\sqrt{3}$$
$$= 7\sqrt{3}$$ ▲

▼   **Example 2** Combine $3\sqrt{8} + 5\sqrt{18}$.

Solution The two radicals do not seem to be similar. We must write each in simplified form before applying the distributive property.

$$
\begin{aligned}
3\sqrt{8} + 5\sqrt{18} &= 3\sqrt{4\cdot 2} + 5\sqrt{9\cdot 2} \\
&= 3\sqrt{4}\sqrt{2} + 5\sqrt{9}\sqrt{2} \\
&= 3\cdot 2\sqrt{2} + 5\cdot 3\sqrt{2} \\
&= 6\sqrt{2} + 15\sqrt{2} \\
&= (6 + 15)\sqrt{2} \\
&= 21\sqrt{2}
\end{aligned}
$$
▲

The result of Example 2 can be generalized to the following rule for sums and differences of radical expressions:

RULE To add or subtract two radical expressions, put each in simplified form and apply the distributive property if possible. We can add only similar radicals. We must write each expression in simplified form for radicals before we can tell if the radicals are similar.

▼   **Example 3** Combine $7\sqrt{75xy^3} - 4y\sqrt{12xy}$, where $x, y \geq 0$.

Solution We write each expression in simplified form and combine similar radicals:

$$
\begin{aligned}
7\sqrt{75xy^3} - 4y\sqrt{12xy} &= 7\sqrt{25y^2}\sqrt{3xy} - 4y\sqrt{4}\sqrt{3xy} \\
&= 35y\sqrt{3xy} - 8y\sqrt{3xy} \\
&= (35y - 8y)\sqrt{3xy} \\
&= 27y\sqrt{3xy}
\end{aligned}
$$
▲

▼   **Example 4** Combine $10\sqrt[3]{8a^4b^2} + 11a\sqrt[3]{27ab^2}$.

Solution Writing each radical in simplified form and combining similar terms, we have

$$
\begin{aligned}
10\sqrt[3]{8a^4b^2} + 11a\sqrt[3]{27ab^2} &= 10\sqrt[3]{8a^3}\sqrt[3]{ab^2} + 11a\sqrt[3]{27}\sqrt[3]{ab^2} \\
&= 20a\sqrt[3]{ab^2} + 33a\sqrt[3]{ab^2} \\
&= 53a\sqrt[3]{ab^2}.
\end{aligned}
$$
▲

▼ **Example 5** Combine $\dfrac{\sqrt{3}}{2} + \dfrac{1}{\sqrt{3}}$.

Solution We begin by writing the second term in simplified form.

$$\dfrac{\sqrt{3}}{2} + \dfrac{1}{\sqrt{3}} - \dfrac{\sqrt{3}}{2} + \dfrac{1}{\sqrt{3}} \cdot \dfrac{\sqrt{3}}{\sqrt{3}}$$

$$= \dfrac{\sqrt{3}}{2} + \dfrac{\sqrt{3}}{3}$$

$$= \left(\dfrac{1}{2} + \dfrac{1}{3}\right)\sqrt{3}$$

The common denominator is 6. Multiplying $\frac{1}{2}$ by $\frac{3}{3}$ and $\frac{1}{3}$ by $\frac{2}{2}$ we have

$$= \left(\dfrac{3}{6} + \dfrac{2}{6}\right)\sqrt{3}$$

$$= \dfrac{5}{6}\sqrt{3}$$

$$= \dfrac{5\sqrt{3}}{6}$$ ▲

Combine the following expressions. (Assume any variables under an even root are positive.)   *Problem Set 5.3*

1. $3\sqrt{5} + 4\sqrt{5}$	2. $6\sqrt{3} - 5\sqrt{3}$
3. $7\sqrt{6} + 9\sqrt{6}$	4. $4\sqrt{2} - 10\sqrt{2}$
5. $3x\sqrt{7} - 4x\sqrt{7}$	6. $6y\sqrt{a} + 7y\sqrt{a}$
7. $5\sqrt[3]{10} - 4\sqrt[3]{10}$	8. $6\sqrt[4]{2} + 9\sqrt[4]{2}$
9. $8\sqrt{6} - 2\sqrt{6} + 3\sqrt{6}$	10. $7\sqrt{7} - \sqrt{7} + 4\sqrt{7}$
11. $3x\sqrt{2} - 4x\sqrt{2} + x\sqrt{2}$	12. $5x\sqrt{6} - 3x\sqrt{6} - 2x\sqrt{6}$
13. $\sqrt{18} + \sqrt{2}$	14. $\sqrt{12} + \sqrt{3}$
15. $\sqrt{20} - \sqrt{80} + \sqrt{45}$	16. $\sqrt{8} - \sqrt{32} - \sqrt{18}$
17. $4\sqrt{8} - 2\sqrt{50} - 5\sqrt{72}$	18. $\sqrt{48} - 3\sqrt{27} + 2\sqrt{75}$
19. $5x\sqrt{8} + 3\sqrt{32x^2} - 5\sqrt{50x^2}$	20. $2\sqrt{50x^2} - 8x\sqrt{18} - 3\sqrt{72x^2}$
21. $5\sqrt[3]{16} - 4\sqrt[3]{54}$	22. $\sqrt[3]{81} + 3\sqrt[3]{24}$
23. $\sqrt[3]{x^4y^2} + 7x\sqrt[3]{xy^2}$	24. $2\sqrt[3]{x^8y^6} - 3y^2\sqrt[3]{8x^8}$
25. $5a^2\sqrt{27ab^3} - 6b\sqrt{12a^5b}$	26. $9a\sqrt{20a^3b^2} + 7b\sqrt{45a^5}$

27. $b\sqrt[3]{24a^5b} + 3a\sqrt[3]{81a^2b^4}$

28. $7\sqrt[3]{a^4b^3c^2} - 6ab\sqrt[3]{ac^2}$

29. $\dfrac{\sqrt{2}}{2} + \dfrac{1}{\sqrt{2}}$

30. $\dfrac{\sqrt{3}}{3} + \dfrac{1}{\sqrt{3}}$

31. $\dfrac{\sqrt{5}}{3} + \dfrac{1}{\sqrt{5}}$

32. $\dfrac{\sqrt{6}}{2} + \dfrac{1}{\sqrt{6}}$

33. $\sqrt{3} - \dfrac{1}{\sqrt{3}}$

34. $\sqrt{5} - \dfrac{1}{\sqrt{5}}$

35. $\dfrac{\sqrt{18}}{6} + \sqrt{\dfrac{1}{2}} + \dfrac{\sqrt{2}}{2}$

36. $\dfrac{\sqrt{12}}{6} + \sqrt{\dfrac{1}{3}} + \dfrac{\sqrt{3}}{3}$

37. $\sqrt{6} - \sqrt{\dfrac{2}{3}}$

38. $\sqrt{15} - \sqrt{\dfrac{3}{5}}$

39. Use the table of powers, roots, and prime factors in the back of the book to find a decimal approximation for $\sqrt{12}$ and for $2\sqrt{3}$.

40. Use the table in the back of the book to find decimal approximations for $\sqrt{50}$ and $5\sqrt{2}$.

41. Use the table in the back of the book to find a decimal approximation for $\sqrt{8} + \sqrt{18}$. Is it equal to the decimal approximations for $\sqrt{26}$ or $\sqrt{50}$?

42. Use the table in the back of the book to find a decimal approximation for $\sqrt{3} + \sqrt{12}$. Is it equal to the decimal approximation for $\sqrt{15}$ or $\sqrt{27}$?

Each statement below is false. Correct the right side of each one.

43. $3\sqrt{2x} + 5\sqrt{2x} = 8\sqrt{4x}$

44. $5\sqrt{3} - 7\sqrt{3} = -2\sqrt{9}$

45. $\sqrt{9 + 16} = 3 + 4$

46. $\sqrt{36 + 64} = 6 + 8$

Review Problems   The problems below review material we covered in Section 3.4. Reviewing these problems will help you understand the next section.

Multiply.

47. $2x(3x - 5)$

48. $5x(4x - 3)$

49. $(a + 5)(2a - 5)$

50. $(3a + 4)(a + 2)$

51. $(3x - 2y)^2$

52. $(2x + 3y)^2$

53. $(x + 2)(x - 2)$

54. $(5x - 7y)(5x + 7y)$

**5.4
Multiplication and
Division of Radical
Expressions**

We use the same process to multiply radical expressions as we have in the past to multiply polynomials.

▼ **Example 1**   Multiply $\sqrt{3}(2\sqrt{6} - 5\sqrt{12})$.

Solution   Applying the distributive property, we have

$$\sqrt{3}(2\sqrt{6} - 5\sqrt{12}) = \sqrt{3}\cdot 2\sqrt{6} - \sqrt{3}\cdot 5\sqrt{12}$$
$$= 2\sqrt{18} - 5\sqrt{36}$$

Writing each radical in simplified form gives

$$2\sqrt{18} - 5\sqrt{36} = 2\sqrt{9}\sqrt{2} - 5\sqrt{36}$$
$$= 6\sqrt{2} - 30 \qquad \blacktriangle$$

▼ **Example 2** Multiply $(\sqrt{3} + \sqrt{5})(4\sqrt{3} - \sqrt{5})$.

Solution The same principle that applies to multiply two binomials applies to this product. We must multiply each term in the first expression by each term in the second one. Any convenient method can be used.

$$\overset{\text{F}}{}\qquad\overset{\text{O}}{}\qquad\overset{\text{I}}{}\qquad\overset{\text{L}}{}$$
$$(\sqrt{3} + \sqrt{5})(4\sqrt{3} - \sqrt{5}) = \sqrt{3}\cdot 4\sqrt{3} - \sqrt{3}\sqrt{5} + \sqrt{5}\cdot 4\sqrt{3} - \sqrt{5}\sqrt{5}$$
$$= 4\cdot 3 - \sqrt{15} + 4\sqrt{15} - 5$$
$$= 12 + 3\sqrt{15} - 5$$
$$= 7 + 3\sqrt{15} \qquad \blacktriangle$$

▼ **Example 3** Expand $(3\sqrt{x} - 2\sqrt{y})^2$.

Solution 1 We write this problem as a multiplication problem and proceed as in Example 2:

$$(3\sqrt{x} - 2\sqrt{y})^2 = (3\sqrt{x} - 2\sqrt{y})(3\sqrt{x} - 2\sqrt{y})$$
$$= 3\sqrt{x}\cdot 3\sqrt{x} - 3\sqrt{x}\cdot 2\sqrt{y} - 2\sqrt{y}\cdot 3\sqrt{x} + 2\sqrt{y}\cdot 2\sqrt{y}$$
$$= 9x - 6\sqrt{xy} - 6\sqrt{xy} + 4y$$
$$= 9x - 12\sqrt{xy} + 4y$$

Solution 2 We can apply the formula for the square of a sum, $(a + b)^2 = a^2 + 2ab + b^2$:

$$(3\sqrt{x} - 2\sqrt{y})^2 = (3\sqrt{x})^2 + 2(3\sqrt{x})(-2\sqrt{y}) + (-2\sqrt{y})^2$$
$$= 9x - 12\sqrt{xy} + 4y \qquad \blacktriangle$$

▼ **Example 4** Multiply $(\sqrt{6} + \sqrt{2})(\sqrt{6} - \sqrt{2})$.

Solution We notice the product is of the form $(a + b)(a - b)$, which always gives the difference of two squares, $a^2 - b^2$:

$$(\sqrt{6} + \sqrt{2})(\sqrt{6} - \sqrt{2}) = (\sqrt{6})^2 - (\sqrt{2})^2$$
$$= 6 - 2$$
$$= 4 \qquad \blacktriangle$$

In Example 4 the two expressions $(\sqrt{6} + \sqrt{2})$ and $(\sqrt{6} - \sqrt{2})$ are called *conjugates*. In general, the conjugate of $\sqrt{a} + \sqrt{b}$ is $\sqrt{a} - \sqrt{b}$. Multiplying conjugates of this form always produces a real number.

Division with radical expressions is the same as rationalizing the denominator. In Section 5.2 we were able to divide $\sqrt{3}$ by $\sqrt{2}$ by rationalizing the denominator:

$$\frac{\sqrt{3}}{\sqrt{2}} = \frac{\sqrt{3}}{\sqrt{2}} \cdot \frac{\sqrt{2}}{\sqrt{2}}$$
$$= \frac{\sqrt{6}}{2}$$

We can accomplish the same result with expressions such as

$$\frac{6}{\sqrt{5} - \sqrt{3}}$$

by multiplying the numerator and denominator by the conjugate of the denominator.

▼ **Example 5**   Divide $\dfrac{6}{\sqrt{5} - \sqrt{3}}$. (Rationalize the denominator.)

Solution   Since the product of two conjugates is a real number, we multiply the numerator and denominator by the conjugate of the denominator:

$$\frac{6}{\sqrt{5} - \sqrt{3}} = \frac{6}{\sqrt{5} - \sqrt{3}} \cdot \frac{(\sqrt{5} + \sqrt{3})}{(\sqrt{5} + \sqrt{3})}$$
$$= \frac{6\sqrt{5} + 6\sqrt{3}}{(\sqrt{5})^2 - (\sqrt{3})^2}$$
$$= \frac{6\sqrt{5} + 6\sqrt{3}}{5 - 3}$$
$$= \frac{6\sqrt{5} + 6\sqrt{3}}{2}$$

$$= \frac{2(3\sqrt{5} + 3\sqrt{3})}{2}$$

$$= 3\sqrt{5} + 3\sqrt{3} \qquad\qquad \blacktriangle$$

Find the following products. (Assume all variables are positive.)   *Problem Set 5.4*

1. $\sqrt{3}(\sqrt{2} - 3\sqrt{3})$
2. $\sqrt{2}(5\sqrt{3} + 4\sqrt{2})$
3. $6\sqrt{6}(2\sqrt{2} + 1)$
4. $7\sqrt{5}(3\sqrt{15} - 2)$
5. $(\sqrt{3} + \sqrt{2})(3\sqrt{3} - \sqrt{2})$
6. $(\sqrt{5} - \sqrt{2})(3\sqrt{5} + 2\sqrt{2})$
7. $(\sqrt{x} + 5)(\sqrt{x} - 3)$
8. $(\sqrt{x} + 4)(\sqrt{x} + 2)$
9. $(3\sqrt{6} + 4\sqrt{2})(\sqrt{6} + 2\sqrt{2})$
10. $(\sqrt{7} - 3\sqrt{3})(2\sqrt{7} - 4\sqrt{3})$
11. $(\sqrt{3} + 4)^2$
12. $(\sqrt{5} - 2)^2$
13. $(\sqrt{x} - 3)^2$
14. $(\sqrt{x} + 4)^2$
15. $(2\sqrt{3} + 3\sqrt{2})^2$
16. $(4\sqrt{6} - 3\sqrt{2})^2$
17. $(2\sqrt{a} - 3\sqrt{b})^2$
18. $(5\sqrt{a} - 2\sqrt{b})^2$
19. $(\sqrt{3} - \sqrt{2})(\sqrt{3} + \sqrt{2})$
20. $(\sqrt{5} - \sqrt{2})(\sqrt{5} + \sqrt{2})$
21. $(2\sqrt{6} - 3)(2\sqrt{6} + 3)$
22. $(3\sqrt{5} - 1)(3\sqrt{5} + 1)$
23. $(\sqrt{a} + 7)(\sqrt{a} - 7)$
24. $(\sqrt{a} + 5)(\sqrt{a} - 5)$
25. $(4\sqrt{7} - 2\sqrt{5})(4\sqrt{7} + 2\sqrt{5})$
26. $(2\sqrt{11} - 5\sqrt{2})(2\sqrt{11} + 5\sqrt{2})$

Rationalize the denominator in each of the following. (Assume all variables are positive.)

27. $\dfrac{2}{\sqrt{3} + \sqrt{2}}$
28. $\dfrac{3}{\sqrt{5} - \sqrt{2}}$
29. $\dfrac{\sqrt{2}}{\sqrt{6} - \sqrt{2}}$
30. $\dfrac{\sqrt{5}}{\sqrt{5} + \sqrt{3}}$
31. $\dfrac{1}{2\sqrt{3} - 1}$
32. $\dfrac{2}{3\sqrt{2} + 1}$
33. $\dfrac{\sqrt{x}}{\sqrt{x} - 3}$
34. $\dfrac{\sqrt{x}}{\sqrt{x} + 2}$
35. $\dfrac{\sqrt{5}}{2\sqrt{5} - 3}$
36. $\dfrac{\sqrt{7}}{3\sqrt{7} - 2}$
37. $\dfrac{3}{\sqrt{x} - \sqrt{y}}$
38. $\dfrac{2}{\sqrt{x} + \sqrt{y}}$

39. $\dfrac{\sqrt{2} + \sqrt{6}}{\sqrt{2} - \sqrt{6}}$

40. $\dfrac{\sqrt{3} - \sqrt{5}}{\sqrt{3} + \sqrt{5}}$

41. $\dfrac{\sqrt{a} + \sqrt{b}}{\sqrt{a} - \sqrt{b}}$

42. $\dfrac{\sqrt{a} - \sqrt{b}}{\sqrt{a} + \sqrt{b}}$

43. $\dfrac{2\sqrt{3} - \sqrt{7}}{3\sqrt{3} + \sqrt{7}}$

44. $\dfrac{5\sqrt{6} + 2\sqrt{2}}{\sqrt{6} - \sqrt{2}}$

45. $\dfrac{3\sqrt{x} + 2}{1 + \sqrt{x}}$

46. $\dfrac{5\sqrt{x} - 1}{2 + \sqrt{x}}$

47. $\dfrac{4\sqrt{3} - 2\sqrt{2}}{3\sqrt{3} - 5\sqrt{2}}$

48. $\dfrac{8\sqrt{5} - 2\sqrt{3}}{2\sqrt{5} + 2\sqrt{3}}$

49. Show that the product $(\sqrt[3]{2} + \sqrt[3]{3})(\sqrt[3]{4} - \sqrt[3]{6} + \sqrt[3]{9})$ is 5. (You may want to look back to Section 3.4 to see how we multiplied a binomial by a trinomial.)

50. Show that the product $(\sqrt[3]{x} + 2)(\sqrt[3]{x^2} - 2\sqrt[3]{x} + 4)$ is $x + 8$.

Each statement below is false. Correct the right side of each one.

51. $5(2\sqrt{3}) = 10\sqrt{15}$

52. $3(2\sqrt{x}) = 6\sqrt{3x}$

53. $(\sqrt{x} + 3)^2 = x + 9$

54. $(\sqrt{x} - 7)^2 = x - 49$

55. $(5\sqrt{3})^2 = 15$

56. $(3\sqrt{5})^2 = 15$

57. If an object is dropped from the top of a 100-ft building, the amount of time t, in seconds, that it takes for the object to be h feet from the ground is given by the formula

$$t = \frac{\sqrt{100 - h}}{4}$$

How long does it take before the object is 50 feet from the ground? How long does it take to reach the ground? (When it is on the ground, h is 0.)

58. Use the formula given in Problem 57 to determine the height from which the ball should be dropped if it is to take exactly 1.25 seconds to hit the ground.

Review Problems   The problems below review material we covered in Section 2.1.

Solve each equation.

59. $3x + 4 = 25$

60. $4x - 7 = 9$

61. $3y - 4 = y + 6$

62. $4y + 7 = y + 2$

63. $4(2a + 1) - (3a - 5) = 24$

64. $5(6a - 3) - (2a - 4) = 45$

This section is concerned with solving equations that involve one or more radicals. The first step in solving an equation that contains a radical is to eliminate the radical from the equation. To do so we need an additional property:

SQUARING PROPERTY OF EQUALITY   If both sides of an equation are squared, the solutions to the original equation are solutions to the resulting equation.

We will never lose solutions to our equations by squaring both sides. We may, however, introduce *extraneous solutions*. Extraneous solutions satisfy the equation obtained by squaring both sides of the original equation, but do not satisfy the original equation.

We know that if two real numbers a and b are equal, then so are their squares:

$$\text{If} \quad a = b$$
$$\text{then} \quad a^2 = b^2$$

On the other hand, extraneous solutions are introduced when we square opposites. That is, even though opposites are not equal, their squares are. For example,

$$5 = -5 \qquad \text{A false statement}$$
$$(5)^2 = (-5)^2 \qquad \text{Square both sides}$$
$$25 = 25 \qquad \text{A true statement}$$

It is because of this that extraneous solutions are sometimes introduced when squaring both sides of an equation.

We are free to square both sides of an equation any time it is convenient. We must be aware, however, that doing so may introduce extraneous solutions. We must, therefore, check all our solutions in the original equation if at any time we square both sides of the original equation.

▼   Example 1   Solve for x: $\sqrt{3x + 4} = 5$.

Solution   We square both sides and proceed as usual:

$$\sqrt{3x + 4} = 5$$
$$(\sqrt{3x + 4})^2 = 5^2$$
$$3x + 4 = 25$$
$$3x = 21$$
$$x = 7$$

Checking $x = 7$ in the original equation, we have

$$\sqrt{3(7) + 4} = 5$$
$$\sqrt{21 + 4} = 5$$
$$\sqrt{25} = 5$$
$$5 = 5$$

The solution $x = 7$ satisfies the original equation.   ▲

▼   **Example 2**   Solve $\sqrt{4x - 7} = -3$.

Solution   There is no solution to this equation since the left side is positive (or zero) for any value of x and the right side is -3. The equation itself is a contradiction; a positive number cannot be equal to -3. Let's look at what would happen if we tried to solve this equation by squaring both sides.

$$\sqrt{4x - 7} = -3$$
$$(\sqrt{4x - 7})^2 = (-3)^2$$
$$4x - 7 = 9$$
$$4x = 16$$
$$x = 4$$

Checking $x = 4$ in the original equation gives

$$\sqrt{4(4) - 7} = -3$$
$$\sqrt{16 - 7} = -3$$
$$\sqrt{9} = -3$$
$$3 = -3$$ ▲

The solution $x = 4$ produces a false statement when checked in the original equation. Since $x = 4$ was the only possible solution, there is no solution to the original equation. The solution set is ∅. The possible solution $x = 4$ is an extraneous solution. It satisfies the equation obtained by squaring both sides of the original equation, but does not satisfy the original equation.

▼   **Example 3**   Solve $\sqrt{5x - 1} + 3 = 7$.

Solution   We must isolate the radical on the left side of the equation. If we attempt to square both sides without doing so, the

resulting equation will also contain a radical. Adding -3 to both sides, we have

$$\sqrt{5x - 1} + 3 = 7$$
$$\sqrt{5x - 1} = 4$$

We can now square both sides and proceed as usual:

$$(\sqrt{5x - 1})^2 = (4)^2$$
$$5x - 1 = 16$$
$$5x = 17$$
$$x = \tfrac{17}{5}$$

Checking $x = \tfrac{17}{5}$, we have

$$\sqrt{5(\tfrac{17}{5}) - 1} + 3 = 7$$
$$\sqrt{17 - 1} + 3 = 7$$
$$4 + 3 = 7$$
$$7 = 7 \qquad \blacktriangle$$

▼   **Example 4**   Solve $\sqrt{3y - 4} - \sqrt{y + 6} = 0$.

Solution We add $\sqrt{y + 6}$ to both sides, square both sides, and solve as usual:

$$\sqrt{3y - 4} - \sqrt{y + 6} = 0$$
$$\sqrt{3y - 4} = \sqrt{y + 6}$$
$$(\sqrt{3y - 4})^2 = (\sqrt{y + 6})^2$$
$$3y - 4 = y + 6$$
$$2y = 10$$
$$y = 5$$

Substituting $y = 5$ into the original equation, we have

$$\sqrt{3(5) - 4} - \sqrt{5 + 6} = 0$$
$$\sqrt{15 - 4} - \sqrt{11} = 0$$
$$\sqrt{11} - \sqrt{11} = 0$$
$$0 = 0 \qquad \blacktriangle$$

It is also possible to raise both sides of an equation to powers greater than 2. We only need to check for extraneous solutions when we raise

both sides of an equation to an even power. Raising both sides of an equation to an odd power will not produce extraneous solutions.

▼   Example 5   Solve $\sqrt[3]{4x + 5} = 3$.

Solution   Cubing both sides we have:

$$(\sqrt[3]{4x + 5})^3 = 3^3$$
$$4x + 5 = 27$$
$$4x = 22$$
$$x = \tfrac{22}{4}$$
$$x = \tfrac{11}{2}$$

We do not need to check $x = \tfrac{11}{2}$ since we raised both sides to an odd power.   ▲

Problem Set 5.5

Solve each of the following equations:

1. $\sqrt{2x + 1} = 3$
2. $\sqrt{3x + 1} = 4$
3. $\sqrt{4x + 1} = -5$
4. $\sqrt{6x + 1} = -5$
5. $\sqrt{2y - 1} = 3$
6. $\sqrt{3y - 1} = 2$
7. $\sqrt{5x - 7} = -1$
8. $\sqrt{8x + 3} = -6$
9. $\sqrt{2x - 3} - 2 = 4$
10. $\sqrt{3x + 1} - 4 = 1$
11. $\sqrt{4a + 1} + 3 = 2$
12. $\sqrt{5a - 3} + 6 = 2$
13. $\sqrt[4]{3x + 1} = 2$
14. $\sqrt[4]{4x + 1} = 3$
15. $\sqrt[3]{2x - 5} = 1$
16. $\sqrt[3]{5x + 7} = 2$
17. $\sqrt[3]{3a + 5} = -3$
18. $\sqrt[3]{2a + 7} = -2$
19. $\sqrt{2x + 4} = \sqrt{1 - x}$
20. $\sqrt{3x + 4} = -\sqrt{2x + 3}$
21. $\sqrt{4a + 7} = -\sqrt{a + 2}$
22. $\sqrt{7a - 1} = \sqrt{2a + 4}$
23. $\sqrt[4]{5x - 8} = \sqrt[4]{4x - 1}$
24. $\sqrt[4]{6x + 7} = \sqrt[4]{x + 5}$
25. $\sqrt{y - 8} = \sqrt{8 - y}$
26. $\sqrt{2y - 5} = \sqrt{5y - 2}$
27. $\sqrt[3]{3x + 5} = \sqrt[3]{5 - 2x}$
28. $\sqrt[3]{4x + 9} = \sqrt[3]{3 - 2x}$
29. $\sqrt[3]{2x + 1} = \sqrt[3]{3x + 2}$
30. $\sqrt[3]{6x - 4} = \sqrt[3]{4x - 3}$
31. $3\sqrt{2x + 1} = 2\sqrt{4x + 7}$
32. $2\sqrt{5x - 1} = 3\sqrt{2x + 4}$
33. $2\sqrt[3]{2y} + 3 = 7$
34. $3\sqrt[3]{2y} + 4 = 1$
35. $5 - 2\sqrt{y} = 2$
36. $7 - 3\sqrt{2y} = 1$
37. $2 - 4\sqrt{5 - x} = -6$
38. $5 - 3\sqrt{6 - x} = -4$

39. The length of time (T) in seconds it takes the pendulum of a grandfather clock to swing through one complete cycle is given by the formula

$$T = 2\pi \sqrt{\frac{L}{32}}$$

where L is the length, in feet, of the pendulum, and π is approximately $\frac{22}{7}$. How long must the pendulum be if one complete cycle takes 2 seconds?

 40. Solve the formula in Problem 39 for L.

41. Solve the formula below for h.

$$t = \frac{\sqrt{100 - h}}{4}$$

42. Solve the formula below for h.

$$t = \sqrt{\frac{2h - 40t}{g}}$$

Review Problems The problems below review material we covered in Section 3.1.

Simplify each expression.

43. $x^3 \cdot x^2$ **44.** $x^2 \cdot x$
45. $(x^4)^{10}$ **46.** $(x^3)^4$
47. $(x^4)^7 \cdot x^2$ **48.** $(x^4)^5 \cdot x$
49. 2^{-5} **50.** 5^{-2}

The equation $x^2 = -9$ has no real solutions since the square of a real number is always positive. We have been unable to work with square roots of negative numbers like $\sqrt{-25}$ and $\sqrt{-16}$ for the same reason. Complex numbers allow us to expand our work with radicals to include square roots of negative numbers and to solve equations like $x^2 = -9$ and $x^2 = -64$. Our work with complex numbers is based on the following definition:

5.6
Addition and
Subtraction of
Complex Numbers

DEFINITION The number i is such that $i = \sqrt{-1}$. (Which is the same as saying $i^2 = -1$.)

The number i is not a real number. The number i can be used to eliminate the negative sign under a square root.

▼ **Example 1** Write (a) $\sqrt{-25}$, (b) $-\sqrt{-49}$, (c) $\sqrt{-12}$, and (d) $-\sqrt{-17}$ in terms of i.

a. $\sqrt{-25} = \sqrt{25(-1)} = \sqrt{25}\sqrt{-1} = 5i$

b. $-\sqrt{-49} = -\sqrt{49(-1)} = -\sqrt{49}\sqrt{-1} = -7i$

c. $\sqrt{-12} = \sqrt{12(-1)} = \sqrt{12}\sqrt{-1} = 2\sqrt{3}i = 2i\sqrt{3}$

d. $-\sqrt{-17} = -\sqrt{17(-1)} = -\sqrt{17}\sqrt{-1} = -\sqrt{17}i = -i\sqrt{17}$ ▲

If we assume all the properties of exponents hold when the base is i, we can write any power of i as either i, -1, $-i$, or 1. Using the fact that $i^2 = -1$, we have

$$i^1 = i$$
$$i^2 = -1$$
$$i^3 = i^2 \cdot i = -1(i) = -i$$
$$i^4 = i^2 \cdot i^2 = -1(-1) = 1$$

Since $i^4 = 1$, i^5 will simplify to i, and we will begin repeating the sequence i, -1, $-i$, 1 as we simplify higher powers of i:

$$i^5 = i^4 \cdot i = 1(i) = i$$
$$i^6 = i^4 \cdot i^2 = 1(-1) = -1$$
$$i^7 = i^4 \cdot i^3 = 1(-i) = -i$$
$$i^8 = i^4 \cdot i^4 = 1(1) = 1$$
$$\vdots$$

These results can be summarized by

$$i = i^1 = i^5 = i^9 \cdots$$
$$-1 = i^2 = i^6 = i^{10} \cdots$$
$$-i = i^3 = i^7 = i^{11} \cdots$$
$$1 = i^4 = i^8 = i^{12} \cdots$$

▼   **Example 2**   Simplify as much as possible:

a. $i^{30} = (i^4)^7 \cdot i^2 = 1(-1) = -1$

b. $i^{21} = (i^4)^5 \cdot i = 1(i) = i$

c. $i^{40} = (i^4)^{10} = 1$ ▲

Complex numbers are defined in terms of the number i.

DEFINITION   A *complex number* is any number that can be put in the form

$$a + bi$$

where a and b are real numbers and $i = \sqrt{-1}$. The form $a + bi$ is

called *standard form* for complex numbers. The number *a* is called the *real part* of the complex number. The number *b* is called the *imaginary part* of the complex number. If $b \neq 0$ we say $a + bi$ is also an *imaginary number*.

▼   **Example 3**

a. The number $3 + 4i$ is a complex number since it has the correct form. The number 3 is the real part and 4 is the imaginary part. It is also an imaginary number since $b \neq 0$.

b. The number $-6i$ is a complex number since it can be written as $0 + (-6i)$. The real part is 0. The imaginary part is -6. It is also an imaginary number since $b \neq 0$.

c. The number 8 is a complex number since it can be written as $8 + 0i$. The real part is 8. The imaginary part is 0.   ▲

From part c in Example 3 it is apparent that all real numbers can be considered complex numbers. The real numbers are a subset of the complex numbers.

Two complex numbers are equal if and only if their real parts are equal and their imaginary parts are equal. That is, for real numbers *a*, *b*, *c*, and *d*,

Equality for Complex Numbers

$$a + bi = c + di \quad \text{if and only if} \quad a = c \quad \text{and} \quad b = d$$

▼   **Example 4**   Find x and y if $3x + 4i = 12 - 8yi$.

Solution   Since the two complex numbers are equal, their real parts are equal and their imaginary parts are equal:

$$\begin{aligned} 3x &= 12 \quad \text{and} \quad 4 = -8y \\ x &= 4 \qquad\qquad\ \ y = -\tfrac{1}{2} \end{aligned}$$ ▲

▼   **Example 5**   Find x and y if $(4x - 3) + 7i = 5 + (2y - 1)i$.

Solution   The real parts are $4x - 3$ and 5. The imaginary parts are 7 and $2y - 1$.

$$\begin{aligned} 4x - 3 &= 5 \quad \text{and} \quad 7 = 2y - 1 \\ 4x &= 8 \qquad\qquad 8 = 2y \\ x &= 2 \qquad\qquad y = 4 \end{aligned}$$ ▲

Addition and
Subtraction of
Complex Numbers

To add two complex numbers, add their real parts and add their imaginary parts. That is, if a, b, c, and d are real numbers, then

$$(a + bi) + (c + di) = (a + c) + (b + d)i$$

If we assume that the commutative, associative, and distributive properties hold for the number i, then the definition of addition is simply an extension of these properties.

We define subtraction in a similar manner. If a, b, c, and d are real numbers, then

$$(a + bi) - (c + di) = (a - c) + (b - d)i$$

▼   Example 6

a. $(3 + 4i) + (7 - 6i) = (3 + 7) + (4 - 6)i$
$$= 10 - 2i$$

b. $(8 - 3i) + (6 - 2i) = (8 + 6) + (-3 - 2)i$
$$= 14 - 5i$$

c. $(7 + 3i) - (5 + 6i) = (7 - 5) + (3 - 6)i$
$$= 2 - 3i$$

d. $(5 - 2i) - (9 - 4i) = (5 - 9) + (-2 + 4)i$
$$= -4 + 2i$$

e. $(3 - 5i) - (4 + 2i) + (6 - 8i) = (3 - 4 + 6) + (-5 - 2 - 8)i$
$$= 5 - 15i$$ ▲

Problem Set 5.6

Write the following in terms of i and simplify as much as possible.

1. $\sqrt{-36}$ 2. $\sqrt{-49}$

3. $-\sqrt{-25}$ 4. $-\sqrt{-81}$

5. $\sqrt{-72}$ 6. $\sqrt{-48}$

7. $-\sqrt{-12}$ 8. $-\sqrt{-75}$

9. $\sqrt{-162}$ 10. $\sqrt{-400}$

Write each of the following as i, -1, $-i$, or 1:

11. i^{17} 12. i^{18}

13. i^{19} 14. i^{20}

15. i^{28} 16. i^{31}

17. i^{26} 18. i^{37}

19. i^{75} 20. i^{42}

Find x and y so each of the following equations is true:

21. $2x + 3yi = 6 - 3i$

22. $4x - 2yi = 4 + 8i$

23. $2 - 5i = -x + 10yi$

24. $4 + 7i = 6x - 14yi$

25. $2x + 10i = -16 - 2yi$

26. $4x - 5i = -2 + 3yi$

27. $4y - 6i = 3 + 5xi$

28. $7y - 3i = 4 + 9xi$

29. $(2x - 4) - 3i = 10 - 6yi$

30. $(4x - 3) - 2i = 8 + yi$

31. $(7x - 1) + 4i = 2 + (5y + 2)i$

32. $(5x + 2) - 7i = 4 + (2y + 1)i$

Combine the following complex numbers:

33. $(2 + 3i) + (3 + 6i)$

34. $(4 + i) + (3 + 2i)$

35. $(3 - 5i) + (2 + 4i)$

36. $(7 + 2i) + (3 - 4i)$

37. $(5 + 2i) - (3 + 6i)$

38. $(6 + 7i) - (4 + i)$

39. $(3 - 5i) - (2 + i)$

40. $(7 - 3i) - (4 + 10i)$

41. $(11 - 6i) - (2 - 4i)$

42. $(10 - 12i) - (6 - i)$

43. $(-3 + 2i) + (6 - 5i)$

44. $(-8 + 3i) + (4 - 8i)$

45. $[(3 + 2i) - (6 + i)] + (5 + i)$

46. $[(4 - 5i) - (2 + i)] + (2 + 5i)$

47. $[(7 - i) - (2 + 4i)] - (6 + 2i)$

48. $[(3 - i) - (4 + 7i)] - (3 - 4i)$

49. $(3 + 2i) - [(3 - 4i) - (6 + 2i)]$

50. $(7 - 4i) - [(-2 + i) - (3 + 7i)]$

51. $(4 - 9i) + [(2 - 7i) - (4 + 8i)]$

52. $(10 - 2i) - [(2 + i) - (3 - i)]$

Review Problems   The problems below review material we covered in Section 5.4. Reviewing these problems will help you with the next section.

Multiply.

53. $\sqrt{2}(\sqrt{3} - \sqrt{2})$

54. $(\sqrt{x} - 4)(\sqrt{x} + 5)$

55. $(\sqrt{x} + 5)^2$

56. $(\sqrt{5} + \sqrt{3})(\sqrt{5} - \sqrt{3})$

Rationalize the denominator.

57. $\dfrac{\sqrt{x}}{\sqrt{x} + 3}$

58. $\dfrac{\sqrt{5} - \sqrt{3}}{\sqrt{5} + \sqrt{3}}$

5.7
Multiplication and
Division of Complex
Numbers

Since complex numbers have the same form as binomials, we find the product of two complex numbers the same way we find the product of two binomials.

▼   **Example 1**   Multiply $(3 - 4i)(2 + 5i)$.

Solution   Multiplying each term in the second complex number by each term in the first, we have

$$(3 - 4i)(2 + 5i) = 3 \cdot 2 + 3 \cdot 5i - 2 \cdot 4i - 4i(5i)$$
$$= 6 + 15i - 8i - 20i^2$$

Combining similar terms and using the fact that $i^2 = -1$, we can simplify as follows:

$$6 + 15i - 8i - 20i^2 = 6 + 7i - 20(-1)$$
$$= 6 + 7i + 20$$
$$= 26 + 7i$$

The product of the complex numbers $3 - 4i$ and $2 + 5i$ is the complex number $26 + 7i$.   ▲

▼   **Example 2**   Multiply $2i(4 - 6i)$.

Solution   Applying the distributive property gives us

$$2i(4 - 6i) = 2i \cdot 4 - 2i(6i)$$
$$= 8i - 12i^2$$
$$= 12 + 8i$$ ▲

▼   **Example 3**   Expand $(3 + 5i)^2$.

Solution   We treat this like the square of a binomial. Remember: $(a + b)^2 = a^2 + 2ab + b^2$.

$$(3 + 5i)^2 = 3^2 + 2(3)(5i) + (5i)^2$$
$$= 9 + 30i + 25i^2$$
$$= 9 + 30i - 25$$
$$= -16 + 30i$$ ▲

▼ **Example 4** Multiply $(2 - 3i)(2 + 3i)$.

Solution This product has the form $(a - b)(a + b)$, which we know results in the difference of two squares, $a^2 - b^2$:

$$
\begin{aligned}
(2 - 3i)(2 + 3i) &= 2^2 - (3i)^2 \\
&= 4 - 9i^2 \\
&= 4 + 9 \\
&= 13
\end{aligned}
$$
▲

The product of the two complex numbers $2 - 3i$ and $2 + 3i$ is the real number 13. The two complex numbers $2 - 3i$ and $2 + 3i$ are called complex conjugates. The fact that their product is a real number is very useful.

DEFINITION The complex numbers $a + bi$ and $a - bi$ are called *complex conjugates*. One important property they have is that their product is the real number $a^2 + b^2$. Here's why:

$$
\begin{aligned}
(a + bi)(a - bi) &= a^2 - (bi)^2 \\
&= a^2 - b^2 i^2 \\
&= a^2 - b^2(-1) \\
&= a^2 + b^2
\end{aligned}
$$

The fact that the product of two complex conjugates is a real number is the key to division with complex numbers.

▼ **Example 5** Divide $\dfrac{2 + i}{3 - 2i}$.

Solution We want a complex number in standard form that is equivalent to the quotient $(2 + i)/(3 - 2i)$. We need to eliminate i from the denominator. Multiplying the numerator and denominator by $3 + 2i$ will give us what we want:

$$
\begin{aligned}
\frac{2 + i}{3 - 2i} &= \frac{2 + i}{3 - 2i} \cdot \frac{(3 + 2i)}{(3 + 2i)} \\
&= \frac{6 + 4i + 3i + 2i^2}{9 - 4i^2} \\
&= \frac{6 + 7i - 2}{9 + 4}
\end{aligned}
$$

$$= \frac{4 + 7i}{13}$$

$$= \tfrac{4}{13} + \tfrac{7}{13}i$$

Dividing the complex number $2 + i$ by $3 - 2i$ gives the complex number $\tfrac{4}{13} + \tfrac{7}{13}i$. The second step in Example 5 is shown for clarity. It takes a while to get used to the idea that $i^2 = -1$. ▲

Here are some further examples illustrating division with complex numbers.

▼ **Example 6**   Divide $\dfrac{7 - 4i}{i}$.

Solution   The conjugate of the denominator is $-i$. Multiplying numerator and denominator by this amount, we have

$$\frac{7 - 4i}{i} = \frac{7 - 4i}{i} \cdot \frac{-i}{-i}$$

$$= \frac{-7i + 4i^2}{-i^2}$$

$$= \frac{-7i + 4(-1)}{-(-1)}$$

$$= -4 - 7i \qquad ▲$$

▼ **Example 7**   Divide $\dfrac{6}{3 - 5i}$.

Solution   The conjugate of $3 - 5i$ is $3 + 5i$.

$$\frac{6}{3 - 5i} = \frac{6}{3 - 5i} \cdot \frac{(3 + 5i)}{(3 + 5i)}$$

$$= \frac{18 + 30i}{9 + 25}$$

$$= \frac{18 + 30i}{34}$$

$$= \tfrac{18}{34} + \tfrac{30}{34}i$$

$$= \tfrac{9}{17} + \tfrac{15}{17}i \qquad ▲$$

Find the following products: *Problem Set 5.7*

1. $3i(4 + 5i)$ 2. $2i(3 + 4i)$
3. $-7i(1 + i)$ 4. $-6i(3 - 8i)$
5. $6i(4 - 3i)$ 6. $11i(2 - i)$
7. $(3 + 2i)(4 + i)$ 8. $(2 - 4i)(3 + i)$
9. $(4 + 9i)(3 - i)$ 10. $(5 - 2i)(1 + i)$
11. $(-3 - 4i)(2 - 5i)$ 12. $(-6 - 2i)(3 - 4i)$
13. $(2 + 5i)^2$ 14. $(3 + 2i)^2$
15. $(1 - i)^2$ 16. $(1 + i)^2$
17. $(3 - 4i)^2$ 18. $(6 - 5i)^2$
19. $(2 + i)(2 - i)$ 20. $(3 + i)(3 - i)$
21. $(6 - 2i)(6 + 2i)$ 22. $(5 + 4i)(5 - 4i)$
23. $(2 + 3i)(2 - 3i)$ 24. $(2 - 7i)(2 + 7i)$
25. $(10 + 8i)(10 - 8i)$ 26. $(11 - 7i)(11 + 7i)$

Find the following quotients. Write all answers in standard form for complex numbers.

27. $\dfrac{2 - 3i}{i}$ 28. $\dfrac{3 + 4i}{i}$

29. $\dfrac{5 + 2i}{-i}$ 30. $\dfrac{4 - 3i}{-i}$

31. $\dfrac{4}{2 - 3i}$ 32. $\dfrac{3}{4 - 5i}$

33. $\dfrac{6}{-3 + 2i}$ 34. $\dfrac{-1}{-2 - 5i}$

35. $\dfrac{2 + 3i}{2 - 3i}$ 36. $\dfrac{4 - 7i}{4 + 7i}$

37. $\dfrac{5 + 4i}{3 + 6i}$ 38. $\dfrac{2 + i}{5 - 6i}$

39. $\dfrac{3 - 7i}{9 - 5i}$ 40. $\dfrac{4 + 10i}{3 + 6i}$

Review Problems The problems that follow review material we covered in Section 4.1.

Reduce each rational expression to lowest terms.

41. $\dfrac{x^2 - 9}{x - 3}$ 42. $\dfrac{x^2 - 5x - 6}{x^2 - 1}$

43. $\dfrac{x^2 - 4x - 12}{x^2 + 8x + 12}$ 44. $\dfrac{6x^2 + 7x - 3}{6x^2 + x - 1}$

45. $\dfrac{x^3 + y^3}{x^2 - y^2}$ 46. $\dfrac{x^3 - 8}{x^2 - 4}$

Examples

Chapter 5   Summary and Review

SQUARE ROOTS [5.1]

1. The number 49 has two square roots, 7 and -7. They are written like this:

$$\sqrt{49} = 7 \qquad -\sqrt{49} = -7$$

Every positive real number x has two square roots. The *positive square root* of x is written $\sqrt{x}$, while the *negative square root* of x is written $-\sqrt{x}$. Both the positive and the negative square roots of x are numbers we square to get x. That is,

$$\left.\begin{array}{c} (\sqrt{x})^2 = x \\ \text{and} \quad (-\sqrt{x})^2 = x \end{array}\right\} \quad \text{for } x \geq 0$$

HIGHER ROOTS [5.1]

2. $\sqrt[3]{8} = 2$
$\sqrt[3]{-27} = -3$

In the expression $\sqrt[n]{a}$, n is the index, a is the *radicand,* and $\sqrt{}$ is the *radical sign*. The expression $\sqrt[n]{a}$ is such that

$$(\sqrt[n]{a})^n = a \qquad a \geq 0 \text{ when } n \text{ is even}$$

FRACTIONAL EXPONENTS [5.1]

3. $25^{1/2} = \sqrt{25} = 5$
$8^{2/3} = (\sqrt[3]{8})^2 = 2^2 = 4$
$9^{3/2} = (\sqrt{9})^3 = 3^3 = 27$

Fractional exponents are used to indicate roots. Using fractional exponents is an alternative to radicals. The relationship between fractional exponents and roots is given by

$$a^{1/n} = \sqrt[n]{a} \qquad a \geq 0 \text{ when } n \text{ is even}$$
$$\text{and}$$
$$a^{m/n} = (a^{1/n})^m = (a^m)^{1/n} \qquad a \geq 0 \text{ when } n \text{ is even}$$

All the properties of exponents developed previously apply to rational exponents.

PROPERTIES OF RADICALS [5.2]

If a and b are nonnegative real numbers whenever n is even, then

1. $\sqrt[n]{ab} = \sqrt[n]{a}\sqrt[n]{b}$

2. $\sqrt[n]{\dfrac{a}{b}} = \dfrac{\sqrt[n]{a}}{\sqrt[n]{b}} \qquad (b \neq 0)$

SIMPLIFIED FORM FOR RADICALS [5.2]

A radical expression is said to be in *simplified form*
   1.   if there is no factor of the radicand that can be writ-
        ten as a power greater than or equal to the index;
   2.   if there are no fractions under the *radical sign*; and
   3.   if there are no radicals in the denominator.

OPERATIONS WITH RADICAL EXPRESSIONS [5.3, 5.4]

The process of combining radical expressions under the four
basic operations—addition, subtraction, multiplication, and
division—is similar to the process of combining polynomials.
   Two radical expressions are *similar* if they have the same
index and the same radicand.

SQUARING PROPERTY OF EQUALITY [5.5]

We may square both sides of an equation any time it is convenient to do so, as long as we check all resulting solutions in
the original equation.

COMPLEX NUMBERS [5.6, 5.7]

A *complex number* is any number that can be put in the form

$$a + bi$$

where a and b are real numbers and $i = \sqrt{-1}$. The *real part*
of the complex number is a, and b is the *imaginary part*.
   If a, b, c, and d are real numbers, then we have the following definitions associated with complex numbers:

   1.   Equality

$$a + bi = c + di \quad \text{if and only if}$$
$$a = c \text{ and } b = d$$

   2.   Addition

$$(a + bi) + (c + di) = (a + c) + (b + d)i$$

   3.   Subtraction

$$(a + bi) - (c + di) = (a - c) + (b - d)i$$

4.
$$\sqrt{\frac{4}{5}} = \frac{\sqrt{4}}{\sqrt{5}}$$
$$= \frac{2}{\sqrt{5}} \cdot \frac{\sqrt{5}}{\sqrt{5}}$$
$$= \frac{2\sqrt{5}}{5}$$

5. $3\sqrt{2} - 7\sqrt{2} = -4\sqrt{2}$
$\sqrt{3}(\sqrt{2} - 5) = \sqrt{6} - 5\sqrt{3}$

6.
$$\sqrt{2x + 1} = 3$$
$$(\sqrt{2x + 1})^2 = 3^2$$
$$2x + 1 = 9$$
$$x = 4$$

7. $3 + 4i$ is a complex number.

$$(3 + 4i) + (2 - 5i) = 5 - i$$

$$(3 + 4i)(2 - 5i)$$
$$= 6 - 15i + 8i - 20i^2$$
$$= 6 - 7i + 20$$
$$= 26 - 7i$$

$$\frac{2}{3 + 4i} = \frac{2}{3 + 4i} \cdot \frac{3 - 4i}{3 - 4i}$$
$$= \frac{6 - 8i}{9 + 16}$$
$$= \frac{6}{25} - \frac{8}{25}i$$

4. Multiplication

$$(a + bi)(c + di) = (ac - bd) + (ad + bc)i$$

5. Conjugates

$$a + bi \quad \text{and} \quad a - bi \text{ are conjugates}$$

COMMON MISTAKES

1. The most common mistake when working with radicals is to assume that the square root of a sum is the sum of the square roots—or:

$$\sqrt{x + y} = \sqrt{x} + \sqrt{y}$$

The problem with this is it just isn't true. If we try it with 16 and 9, the mistake becomes obvious:

$$\sqrt{16 + 9} \overset{?}{=} \sqrt{16} + \sqrt{9}$$
$$\sqrt{25} \overset{?}{=} 4 + 3$$
$$5 \neq 7$$

2. A common mistake when working with complex numbers is to mistake i for -1. The letter i is not -1, it is the square root of -1. That is, $i = \sqrt{-1}$.

**Chapter 5
Test**

Write with radical notation:

1. $8^{3/5}$ 2. $17^{4/3}$

Write with rational exponents:

3. $\sqrt[4]{3x^3}$ 4. $\sqrt[3]{7x^2}$

Simplify each of the following:

5. $27^{-2/3}$ 6. $\left(\frac{25}{49}\right)^{-1/2}$

7. $a^{3/4} \cdot a^{-1/3}$ 8. $\dfrac{(x^{2/3}y^{-3})^{1/2}}{(x^{3/4}y^{1/2})^{-1}}$

Write in simplified form:

9. $\sqrt{125x^3y^5}$ 10. $\sqrt[3]{40x^7y^8}$

11. $\sqrt{\frac{2}{3}}$ 12. $\sqrt{\dfrac{12a^4b^3}{5c}}$

Combine:

13. $3\sqrt{12} - 4\sqrt{27}$ 14. $2\sqrt[3]{24a^3b^3} - 5a\sqrt[3]{3b^3}$

Multiply:

15. $(\sqrt{7} - \sqrt{2})(3\sqrt{7} + 2\sqrt{2})$ **16.** $(3\sqrt{2} - \sqrt{3})^2$

Rationalize the denominator:

17. $\dfrac{5}{\sqrt{3} - 1}$ **18.** $\dfrac{\sqrt{5} - \sqrt{2}}{\sqrt{5} + \sqrt{2}}$

Solve for x:

19. $\sqrt{5x - 1} = 7$ **20.** $\sqrt{3x + 2} + 4 = 0$

21. $\sqrt[3]{2x + 7} = -1$ **22.** $\sqrt[4]{3x + 1} = -2$

Solve for x and y so that each of the following equations is true:

23. $3x - 4i = 2 - 8yi$

24. $(2x + 5) - 4i = 6 - (y - 3)i$

Perform the indicated operations:

25. $(3 - 7i) - (4 + 2i)$

26. $(3 + 2i) - [(7 - i) - (4 + 3i)]$

27. $(2 - 3i)(4 + 3i)$ **28.** $(5 - 4i)^2$

29. $\dfrac{3 + 2i}{i}$ **30.** $\dfrac{5}{2 - 3i}$

31. $\dfrac{2 - 3i}{2 + 3i}$

32. Show that i^{38} can be written as -1.

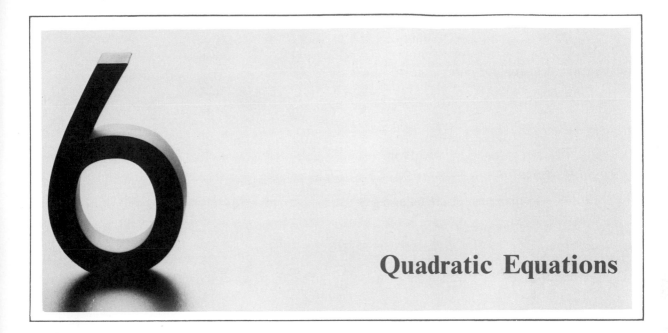

Quadratic Equations

To the student:

If an object is thrown straight up into the air with an initial velocity of 32 feet/second, and we neglect the friction of the air on the object, then its height h above the ground, t seconds later, can be found by using the equation

$$h = 32t - 16t^2$$

Notice that the height depends only on t. The height of the object does not depend on the size or weight of the object. If we neglect the resistance of air on the object, then any object, whether it is a golf ball or a bowling ball, that is thrown into the air with an initial velocity of 32 feet/second, will reach the same height. If we want to find how long it takes the object to hit the ground, we let $h = 0$ (it is 0 feet above the ground when it hits the ground) and solve for t in

$$0 = 32t - 16t^2$$

This last equation is called a quadratic equation because it contains a polynomial of degree 2. Until now we have solved only first-degree equations. This chapter is about solving second-degree equations in one

variable: quadratic equations. The chapter begins with three basic methods of solving quadratic equations: factoring, completing the square, and using the quadratic formula. We will also include applications of quadratic equations and some additional topics on the properties of the solutions to quadratic equations. To be successful in this chapter you should have a working knowledge of factoring, binomial squares, square roots, and complex numbers.

We are going to combine our ability to solve first-degree equations with our knowledge of factoring to solve quadratic equations.

6.1 Solving Quadratic Equations by Factoring

DEFINITION   Any equation that can be written in the form

$$ax^2 + bx + c = 0$$

where a, b, and c are constants and a is not 0 ($a \neq 0$), is called a *quadratic equation*. The form $ax^2 + bx + c = 0$ is called *standard form* for quadratic equations.

Each of the following is a quadratic equation:

$$2x^2 = 5x + 3 \qquad 5x^2 = 75 \qquad 4x^2 - 3x + 2 = 0$$

The third equation is clearly a quadratic equation because it is written in the form $ax^2 + bx + c = 0$. (Notice in this case that a is 4, b is -3, and c is 2.) The first two are also quadratic equations because both could be written in the form $ax^2 + bx + c = 0$ by using the addition property of equality to rearrange terms.   ▲

NOTATION   For a quadratic equation written in standard form, the first term, ax^2, is called the *quadratic term;* the second term, bx, is the *linear term;* and the last term, c, is called the *constant term.*

In the past we have noticed that the number 0 is a special number. That is, 0 has some unique properties. In some situations it does not behave like other numbers. For example, division by 0 does not make sense, whereas division by all other real numbers does. Note also that 0 is the only number without a reciprocal. There is another property of 0 that is the key to solving quadratic equations. It is called the *zero-factor property* and we state it as a theorem.

THEOREM 6.1 (ZERO-FACTOR PROPERTY)   For all real numbers r and s,

$$r \cdot s = 0 \quad \text{if and only if} \quad r = 0 \quad \text{or} \quad s = 0 \quad \text{(or both)}$$

What Theorem 6.1 says in words is that every time we multiply two numbers and the product (answer) is 0, then one or both of the original numbers must have been 0.

▼   **Example 1**   Solve $x^2 - 2x - 24 = 0$.

Solution   We begin by factoring the left side as $(x - 6)(x + 4)$ and get

$$(x - 6)(x + 4) = 0$$

Now both $(x - 6)$ and $(x + 4)$ represent real numbers. We notice that their product is 0. By the zero-factor property, one or both of them must be 0:

$$x - 6 = 0 \text{ or } x + 4 = 0$$

We have used factoring and the zero-factor property to rewrite our original second-degree equation as two first-degree equations connected by the word *or*. Completing the solution, we solve the two first-degree equations:

$$x - 6 = 0 \text{ or } x + 4 = 0$$
$$x = 6 \text{ or } \qquad x = -4$$

Our solution set is $\{6, -4\}$. Both numbers satisfy the original quadratic equation $x^2 - 2x - 24 = 0$. We can check this as follows:

$$
\begin{array}{rl}
\text{When} & x = 6 \\
\text{the equation} & x^2 - 2x - 24 = 0 \\
\text{becomes} & 6^2 - 2(6) - 24 = 0 \\
& 36 - 12 - 24 = 0 \\
& 0 = 0
\end{array}
$$

$$
\begin{array}{rl}
\text{When} & x = -4 \\
\text{the equation} & x^2 - 2x - 24 = 0 \\
\text{becomes} & (-4)^2 - 2(-4) - 24 = 0 \\
& 16 + 8 - 24 = 0 \\
& 0 = 0
\end{array}
$$

In both cases the result is a true statement, which means that both 6 and -4 are solutions to the original equation.   ▲

▼ **Example 2**   Solve $2x^2 = 5x + 3$.

Solution   We begin by adding $-5x$ and -3 to both sides in order to rewrite the equation in standard form:

$$2x^2 - 5x - 3 = 0$$

We then factor the left side and use the zero-factor property to set each factor to zero:

$$(2x + 1)(x - 3) = 0 \qquad \text{Factor}$$
$$2x + 1 = 0 \quad \text{or} \quad x - 3 = 0 \qquad \text{Zero-factor property}$$

Solving each of the resulting first-degree equations, we have

$$x = -\tfrac{1}{2} \quad \text{or} \quad x = 3$$

The solution set is $\{-\tfrac{1}{2}, 3\}$.   ▲

To generalize the above example, here are the steps used in solving a quadratic equation by factoring:

Step 1.   Write the equation in standard form.
Step 2.   Factor the left side.
Step 3.   Use the zero-factor property to set each factor equal to 0.
Step 4.   Solve the resulting first-degree equations.

▼ **Example 3**   Solve $x^2 = 3x$.

Solution   We begin by writing the equation in standard form and factoring:

$$x^2 = 3x$$
$$x^2 - 3x = 0 \qquad \text{Standard form}$$
$$x(x - 3) = 0 \qquad \text{Factor}$$

Using the zero-factor property to set each factor to 0, we have

$$x = 0 \quad \text{or} \quad x - 3 = 0$$
$$x = 3$$

The solution set is $\{0, 3\}$.   ▲

▼ **Example 4**   Solve $(x - 2)(x + 1) = 4$.

Solution We begin by multiplying the two factors on the left side. (Notice that it would be incorrect to set each of the factors on the left side equal to 4. The fact that the product is 4 does not imply that either of the factors must be 4.)

$$(x - 2)(x + 1) = 4$$
$$x^2 - x - 2 = 4 \qquad \text{Multiply the left side}$$
$$x^2 - x - 6 = 0 \qquad \text{Standard form}$$
$$(x - 3)(x + 2) = 0 \qquad \text{Factor}$$
$$x - 3 = 0 \quad \text{or} \quad x + 2 = 0 \qquad \text{Zero-factor property}$$
$$x = 3 \quad \text{or} \qquad x = -2$$

The solution set is $\{3, -2\}$.   ▲

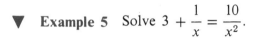

▼ **Example 5** Solve $3 + \dfrac{1}{x} = \dfrac{10}{x^2}$.

Solution To clear the equation of denominators we multiply both sides by x^2:

$$3(x^2) + \left(\frac{1}{x}\right)(x^2) = \left(\frac{10}{x^2}\right)(x^2)$$
$$3x^2 + x = 10$$

Rewrite in standard form and solve:

$$3x^2 + x - 10 = 0$$
$$(3x - 5)(x + 2) = 0$$
$$3x - 5 = 0 \quad \text{or} \quad x + 2 = 0$$
$$x = \tfrac{5}{3} \quad \text{or} \qquad x = -2$$

The solution set is $\{-2, \tfrac{5}{3}\}$. Both solutions check in the original equation. Remember: We have to check *all solutions* any time we multiply both sides of the equation by an expression that contains the variable.   ▲

▼ **Example 6** Solve $\sqrt{6x^2 + 5x} = 5$.

Solution To eliminate the radical we begin by squaring both sides of the equation. We then write the resulting equation in standard form and solve as before:

$$(\sqrt{6x^2 + 5x})^2 = 5^2 \qquad \text{Square both sides}$$
$$6x^2 + 5x = 25 \qquad \text{Simplify}$$
$$6x^2 + 5x - 25 = 0 \qquad \text{Standard form}$$
$$(3x - 5)(2x + 5) = 0 \qquad \text{Factor}$$
$$3x - 5 = 0 \quad \text{or} \quad 2x + 5 = 0 \qquad \text{Zero-factor property}$$
$$x = \frac{5}{3} \quad \text{or} \qquad x = -\frac{5}{2}$$

Since we raised both sides of the original equation to an even power, we have the possibility that one or both of the solutions are extraneous. We must check each solution in the original equation.

$$\text{When} \qquad\qquad\qquad\qquad x = \frac{5}{3},$$

$$\text{the equation} \quad \sqrt{6x^2 + 5x} \qquad\quad = 5$$

$$\text{becomes} \quad \sqrt{6\left(\frac{25}{9}\right) + 5\left(\frac{5}{3}\right)} = 5$$

$$\text{or} \quad \sqrt{\frac{50}{3} + \frac{25}{3}} \qquad = 5$$

$$\sqrt{\frac{75}{3}} \qquad\qquad = 5$$

$$\sqrt{25} \qquad\qquad = 5$$

$$5 \qquad\qquad\quad = 5$$

$$\text{When} \qquad\qquad\qquad\qquad x = -\frac{5}{2},$$

$$\text{the equation} \quad \sqrt{6x^2 + 5x} \qquad\quad = 5$$

$$\text{becomes} \quad \sqrt{6\left(\frac{25}{4}\right) + 5\left(-\frac{5}{2}\right)} = 5$$

$$\text{or} \quad \sqrt{\frac{75}{2} - \frac{25}{2}} \qquad = 5$$

$$\sqrt{\frac{50}{2}} \qquad\qquad = 5$$

$$\sqrt{25} \qquad\qquad = 5$$

$$5 \qquad\qquad\quad = 5$$

Since both checks result in a true statement, neither of the two possible solutions is extraneous.
The solution set is $\{\frac{5}{3}, -\frac{5}{2}\}$. ▲

Problem Set 6.1

The following quadratic equations are in standard form. Factor the left side and solve:

1. $x^2 - 5x - 6 = 0$
2. $x^2 + 5x - 6 = 0$
3. $x^2 - 5x + 6 = 0$
4. $x^2 + 5x + 6 = 0$
5. $3y^2 + 11y - 4 = 0$
6. $3y^2 - y - 4 = 0$
7. $6x^2 - 13x + 6 = 0$
8. $9x^2 + 6x - 8 = 0$
9. $t^2 - 25 = 0$
10. $4t^2 - 49 = 0$

Write each of the following in standard form and solve for the indicated variable:

11. $x^2 = 4x + 21$
12. $x^2 = -4x + 21$
13. $2y^2 - 20 = -3y$
14. $3y^2 + 10 = 17y$
15. $9x^2 - 12x = 0$
16. $4x^2 + 4x = 0$
17. $2r + 1 = 15r^2$
18. $2r - 1 = -8r^2$
19. $9a^2 = 16$
20. $16a^2 = 25$
21. $-10x = x^2$
22. $8x = x^2$
23. $(x + 6)(x - 2) = -7$
24. $(x - 7)(x + 5) = -20$
25. $(y - 4)(y + 1) = -6$
26. $(y - 6)(y + 1) = -12$
27. $(x + 1)^2 = 3x + 7$
28. $(x + 2)^2 = 9x$
29. $(2r + 3)(2r - 1) = -(3r + 1)$
30. $(3r + 2)(r - 1) = -(7r - 7)$

Multiply each of the following by its least common denominator and solve for the indicated variable:

31. $1 - \dfrac{1}{x} = \dfrac{12}{x^2}$
32. $2 + \dfrac{5}{x} = \dfrac{3}{x^2}$
33. $x - \dfrac{4}{3x} = -\dfrac{1}{3}$
34. $\dfrac{x}{2} - \dfrac{4}{x} = -\dfrac{7}{2}$
35. $1 = \dfrac{9}{y^2}$
36. $1 = \dfrac{4}{x^2}$
37. $\dfrac{6}{(r - 1)(r + 1)} - \dfrac{1}{2} = \dfrac{1}{(r + 1)}$
38. $2 + \dfrac{5}{x - 1} = \dfrac{12}{(x - 1)^2}$

Square both sides of the following equations and solve for the indicated variable:

39. $\sqrt{6x^2 + 5x} = 2$
40. $\sqrt{2x^2 - 5x} = 5$
41. $\sqrt{y + 2} = y - 4$
42. $\sqrt{11 - 5x} = x - 3$
43. $\sqrt{5y + 1} = 11 - 5y$
44. $4\sqrt{x + 2} = x + 5$
45. $\sqrt{3x^2 + 4x} - 2 = 0$
46. $\sqrt{2x^2 + 3x} - 3 = 0$

47. In the introduction to this chapter we said the height h of an object thrown straight up into the air with an initial velocity of 32 feet/second could be found by using the equation $h = 32t - 16t^2$. When does the object hit the ground?

48. An object is tossed straight up with an initial velocity of 64 feet/second. The equation that gives the height at time t is

$$h = 64t - 16t^2.$$

When is the object on the ground?

Review Problems The problems below review material we covered in Sections 3.4 and 3.7. Reviewing these problems will help you understand the next section.

Multiply.

49. $(x + 3)^2$ **50.** $(x - 5)^2$

51. $(x - 4)^2$ **52.** $(x + 6)^2$

Factor.

53. $x^2 - 6x + 9$ **54.** $x^2 + 10x + 25$

55. $x^2 + 4x + 4$ **56.** $x^2 - 16x + 64$

In this section we will produce another method of solving quadratic equations. The method is called *completing the square*. Completing the square on a quadratic equation always allows us to obtain solutions, regardless of whether or not the equation can be factored.

Consider the equation

6.2 Completing the Square

$$x^2 = 16$$

We could solve it by writing it in standard form, factoring the left side, and proceeding as we did in the last section. However, we can shorten our work considerably if we simply notice that x must be either the positive square root of 16 or the negative square root of 16. That is,

$$\text{If} \quad x^2 = 16,$$
$$\text{then} \quad x = \sqrt{16} \quad \text{or} \quad x = -\sqrt{16}$$
$$x = 4 \qquad \text{or} \quad x = -4$$

We can generalize this result into a theorem as follows:

THEOREM 6.2 If $a^2 = b$ where b is a real number, then $a = \sqrt{b}$ or $a = -\sqrt{b}$.

NOTATION The expression $a = \sqrt{b}$ or $a = -\sqrt{b}$ can be written in shorthand form as $a = \pm\sqrt{b}$. The symbol $\pm$ is read "plus or minus."

We can apply Theorem 1 to some fairly complicated quadratic equations.

▼ **Example 1**   Solve $(2x - 3)^2 = 25$.

Solution $(2x - 3)^2 = 25$

$2x - 3 = \pm\sqrt{25}$ Theorem 6.2

$2x - 3 = \pm 5$ $\sqrt{25} = 5$

$2x = 3 \pm 5$ Add 3 to both sides

$x = \dfrac{3 \pm 5}{2}$ Divide both sides by 2

The last equation can be written as two separate statements:

$$x = \frac{3 + 5}{2} \quad \text{or} \quad x = \frac{3 - 5}{2}$$

$$x = 4 \qquad \text{or} \quad x = -1$$

The solution set is $\{4, -1\}$.   ▲

▼ **Example 2**   Solve for x: $(3x - 1)^2 = -12$.

Solution

$(3x - 1)^2 = -12$

$3x - 1 = \pm\sqrt{-12}$ Theorem 6.2

$3x - 1 = \pm 2i\sqrt{3}$ $\sqrt{-12} = i\sqrt{12} = \sqrt{4}\,i\sqrt{3} = 2i\sqrt{3}$

$3x = 1 \pm 2i\sqrt{3}$ Add 1 to both sides

$x = \dfrac{1 \pm 2i\sqrt{3}}{3}$ Divide both sides by 3

The solution set is

$$\left\{ \frac{1 + 2i\sqrt{3}}{3}, \ \frac{1 - 2i\sqrt{3}}{3} \right\}$$

Both solutions are complex. Although the arithmetic is somewhat more complicated, we check each solution in the usual manner.

Here is a check of the first solution:

$$\text{When} \quad x = \frac{1 + 2i\sqrt{3}}{3},$$

the equation $(3x - 1)^2 = -12$

becomes $\left(3 \cdot \dfrac{1 + 2i\sqrt{3}}{3} - 1\right)^2 = -12$

or $(1 + 2i\sqrt{3} - 1)^2 = -12$

$$(2i\sqrt{3})^2 = -12$$
$$4 \cdot 3 \cdot i^2 = -12$$
$$12(-1) = -12$$
$$-12 = -12 \qquad \blacktriangle$$

The method of completing the square is simply a way of transforming any quadratic equation into an equation of the form found in the preceding two examples.

The key to understanding the method of completing the square lies in recognizing the relationship between the last two terms of any perfect square trinomial whose leading coefficient is 1.

Consider the following list of perfect square trinomials and their corresponding binomial squares:

$$x^2 - 6x + 9 = (x - 3)^2$$
$$x^2 + 8x + 16 = (x + 4)^2$$
$$x^2 - 10x + 25 = (x - 5)^2$$
$$x^2 + 12x + 36 = (x + 6)^2$$

In each case the leading coefficient is 1. A more important observation comes from noticing the relationship between the linear and constant terms (middle and last terms) in each trinomial. Observe that the constant term in each case is the square of half the coefficient of x in the middle term. For example, in the last expression, the constant term, 36, is the square of half of 12, where 12 is the coefficient of x in the middle term. (Notice also that the second terms in all the binomials on the right side are half the coefficients of the middle terms of the trinomials on the left side.) We can use these observations to build our own perfect square trinomials, and in doing so, solve some quadratic equations. Consider the following equation:

$$x^2 + 6x = 3$$

We can think of the left side as having the first two terms of a perfect

square trinomial. We need only add the correct constant term. If we take half the coefficient of x, we get 3. If we then square this quantity, we have 9. Adding the 9 to both sides, the equation becomes

$$x^2 + 6x + 9 = 3 + 9$$

The left side is the perfect square $(x + 3)^2$; the right side is 12:

$$(x + 3)^2 = 12$$

The equation is now in the correct form. We can apply Theorem 6.2 and finish the solution:

$$
\begin{aligned}
(x + 3)^2 &= 12 \\
x + 3 &= \pm\sqrt{12} \qquad \text{Theorem 6.2} \\
x + 3 &= \pm 2\sqrt{3} \\
x &= -3 \pm 2\sqrt{3}
\end{aligned}
$$

The solution set is $\{-3 + 2\sqrt{3}, -3 - 2\sqrt{3}\}$. The method just used is called *completing the square,* since we complete the square on the left side of the original equation by adding the appropriate constant term.

▼   **Example 3** Solve by completing the square: $x^2 + 5x - 2 = 0$.

Solution We must begin by adding 2 to both sides. (The left side of the equation, as it is, is not a perfect square because it does not have the correct constant term. We will simply "move" that term to the other side and use our own constant term.)

$$x^2 + 5x = 2$$

We complete the square by adding the square of half the coefficient of the linear term to both sides:

$$x^2 + 5x + \frac{25}{4} = 2 + \frac{25}{4} \qquad \text{Half of 5 is } \tfrac{5}{2}, \text{ the square of which is } \tfrac{25}{4}$$

$$\left(x + \frac{5}{2}\right)^2 = \frac{33}{4}$$

$$x + \frac{5}{2} = \pm\sqrt{\frac{33}{4}} \qquad \text{Theorem 6.2}$$

$$x + \frac{5}{2} = \pm\frac{\sqrt{33}}{2} \qquad \text{Simplify the radical}$$

$$x = -\frac{5}{2} \pm \frac{\sqrt{33}}{2} \qquad \text{Add } -\frac{5}{2} \text{ to both sides}$$

$$x = \frac{-5 \pm \sqrt{33}}{2}$$

The solution set is $\left\{ \dfrac{-5 + \sqrt{33}}{2}, \dfrac{-5 - \sqrt{33}}{2} \right\}$

▲

▼ **Example 4** Solve for x: $3x^2 - 8x + 7 = 0$.

Solution

$$3x^2 - 8x + 7 = 0$$
$$3x^2 - 8x = -7 \qquad \text{Add } -7 \text{ to both sides}$$

We cannot complete the square on the left side because the leading coefficient is not 1. We take an extra step and divide both sides by 3:

$$\frac{3x^2}{3} - \frac{8x}{3} = -\frac{7}{3}$$

$$x^2 - \frac{8}{3}x = -\frac{7}{3}$$

Half of $\frac{8}{3}$ is $\frac{4}{3}$, the square of which is $\frac{16}{9}$.

$$x^2 - \frac{8}{3}x + \frac{16}{9} = -\frac{7}{3} + \frac{16}{9} \qquad \text{Add } \frac{16}{9} \text{ to both sides}$$

$$\left(x - \frac{4}{3}\right)^2 = -\frac{5}{9} \qquad \text{Simplify right side}$$

$$x - \frac{4}{3} = \pm \frac{i\sqrt{5}}{3} \qquad \text{Theorem 6.2}$$

$$x = \frac{4}{3} \pm \frac{i\sqrt{5}}{3} \qquad \text{Add } \frac{4}{3} \text{ to both sides}$$

$$x = \frac{4 \pm i\sqrt{5}}{3}$$

The solution set is $\left\{ \dfrac{4 + i\sqrt{5}}{3}, \dfrac{4 - i\sqrt{5}}{3} \right\}$

▲

To Solve a Quadratic
Equation by
Completing the Square

To summarize the method used in the preceding two examples, we list the following steps:

Step 1.   Write the equation in the form $ax^2 + bx = c$.

Step 2.   If the leading coefficient is not 1, divide both sides by the coefficient so that the resulting equation has a leading coefficient of 1. That is, if $a \neq 1$, then divide both sides by a.

Step 3.   Add the square of half the coefficient of the linear term to both sides of the equation.

Step 4.   Write the left side of the equation as the square of a binomial and simplify the right side if possible.

Step 5.   Apply Theorem 6.2 and solve as usual.

Problem Set 6.2

Solve the following by applying Theorem 6.2.

1. $x^2 = 25$
2. $x^2 = 16$
3. $a^2 = -9$
4. $a^2 = -49$
5. $y^2 = \frac{3}{4}$
6. $y^2 = \frac{5}{9}$
7. $x^2 - 5 = 0$ (Add 5 to both sides first.)
8. $x^2 - 7 = 0$
9. $x^2 + 12 = 0$
10. $x^2 + 8 = 0$
11. $4a^2 - 45 = 0$
12. $9a^2 - 20 = 0$
13. $(x - 5)^2 = 9$
14. $(x + 2)^2 = 16$
15. $(2y - 1)^2 = 25$
16. $(3y + 7)^2 = 1$
17. $(2a + 3)^2 = -9$
18. $(3a - 5)^2 = -49$
19. $(5x + 2)^2 = -8$
20. $(6x - 7)^2 = -75$

Copy each of the following and fill in the blanks so that the left side of each is a perfect square trinomial. That is, complete the square.

21. $x^2 + 12x + \underline{} = (x + \underline{})^2$
22. $x^2 + 6x + \underline{} = (x + \underline{})^2$
23. $x^2 - 4x + \underline{} = (x - \underline{})^2$
24. $x^2 - 2x + \underline{} = (x - \underline{})^2$
25. $a^2 - 10a + \underline{} = (a - \underline{})^2$
26. $a^2 - 8a + \underline{} = (a - \underline{})^2$
27. $x^2 + 5x + \underline{} = (x + \underline{})^2$
28. $x^2 + 3x + \underline{} = (x + \underline{})^2$
29. $y^2 - 7y + \underline{} = (y - \underline{})^2$
30. $y^2 - y + \underline{} = (y - \underline{})^2$

Solve each of the following quadratic equations by completing the square:

31. $x^2 + 4x = 12$
32. $x^2 - 2x = 8$
33. $x^2 + 12x = -27$
34. $x^2 - 6x = 16$
35. $a^2 - 2a + 5 = 0$
36. $a^2 + 10a + 22 = 0$
37. $y^2 - 8y + 1 = 0$
38. $y^2 + 6y - 1 = 0$
39. $x^2 - 5x - 3 = 0$
40. $x^2 - 5x - 2 = 0$

41. $2x^2 - 4x - 8 = 0$ **42.** $3x^2 - 9x - 12 = 0$
43. $3t^2 - 8t + 1 = 0$ **44.** $5t^2 + 12t - 1 = 0$
45. $4x^2 - 3x + 5 = 0$ **46.** $7x^2 - 5x + 2 = 0$

For each of the following equations, multiply both sides by the least common denominator. Solve the resulting equation.

47. $3 + \dfrac{2}{x} = \dfrac{4}{x^2}$ **48.** $5 - \dfrac{1}{x} = \dfrac{3}{x^2}$

49. $2t - 3 = \dfrac{5}{t}$ **50.** $3t + 1 = \dfrac{2}{t}$

51. $\dfrac{1}{y - 1} + \dfrac{1}{y + 1} = 1$ **52.** $\dfrac{2}{y + 2} + \dfrac{3}{y - 2} = 1$

53. $\dfrac{10}{x^2} = \dfrac{-3}{x}$ **54.** $\dfrac{5}{x} = \dfrac{4}{x^2}$

55. Check the solution $x = -2 + 3\sqrt{2}$ in the equation $(x + 2)^2 = 18$.
56. Check the solution $x = 2 - 5\sqrt{2}$ in the equation $(x - 2)^2 = 50$.
57. Check the solution $x = 3 - 2\sqrt{3}$ in the equation $x^2 - 6x - 3 = 0$.
58. Check the solution $x = -2 + 3\sqrt{2}$ in the equation $x^2 + 4x - 14 = 0$.
59. The table at the back of the book gives the decimal approximation for $\sqrt{5}$ as 2.236. Use this number to find a decimal approximation for $\dfrac{2 + \sqrt{5}}{2}$ and $\dfrac{2 - \sqrt{5}}{2}$.

60. A decimal approximation for $\sqrt{13}$ is 3.606. Use this number to find decimal approximations for $\dfrac{-3 + \sqrt{13}}{2}$ and $\dfrac{-3 - \sqrt{13}}{2}$.

Review Problems   The problems below review material we have covered previously in a number of sections.

Let $a = 2$, $b = -3$, and $c = -1$ in each expression below, and then simplify.

61. b^2 **62.** $4ac$
63. $b^2 - 4ac$ **64.** $\sqrt{b^2 - 4ac}$

65. $-b + \sqrt{b^2 - 4ac}$ **66.** $\dfrac{-b + \sqrt{b^2 - 4ac}}{2a}$

In this section we will use the method of completing the square from the preceding section to derive the quadratic formula. The quadratic formula is a very useful tool in mathematics. It allows us to solve all types of quadratic equations.

**6.3
The Quadratic
Formula**

THEOREM 6.3 (THE QUADRATIC THEOREM) For any quadratic equation in the form $ax^2 + bx + c = 0$, where $a \neq 0$, the two solutions are

$$x = \frac{-b + \sqrt{b^2 - 4ac}}{2a} \quad \text{and} \quad x = \frac{-b - \sqrt{b^2 - 4ac}}{2a}$$

Proof We will prove the quadratic theorem by completing the square on $ax^2 + bx + c = 0$.

$$ax^2 + bx + c = 0$$
$$ax^2 + bx \quad = -c \qquad \text{Add } -c \text{ to both sides}$$
$$x^2 + \frac{b}{a}x \quad = -\frac{c}{a} \qquad \text{Divide both sides by } a$$

To complete the square on the left side we add the square of 1/2 of b/a to both sides. (1/2 of b/a is $b/2a$.)

$$x^2 + \frac{b}{a}x + \left(\frac{b}{2a}\right)^2 = -\frac{c}{a} + \left(\frac{b}{2a}\right)^2$$

We now simplify the right side as a separate step. We square the second term and combine the two terms by writing each with the least common denominator $4a^2$:

$$-\frac{c}{a} + \left(\frac{b}{2a}\right)^2 = -\frac{c}{a} + \frac{b^2}{4a^2} = \frac{4a}{4a}\left(\frac{-c}{a}\right) + \frac{b^2}{4a^2} = \frac{-4ac + b^2}{4a^2}$$

It is convenient to write this last expression as

$$\frac{b^2 - 4ac}{4a^2}$$

Continuing with the proof, we have

$$x^2 + \frac{b}{a}x + \left(\frac{b}{2a}\right)^2 = \frac{b^2 - 4ac}{4a^2}$$

$$\left(x + \frac{b}{2a}\right)^2 = \frac{b^2 - 4ac}{4a^2} \qquad \text{Write left side as a binomial square}$$

$$x + \frac{b}{2a} = \pm\frac{\sqrt{b^2 - 4ac}}{2a} \qquad \text{Theorem 6.2}$$

$$x = -\frac{b}{2a} \pm \frac{\sqrt{b^2 - 4ac}}{2a} \qquad \text{Add } -\frac{b}{2a} \text{ to both sides}$$

$$x = \frac{-b \pm \sqrt{b^2 - 4ac}}{2a}$$

which is equivalent to

$$x = \frac{-b + \sqrt{b^2 - 4ac}}{2a} \quad \text{or} \quad x = \frac{-b - \sqrt{b^2 - 4ac}}{2a}$$

Our proof is now complete. What we have is this: If our equation is in the form $ax^2 + bx + c = 0$ (standard form), where $a \neq 0$, the two solutions are always given by the formula

$$x = \frac{-b \pm \sqrt{b^2 - 4ac}}{2a}$$

This formula is known as the *quadratic formula*. If we substitute the coefficients a, b, and c of any quadratic equation in standard form in the formula, we need only perform some basic arithmetic to arrive at the solution set.

▼   **Example 1**   Use the quadratic formula to solve $6x^2 + 7x - 5 = 0$.

Solution   Using the coefficients $a = 6$, $b = 7$, and $c = -5$ in the formula

$$x = \frac{-b \pm \sqrt{b^2 - 4ac}}{2a}$$

we have

$$x = \frac{-7 \pm \sqrt{49 - 4(6)(-5)}}{2(6)}$$

or

$$x = \frac{-7 \pm \sqrt{49 + 120}}{12}$$

$$= \frac{-7 \pm \sqrt{169}}{12}$$

$$= \frac{-7 \pm 13}{12}$$

We separate the last equation into the two statements

$$x = \frac{-7 + 13}{12} \quad \text{or} \quad x = \frac{-7 - 13}{12}$$

$$x = \frac{1}{2} \quad \text{or} \quad x = -\frac{5}{3}$$

The solution set is $\{\frac{1}{2}, -\frac{5}{3}\}$.

▼   **Example 2** Solve $\dfrac{x^2}{3} - x = -\dfrac{1}{2}$.

Solution Multiplying through by 6 and writing the result in standard form, we have

$$2x^2 - 6x + 3 = 0$$

In this case $a = 2$, $b = -6$, and $c = 3$. The two solutions are given by

$$x = \dfrac{-(-6) \pm \sqrt{36 - 4(2)(3)}}{2(2)}$$

$$= \dfrac{6 \pm \sqrt{12}}{4}$$

$$= \dfrac{6 \pm 2\sqrt{3}}{4}$$

We can reduce this last expression to lowest terms by dividing the numerator and denominator by 2:

$$x = \dfrac{3 \pm \sqrt{3}}{2}$$

▲

▼   **Example 3** Solve $\sqrt{x + 1} + \sqrt{2x} = 1$.

Solution This equation has two separate terms involving radical signs. In this situation it is usually best to separate the radical terms on opposite sides of the equal sign. [*Note:* If we were to square both sides of the equation in its present form, we would not get $(x + 1) + (2x)$ for the left side. The square of the left side is $(\sqrt{x + 1} + \sqrt{2x})^2 = (x + 1) + 2\sqrt{x + 1}\sqrt{2x} + 2x$.]
    Adding $-\sqrt{2x}$ to both sides, we have

$$\sqrt{x + 1} = 1 - \sqrt{2x}$$

Squaring both sides gives

$x + 1 = 1 - 2\sqrt{2x} + 2x$	Recall: $(a + b)^2 = a^2 + 2ab + b^2$
$-x = -2\sqrt{2x}$	Add $-2x$ and -1 to both sides
$x^2 = 4(2x)$	Square both sides
$x^2 - 8x = 0$	Standard form

The coefficients are $a = 1$, $b = -8$, and $c = 0$.

$$x = \frac{-(-8) \pm \sqrt{64 - 4(1)(0)}}{2(1)}$$

$$x = \frac{8 \pm 8}{2}$$

$$x = 8 \quad \text{or} \quad x = 0$$

Since we squared both sides, we have the possibility that one or both of the solutions are extraneous. We must check each one in the original equation:

When $x = 8$,

we have $\sqrt{8 + 1} + \sqrt{2 \cdot 8} = 1$

$\sqrt{9} + \sqrt{16} = 1$

$3 + 4 = 1$

$7 = 1$

which implies $x = 8$

is extraneous

When $x = 0$,

we have $\sqrt{0 + 1} + \sqrt{2(0)} = 1$

$\sqrt{1} + \sqrt{0} = 1$

$1 + 0 = 1$

$1 = 1$

which implies $x = 0$

is a solution   ▲

Some of the examples in this section could have been solved by simply factoring. The same will be true of some of the problems in the problem set. It is a good idea to use the quadratic formula to solve these problems, even though factoring may be faster. The reason is that we need to be able to understand and use the quadratic formula. The more problems we work, the better we will understand the formula and the longer we will remember it.

Use the quadratic formula to solve each of the following: *Problem Set 6.3*

1. $x^2 + 5x + 6 = 0$
2. $x^2 + 5x - 6 = 0$
3. $a^2 - 4a + 1 = 0$
4. $a^2 + 4a + 1 = 0$
5. $x^2 - 3x + 2 = 0$
6. $x^2 + x - 2 = 0$
7. $\dfrac{x^2}{2} + 1 = \dfrac{2x}{3}$
8. $\dfrac{x^2}{2} + \dfrac{2}{3} = -\dfrac{2x}{3}$
9. $y^2 - 5y = 0$
10. $2y^2 + 10y = 0$
11. $\dfrac{2t^2}{3} - t = -\dfrac{1}{6}$
12. $\dfrac{t^2}{3} - \dfrac{t}{2} = -\dfrac{3}{2}$
13. $x^2 + 6x - 8 = 0$
14. $2x^2 - 3x + 5 = 0$
15. $2r^2 + 5r + 3 = 0$
16. $3r^2 - 4r - 5 = 0$
17. $2x + 3 = -2x^2$
18. $2x - 3 = 3x^2$
19. $x^2 - 2x + 1 = 0$
20. $x^2 - 6x + 9 = 0$
21. $2x^2 - 5 = 2x$
22. $7x^2 - 8 = 3x$

23. $3r^2 = r - 4$ 24. $5r^2 = 8r + 2$
25. $(x - 3)(x - 5) = 1$ 26. $(x - 3)(x + 1) = -6$
27. $(2x + 3)(x + 4) = 1$ 28. $(3x - 5)(2x + 1) = 4$

29. $\dfrac{x^2}{3} - \dfrac{5x}{6} = \dfrac{1}{2}$ 30. $\dfrac{x^2}{6} + \dfrac{5}{6} = -\dfrac{x}{3}$

Solve the following equations by first clearing each of the radicals:

31. $\sqrt{x + 5} = \sqrt{x} + 1$ 32. $\sqrt{x - 2} = 2 - \sqrt{x}$
33. $\sqrt{x + 4} = 2 - \sqrt{2x}$ 34. $\sqrt{5y + 1} = 1 + \sqrt{3y}$
35. $\sqrt{y + 21} + \sqrt{y} = 7$ 36. $\sqrt{y - 3} - \sqrt{y} = -1$
37. $\sqrt{2(x + 2)} = \sqrt{x + 3} + 1$ 38. $\sqrt{2x - 1} = \sqrt{x - 4} + 2$
39. $\sqrt{y + 9} - \sqrt{y - 6} = 3$ 40. $\sqrt{y + 7} - \sqrt{y + 2} = 1$

41. Solve $2x^3 + 2x^2 + 3x = 0$ by first factoring out the common factor x, and then using the quadratic formula. There are three solutions.

42. Solve $6x^3 - 4x^2 + 6x = 0$ by first factoring out the greatest common factor, and then applying the quadratic formula. There are three solutions.

43. One solution to a quadratic equation is $\dfrac{-3 + 2i}{5}$. What do you think the other solution is?

44. One solution to a quadratic equation is $\dfrac{-2 + 3i\sqrt{2}}{5}$. What is the other solution?

45. A manufacturer can produce x items at a total cost of $C = -x^2 + 40x$. He sells each item for \$11.00, so his total revenue for selling x items is $R = 11x$. The manufacturer will break even when his total revenue is equal to his total cost—that is, when $R = C$. How many items must he sell to break even?

46. If the cost to produce x items is $c = -x^2 + 100x$, while the revenue for x items is $R = 15x$, how many items must be sold in order to break even?

Review Problems   The problems below review material we covered in Section 5.1.

Simplify each expression.

47. $25^{1/2}$ 48. $8^{1/3}$
49. $9^{3/2}$ 50. $16^{3/4}$
51. $8^{-2/3}$ 52. $4^{-3/2}$
53. $\left(\frac{25}{36}\right)^{-1/2}$ 54. $\left(\frac{49}{16}\right)^{-1/2}$

6.4
**Equations Quadratic
in Form**

We are now in a position to put our knowledge of quadratic equations to work to solve a variety of equations. Although, at first, the equations we will solve in this section may not look like quadratic equa-

tions, each of them has the *form* of a quadratic equation. Let's look at an example.

▼ **Example 1**   Solve $(x + 3)^2 - 2(x + 3) - 8 = 0$

Solution   We can see that this equation is quadratic in form by replacing $x + 3$ with another variable, say y. Replacing $x + 3$ with y we have

$$y^2 - 2y - 8 = 0.$$

We can solve this equation by factoring the left side and then setting each factor to 0.

$$
\begin{array}{ll}
y^2 - 2y - 8 = 0 & \\
(y - 4)(y + 2) = 0 & \text{Factor} \\
y - 4 = 0 \text{ or } y + 2 = 0 & \text{Set factors to 0} \\
y = 4 \text{ or } \quad y = -2 &
\end{array}
$$

Since our original equation was written in terms of the variable x, we would like our solutions in terms of x also. Replacing y with $x + 3$, and then solving for x we have

$$
\begin{array}{ll}
x + 3 = 4 \quad \text{or} \quad x + 3 = -2 \\
x = 1 \quad \text{or} \qquad x = -5.
\end{array}
$$

The solutions to our original equation are 1 and -5.

The method we have just shown lends itself well to other types of equations that are quadratic in form, as we will see. In this example, however, there is another method that works just as well. Let's solve our original equation again, but this time, let's begin by expanding $(x + 3)^2$ and $2(x + 3)$.

$$
\begin{array}{ll}
(x + 3)^2 - 2(x + 3) - 8 = 0 & \\
x^2 + 6x + 9 - 2x - 6 - 8 = 0 & \text{Multiply} \\
x^2 + 4x - 5 = 0 & \text{Combine similar terms} \\
(x - 1)(x + 5) = 0 & \text{Factor} \\
x - 1 = 0 \quad \text{or} \quad x + 5 = 0 & \text{Set factors to 0} \\
x = 1 \quad \text{or} \qquad x = -5 &
\end{array}
$$

As you can see, either method produces the same result.   ▲

▼ **Example 2**   Solve $4x^4 + 7x^2 = 2$

Solution   This equation is quadratic in x^2. We can make it easier to

look at by using the substitution $y = x^2$. (The choice of the letter y is arbitrary. We could just as easily use the substitution $m = x^2$.) Making the substitutions $y = x^2$ and then solving the resulting equation we have

$$4y^2 + 7y = 2$$
$$4y^2 + 7y - 2 = 0 \qquad \text{Standard form}$$
$$(4y - 1)(y + 2) = 0 \qquad \text{Factor}$$
$$4y - 1 = 0 \quad \text{or} \quad y + 2 = 0 \qquad \text{Set factors to 0}$$
$$y = \tfrac{1}{4} \quad \text{or} \qquad y = -2$$

Now we replace y with x^2 in order to solve for x

$$x^2 = \tfrac{1}{4} \qquad \text{or} \quad x^2 = -2$$
$$x = \pm\sqrt{\tfrac{1}{4}} \quad \text{or} \quad x = \pm\sqrt{-2} \qquad \text{Theorem 6.2}$$
$$x = \pm\tfrac{1}{2} \quad \text{or} \quad x = \pm i\sqrt{2}$$

The solution set is $\{\tfrac{1}{2}, -\tfrac{1}{2}, i\sqrt{2}, -i\sqrt{2}\}$.   ▲

▼   **Example 3** Solve $2a^{2/3} - 11a^{1/3} + 12 = 0$

Solution Since $a^{2/3} = (a^{1/3})^2$, this equation is quadratic in $a^{1/3}$. Let's replace $a^{1/3}$ with y and solve for y.

When $y = a^{1/3}$
the equation $2a^{2/3} - 11a^{1/3} + 12 = 0$
becomes $2y^2 - 11y + 12 = 0$
$$(2y - 3)(y - 4) = 0 \qquad \text{Factor}$$
$$2y - 3 = 0 \quad \text{or} \quad y - 4 = 0 \qquad \text{Set factors to 0}$$
$$y = \tfrac{3}{2} \quad \text{or} \qquad y = 4$$

Now we replace y with $a^{1/3}$ and solve for a.

$$a^{1/3} = \tfrac{3}{2} \quad \text{or} \quad a^{1/3} = 4$$
$$a = \tfrac{27}{8} \quad \text{or} \qquad a = 64 \qquad \text{Cube both sides}$$
$$\text{of each equation.} \qquad ▲$$

▼   **Example 4** Solve for x: $x + \sqrt{x} - 6 = 0$

Solution To see that this equation is quadratic in form, we have to notice that $(\sqrt{x})^2 = x$. That is, the equation can be rewritten as

$$(\sqrt{x})^2 + \sqrt{x} - 6 = 0$$

Replacing $\sqrt{x}$ with y and solving as usual we have

$$y^2 + y - 6 = 0$$
$$(y + 3)(y - 2) = 0$$
$$y + 3 = 0 \quad \text{or} \quad y - 2 = 0$$
$$y = -3 \quad \text{or} \quad y = 2$$

Again, to find x we replace y with $\sqrt{x}$ and solve.

$$\sqrt{x} = -3 \quad \text{or} \quad \sqrt{x} = 2$$
$$x = 9 \qquad\qquad x = 4 \qquad \text{Square both sides}$$
$$\text{of each equation.}$$

Since we squared both sides of each equation, we have the possibility of obtaining extraneous solutions. We have to check both solutions in our original equation.

When
the equation
becomes

$$x = 9$$
$$x + \sqrt{x} - 6 = 0$$
$$9 + \sqrt{9} - 6 = 0$$
$$9 + 3 - 6 = 0$$
$$6 = 0$$

This implies 9 is extraneous.

When
the equation
becomes

$$x = 4$$
$$x + \sqrt{x} - 6 = 0$$
$$4 + \sqrt{4} - 6 = 0$$
$$4 + 2 - 6 = 0$$
$$0 = 0$$

This means 4 is a solution.

The only solution to the equation $x + \sqrt{x} - 6 = 0$ is $x = 4$. ▲

As a final note, we should mention that the two possible solutions, 9 and 4, to the equation in Example 4 can be obtained by another method. Instead of substituting for $\sqrt{x}$, we can isolate it on one side of the equation and then square both sides to clear the equation of radicals.

$$x + \sqrt{x} - 6 = 0$$
$$\sqrt{x} = -x + 6 \qquad\qquad \text{Isolate } \sqrt{x}$$
$$x = x^2 - 12x + 36 \qquad \text{Square both sides}$$
$$0 = x^2 - 13x + 36 \qquad \text{Add } -x \text{ to both sides}$$
$$0 = (x - 4)(x - 9) \qquad \text{Factor}$$
$$x - 4 = 0 \quad \text{or} \quad x - 9 = 0$$
$$x = 4 \quad \text{or} \quad x = 9$$

We obtain the same two possible solutions. Since we squared both sides of the equation to find them, we would have to check each one in the original equation. As was the case in Example 4, only $x = 4$ is a solution, $x = 9$ is extraneous.

Problem Set 6.4

Solve each equation.

1. $(x - 3)^2 + 3(x - 3) + 2 = 0$
2. $(x + 4)^2 - (x + 4) - 6 = 0$
3. $2(x + 4)^2 + 5(x + 4) - 12 = 0$
4. $3(x - 5)^2 + 14(x - 5) - 5 = 0$
5. $x^4 - 6x^2 - 27 = 0$
6. $x^4 + 2x^2 - 8 = 0$
7. $x^4 + 9x^2 = -20$
8. $x^4 - 11x^2 = -30$
9. $(2a - 3)^2 - 9(2a - 3) = -20$
10. $(3a - 2)^2 + 2(3a - 2) = 3$
11. $2(4a + 2)^2 = 3(4a + 2) + 20$
12. $6(2a + 4)^2 = (2a + 4) + 2$
13. $6t^4 = -t^2 + 5$
14. $3t^4 = -2t^2 + 8$
15. $9x^4 - 49 = 0$
16. $25x^4 - 9 = 0$

Solve each of the following equations. Remember. if you square both sides of an equation in the process of solving it, you have to check all solutions in the original equation.

17. $x^{2/3} + x^{1/3} - 6 = 0$
18. $x^{2/3} + 5x^{1/3} + 6 = 0$
19. $9a^{2/3} + 12a^{1/3} = -4$
20. $12a^{2/3} - 24a^{1/3} = -9$
21. $x - 7\sqrt{x} + 10 = 0$
22. $x - 6\sqrt{x} + 8 = 0$
23. $t - 2\sqrt{t} - 15 = 0$
24. $t - 3\sqrt{t} - 10 = 0$
25. $2x^{2/5} - 3x^{1/5} - 2 = 0$
26. $2x^{2/5} + 5x^{1/5} - 3 = 0$
27. $6x + 11\sqrt{x} = 35$
28. $2x + \sqrt{x} = 15$
29. $(a - 2) - 11\sqrt{a - 2} + 30 = 0$

30. $(a - 3) - 9\sqrt{a - 3} + 20 = 0$
31. $(x - 2)^{2/3} - 3(x - 2)^{1/3} + 2 = 0$
32. $(x - 3)^{2/3} - 5(x - 3)^{1/3} + 6 = 0$
33. $(2x + 1) - 8\sqrt{2x + 1} + 15 = 0$
34. $(2x - 3) - 7\sqrt{2x - 3} + 12 = 0$
35. Solve $x^3 - 8 = 0$ by factoring $x^3 - 8$ and then setting each factor to 0. (You will have to use the quadratic formula on the second factor.) There are three solutions.
36. Solve $x^3 - 27 = 0$ by factoring $x^3 - 27$ and setting each factor to 0.

37. We can solve the equation $x^2 + 2xy + y^2 - 4 = 0$ for x if we look at it this way:

$$1 \cdot x^2 + (2y)x + (y^2 - 4) = 0.$$

Looking at it this way allows us to use the quadratic formula with $a = 1$, $b = 2y$, and $c = y^2 - 4$. Substitute these values into the quadratic formula to solve for x.

38. Solve the equation $x^2 - 2xy + y^2 + 4 = 0$ by first writing it as

$$1 \cdot x^2 - (2y)x + (y^2 + 4) = 0$$

and then using the quadratic formula.

Review Problems The problems below review material we covered in Section 2.6. They are taken from the book *A First Course in Algebra,* written by Wallace C. Boyden and published by Silver, Burdett and Company in 1894.

39. A man bought 12 pairs of boots and 6 suits of clothes for $168. If a suit of clothes cost $2 less than four times as much as a pair of boots, what was the price of each?

40. A farmer pays just as much for 4 horses as he does for 6 cows. If a cow costs 15 dollars less than a horse, what is the cost of each?

41. Two men whose wages differ by 8 dollars receive both together $44 per month. How much does each receive?

42. Mr. Ames builds three houses. The first cost $2000 more than the second, and the third twice as much as the first. If they all together cost $18,000, what was the cost of each house?

We will use the same four steps in solving word problems in this section that we have used in the past. The most important part of solving word problems is finding an equation that describes the situation. In this section, the equation will be second-degree. We can solve the equations by any convenient method.

**6.5
Word Problems**

▼ **Example 1** The sum of the squares of two consecutive integers is 25. Find the two integers.

Solution Let $x =$ the first integer; then $x + 1 =$ the next consecutive integer. The sum of the squares of x and $x + 1$ is 25.

$$
\begin{aligned}
x^2 + (x + 1)^2 &= 25 \\
x^2 + x^2 + 2x + 1 &= 25 \\
2x^2 + 2x - 24 &= 0 \\
x^2 + x - 12 &= 0 \qquad \text{Divide both sides by 2} \\
(x + 4)(x - 3) &= 0 \\
x = -4 \quad &\text{or} \quad x = 3
\end{aligned}
$$

These are the possible values for the first integer:

$$\text{If} \quad x = -4 \qquad\qquad \text{If} \quad x = 3$$
$$\text{then} \quad x + 1 = -3 \qquad \text{then} \quad x + 1 = 4$$

There are two pairs of consecutive integers, the sum of whose squares is 25. They are $\{-4, -3\}$ and $\{3, 4\}$.   ▲

Another application of quadratic equations involves the Pythagorean theorem, an important theorem from geometry. The theorem gives the relationship between the sides of any right triangle (a triangle with a 90° angle). We state it here without proof.

PYTHAGOREAN THEOREM In any right triangle, the square of the longest side (hypotenuse) is equal to the sum of the squares of the other two sides (legs).

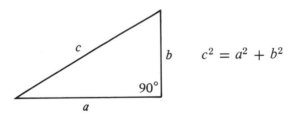

$$c^2 = a^2 + b^2$$

▼ **Example 2** The lengths of the three sides of a right triangle are given by three consecutive integers. Find the lengths of the three sides.

Solution Let x = first integer (shortest side)
Then $x + 1$ = next consecutive integer
        $x + 2$ = last consecutive integer (longest side)

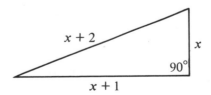

By the Pythagorean theorem, we have

$$(x + 2)^2 = (x + 1)^2 + x^2$$
$$x^2 + 4x + 4 = x^2 + 2x + 1 + x^2$$
$$x^2 - 2x - 3 = 0$$
$$(x - 3)(x + 1) = 0$$
$$x = 3 \quad \text{or} \quad x = -1$$

The shortest side is 3. The other two sides are 4 and 5.   ▲

▼ **Example 3** If an object is thrown downward with an initial velocity of 20 feet/second, the distance (s) it travels in an amount of time t is given by the equation $s = 20t + 16t^2$. (This is because of the acceleration of the object due to gravity. The equation does not take into account the force of friction on the object due to its falling through air.) How long does it take the object to fall 40 feet?

Solution In this example, the equation that describes the situation is given. We simply let $s = 40$ and solve for t:

When $s = 40$,
the equation $s = 20t + 16t^2$
becomes $40 = 20t + 16t^2$
or $16t^2 + 20t - 40 = 0$
$4t^2 + 5t - 10 = 0$ Divide by 4

Using the quadratic formula, we have

$$t = \frac{-5 \pm \sqrt{25 - 4(4)(-10)}}{2(4)}$$

$$= \frac{-5 \pm \sqrt{185}}{8}$$

$$t = \frac{-5 + \sqrt{185}}{8} \quad \text{or} \quad t = \frac{-5 - \sqrt{185}}{8}$$

The second solution is impossible since it is a negative number and t must be positive.

It takes

$$t = \frac{-5 + \sqrt{185}}{8}$$

or approximately

$$\frac{-5 + 13.60}{8} = 1.08 \text{ seconds}$$

for the object to fall 40 feet.   ▲

▼ **Example 4**   The current of a river is 3 miles/hour. It takes a motorboat a total of 3 hours to travel 12 miles upstream and return 12 miles downstream. What is the speed of the boat in still water?

Solution   Let x = the speed of the boat in still water. The basic equation that gives the relationship between distance, rate, and time is $d = rt$.

It is usually helpful to summarize rate problems like this by using a table:

	Distance	Rate	Time
Upstream			
Downstream			

We fill in as much of the table as possible using the information given in the problem. For instance, since we let x = the speed of the boat in still water, the rate upstream (against the current) must be $x - 3$. The rate downstream (with the current) is $x + 3$.

	d	r	t
Upstream	12	$x - 3$	
Downstream	12	$x + 3$	

The last two boxes can be filled in using the relationship $d = r \cdot t$. Since the boxes correspond to t, we solve $d = r \cdot t$ for t and get

$$t = \frac{d}{r}$$

Completing the table, we have

	d	r	t
Upstream	12	$x - 3$	$12/(x - 3)$
Downstream	12	$x + 3$	$12/(x + 3)$

The total time for the trip up and back is 3 hours:

$$\text{Time upstream} + \text{Time downstream} = \text{Total time}$$

$$\frac{12}{x - 3} \quad + \quad \frac{12}{x + 3} \quad = \quad 3$$

Multiplying both sides by $(x - 3)(x + 3)$, we have

$$12(x + 3) + 12(x - 3) = 3(x^2 - 9)$$
$$12x + 36 + 12x - 36 = 3x^2 - 27$$
$$3x^2 - 24x - 27 = 0$$
$$x^2 - 8x - 9 = 0 \qquad\qquad \text{Divide both sides by 3}$$
$$(x - 9)(x + 1) = 0$$
$$x = 9 \quad \text{or} \quad x = -1$$

The speed of the motorboat in still water is 9 miles/hour.

---

1. The sum of the squares of two consecutive odd integers is 34. Find the two integers.

2. The sum of the squares of two consecutive even integers is 100. Find the two integers.

3. The square of the sum of two consecutive integers is 81. Find the two integers.

4. Find two consecutive even integers whose sum squared is 100.

5. One integer is 3 more than twice another. Their product is 65. Find the two integers.

6. The product of two integers is 150. One integer is 5 less than twice the other. Find the integers.

7. The sum of a number and its reciprocal is $\frac{10}{3}$. Find the number.

8. The sum of a number and twice its reciprocal is $\frac{27}{5}$. Find the number.

9. The sum of the reciprocals of two consecutive integers is $\frac{7}{12}$. Find the two integers.

10. Find two consecutive even integers, the sum of whose reciprocals is $\frac{3}{4}$.

Problem Set 6.5

11. The sum of a number and its positive square root is 6. Find the number.
12. The difference of a number and twice its positive square root is 15. Find the number.
13. The lengths of the three sides of a right triangle are given by three consecutive even integers. Find the lengths of the three sides.
14. The longest side of a right triangle is twice the shortest side. The third side measures 6 inches. Find the length of the shortest side.
15. One leg of a right triangle is 3 times the other leg. The hypotenuse is $2\sqrt{10}$ centimeters. What are the lengths of the legs?
16. The longest side of a right triangle is 2 less than twice the shortest side. The third side is 2 more than the shortest side. Find the length of all three sides.
17. An object is thrown downward with an initial velocity of 5 feet/second. The relationship between the distance (s) it travels and time (t) is given by $s = 5t + 16t^2$. How long does it take the object to fall 74 feet?
18. The distance an object falls from rest is given by the equation $s = 16t^2$, where s = distance and t = time. How long does it take an object dropped from a 100-foot cliff to hit the ground?
19. An object is thrown upward with an initial velocity of 20 feet/second. The equation that gives the height (h) of the object at any time (t) is $h = 20t - 16t^2$. At what times will the object be 4 feet off the ground?
20. An object is propelled upward with an initial velocity of 32 feet/second from a height of 16 feet above the ground. The equation giving the object's height (h) at any time (t) is $h = 16 + 32t - 16t^2$. Does the object ever reach a height of 32 feet?
21. The current of a river is 2 miles/hour. A boat travels to a point 8 miles upstream and back again in 3 hours. What is the speed of the boat in still water?
22. A boat travels 15 miles/hour in still water. It takes twice as long for the boat to go 20 miles upstream as it does to go downstream. Find the speed of the current.
23. A man can ride his bicycle 3 times as fast as he can walk. He rode his bike for 15 miles, then walked an additional 2.5 miles. The total time for the trip was $1\frac{1}{2}$ hours. How fast does he ride his bicycle?
24. Train A can travel the 50 miles between the farm and the city 10 miles/hour faster than train B. If a person takes train A to the city and train B back to the farm, the total trip takes $2\frac{1}{4}$ hours. Find the speed of each train.

Review Problems   The problems below review material we covered in Sections 2.2 and 4.4. Reviewing these problems will help you understand the next section.

Solve each inequality and graph the solution set.

25. $2x + 1 < 0$ or $3x - 2 > 0$
26. $3x - 4 < 0$ or $2x - 8 > 0$

27. $x + 2 \geq 0$ and $x - 4 \leq 0$
28. $x + 4 \geq 0$ and $x - 2 \leq 0$

Add or subtract as indicated.

29. $\dfrac{2}{x - 3} + 1$ 30. $\dfrac{3}{x + 2} + 1$

31. $\dfrac{3}{x - 2} - \dfrac{2}{x - 3}$ 32. $\dfrac{4}{x + 3} - \dfrac{3}{x + 2}$

Quadratic inequalities in one variable are inequalities of the form

6.6
More on Inequalities

$$ax^2 + bx + c < 0$$
$$ax^2 + bx + c \leq 0$$
$$ax^2 + bx + c > 0$$
$$ax^2 + bx + c \geq 0$$

where a, b, and c are constants, with $a \neq 0$. The technique we will use to solve inequalities of this type involves graphing. Suppose, for example, we wish to find the solution set for the inequality $x^2 - x - 6 > 0$. We begin by factoring the left side to obtain

$$(x - 3)(x + 2) > 0$$

We have two real numbers $x - 3$ and $x + 2$ whose product $(x - 3) \cdot (x + 2)$ is greater than zero. That is, their product is positive. The only way the product can be positive is either if both factors, $(x - 3)$ and $(x + 2)$, are positive or if they are both negative. To help visualize where $x - 3$ is positive and where it is negative, we draw a real number line and label it accordingly:

Here is a similar diagram showing where the factor $x + 2$ is positive and where it is negative:

Drawing the two number lines together and eliminating the unnecessary numbers, we have

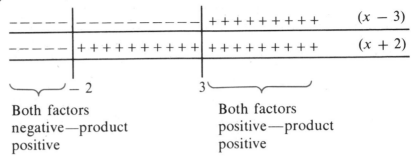

Since the original inequality indicates that the product $(x - 3)(x + 2)$ is positive (greater than 0), the solution set will consist of those regions where both factors are positive or both factors are negative. Here is the solution set and its graph.

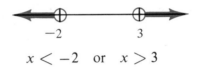

$$x < -2 \quad \text{or} \quad x > 3$$

▼ **Example 1**   Solve for x: $x^2 - 2x - 8 \le 0$.

Solution   We begin by factoring:

$$x^2 - 2x - 8 \le 0$$
$$(x - 4)(x + 2) \le 0$$

The product $(x - 4)(x + 2)$ is negative or zero. The factors must have opposite signs. We draw a diagram showing where each factor is positive and where each factor is negative:

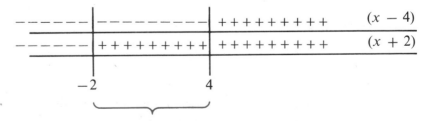

Factors have different signs—
their product is negative

From the diagram we have the graph of the solution set:

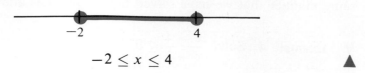

$$-2 \leq x \leq 4 \qquad \blacktriangle$$

▼   **Example 2**   Solve for x: $6x^2 - x \geq 2$.

Solution

$$6x^2 - x \geq 2$$
$$6x^2 - x - 2 \geq 0 \qquad \leftarrow \text{Standard form}$$
$$(3x - 2)(2x + 1) \geq 0$$

The product is positive, so the factors must agree in sign. Here is the diagram showing where that occurs:

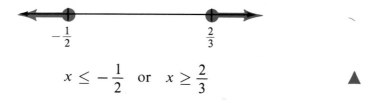

Since the factors agree in sign below $-\frac{1}{2}$ and above $\frac{2}{3}$, the graph of the solution set is

$$x \leq -\frac{1}{2} \quad \text{or} \quad x \geq \frac{2}{3} \qquad \blacktriangle$$

▼   **Example 3**   Solve $x^2 - 6x + 9 \geq 0$.

Solution

$$x^2 - 6x + 9 \geq 0$$
$$(x - 3)^2 \geq 0$$

This is a special case in which both factors are the same. Since $(x - 3)^2$ is always positive or zero, the solution set is all real numbers. That is, any real number that is used in place of x in the original inequality will produce a true statement. $\blacktriangle$

Inequalities that involve rational expressions can be solved in the same manner that we have solved the inequalities above.

▼   **Example 4**   Solve $\dfrac{x-4}{x+1} \leq 0$.

Solution   The inequality indicates that the quotient of $(x-4)$ and $(x+1)$ is negative or 0 (less than or equal to 0). We can use the same reasoning we used to solve the first three examples, because quotients are positive or negative under the same conditions that products are positive or negative. Here is the diagram that shows when each factor is positive and where each factor is negative.

$$
\begin{array}{ccccc}
---- & |++++| & ++++ & \quad (x+1) \\[4pt]
\hline
---- & |----| & ++++ & \quad (x-4) \\[4pt]
\hline
 & -1 & \quad 4 &
\end{array}
$$

Factors have different signs—
quotient is negative.

The region between -1 and 4 is where the solutions lie, since the original inequality indicates the quotient $\dfrac{x-4}{x+1}$ is negative, and between -1 and 4 the factors have opposite signs, making the quotient negative.
The solution set and its graph are:

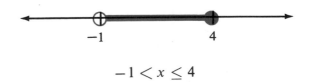

$$-1 < x \leq 4$$

Notice the left end point is open—that is, not included in the solution set—because $x = -1$ would make the original inequality undefined. It is important to check all end points of solution sets to inequalities that involve rational expressions.   ▲

▼   **Example 5**   Solve $\dfrac{3}{x-2} - \dfrac{2}{x-3} > 0$.

Solution   We begin by adding the two rational expressions on the left side. The common denominator is $(x - 2)(x - 3)$.

$$\frac{3}{x - 2} \cdot \frac{(x - 3)}{(x - 3)} - \frac{2}{x - 3} \cdot \frac{(x - 2)}{(x - 2)} > 0$$

$$\frac{3x - 9 - 2x + 4}{(x - 2)(x - 3)} > 0$$

$$\frac{x - 5}{(x - 2)(x - 3)} > 0$$

This time the quotient involves three factors. Here is the diagram that shows the signs of the three factors.

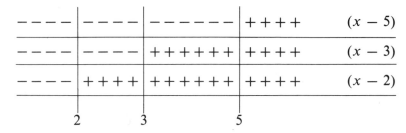

The original inequality indicates that the quotient is positive. In order for this to happen, of the three factors either all must be positive, or exactly two must be negative. Looking back to our diagram, we see the regions that satisfy these conditions are between 2 and 3 or above 5. Here is our solution set.

$$2 < x < 3 \quad \text{or} \quad x > 5 \qquad \blacktriangle$$

Solve each of the following inequalities and graph the solution set.   *Problem Set 6.6*

1. $x^2 + x - 6 > 0$
2. $x^2 + x - 6 < 0$
3. $x^2 - x - 12 \leq 0$
4. $x^2 - x - 12 \geq 0$
5. $x^2 + 5x \geq -6$
6. $x^2 - 5x > 6$
7. $6x^2 < 5x - 1$
8. $4x^2 \geq -5x + 6$
9. $x^2 - 9 < 0$
10. $x^2 - 16 \geq 0$
11. $4x^2 - 9 \geq 0$
12. $9x^2 - 4 < 0$
13. $2x^2 - x - 3 < 0$
14. $3x^2 + x - 10 \geq 0$

15. $x^2 - 4x + 4 \geq 0$

16. $x^2 - 4x + 4 < 0$

17. $x^2 - 10x + 25 < 0$

18. $x^2 - 10x + 25 > 0$

19. $(x - 2)(x - 3)(x - 4) > 0$

20. $(x - 2)(x - 3)(x - 4) < 0$

21. $(x + 1)(x + 2)(x + 3) \leq 0$

22. $(x + 1)(x + 2)(x + 3) \geq 0$

23. $\dfrac{x - 4}{x + 1} > 0$

24. $\dfrac{x + 1}{x - 4} \leq 0$

25. $\dfrac{x + 5}{x + 3} \leq 0$

26. $\dfrac{x + 2}{x + 5} > 0$

27. $\dfrac{2x - 3}{4x + 1} < 0$

28. $\dfrac{3x + 2}{4x - 1} \geq 0$

29. $\dfrac{3}{x - 2} - \dfrac{2}{x - 3} \leq 0$

30. $\dfrac{4}{x + 3} - \dfrac{3}{x + 2} > 0$

31. $\dfrac{2}{x - 3} + 1 > 0$

32. $\dfrac{3}{x + 2} + 1 \leq 0$

33. Why can't we solve the inequality $\dfrac{x - 3}{x} < 0$ by simply multiplying both sides by x to clear it of fractions?

34. Why can't we solve the inequality $x(x + 3) < 0$ by dividing both sides by x?

Review Problems   The problems below review material we covered in Section 5.2.

Write each radical in simplified form.

35. $\sqrt{48}$

36. $\sqrt{50}$

37. $\sqrt{18x^3}$

38. $\sqrt{12x^5}$

39. $\sqrt[3]{8x^3y^4}$

40. $\sqrt[3]{27x^4y^3}$

41. $\sqrt{\frac{2}{3}}$

42. $\sqrt{\frac{3}{5}}$

Examples

Chapter 6   Summary and Review

THEOREM 6.1 (ZERO-FACTOR PROPERTY) [6.1]

For all real numbers r and s, if $r \cdot s = 0$, then $r = 0$ or $s = 0$ (or both).

1. Solve $x^2 - 5x = -6$.

$$x^2 - 5x + 6 = 0$$
$$(x - 3)(x - 2) = 0$$
$$x - 3 = 0 \text{ or } x - 2 = 0$$
$$x = 3 \text{ or } \quad x = 2$$

TO SOLVE A QUADRATIC EQUATION BY FACTORING [6.1]

Step 1.   Write the equation in standard form:
$$ax^2 + bx + c = 0, \quad a \neq 0.$$
Step 2.   Factor the left side.

Step 3. Use the zero-factor property to set each factor equal to zero.

Step 4. Solve the resulting first-degree equations.

THEOREM 6.2 [6.2]

If $a^2 = b$, where b is a real number, then

$$a = \sqrt{b} \quad \text{or} \quad a = -\sqrt{b} \quad \text{or} \quad a = \pm\sqrt{b}.$$

TO SOLVE A QUADRATIC EQUATION BY COMPLETING THE SQUARE [6.2]

Step 1. Write the equation in the form $ax^2 + bx = c$.

Step 2. If $a \neq 1$, divide through by the constant a so the coefficient of x^2 is 1.

Step 3. Complete the square on the left side by adding the square of $\frac{1}{2}$ the coefficient of x to both sides.

Step 4. Write the left side of the equation as the square of a binomial. Simplify the right side if possible.

Step 5. Apply Theorem 6.2 and solve as usual.

2. Solve $x^2 - 6x - 6 = 0$.

$$x^2 - 6x = 6$$
$$x^2 - 6x + 9 = 6 + 9$$
$$(x - 3)^2 = 15$$
$$x - 3 = \pm\sqrt{15}$$
$$x = 3 \pm \sqrt{15}$$

THEOREM 6.3 (THE QUADRATIC THEOREM) [6.3]

For any quadratic equation in the form $ax^2 + bx + c = 0$, $a \neq 0$, the two solutions are

$$x = \frac{-b \pm \sqrt{b^2 - 4ac}}{2a}.$$

This last expression is known as the *quadratic formula.*

3. If $2x^2 + 3x - 4 = 0$, then

$$x = \frac{-3 \pm \sqrt{9 - 4(2)(-4)}}{2(2)}$$
$$= \frac{-3 \pm \sqrt{41}}{4}$$

EQUATIONS QUADRATIC IN FORM [6.4]

There are a variety of equations whose form is quadratic. We solve most of them by making a substitution so the equation becomes quadratic, and then solving that equation by factoring or the quadratic formula. For example

The equation	*is quadratic in*
$(2x - 3)^2 + 5(2x - 3) - 6 = 0$	$2x - 3$
$4x^4 - 7x^2 - 2 = 0$	x^2
$2x - 7\sqrt{x} + 3 = 0$	$\sqrt{x}$

4. The equation $x^{2/3} - 5x^{1/3} - 6 = 0$ is quadratic in $x^{1/3}$. Letting $y = x^{1/3}$ we have

$$y^2 - 5y - 6 = 0$$
$$(y - 6)(y + 1) = 0$$
$$y = 6 \quad \text{or} \quad y = -1$$

Resubstituting $x^{1/3}$ for y we have

$$x^{1/3} = 6 \quad \text{or} \quad x^{1/3} = -1$$
$$x = 216 \quad \text{or} \quad x = -1$$

5. Solve $x^2 - 2x - 8 > 0$
We factor and draw the sign diagram.

$$(x - 4)(x + 2) > 0$$

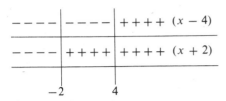

-2 4

A positive product means the solution set is

-2 4

QUADRATIC INEQUALITIES [6.6]

We solve quadratic inequalities by manipulating the inequality to get 0 on the right side and then factoring the left side. We then make a diagram that indicates where the factors are positive and where they are negative. From this sign diagram and the original inequality we graph the appropriate solution set.

INEQUALITIES INVOLVING RATIONAL EXPRESSIONS [6.6]

We manipulate these inequalities to get 0 on the right side and a single rational expression in factored form on the left side. We then draw a sign diagram for the factors and proceed as we did with quadratic inequalities.

COMMON MISTAKES

1. Attempting to apply the zero-factor property to numbers other than 0. For example, consider the equation

$$(x + 2)(x - 3) = 7$$

The mistake takes place when we try to solve it by setting each factor equal to 7.

$$x + 2 = 7 \quad \text{or} \quad x - 3 = 7$$
$$x = 5 \quad \text{or} \quad x = 10$$

Neither of these two numbers is a solution to the original equation. The mistake arises when we assume that since the product of $(x + 2)$ and $(x - 3)$ is 7, one of the two factors must also be 7.

2. When both sides of an equation are squared in the process of solving the equation, a common mistake occurs when the resulting solutions are not checked in the original equation. Remember, every time we square both sides of an equation, there is the possibility we have introduced an extraneous root.

3. When squaring a quantity that has two terms involving radicals, it is a common mistake to omit the middle term in the result. For example,

$$(\sqrt{x + 3} + \sqrt{2x})^2 = (x + 3) + (2x)$$

is a common mistake. It should look like this:

$$(\sqrt{x + 3} + \sqrt{2x})^2 = (x + 3) + 2\sqrt{2x}\sqrt{x + 3} + (2x)$$

Remember: $(a + b)^2 = a^2 + 2ab + b^2$.

Solve the following by factoring:

1. $x^2 - 4x = 21$ 2. $25x^2 - 81 = 0$
3. $4x^2 = 6 - 5x$

Solve by completing the square:

4. $x^2 - 4x = -2$ 5. $2x^2 - 8x = 5$
6. $3y^2 + 8y + 5 = 0$

Solve by using the quadratic formula:

7. $2x^2 - x = 3$ 8. $x^2 - 7x - 7 = 0$

Solve by any method:

9. $4x^4 - 7x^2 - 2 = 0$ 10. $\sqrt{2x + 5} = x + 1$

11. $\sqrt{2x + 1} = 1 + \sqrt{x}$ 12. $\dfrac{1}{x - 1} = \dfrac{5}{4} - \dfrac{1}{x - 4}$

13. $x^{2/3} - 3x^{1/3} + 2 = 0$
14. $(2x + 1)^2 - 5(2x + 1) + 6 = 0$
15. $2x - 7\sqrt{x} + 3 = 0$
16. Are $x = (3 + i)/5$ and $x = (3 - i)/5$ solutions to $5x^2 - 6x + 2 = 0$?
17. One number is 1 less than twice another. The sum of their squares is 34.
     Find the two numbers.
18. A motorboat travels at 4 miles/hour in still water. It goes 12 miles
     upstream and 12 miles back again in a total of 8 hours. Find the speed of
     the current of the river.

Solve each inequality and graph the solution set.

19. $2x^2 + 5x > 3$ 20. $\dfrac{(x - 3)}{(x + 2)(x - 5)} \leq 0$

7
Linear Equations and Inequalities

To the student:

As we have said before, mathematics is a language that can describe certain aspects of our world better than English. One important aspect of the world is the idea of a path, track, orbit, or course. Some simple paths we are familiar with are circles, cloverleaf interchanges on a highway, the sloping straight line of a sewer pipe, and the elliptical route taken by a satellite. Mathematics can be used to describe these paths very accurately. The simplest of paths—a straight line—can be described by an equation such as $3x - 2y = 5$, which we call a linear equation in two variables. In this chapter we will concern ourselves with linear equations (and inequalities) in two variables.

To make a successful attempt at Chapter 7 you should be familiar with the concepts developed in Chapter 2. The main concept is how to solve a linear equation in one variable.

7.1 Graphing in Two Dimensions

Up to this point we have always considered equations with one variable. Solution sets for these equations are sets of single numbers.

Let us now consider the equation $2x - y = 5$. The equation contains two variables. A solution, therefore, must be in the form of a pair of

numbers, one for x and one for y, that make the equation a true statement. One pair of numbers that works is $x = 3$ and $y = 1$, because when we substitute them for x and y in the equation we get a true statement. That is,

$$2(3) - 1 = 5$$
$$5 = 5 \qquad \leftarrow \text{A true statement}$$

The pair of numbers $x = 3$ and $y = 1$ can be written as $(3, 1)$. This is called an *ordered pair,* because it is a pair of numbers written in a specific order. The first number in the ordered pair is always associated with the variable x, the second number with the variable y. The first number is called the *x-coordinate* (or x-component) of the ordered pair and the second number is called the *y-coordinate* (or y-component) of the ordered pair.

With equations that contain *one* variable, solution sets are graphed on the real number line. Solution sets for equations in two variables consist of ordered pairs of numbers which we will graph on a rectangular (cartesian) coordinate system.

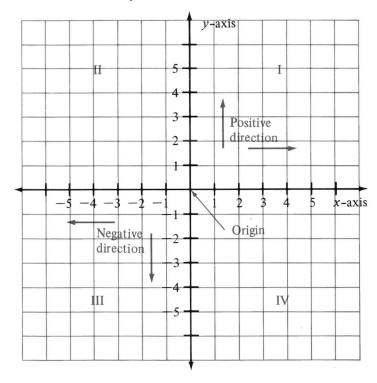

Figure 7-1

A *rectangular coordinate system* is made by drawing two real number lines at right angles to each other. The two number lines, called *axes,* cross each other at 0. This point is called the *origin.* Positive directions are to the right and up. Negative directions are down and to the left. The rectangular coordinate system is shown in Figure 7-1.

The horizontal number line is called the *x-axis* and the vertical number line is called the *y-axis.* The two number lines divide the coordinate system into four quadrants which we number I through IV in a counterclockwise direction. Points on the axes are not considered as being in any quadrant.

Graphing Ordered Pairs

To graph the ordered pair (*a, b*) on a rectangular coordinate system, we start at the origin and move *a* units right or left (right if *a* is positive, left if *a* is negative). Then we move *b* units up or down (up if *b* is positive and down if *b* is negative). The point where we end is the graph of the ordered pair (*a, b*).

▼   Example 1   Plot (graph) the ordered pairs (2, 5), (−2, 5), (−2, −5), and (2, −5).

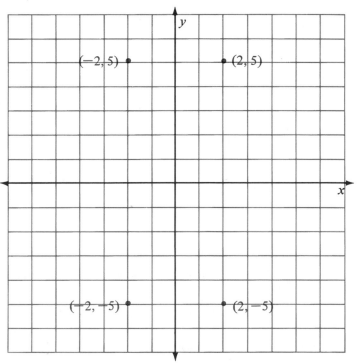

Figure 7-2

Solution To graph the ordered pair (2, 5), we start at the origin and move 2 units to the right, then 5 units up. We are now at the point whose coordinates are (2, 5). We graph the other three ordered pairs in a similar manner (see Figure 7-2). ▲

From Example 1 we see that any point in quadrant I has both its x- and y- coordinates positive $(+, +)$. Points in quadrant II have negative x-coordinates and positive y-coordinates $(-, +)$. In quadrant III both coordinates are negative $(-, -)$. In quadrant IV the form is $(+, -)$.

Every point has a unique set of coordinates and every ordered pair (a, b) has as its graph a point in the coordinate plane. That is, for every point there corresponds an ordered pair (a, b), and for every ordered pair (a, b) there corresponds a point. We say there is a *one-to-one correspondence* between points on the coordinate system and ordered pairs (a, b) where a and b are real numbers.

▼ **Example 2** Graph the ordered pairs $(1, -3)$, $(\frac{1}{2}, 2)$, $(3, 0)$, $(0, -2)$, $(-1, 0)$ and $(0, 5)$.

Solution

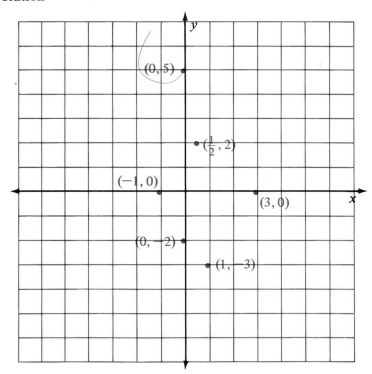

Figure 7-3   ▲

From Example 2 we see that any point on the x-axis has a y-coordinate of 0 (it has no vertical displacement), and any point on the y-axis has an x-coordinate of 0 (no horizontal displacement).

We will now turn our attention to graphing some straight lines.

DEFINITION Any equation that can be put in the form $ax + by = c$, where a, b, and c are real numbers and a and b are not both 0, is called a *linear equation* in two variables. The graph of any equation of this form is a straight line (that is why these equations are called "linear"). The form $ax + by = c$ is called *standard form*.

To graph a linear equation in two variables, we simply graph its solution set. That is, we will draw a line through all the points whose coordinates satisfy the equation.

▼   **Example 3**   Graph $y = 2x - 3$.

Solution   Since $y = 2x - 3$ can be put in the form $ax + by = c$, it is a linear equation in two variables. Hence, the graph of its solution set is a straight line. We can find some specific solutions by substituting numbers for x and then solving for the corresponding values of y. We are free to choose any convenient numbers for x, so let's use the numbers -2, 0, and 2:

$$\begin{aligned} \text{When} \quad & x = -2, \\ \text{the equation} \quad & y = 2x - 3 \\ \text{becomes} \quad & y = 2(-2) - 3 \\ & y = -7 \end{aligned}$$

The ordered pair $(-2, -7)$ is a solution.

$$\begin{aligned} \text{When} \quad & x = 0, \\ \text{we have} \quad & y = 2(0) - 3 \\ & y = -3 \end{aligned}$$

The ordered pair $(0, -3)$ is also a solution.

$$\begin{aligned} \text{Using} \quad & x = 2, \\ \text{we have} \quad & y = 2(2) - 3 \\ & y = 1 \end{aligned}$$

The ordered pair (2, 1) is another solution.

Graphing these three ordered pairs and drawing a line through them, we have the graph of $y = 2x - 3$:

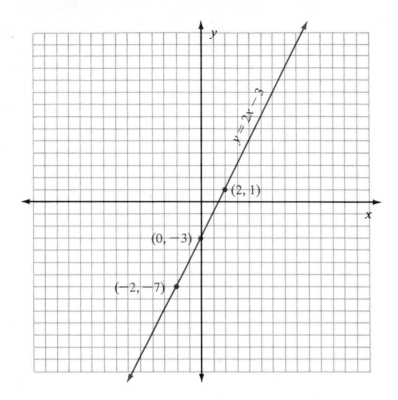

Figure 7-4

There is a one-to-one correspondence between points on the line and solutions to the equation $y = 2x - 3$. ▲

It actually takes only two points to determine a straight line. We have included a third point in the example above for insurance. If all three points do not line up in a straight line, we have made a mistake.

Two important points on the graph of a straight line, if they exist, are the points where the graph crosses the axes.

DEFINITION The *x-intercept* of the graph of an equation is the *x*-coordinate of the point where the graph crosses the *x*-axis. The *y-intercept* is defined similarly.

Since any point on the x-axis has a y-coordinate of 0, we can find the x-intercept by letting $y = 0$ and solving the equation for x. We find the y-intercept by letting $x = 0$ and solving for y.

▼   **Example 4**   Find the x- and y-intercepts for $2x + 3y = 6$; then graph the solution set.

Solution   To find the y-intercept we let $x = 0$.

$$\text{When} \qquad x = 0,$$
$$\text{we have} \quad 2(0) + 3y = 6$$
$$3y = 6$$
$$y = 2$$

The y-intercept is 2, and the graph crosses the y-axis at the point $(0, 2)$.

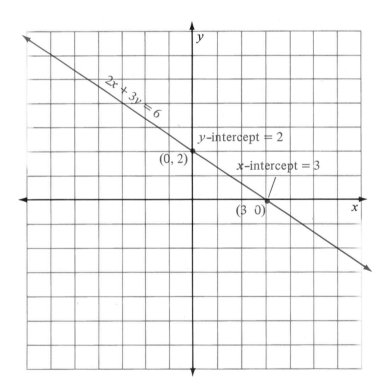

Figure 7-5

$$\text{When} \qquad y = 0$$
$$\text{we have} \quad 2x + 3(0) = 6$$
$$x = 3$$

The x-intercept is 3, so the graph crosses the x-axis at the point (3, 0). We use these results to graph the solution set for $2x + 3y = 6$ (see Figure 7-5).   ▲

Graphing straight lines by finding the intercepts works best when the coefficients of x and y are factors of the constant term, as was the case in Example 4.

▼   **Example 5**   Graph the line $x = 3$ and the line $y = -2$.

Solution   The line $x = 3$ is the set of all points whose x-coordinate is 3. The variable y does not appear in the equation, so the y-coordinates can be any number.

The line $y = -2$ is the set of all points whose y-coordinate is -2. The x-coordinate can be any number.

Here are the graphs:

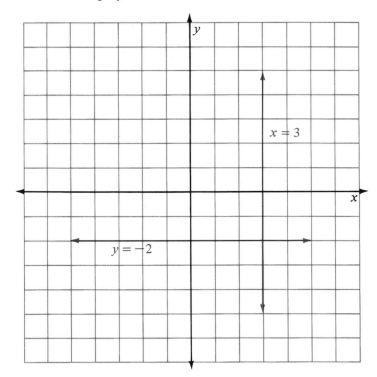

Figure 7-6   ▲

Problem Set 7.1

Graph each of the following ordered pairs on a rectangular coordinate system:

1. $(1, 2)$ 2. $(-1, 2)$
3. $(-1, -2)$ 4. $(1, -2)$
5. $(3, 4)$ 6. $(-3, -4)$
7. $(5, 0)$ 8. $(0, -3)$
9. $(0, 2)$ 10. $(4, 0)$
11. $(-5, -5)$ 12. $(-4, -1)$
13. $(\frac{1}{2}, 2)$ 14. $(3, \frac{1}{4})$
15. $(5, -2)$ 16. $(0, 4)$

Give the coordinates of each of the following points:

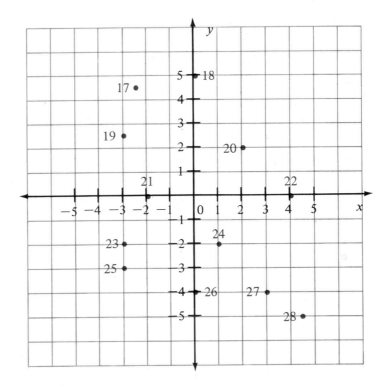

Graph each of the following linear equations by first finding the intercepts:

29. $2x - 3y = 6$ 30. $3x - 2y = 6$
31. $y + 2x = 4$ 32. $y - 2x = 4$
33. $4x - 5y = 20$ 34. $4x + 5y = 20$

35. $2x - y = 8$ 36. $x - 2y = 8$
37. $7x - 2y = 14$ 38. $2x + 7y = 14$
39. $3x + 5y = 15$ 40. $5x - 3y = 15$
41. $2x + 5y = 10$ 42. $2x - 5y = 10$

Graph each of the following straight lines:

43. $y = 2x + 3$ 44. $y = 3x - 2$
45. $2y = 4x - 8$ 46. $2y = 4x + 8$
47. $x - y = 5$ 48. $x + y = 6$
49. $2x - y = 3$ 50. $3x - y = 2$
51. $y = 5$ 52. $y = -3$
53. $x = -7$ 54. $x = 2$
55. $y = 4x - 1$ 56. $y = 3x + 3$
57. $y = 5x + 1$ 58. $y = 4x + 3$
59. $y = \frac{1}{2}x + 1$ 60. $y = \frac{1}{3}x + 1$
61. $y = -\frac{1}{2}x + 2$ 62. $y = -\frac{1}{3}x + 2$

63. If the perimeter of a rectangle is 12 meters, then the relationship between
     the length l and width w can be written

$$2l + 2w = 12$$

Graph this equation on a rectangular coordinate system in which the
horizontal axis is labeled l and the vertical axis is labeled w.

64. The perimeter of a rectangle is 10 inches. Graph the equation that de-
     scribes the relationship between the length l and the width w.

65. Complete the following ordered pairs so they are solutions to $y = |x + 2|$.
     Then graph the equation by connecting the points in the way that makes
     the most sense to you.
     $(-5,\ )$ $(-4,\ )$ $(-3,\ )$ $(-2,\ )$ $(-1,\ )$ $(0,\ )$ $(1,\ )$

66. Complete each ordered pair below so they are solutions to $y = |x - 2|$.
     Then use these points to graph $y = |x - 2|$.
     $(-1,\ )$ $(0,\ )$ $(1,\ )$ $(2,\ )$ $(3,\ )$ $(4,\ )$ $(5,\ )$

Review Problems   The problems below review material we covered in Section
4.6.

Solve each equation.

67. $\dfrac{a + 8}{5} = \dfrac{5}{4}$ 68. $\dfrac{a - 5}{2} = -4$

69. $\dfrac{8}{x + 2} = -\dfrac{8}{3}$ **70.** $\dfrac{-11}{x - 4} = -\dfrac{7}{3}$

71. $\dfrac{y^2 - 6y}{4} = -2$ **72.** $\dfrac{y^2 - y}{6} = 1$

7.2
The Slope of a Line

In defining the slope of a straight line, we are looking for a number to associate with a straight line that does two things. First of all, we want the slope of a line to measure the "steepness" of the line. That is, in comparing two lines, the slope of the steeper line should have the larger numerical value. Secondly, we want a line that *rises* going from left to right to have a *positive* slope. We want a line that *falls* going from left to right to have a *negative* slope. (A line that neither rises nor falls going from left to right must, therefore, have 0 slope.)

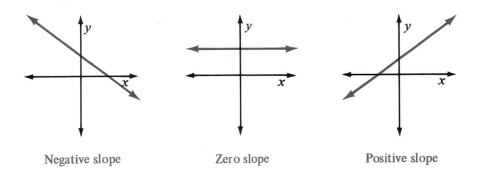

Negative slope   Zero slope   Positive slope

Geometrically, we can define the *slope* of a line as the ratio of the vertical change to the horizontal change encountered when moving from one point to another on the line. The vertical change is sometimes called the *rise*. The horizontal change is called the *run*.

▼   **Example 1**   Find the slope of the line $y = 2x - 3$.

Solution   In order to use our geometric definition, we first graph $y = 2x - 3$. We then pick any two convenient points and find the ratio of rise to run.

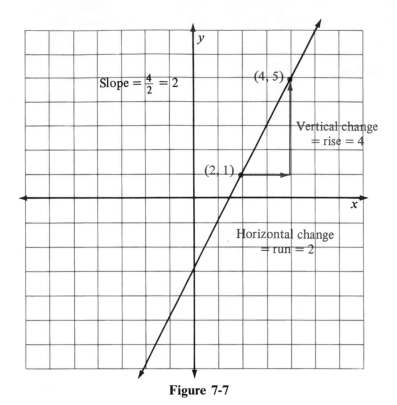

Figure 7-7

Our line has a slope of 2.   ▲

Notice that we can measure the vertical change by subtracting the y-coordinates of the two points shown: $5 - 1 = 4$. The horizontal change is the difference of the x-coordinates: $4 - 2 = 2$. This gives us a second way of defining the slope of a line. Algebraically, we say the slope of a line between two points whose coordinates are given is the ratio of the difference in the y-coordinates to the difference in the x-coordinates. We can summarize the above discussion by formalizing our definition for slope.

DEFINITION   The slope of the line between two points (x_1, y_1) and (x_2, y_2) is given by

$$\text{Slope} = m = \frac{\text{rise}}{\text{run}} = \frac{y_2 - y_1}{x_2 - x_1}$$

The letter m is usually used to designate slope. Our definition includes both the geometric form (rise/run) and the algebraic form $(y_2 - y_1)/(x_2 - x_1)$.

▼ **Example 2** Find the slope of the line through $(-2, -3)$ and $(-5, 1)$.

Solution

$$m = \frac{y_2 - y_1}{x_2 - x_1} = \frac{1 - (-3)}{-5 - (-2)} = \frac{4}{-3} = -\frac{4}{3}$$

Looking at the graph of the line between the two points, we can see our geometric approach does not conflict with our algebraic approach:

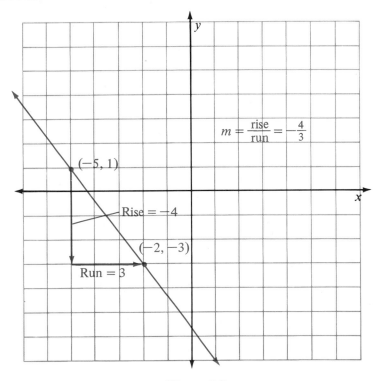

$$m = \frac{\text{rise}}{\text{run}} = -\frac{4}{3}$$

$(-5, 1)$

Rise $= -4$

$(-2, -3)$

Run $= 3$

Figure 7-8

We should note here that it does not matter which ordered pair we call (x_1, y_1) and which we call (x_2, y_2). If we were to reverse the order of subtraction of both the x- and y-coordinates in the above example, we would have

$$m = \frac{-3 - 1}{-2 - (-5)} = \frac{-4}{3} = -\frac{4}{3}$$

which is the same as our previous result.   ▲

▼ **Example 3** Find the slope of the line containing $(3, -1)$ and $(3, 4)$.

Solution Using the definition for slope, we have

$$m = \frac{-1 - 4}{3 - 3} = \frac{-5}{0}$$

The expression $-5/0$ is undefined. That is, there is no real number to associate with it. In this case, we say the line has *no slope*.

The graph of our line is as follows:

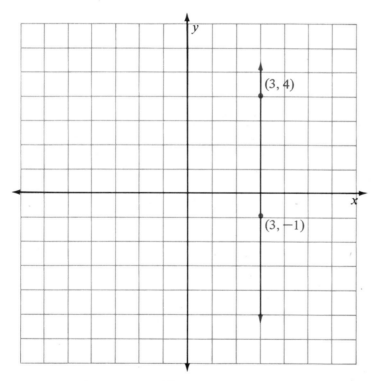

Figure 7-9

Our line with no slope is a vertical line. All vertical lines have no slope. (And all horizontal lines, as was mentioned earlier, have 0 slope.)   ▲

In geometry we call lines in the same plane that never intersect parallel. In order for two lines to be nonintersecting, they must rise or fall at the same rate. That is, the ratio of vertical change to horizontal change must

Slope of Parallel and Perpendicular Lines

be the same for each line. In other words, two lines are *parallel* if and only if they have the *same slope*.

Although it is not as obvious, it is also true that two nonvertical lines are *perpendicular* if and only if the *product of their slopes is* -1. This is the same as saying their slopes are negative reciprocals.

We can state these facts with symbols as follows.

If line l_1 has slope m_1, and line l_2 has slope m_2, then

$$l_1 \text{ and } l_2 \text{ are parallel} \Leftrightarrow m_1 = m_2$$

and

$$l_1 \text{ and } l_2 \text{ are perpendicular} \Leftrightarrow m_1 \cdot m_2 = -1$$

$$\left(\text{or } m_1 = \frac{-1}{m_2} \right)$$

To clarify this, if a line has a slope of $\frac{2}{3}$, then any line parallel to it has a slope of $\frac{2}{3}$. Any line perpendicular to it has a slope of $-\frac{3}{2}$ (the negative reciprocal of $\frac{2}{3}$).

▼   **Example 4**   Find a if the line through $(3, a)$ and $(-2, -8)$ is perpendicular to a line with slope $-\frac{4}{5}$.

Solution   The slope of the line through the two points is

$$m = \frac{a - (-8)}{3 - (-2)} = \frac{a + 8}{5}$$

Since the line through the two points is perpendicular to a line with slope $-\frac{4}{5}$, we can also write its slope as $\frac{5}{4}$:

$$\frac{a + 8}{5} = \frac{5}{4}$$

Multiplying both sides by 20, we have

$$4(a + 8) = 5 \cdot 5$$
$$4a + 32 = 25$$
$$4a = -7$$
$$a = -\frac{7}{4}$$ ▲

Find the slope of each of the following lines from the given graph: *Problem Set 7.2*

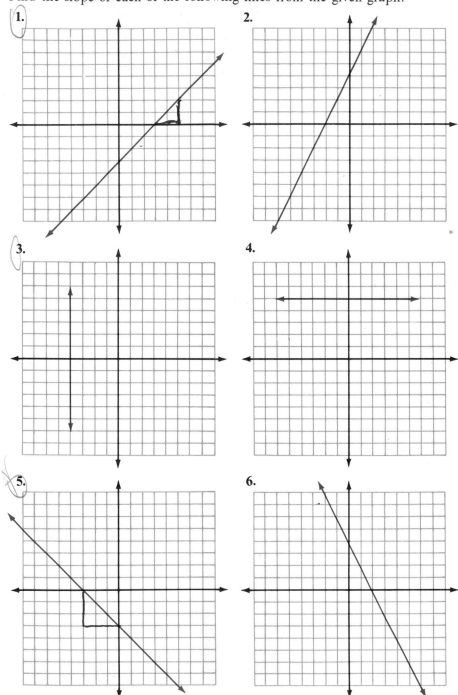

Find the slope of the line through the following pairs of points:

7. (4, 1), (2, 5)   8. (7, 3), (2, −2)
9. (5, −1), (4, 3)   10. (2, −3), (5, 6)
11. (1, −2), (3, −5)   12. (4, −1), (5, −3)
13. (1, 7), (1, 2)   14. (−3, −4), (−3, −1)
15. (5, 4), (−1, 4)   16. (2, −1), (6, −1)
17. $(\frac{1}{2}, 3), (\frac{2}{3}, -\frac{1}{4})$ 18. $(\frac{3}{4}, \frac{5}{6}), (-\frac{1}{2}, \frac{1}{3})$

Solve for the indicated variable if the line through the two given points has the given slope:

19. (5, a), (4, 2); $m = 3$ 20. (3, a), (1, 5); $m = -4$
21. (2, 6), (3, y); $m = -7$ 22. (−4, 9), (−5, y); $m = 3$
23. (−2, −3), (x, 5); $m = -\frac{8}{3}$ 24. (4, 9), (x, −2); $m = -\frac{7}{3}$
25. (x, 1), (8, 9); $m = \frac{1}{2}$ 26. (x, 4), (1, −3); $m = \frac{2}{3}$

27. Find the slope of any line parallel to the line through ($\sqrt{2}$, 3) and $(-\sqrt{8}, 1)$.
28. Find the slope of any line parallel to the line through (2, $\sqrt{27}$) and $(5, -\sqrt{3})$.
29. Line l contains the points $(5\sqrt{2}, -6)$ and $(\sqrt{50}, 2)$. Give the slope of any line perpendicular to l.
30. Line l contains points $(3\sqrt{6}, 4)$ and $(-3\sqrt{24}, 1)$. Give the slope of any line perpendicular to l.
31. Line l has a slope of $\frac{2}{3}$. A horizontal change of 12 will always be accompanied by how much of a vertical change?
32. For any line with slope $\frac{4}{5}$, a vertical change of 8 is always accompanied by how much of a horizontal change?
33. The line through $(2, y^2)$ and $(1, y)$ is perpendicular to a line with slope $-\frac{1}{6}$. What are the possible values for y?
34. The line through $(7, y^2)$ and $(3, 6y)$ is parallel to a line with slope -2. What are the possible values for y?
35. A pile of sand at a construction site is in the shape of a cone. If the slope of the side of the pile is $\frac{2}{3}$ and the pile is 8 feet high, how wide is the diameter of the base of the pile?
36. The slope of the sides of one of the Great Pyramids in Egypt is 13/10. If the base of the pyramid is 750 feet, how tall is the pyramid?
37. Graph the lines $y = 2x + 3$ and $y = 2x - 1$ on the same coordinate system. For each one, use the graph to name the slope and y-intercept.
38. Graph the lines $y = 2x + 1$ and $y = -2x + 1$ on the same coordinate system. For each one, use the graph to name the slope and y-intercept.

Review Problems   The problems below review material we covered in Section 2.5. Reviewing these problems will help you with some parts of the next section.

Solve each formula for y.

39. $3x - 2y = 6$	**40.** $2x + 3y = 6$
41. $2x + 3y = 5$	**42.** $3x + 2y = 5$
43. $y + 2 = 5(x - 3)$	**44.** $y - 4 = 2(x + 3)$
45. $y - 5 = \frac{4}{5}(x - 5)$	**46.** $y - 3 = -\frac{1}{2}(x + 2)$

In the first section of this chapter we defined the y-intercept of a line to be the y-coordinate of the point where the graph crosses the y-axis. We can use this definition, along with the definition of slope from the preceding section, to derive the slope-intercept form of the equation of a straight line.

Suppose line l has slope m and y-intercept b. What is the equation of l?

Since the y-intercept is b, we know that the point $(0, b)$ is on the line. If (x, y) is any other point on l, then using the definition for slope, we have

$$\frac{y - b}{x - 0} = m \qquad \text{Definition of slope}$$

$$y - b = mx \qquad \text{Multiply both sides by } x$$

$$y = mx + b \qquad \text{Add } b \text{ to both sides}$$

This last equation is known as the *slope-intercept form* of the equation of a straight line.

The equation of any line with slope m and y-intercept b is given by

$$y = mx + b$$

Slope ↗   ↑ y-intercept

This form of the equation of a straight line is very useful. When the equation is in this form, the *slope* of the line is always the *coefficient of x*, and the *y-intercept* is always the *constant term*. Both slope and y-intercept are easy to identify.

▼   **Example 1**   Write the equation of the line with slope -2 and y-intercept 3.

Solution   Since $m = -2$ and $b = 3$, we have

7.3
The Equation of a
Straight Line

Slope-Intercept Form

$$y = mx + b$$
$$\text{or} \quad y = -2x + 3$$

It is just as easy as it seems. The equation of the line with slope -2 and y-intercept 3 is $y = -2x + 3$. ▲

▼ **Example 2**   Give the slope and y-intercept for the line $2x - 3y = 5$.

Solution   To use the slope-intercept form we must solve the equation for y in terms of x:

$$2x - 3y = 5$$
$$-3y = -2x + 5 \qquad \text{Add } -2x \text{ to both sides}$$
$$y = \tfrac{2}{3}x - \tfrac{5}{3} \qquad \text{Divide by } -3$$

The last equation has the form $y = mx + b$. The slope must be $m = \tfrac{2}{3}$, and the y-intercept is $b = -\tfrac{5}{3}$. ▲

▼ **Example 3**   Graph the equation $2x + 3y = 6$.

Solution   Although we could graph this equation using the methods developed in Section 7.1 (by finding ordered pairs that are solutions to the equation and drawing a line through their graphs), it is sometimes easier to graph a line using the slope-intercept form of the equation.

Solving the equation for y, we have

$$2x + 3y = 6$$
$$3y = -2x + 6 \qquad \text{Add } -2x \text{ to both sides.}$$
$$y = -\tfrac{2}{3}x + 2 \qquad \text{Divide by 3.}$$

The slope is $m = -\tfrac{2}{3}$ and the y-intercept is $b = 2$. Therefore, the point $(0, 2)$ is on the graph and the ratio rise/run going from $(0, 2)$ to any other point on the line is $-\tfrac{2}{3}$. If we start at $(0, 2)$ and move 2 units up (that's a rise of 2) and 3 units to the left (a run of -3), we will be at another point on the graph. (We could also go down 2 units and right 3 units and also be assured of ending up at another point on the line, since $2/-3$ is the same as $-2/3$.)

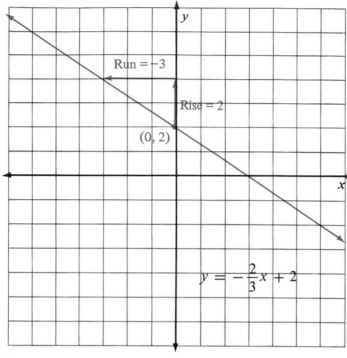

Figure 7-10   ▲

A second useful form of the equation of a straight line is the point-slope form.

Let line l contain the point (x_1, y_1) and have slope m. If (x, y) is any other point on l, then by the definition of slope we have

$$\frac{y - y_1}{x - x_1} = m$$

Multiplying both sides by $(x - x_1)$ gives us

$$(x - x_1) \cdot \frac{y - y_1}{x - x_1} = m(x - x_1)$$

$$y - y_1 = m(x - x_1)$$

This last equation is known as the *point-slope form* of the equation of a straight line.   *Point-Slope Form*

The equation of the line through (x_1, y_1) with slope m is given by
$$y - y_1 = m(x - x_1)$$
This form of the equation of a straight line is also very easy to work with.

▼   **Example 4** Find the equation of the line with slope 5 that contains the point $(3, -2)$.

   Solution   We have
$$
\begin{aligned}
(x_1, y_1) &= (3, -2) \quad \text{and} \quad m = 5 \\
y - y_1 &= m(x - x_1) \\
y + 2 &= 5(x - 3) \\
y + 2 &= 5x - 15 \\
y &= 5x - 17
\end{aligned}
$$

▲

▼   **Example 5** Find the equation of the line through $(-3, 1)$ and $(2, 5)$.

   Solution   We begin by using the two given points to find the slope of the line:

$$m = \frac{5 - 1}{2 - (-3)} = \frac{4}{5}$$

We can use either of the two given points in the point-slope form. Let's choose $(x_1, y_1) = (2, 5)$:

$$
\begin{aligned}
y - y_1 &= m(x - x_1) \\
y - 5 &= \tfrac{4}{5}(x - 2) \\
5y - 25 &= 4x - 8 \\
5y &= 4x + 17 \\
y &= \tfrac{4}{5}x + \tfrac{17}{5}
\end{aligned}
$$

▲

▼   **Example 6** Give the equation of the line through $(-1, 4)$ whose graph is perpendicular to the graph of $2x - y = -3$.

   Solution   To find the slope of $2x - y = -3$, we solve for y:
$$
\begin{aligned}
2x - y &= -3 \\
y &= 2x + 3
\end{aligned}
$$

   The slope of this line is 2. The line we are interested in is perpendicular to the line with slope 2 and must, therefore, have a slope of $-\tfrac{1}{2}$.

   Using $(x_1, y_1) = (-1, 4)$ and $m = -\tfrac{1}{2}$, we have

$$
\begin{aligned}
y - y_1 &= m(x - x_1) \\
y - 4 &= -\tfrac{1}{2}(x + 1) \\
2y - 8 &= -x - 1
\end{aligned}
$$

$$2y = -x + 7$$
$$y = -\tfrac{1}{2}x + \tfrac{7}{2}$$

Our answer is in slope-intercept form. If we were to write it in standard form, we would have

$$x + 2y = 7 \qquad\qquad \blacktriangle$$

As a final note, we should mention again that all horizontal lines have equations of the form $y = b$ and slopes of 0. Vertical lines have no slope and have equations of the form $x = a$. These two special cases do not lend themselves to either the slope-intercept form or the point-slope form (although the equation $y = b$ could be written in slope-intercept form as $y = 0x + b$).

Give the equation of the line with the following slope and y-intercept:

Problem Set 7.3

1. $m = 2, b = 3$
2. $m = -4, b = 2$
3. $m = 1, b = -5$
4. $m = -5, b = -3$
5. $m = \tfrac{1}{2}, b = \tfrac{3}{2}$
6. $m = \tfrac{2}{3}, b = \tfrac{5}{6}$
7. $m = 0, b = 4$
8. $m = 0, b = -2$
9. $m = -\sqrt{2}, b = 3\sqrt{2}$
10. $m = 2\sqrt{5}, b = -\sqrt{5}$

Give the slope and y-intercept for each of the following equations. Sketch the graph using the slope and y-intercept. Give the slope of any line perpendicular to the given line.

11. $y = 3x - 2$
12. $y = 2x + 3$
13. $2x - y = 4$
14. $3x + y = -2$
15. $2x - 3y = 12$
16. $3x - 2y = 12$
17. $4x + 5y = 20$
18. $5x - 4y = 20$
19. $3x - 5y = 10$
20. $4x - 3y = -9$

Find the equation of the line that contains the given point and has the given slope:

21. $(1, 3); m = 2$
22. $(3, -2); m = 4$
23. $(5, -1); m = -3$
24. $(6, -4); m = -2$
25. $(-2, 3); m = -\tfrac{1}{2}$
26. $(-1, -1); m = \tfrac{1}{4}$
27. $(\tfrac{1}{2}, -\tfrac{3}{2}); m = \tfrac{2}{3}$
28. $(\tfrac{2}{3}, -\tfrac{3}{4}); m = -\tfrac{3}{2}$

Find the equation of the line that contains the given pair of points:

29. $(2, 3), (1, 5)$
30. $(4, 6), (2, -2)$
31. $(-5, -8), (0, 2)$
32. $(2, -7), (-6, -3)$
33. $(3, -2), (1, 5)$
34. $(-4, 1), (-2, 4)$
35. $(0, 5), (-3, 0)$
36. $(0, -7), (4, 0)$
37. $(3, 5), (-2, 5)$
38. $(-8, 2), (4, 2)$

39. $(5, -1), (5, 4)$ **40.** $(-2, 1), (-2, 3)$

41. Give the slope and y-intercept, and sketch the graph, of $y = -2$.

42. Give the slope and y-intercept, and sketch the graph, of $y = 3\sqrt{2}$.

43. For the line $x = \sqrt{5}$ sketch the graph, give the slope, and name any intercepts.

44. For the line $x = -3$ sketch the graph, give the slope, and name any intercepts.

45. Find the equation of the line parallel to the graph of $3x - y = 5$ that contains the point $(-1, 4)$.

46. Find the equation of the line parallel to the graph of $2x - 4y = 5$ that contains the point $(0, 3)$.

47. Line l is perpendicular to the graph of $2x - 5y = 10$ and contains the point $(-4, -3)$. Find the equation for l.

48. Line l is perpendicular to the graph of $-3x - 5y = 2$ and contains the point $(2, -6)$. Find the equation for l.

49. Give the equation of the line perpendicular to $y = -4x + 2$ that has an x-intercept of -1.

50. Write the equation of the line parallel to the graph of $7x - 2y = 14$ that has an x-intercept of 5.

51. Give the equation of the line with x-intercept 3 and y-intercept -2.

52. Give the equation of the line with x-intercept $\frac{1}{2}$ and y-intercept $-\frac{1}{4}$.

53. Sound travels at approximately 1100 feet/second. It is for this reason that people who observe lightning strike the earth do not hear the sound (thunder) associated with the strike until after they have seen the flash. If we let t represent time (in seconds) and d represent distance (in feet), then we can write a linear equation in d and t that gives the relationship between the distance between us and the lightning strike and the time that elapses before we hear the thunder. If $d = 1100$ when $t = 1$, and $d = 2200$ when $t = 2$, find

  a. The equation that describes the relationship between d and t.
  b. How far away a lightning strike is if it takes 4 seconds before thunder is heard.
  c. How long it will be before we hear a lightning strike that is 1 mile (5280 feet) away.

54. The formula $F = \frac{9}{5}C + 32$ gives the relationship between the Fahrenheit and Celsius temperature scales. This formula can be derived experimentally. Suppose you are working in a laboratory and find that the temperature at which water freezes is $32°$ Fahrenheit and $0°$ Celsius, ($C = 0$ when $F = 32$) and that water boils at $212°$ Fahrenheit and $100°$ Celsius ($C = 100$ when $F = 212$).

  a. Give two ordered pairs of the form (C, F) that summarize the data on freezing and boiling temperatures.

   b. Use the ordered pairs found in part **a** to derive the formula
$F = \frac{9}{5}C + 32$.
   c. Use the formula found in part **b** to predict the Fahrenheit temperature reading of a liquid with a temperature of $50°$ Celsius.

55. Give the slope and y-intercept of the line $ax + by = c$.
56. Give the x-intercept of the line $y = mx + b$.

Review Problems   The problems below review material we covered in Section
5.4.

Multiply.

57. $(\sqrt{x} - 3)(\sqrt{x} + 5)$ **58.** $(\sqrt{x} - 2)(\sqrt{x} - 3)$
59. $(\sqrt{5} - 2)^2$ **60.** $(\sqrt{3} + 2)^2$

Rationalize the denominator.

61. $\dfrac{\sqrt{x}}{\sqrt{x} + 2}$ **62.** $\dfrac{\sqrt{x}}{\sqrt{x} - 3}$

63. $\dfrac{\sqrt{5} + \sqrt{2}}{\sqrt{5} - \sqrt{2}}$ **64.** $\dfrac{\sqrt{7} - \sqrt{3}}{\sqrt{7} + \sqrt{3}}$

A linear inequality in two variables is any expression that can be put in
the form

$$ax + by < c$$

where a, b, and c are real numbers (a and b not both 0). The inequality
symbol can be any one of the following four: $<, \le, >, \ge$.
   Some examples of linear inequalities are

$$2x + 3y < 6 \qquad y \ge 2x + 1 \qquad x - y \le 0$$

   Although not all of the above have the form $ax + by < c$, each one
can be put in that form.

   The solution set for a linear inequality is a section of the coordinate
plane. The boundary for the section is found by replacing the inequality
symbol with an equal sign and graphing the resulting equation. The
boundary is included in the solution set (and represented with a solid
line) if the inequality symbol used originally is $\le$ or $\ge$. The boundary
is not included (and is represented with a dotted line) if the original
symbol is $<$ or $>$.
   Let's look at some examples.

7.4

**Linear Inequalities in
Two Variables**

▼   Example 1   Graph the solution set for $x + y \leq 4$.

Solution   The boundary for the graph is the graph of $x + y = 4$;
the x- and y-intercepts are both 4. (Remember, the x-intercept is
found by letting $y = 0$, and the y-intercept by letting $x = 0$.) The
boundary is included in the solution set because the inequality
symbol is $\leq$.

Here is the graph of the boundary:

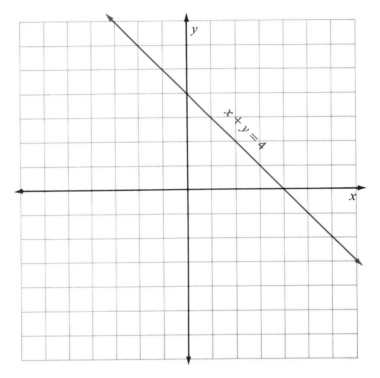

Figure 7-11

The boundary separates the coordinate plane into two sections or
regions—the region above the boundary and the region below the
boundary. The solution set for $x + y \leq 4$ is one of these two regions
along with the boundary. To find the correct region, we simply
choose any convenient point that is *not* on the boundary. We then
substitute the coordinates of the point into the original inequality
$x + y \leq 4$. If the point we choose satisfies the inequality, then it is
a member of the solution set, and we can assume that all points
on the same side of the boundary as the chosen point are also in
the solution set. If the coordinates of our point do not satisfy the ori-

ginal inequality, then the solution set lies on the other side of the boundary.

In this example, a convenient point off the boundary is the origin.

$$\begin{aligned} \text{Substituting} \quad & (0, 0) \\ \text{into} \quad & x + y \le 4 \\ \text{gives us} \quad & 0 + 0 \le 4 \\ & 0 \le 4 \quad \leftarrow \text{A true statement} \end{aligned}$$

Since the origin is a solution to the inequality $x + y \le 4$, and the origin is below the boundary, all other points below the boundary are also solutions.

Here is the graph of $x + y \le 4$:

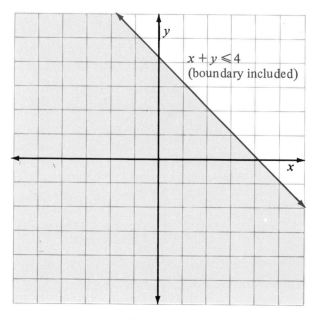

Figure 7-12

The region above the boundary is described by the inequality $x + y > 4$.

▼ **Example 2** Graph all linear inequalities with boundary $x + y = 4$.

Solution   Extending the results from Example 1, we have four different graphs with boundary $x + y = 4$. We have the region above the boundary and the region below the boundary. The boundary can be included in the solution set or not included in the solution set (see Figures 7-13 through 7-16).

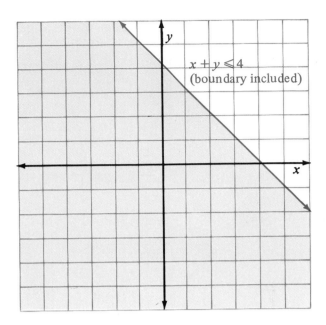

Figure 7-13

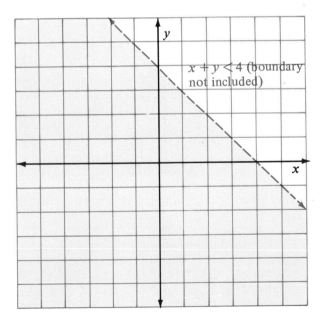

Figure 7-14

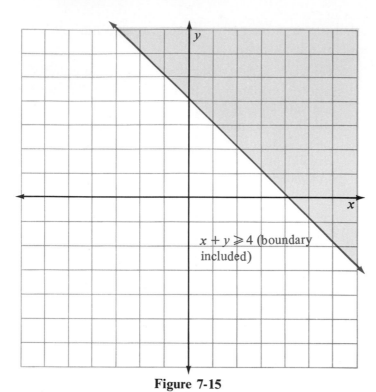

Figure 7-15

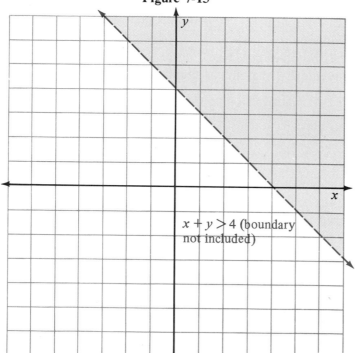

Figure 7-16

Here is a list of steps to follow when graphing the solution set for linear inequalities in two variables:

Step 1. Replace the inequality symbol with an equal sign. The resulting equation represents the boundary for the solution set.

Step 2. Graph the boundary found in step 1 using a *solid line* if the boundary is included in the solution set (that is, if the original inequality symbol was either $\leq$ or $\geq$). Use a *broken line* to graph the boundary if it is *not* included in the solution set. (It is not included if the original inequality was either $<$ or $>$.)

Step 3. Choose any convenient point not on the boundary and substitute the coordinates into the *original* inequality. If the resulting statement is *true*, the graph lies on the *same* side of the boundary as the chosen point. If the resulting statement is *false*, the solution set lies on the *opposite* side of the boundary.

▼ **Example 3** Graph the solution set for $y < 2x - 3$.

Solution The boundary is the graph of $y = 2x - 3$: a line with slope 2 and y-intercept -3. The boundary is not included since the original inequality symbol is $<$. Therefore, we use a broken line to represent the boundary (see Figure 7-17).

A convenient test point is again the origin:

$$\begin{array}{ll} \text{Using} & (0, 0) \\ \text{in} & y < 2x - 3, \\ \text{we have} & 0 < 2(0) - 3 \\ & 0 < -3 \qquad \leftarrow \text{A false statement} \end{array}$$

Since our test point gives us a false statement and it lies above the boundary, the solution set must lie on the other side of the boundary.

The complete graph is shown in Figure 7-18.   ▲

▼ **Example 4** Graph the solution set for $x \leq 5$.

Solution The boundary is $x = 5$, which is a vertical line. All points to the left have x-coordinates less than 5 and all points to the right have x-coordinates greater than 5 (see Figure 7-19).   ▲

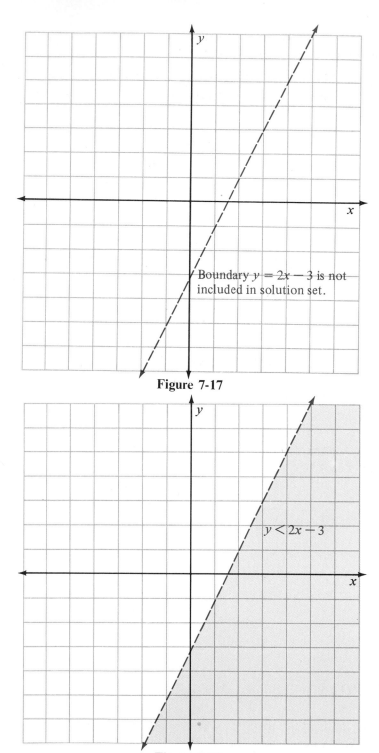

Boundary $y = 2x - 3$ is not included in solution set.

Figure 7-17

$y < 2x - 3$

Figure 7-18

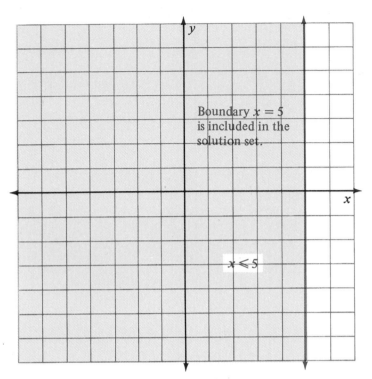

Boundary $x = 5$
is included in the
solution set.

$x \leqslant 5$

Figure 7-19

Problem Set 7.4

Graph the solution set for each of the following:

1. $x + y < 5$
3. $x - y \geq -3$
5. $2x + 3y \leq 6$
7. $x - 2y < 4$
9. $2x + y < 5$
11. $3x + 5y \geq 7$
13. $y < 2x - 1$
15. $y \geq -3x - 4$
17. $x > 3$
19. $y \leq 4$

2. $x + y \leq 5$
4. $x - y > -3$
6. $2x - 3y > -6$
8. $x + 2y > -4$
10. $2x + y < -5$
12. $3x - 5y > 7$
14. $y \geq 2x - 1$
16. $y < -3x + 4$
18. $x \leq -2$
20. $y > -5$

Find all graphs with the following boundaries. For each graph give the corresponding inequality. (There are four graphs for each boundary.)

21. $4x + 5y = 20$
23. $y = 2x + 5$

22. $3x + 4y = 12$
24. $y = 3x - 2$

25. Give the inequality whose graph lies above a boundary with slope -2 and y-intercept 4, if the boundary is included.

26. Give the inequality whose graph lies above a boundary with slope $+3$ and y-intercept -1, if the boundary is not included.
27. Give the inequality whose graph lies below a boundary with x-intercept 2 and y-intercept -3, if the boundary is not included.
28. Give the inequality whose graph lies below a boundary with x-intercept -3 and y-intercept 2, if the boundary is included.
29. Graph the inequality $y \leq |x|$ by first graphing the boundary $y = |x|$ using $x = -3, -2, -1, 0, 1, 2,$ and 3.
30. Graph the inequality $y > |x|$.
31. Graph the inequalities $x > -2$ and $x < 3$ on the same coordinate system. Write a continued inequality that gives the intersection of the two graphs.
32. Graph the inequalities $y > 0$ and $y < 4$ on the same coordinate system. Write a continued inequality that gives the intersection of these two graphs.

Review Problems   The problems below review material we covered in Sections 5.5 and 6.1.

Solve each equation.

33.	$\sqrt{3x + 4} = 5$	34.	$\sqrt{3x + 1} = -4$
35.	$\sqrt{2x - 1} - 3 = 2$	36.	$\sqrt{3x + 6} + 2 = 5$
37.	$\sqrt{x + 15} = x + 3$	38.	$\sqrt{x + 3} = x - 3$
39.	$\sqrt{2x + 9} = x + 5$	40.	$\sqrt{2x + 13} = x + 7$

There are two main types of variation—direct variation and inverse variation. Variation problems are most common in the sciences, particularly in chemistry and physics. They are also found in nonscience subjects such as business and economics. For the most part, variation problems are just a matter of translating specific English phrases into algebraic equations.

**7.5
Variation**

We say the variable y varies *directly* with the variable x if y increases as x increases (which is the same as saying y decreases as x decreases). A change in one variable produces a corresponding and similar change in the other variable. Algebraically, the relationship is written $y = Kx$ where K is a nonzero constant called the *constant of variation* (proportionality constant).

Direct Variation

Another way of saying y varies directly with x is to say y is *directly proportional* to x.

Study the following list. It gives the mathematical equivalent of some direct-variation statements.

English phrase	Algebraic equation
y varies directly with x	$y = Kx$
s varies directly with the square of t	$s = Kt^2$
y is directly proportional to the cube of z	$y = Kz^3$
u is directly proportional to the square root of v	$u = K\sqrt{v}$

Here are some sample problems involving direct-variation statements.

▼ **Example 1** y varies directly with x. If y is 15 when x is 5, find y when x is 7.

Solution The first sentence gives us the general relationship between x and y. The equation equivalent to the statement "y varies directly with x" is

$$y = Kx$$

The first part of the second sentence in our example gives us the information necessary to evaluate the constant K:

$$
\begin{aligned}
\text{When} \quad & y = 15 \\
\text{and} \quad & x = 5, \\
\text{the equation} \quad & y = Kx \\
\text{becomes} \quad & 15 = K \cdot 5 \\
\text{or} \quad & K = 3
\end{aligned}
$$

The equation can now be written specifically as

$$y = 3x$$

Letting $x = 7$, we have

$$
\begin{aligned}
y &= 3 \cdot 7 \\
y &= 21
\end{aligned}
$$

▲

Almost every problem involving variation can be solved by the above procedure. That is, begin by writing a general variation equation that gives the relationship between the variables. Next, use the information in the problem to evaluate the constant of variation. This results in a specific variation equation. Finally, use the specific equation and the rest of the information in the problem to find the quantity asked for.

▼ **Example 2** The distance a body falls from rest toward the earth is

directly proportional to the square of the time it has been falling. If a body falls 64 feet in 2 seconds, how far will it fall in 3.5 seconds?

Solution We will let d = distance and t = time. Since distance is directly proportional to the square of time, we have

$$d = Kt^2$$

Next we evaluate the constant K:

$$
\begin{aligned}
\text{When} \quad & t = 2 \\
\text{and} \quad & d = 64, \\
\text{the equation} \quad & d = Kt^2 \\
\text{becomes} \quad & 64 = K(2)^2 \\
\text{or} \quad & 64 = 4K \\
\text{and} \quad & K = 16
\end{aligned}
$$

Specifically, then, the relationship between d and t is

$$d = 16t^2$$

Finally, we find d when $t = 3.5$:

$$
\begin{aligned}
d &= 16(3.5)^2 \\
d &= 16(12.25) \\
d &= 196
\end{aligned}
$$

After 3.5 seconds the body will have fallen 196 feet.   ▲

We should note in the above example that the equation $d = 16t^2$ is a specific example of the equation used in physics to find the distance a body falls from rest after a time t. In physics, the equation is $d = \frac{1}{2}gt^2$, where g is the acceleration of gravity. On earth the acceleration of gravity is $g = 32$ ft/sec². The equation, then, is $d = \frac{1}{2} \cdot 32t^2$ or $d = 16t^2$.

Inverse Variation

If two variables are related so that an *increase* in one produces a proportional *decrease* in the other, then the variables are said to *vary inversely*. If y varies inversely with x, then $y = K\dfrac{1}{x}$ or $y = \dfrac{K}{x}$. We can also say y is inversely proportional to x. The constant K is again called the *constant of variation or proportionality constant.*

The following list gives the relationship between some inverse-variation statements and their corresponding equations:

English phrase	Algebraic equation
y is inversely proportional to x	$y = \dfrac{K}{x}$
s varies inversely with the square of t	$s = \dfrac{K}{t^2}$
y is inversely proportional to x^4	$y = \dfrac{K}{x^4}$
z varies inversely with the cube root of t	$z = \dfrac{K}{\sqrt[3]{t}}$

The procedure used to solve inverse-variation problems is the same as that used to solve direct-variation problems.

▼   **Example 3** y varies inversely with the square of x. If y is 4 when x is 5, find y when x is 10.

Solution   Since y is inversely proportional to the square of x, we can write

$$y = \frac{K}{x^2}$$

Evaluating K using the information given, we have

$$\text{When} \quad x = 5$$
$$\text{and} \quad y = 4,$$
$$\text{the equation} \quad y = \frac{K}{x^2}$$
$$\text{becomes} \quad 4 = \frac{K}{5^2}$$
$$\text{or} \quad 4 = \frac{K}{25}$$
$$\text{and} \quad K = 100$$

Now we write the equation again as

$$y = \frac{100}{x^2}$$

We finish by substituting $x = 10$ into the last equation:

$$y = \frac{100}{10^2}$$

$$y = \frac{100}{100}$$

$$y = 1 \qquad \blacktriangle$$

▼   **Example 4**   The volume of a gas is inversely proportional to the pressure of the gas on its container. If a pressure of 48 pounds per square inch corresponds to a volume of 50 cubic feet, what pressure is needed to produce a volume of 100 cubic feet?

Solution   We can represent volume with V and pressure by P:

$$V = \frac{K}{P}$$

Using $P = 48$ and $V = 50$, we have

$$50 = \frac{K}{48}$$

$$K = 50(48)$$

$$K = 2400$$

The equation that describes the relationship between P and V is

$$V = \frac{2400}{P}$$

Substituting $V = 100$ into this last equation, we get

$$100 = \frac{2400}{P}$$

$$100P = 2400$$

$$P = \frac{2400}{100}$$

$$P = 24$$

A volume of 100 cubic feet is produced by a pressure of 24 pounds per square inch.   ▲

The relationship between pressure and volume as given in the example above is known in chemistry as Boyle's law.

Many times relationships among different quantities are described in terms of more than two variables. If the variable y varies directly with *two* other variables, say x and z, then we say y varies *jointly* with x and z.

Joint Variation and Other Variation Combinations

In addition to joint variation, there are many other combinations of direct and inverse variation involving more than two variables. The following table is a list of some variation statements and their equivalent mathematical form:

English phrase	Algebraic equation
y varies jointly with x and z	$y = Kxz$
z varies jointly with r and the square of s	$z = Krs^2$
V is directly proportional to T and inversely proportional to P	$V = \dfrac{KT}{P}$
F varies jointly with m_1 and m_2 and inversely with the square of r	$F = \dfrac{Km_1 \cdot m_2}{r^2}$

▼   **Example 5** y varies jointly with x and the square of z. When x is 5 and z is 3, y is 180. Find y when x is 2 and z is 4.

Solution   The general equation is given by

$$y = Kxz^2$$

Substituting $x = 5$, $z = 3$, and $y = 180$, we have

$$180 = K(5)(3)^2$$
$$180 = 45K$$
$$K = 4$$

The specific equation is

$$y = 4xz^2$$

When $x = 2$ and $z = 4$, the last equation becomes

$$y = 4(2)(4)^2$$
$$y = 128$$ ▲

▼   **Example 6** In electricity, the resistance of a cable is directly proportional to its length and inversely proportional to the square of the diameter. If a 100-foot cable 0.5 inch in diameter has a resistance of 0.2 ohm, what will be the resistance of a cable made from the same material if it is 200 feet long with a diameter of 0.25 inch?

Solution   Let R = resistance, l = length, and d = diameter. The equation is

$$R = \frac{Kl}{d^2}$$

When $R = 0.2$, $l = 100$, and $d = 0.5$, the equation becomes

$$0.2 = \frac{K(100)}{(0.5)^2}$$

or $K = 0.0005$

Using this value of K in our original equation, the result is

$$R = \frac{0.0005l}{d^2}$$

When $l = 200$ and $d = 0.25$, the equation becomes

$$R = \frac{0.0005(200)}{(0.25)^2}$$

$$R = 1.6$$

The resistance is 1.6 ohms.   ▲

For the following problems, y varies directly with x: *Problem Set 7.5*

1. If y is 10 when x is 2, find y when x is 6.
2. If y is 20 when x is 5, find y when x is 3.
3. If y is -32 when x is 4, find x when y is -40.
4. If y is -50 when x is 5, find x when y is -70.
5. If y is 33 when x is 8, find y when x is 11.
6. If y is 26 when x is 3, find y when x is 4.

For the following problems, r is inversely proportional to s:

7. If r is -3 when s is 4, find r when s is 2.
8. If r is -10 when s is 6, find r when s is -5.
9. If r is 8 when s is 3, find s when r is 48.
10. If r is 12 when s is 5, find s when r is 30.
11. If r is 15 when s is 4, find r when s is 5.
12. If r is 21 when s is 10, find r when s is 8.

For the following problems, d varies directly with the square root of r:

13. If $d = 10$, when $r = 25$, find d when $r = 16$.
14. If $d = 12$, when $r = 36$, find d when $r = 49$.
15. If $d = 10\sqrt{2}$, when $r = 50$, find d when $r = 8$.
16. If $d = 6\sqrt{3}$, when $r = 12$, find d when $r = 75$.

For the following problems, y varies inversely with the square of x:

17. If $y = 45$ when $x = 3$, find y when x is 5.
18. If $y = 12$ when $x = 2$, find y when x is 6.
19. If $y = 10$ when $x = 6$, find x when y is 20.
20. If $y = 13$ when $x = 4$, find x when y is 30.

For the following problems, z varies jointly with x and the square of y:

21. If z is 54 when x and y are 3, find z when $x = 2$ and $y = 4$.
22. If z is 80 when x is 5 and y is 2, find z when $x = 2$ and $y = 5$.
23. If z is 64 when $x = 1$ and $y = 4$, find x when $z = 32$ and $y = 1$.
24. If z is 27 when $x = 6$ and $y = 3$, find x when $z = 50$ and $y = 4$.
25. The length a spring stretches is directly proportional to the force applied. If a force of 5 pounds stretches a spring 3 inches, how much force is necessary to stretch the same spring 10 inches?
26. The weight of a certain material varies directly with the surface area of that material. If 8 square feet weighs half a pound, how much will 10 square feet weigh?
27. The volume of a gas is inversely proportional to the pressure. If a pressure of 36 pounds per square inch corresponds to a volume of 25 cubic feet, what pressure is needed to produce a volume of 75 cubic feet?
28. The frequency of an electromagnetic wave varies inversely with the length. If a wave of length 200 meters has a frequency of 800 kilocycles per second, what frequency will be associated with a wave of length 500 meters?
29. The surface area of a hollow cylinder varies jointly with the height and radius of the cylinder. If a cylinder with radius 3 inches and height 5 inches has a surface area of 94 square inches, what is the surface area of a cylinder with radius 2 inches and height 8 inches?
30. The capacity of a cylinder varies jointly with the height and the square of the radius. If a cylinder with radius of 3 cm (centimeters) and a height of 6 cm has a capacity of 3 cm^3 (cubic centimeters), what will be the capacity of a cylinder with radius 4 cm and height 9 cm?
31. The resistance of a wire varies directly with the length and inversely with the square of the diameter. If 100 feet of wire with diameter 0.01 inch has a resistance of 10 ohms, what is the resistance of 60 feet of the same type of wire if its diameter is 0.02 inch?
32. The volume of a gas varies directly with its temperature and inversely with the pressure. If the volume of a certain gas is 30 cubic feet at a temperature of 300°K and a pressure of 20 pounds per square inch, what is the volume of the same gas at 340°K when the pressure is 30 pounds per square inch?

Review Problems   The problems below review material we covered in Sections 5.6 and 5.7.

Simplify:

33. $(11 - 6i) - (2 - 4i)$ **34.** $(5 + 8i) - (3 - 4i)$

Multiply:

35. $(2 + 3i)(4 - i)$ **36.** $(3 - 5i)(2 + i)$
37. $(3 - 2i)^2$ **38.** $(4 + 5i)^2$

Divide:

39. $\dfrac{2 + 3i}{2 - 3i}$ **40.** $\dfrac{2 - 3i}{2 + 3i}$

Chapter 7   Summary and Review

Examples

LINEAR EQUATIONS IN TWO VARIABLES [7.1]

A linear equation in two variables is any equation that can be put in the form $ax + by = c$. The graph of every linear equation is a straight line.

1. The equation $3x + 2y = 6$ is an example of a linear equation in two variables.

INTERCEPTS [7.1]

The x-intercept of an equation is the x-coordinate of the point where the graph crosses the x-axis. The y-intercept is the y-coordinate of the point where the graph crosses the y-axis. We find the y-intercept by substituting $x = 0$ into the equation and solving for y. The x-intercept is found by letting $y = 0$ and solving for x.

2. To find the x-intercept for $3x + 2y = 6$ we let $y = 0$ and get

$$3x = 6$$
$$x = 2$$

In this case the x-intercept is 2, and the graph crosses the x-axis at $(2, 0)$.

THE SLOPE OF A LINE [7.2]

The slope of the line containing points (x_1, y_1) and (x_2, y_2) is given by

$$\text{Slope} = m = \frac{\text{rise}}{\text{run}} = \frac{y_2 - y_1}{x_2 - x_1}.$$

Horizontal lines have 0 slope, and vertical lines have no slope.

Parallel lines have equal slopes, and perpendicular lines have slopes which are negative reciprocals.

3. The slope of the line through $(6, 9)$ and $(1, -1)$ is

$$m = \frac{9 - (-1)}{6 - 1} = \frac{10}{5} = 2$$

4. The equation of the line with slope 5 and y-intercept 3 is

$$y = 5x + 3$$

5. The equation of the line through $(3, 2)$ with slope -4 is

$$y - 2 = -4(x - 3)$$

which can be simplified to

$$y = -4x + 14$$

6. The graph of $x - y \leq 3$ is

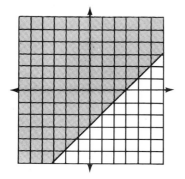

7. If y varies directly with x then

$$y = Kx$$

If also, y is 18 when x is 6, then

$$18 = K \cdot 6$$
or$$\qquad K = 3$$

So the equation can be written more specifically as

$$y = 3x$$

If we want to know what y is when x is 4, we simply substitute:

$$\begin{aligned} y &= 3 \cdot 4 \\ &= 12 \end{aligned}$$

THE SLOPE-INTERCEPT FORM OF A STRAIGHT LINE [7.3]

The equation of a line with slope m and y-intercept b is given by

$$y = mx + b.$$

THE POINT-SLOPE FORM OF A STRAIGHT LINE [7.3]

The equation of a line through (x_1, y_1) that has a slope of m can be written as

$$y - y_1 = m(x - x_1).$$

LINEAR INEQUALITIES IN TWO VARIABLES [7.4]

An inequality of the form $ax + by < c$ is a linear inequality in two variables. The equation for the boundary of the solution set is given by $ax + by = c$. (This equation is found by simply replacing the inequality symbol with an equal sign.)

To graph a linear inequality, first graph the boundary. Next, choose any point not on the boundary and substitute its coordinates into the original inequality. If the resulting statement is true, the graph lies on the same side of the boundary as the test point. A false statement indicates that the solution set lies on the other side of the boundary.

VARIATION [7.5]

If y varies directly with x (y is directly proportional to x), then we say:

$$y = Kx.$$

If y varies inversely with x (y is inversely proportional to x), then we say:

$$y = \frac{K}{x}.$$

If z varies jointly with x and y (z is directly proportional to both x and y), then we say:

$$z = Kxy.$$

In each case, K is called the constant of variation.

For each of the following straight lines, identify the x-intercept, y-intercept, and slope, and sketch the graph:

1. $2x + y = 6$
2. $3x - 2y = 5$
3. $y = -2x - 3$
4. $y = \frac{3}{2}x + 4$
5. $x = -2$
6. $y = 3$

Graph the following linear inequalities:

7. $x + y \geq 4$
8. $3x - 4y < 12$
9. $y \leq -x + 2$
10. $x > 5$

11. Give the equation of the line through $(-1, 3)$ that has slope $m = 2$.
12. Give the equation of the line through $(-3, 2)$ and $(4, -1)$.
13. Line l contains the point $(5, -3)$ and has a graph parallel to the graph of $2x - 5y = 10$. Find the equation for l.
14. Line l contains the point $(-1, -2)$ and has a graph perpendicular to the graph of $y = 3x - 1$. Find the equation for l.
15. Give the equation of the vertical line through $(4, -7)$.
16. Quantity y varies directly with the square of x. If y is 50 when x is 5, find y when x is 3.
17. Quantity z varies jointly with x and the square root of y. If z is 15 when x is 5 and y is 36, find z when x is 2 and y is 25.
18. The maximum load (L) a horizontal beam can safely hold varies jointly with the width (w) and the square of the depth (d) and inversely with the length (l). If a 10-foot beam with width 3 and depth 4 will safely hold up to 800 pounds load, how many pounds will a 12-foot beam with width 3 and depth 4 hold?

Systems of
Linear Equations

To the student:

In Chapter 7 we did some work with linear equations in two variables. In this chapter we will extend our work with linear equations to include systems of linear equations in two and three variables.

Systems of linear equations are used extensively in many different disciplines. Systems of linear equations can be used to solve multiple-loop circuit problems in electronics, kinship patterns in anthropology, genetics problems in biology, and profit-and-cost problems in economics. There are many other applications as well.

We will begin this chapter by looking at three different methods of solving linear systems in two variables. We will then extend two of these methods to include solutions to systems in three variables. A fourth method of solving linear systems involves what are known as determinants. The chapter ends with a look at some word problems whose solutions depend on linear systems.

To be successful in this chapter you should be familiar with the concepts in Chapter 7 as well as the process of solving a linear equation in one variable.

In Chapter 7 we found the graph of an equation of the form $ax + by = c$ to be a straight line. Since the graph is a straight line, the equation is said to be a linear equation. Two linear equations considered together form a *linear system* of equations. For example,

$$3x - 2y = 6$$
$$2x + 4y = 20$$

is a linear system. The solution set to the system is the set of all ordered pairs that satisfy both equations. If we graph each equation on the same set of axes, we can see the solution set (see Figure 8-1).

The point $(4, 3)$ lies on both lines and therefore must satisfy both equations. It is obvious from the graph that it is the only point that does so. The solution set for the system is $\{(4, 3)\}$.

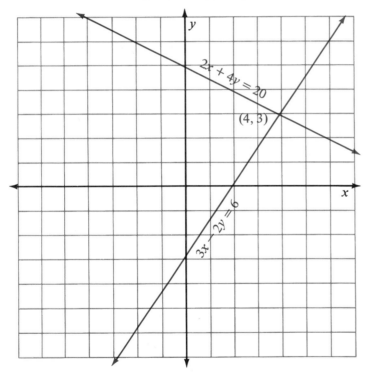

Figure 8-1

More generally, if $a_1x + b_1y = c_1$ and $a_2x + b_2y = c_2$ are linear equations, then the solution set for the system

$$a_1x + b_1y = c_1$$
$$a_2x + b_2y = c_2$$

can be illustrated through one of the following graphs:

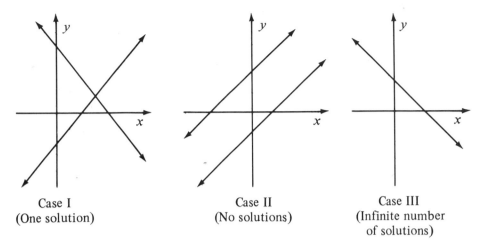

| Case I | Case II | Case III |
| (One solution) | (No solutions) | (Infinite number of solutions) |

Case I. The two lines intersect at one and only one point. The coordinates of the point give the solution to the system. This is what usually happens.

Case II. The lines are parallel and therefore have no points in common. The solution set to the system is $\varnothing$. In this case, we say the equations are *inconsistent*.

Case III. The lines coincide. That is, their graphs represent the same line. The solution set consists of all ordered pairs that satisfy either equation. In this case, the equations are said to be *dependent*.

In the beginning of this section we found the solution set for the system

$$3x - 2y = 6$$
$$2x + 4y = 20$$

by graphing each equation and then reading the solution set from the graph. Solving a system of linear equations by graphing is the least accurate method. If the coordinates of the point of intersection are not integers, it can be very difficult to read the solution set from the graph. There is another method of solving a linear system that does not depend on the graph. It is called the *elimination method* and depends on the following theorem.

THEOREM 8.1   If the ordered pair (x, y) satisfies the equations in the system

$$a_1 x + b_1 y = c_1$$
$$a_2 x + b_2 y = c_2$$

then it will also satisfy the equation

$$(a_1 + a_2)x + (b_1 + b_2)y = c_1 + c_2$$

Although Theorem 8.1 may look complicated, it simply indicates that the single equation that results from adding the left and right sides of the equations in our system has the same solution as the original system.

Here are some examples illustrating how we use Theorem 8.1 to find the solution set for a system of linear equations.

▼ **Example 1** Solve the system

$$4x + 3y = 10$$
$$2x + y = 4$$

Solution If we multiply the bottom equation by -3, the coefficients of y in the resulting equation and the top equation will be opposites:

$$
\begin{array}{ll}
4x + 3y = 10 & \xrightarrow{\text{No change}} \quad 4x + 3y = 10 \\
2x + y = 4 & \xrightarrow{\text{Multiply by } -3} \quad -6x - 3y = -12
\end{array}
$$

Adding the left and right sides of the resulting equations, according to Theorem 8.1, we have

$$
\begin{array}{r}
4x + 3y = 10 \\
-6x - 3y = -12 \\
\hline
-2x = -2
\end{array}
$$

The result is a linear equation in one variable. We have eliminated the variable y from the equations. (It is for this reason we call this method of solving a linear system the *elimination method*.) Solving $-2x = -2$ for x, we have

$$x = 1$$

This is the x-coordinate of the solution to our system. To find the y-coordinate, we substitute $x = 1$ into any of the equations containing both the variables x and y. Let's try the second equation in

our original system:

$$2(1) + y = 4$$
$$2 + y = 4$$
$$y = 2$$

This is the y-coordinate of the solution to our system. The ordered pair $(1, 2)$ is the solution to the system.   ▲

▼ **Example 2** Solve the system

$$3x - 5y = -2$$
$$2x - 3y = 1$$

Solution We can eliminate either variable. Let's decide to eliminate the variable x. We can do so by multiplying the top equation by 2 and the bottom equation by -3, and then adding the left and right sides of the resulting equations:

$$
\begin{array}{l}
3x - 5y = -2 \quad \xrightarrow{\text{Multiply by 2}} \quad 6x - 10y = -4 \\
2x - 3y = 1 \quad \xrightarrow[\text{Multiply by } -3]{} \quad -6x + 9y = -3 \\
\hline
\qquad\qquad\qquad\qquad\qquad\qquad\qquad -y = -7 \\
\qquad\qquad\qquad\qquad\qquad\qquad\qquad\quad y = 7
\end{array}
$$

The y-coordinate of the solution to the system is 7. Substituting this value of y into any of the equations with both x- and y-variables gives $x = 11$. The solution to the system is $(11, 7)$. It is the only ordered pair that satisfies both equations.   ▲

▼ **Example 3** Solve the system

$$5x - 2y = 1$$
$$-10x + 4y = 3$$

Solution We can eliminate y by multiplying the first equation by 2 and adding the result to the second equation:

$$
\begin{array}{l}
5x - 2y = 1 \quad \xrightarrow{\text{Multiply by 2}} \quad 10x - 4y = 2 \\
-10x + 4y = 3 \quad \xrightarrow[\text{No change}]{} \quad -10x + 4y = 3 \\
\hline
\qquad\qquad\qquad\qquad\qquad\qquad\qquad\qquad 0 = 5
\end{array}
$$

The result is the false statement $0 = 5$, which indicates there is no solution to the system. The equations are said to be *inconsistent;*

their graphs are parallel lines. Whenever both variables have been eliminated and the resulting statement is false, the solution set for the system will be $\varnothing$.   ▲

▼ **Example 4** Solve the system

$$4x - 3y = 2$$
$$8x - 6y = 4$$

Solution Multiplying the top equation by -2 and adding, we can eliminate the variable x:

$$
\begin{array}{lcl}
4x - 3y = 2 & \xrightarrow{\text{Multiply by } -2} & -8x + 6y = -4 \\
8x - 6y = 4 & \xrightarrow[\text{No change}]{} & \underline{8x - 6y = 4} \\
& & 0 = 0
\end{array}
$$

Both variables have been eliminated and the resulting statement $0 = 0$ is true. In this case the lines coincide and the equations are said to be *dependent*. The solution set consists of all ordered pairs that satisfy either equation. We can write the solution set as $\{(x, y)|4x - 3y = 2\}$ or $\{(x, y)|8x - 6y = 4\}$.   ▲

The last two examples illustrate the two special cases in which the graphs of the equations in the system either coincide or are parallel. In both cases the left-hand sides of the equations were multiples of one another. In the case of the dependent equations, the right-hand sides were also multiples. We can generalize these observations as follows:
The equations in the system

$$a_1 x + b_1 y = c_1$$
$$a_2 x + b_2 y = c_2$$

will be inconsistent (their graphs are parallel lines) if

$$\frac{a_1}{a_2} = \frac{b_1}{b_2} \neq \frac{c_1}{c_2}$$

and will be dependent (their graphs will coincide) if

$$\frac{a_1}{a_2} = \frac{b_1}{b_2} = \frac{c_1}{c_2}$$

We end this section by considering another method of solving a linear system. The method is called the *substitution* method and is shown in the following example.

▼ **Example 5** Solve the system

$$2x - 3y = -6$$
$$y = 3x - 5$$

Solution The second equation tells us y *is* $3x - 5$. Substituting the expression $3x - 5$ for y in the first equation, we have

$$2x - 3(3x - 5) = -6$$

The result of the substitution is the elimination of the variable y. Solving the resulting linear equation in x as usual, we have

$$2x - 9x + 15 = -6$$
$$-7x + 15 = -6$$
$$-7x = -21$$
$$x = 3$$

Putting $x = 3$ into the second equation in the original system, we have

$$y = 3(3) - 5$$
$$= 9 - 5$$
$$= 4$$

The solution to the system is $(3, 4)$.   ▲

▼ **Example 6** Solve by substitution:

$$2x + 3y = 5$$
$$x - 2y = 6$$

Solution In order to use the substitution method we must solve one of the two equations for x or y. We can solve for x in the second equation by adding $2y$ to both sides:

$$x - 2y = 6$$
$$x = 2y + 6 \qquad \text{Add } 2y \text{ to both sides}$$

Substituting the expression $2y + 6$ for x in the first equation of our system, we have

$$2(2y + 6) + 3y = 5$$
$$4y + 12 + 3y = 5$$
$$7y + 12 = 5$$
$$7y = -7$$
$$y = -1$$

Using $y = -1$ in either equation in the original system, we find $x = 4$. The solution is $(4, -1)$.   ▲

Solve each system by graphing both equations on the same set of axes and then reading the solution from the graph:

Problem Set 8.1

1. $x + y = 3$
   $x - y = 3$

2. $x + y = -2$
   $x - y = 6$

3. $4x + 5y = 3$
   $2x - 5y = 9$

4. $6x - y = 4$
   $2x + y = 4$

5. $x + y = 4$
   $-2x - 2y = 3$

6. $2x - y = 5$
   $4x - 2y = 10$

7. $y = 2x + 3$
   $y = 3x + 4$

8. $y = 3x - 1$
   $y = 2x + 1$

Solve each of the following systems by the elimination method:

9. $x + y = 5$
   $3x - y = 3$

10. $x - y = 4$
    $-x + 2y = -3$

11. $3x + y = 4$
    $4x + y = 5$

12. $6x - 2y = -10$
    $6x + 3y = -15$

13. $3x - 2y = 6$
    $6x - 4y = 12$

14. $4x + 5y = -3$
    $-8x - 10y = 3$

15. $3x - 5y = 7$
    $-x + y = -1$

16. $4x + 2y = 32$
    $x + y = -2$

17. $x + 2y = 0$
    $2x - y = 0$

18. $x + 3y = 9$
    $2x - y = 4$

19. $2x + y = 5$
    $5x + 3y = 11$

20. $5x + 2y = 11$
    $7x + y = 10$

21. $4x + 3y = 14$
    $9x - 2y = 14$

22. $7x - 6y = 13$
    $6x - 5y = 11$

23. $2x - 5y = 3$
    $-4x + 10y = 3$

24. $3x - 2y = 1$
    $-6x + 4y = -2$

25. $2x + 3y = 8$
    $3x - 4y = -5$

26. $3x - 4y = -1$
    $6x - 5y = 10$

27. $\frac{1}{2}x + \frac{1}{3}y = 13$
    $\frac{1}{5}x + \frac{1}{8}y = 5$

28. $\frac{1}{2}x + \frac{1}{3}y = \frac{2}{3}$
    $\frac{1}{3}x + \frac{1}{5}y = \frac{7}{15}$

29. $\frac{1}{3}x + \frac{1}{5}y = 2$
    $\frac{1}{3}x - \frac{1}{2}y = -\frac{1}{3}$

30. $\frac{1}{2}x - \frac{1}{3}y = \frac{5}{6}$
    $-\frac{1}{5}x + \frac{1}{4}y = -\frac{9}{20}$

Solve each of the following systems by the substitution method:

31. $y = x + 3$
    $x + y = 3$

32. $y = x - 5$
    $2x - 6y = -2$

33. $x - y = 4$
    $2x - 3y = 6$

34. $x + y = 3$
    $2x + 3y = -4$

35. $y = 3x - 2$
    $y = 4x - 4$
37. $7x - y = 24$
    $x = 2y + 9$
39. $2x - y = 5$
    $4x - 2y = 10$

36. $y = 5x - 2$
    $y = -2x + 5$
38. $3x - y = -8$
    $y = 6x + 3$
40. $5x - 4y = 3$
    $-10x + 8y = -6$

Solve each of the following systems by letting $a = 1/x$ and $b = 1/y$. Solve by any method you choose.

41. $\dfrac{2}{x} - \dfrac{1}{y} = 1$

$\dfrac{1}{x} + \dfrac{1}{y} = 2$

Hint: If we let $a = 1/x$ and $b = 1/y$, then this system becomes

$$2a - b = 1$$
$$a + b = 2$$

which can be solved very simply by either the elimination method or the substitution method.

42. $\dfrac{1}{x} + \dfrac{3}{y} = 9$

$\dfrac{1}{x} - \dfrac{1}{y} = 1$

43. $\dfrac{2}{x} - \dfrac{3}{y} = 3$

$\dfrac{2}{x} + \dfrac{2}{y} = 8$

44. $\dfrac{2}{x} - \dfrac{5}{y} = 10$

$\dfrac{3}{x} + \dfrac{1}{y} = 15$

45. Multiply both sides of the second equation in the following system by 100 and then solve as usual.

$$x + y = 10,000$$
$$.06x + .05y = 560$$

46. Multiply both sides of the second equation in the following system by 10 and then solve as usual.

$$x + y = 12$$
$$.20x + .50y = .30(12)$$

47. What value of c will make the following system a dependent system? (One in which the lines coincide.)

$$6x - 9y = 3$$
$$4x - 6y = c$$

48. What value of c will make the following system a dependent system?

$$5x - 7y = c$$
$$-15x + 21y = 9$$

49. One telephone company charges 25 cents for the first minute and 15 cents for each additional minute for a certain long-distance phone call. If the number of additional minutes after the first minute is x and the cost, in cents, for the call is y, then the equation that gives the total cost, in cents, for the call is $y = 15x + 25$.

   a. If a second phone company charges 22 cents for the first minute and 16 cents for each additional minute, write the equation that gives the total cost (y) of a call in terms of the number of additional minutes (x).

   b. After how many additional minutes will the two companies charge an equal amount? (What is the x-coordinate of the point of intersection of the two lines?)

50. In a certain city a taxi ride costs 75¢ for the first $\frac{1}{7}$ of a mile and 10¢ for every additional $\frac{1}{7}$ of a mile after the first seventh. If x is the number of additional sevenths of a mile, then the total cost y of a taxi ride is

$$y = 10x + 75$$

   a. How much does it cost to ride a taxi for 10 miles in this city?

   b. Suppose a taxi ride in another city costs 50¢ for the first $\frac{1}{7}$ of a mile, and 15¢ for each additional $\frac{1}{7}$ of a mile. Write an equation that gives the total cost y, in cents, to ride x sevenths of a mile past the first seventh, in this city.

   c. Solve the two equations given above simultaneously (as a system of equations) and explain in words what your solution represents.

Review Problems The problems below review material we covered in Section 5.3.

Combine the following radicals. (Assume all variables are positive.)

51. $5\sqrt{3} + 2\sqrt{3}$ **52.** $8\sqrt{2} - 6\sqrt{2}$

53. $2x\sqrt{5} - 7x\sqrt{5}$ **54.** $5x\sqrt{7} - 3x\sqrt{7}$

55. $3\sqrt{8} + 5\sqrt{18}$ **56.** $5\sqrt{12} + 3\sqrt{27}$

57. $5\sqrt[3]{16} - 4\sqrt[3]{54}$ **58.** $\sqrt[3]{81} + 3\sqrt[3]{24}$

A solution to an equation in three variables such as

$$2x + y - 3z = 6$$

is an ordered triple of numbers (x, y, z). For example, the ordered triples $(0, 0, -2)$, $(2, 2, 0)$, and $(0, 9, 1)$ are solutions to the equation $2x + y - 3z = 6$, since they produce a true statement when their coordinates are replaced for x, y, and z in the equation.

8.2
Systems of Linear Equations in Three Variables

In this section we are concerned with solution sets to systems of linear equations in three variables.

DEFINITION   The solution set for a system of three linear equations in three variables is the set of ordered triples that satisfy all three equations.

We can solve a system in three variables by methods similar to those developed in the last section.

▼   **Example 1**   Solve the system

$$\begin{aligned}
x + y + z &= 6 \qquad &(1) \\
2x - y + z &= 3 \qquad &(2) \\
x + 2y - 3z &= -4 \qquad &(3)
\end{aligned}$$

Solution   We want to find the ordered triple (x, y, z) that satisfies all three equations. We have numbered the equations so it will be easier to keep track of where they are and what we are doing.

There are many ways to proceed. The main idea is to take two different pairs of equations and eliminate the same variable from each pair. We begin by adding equations (1) and (2) to eliminate the y-variable. The resulting equation is numbered (4):

$$\begin{array}{lr}
x + y + z = 6 & (1) \\
\underline{2x - y + z = 3} & (2) \\
3x \quad\;\; + 2z = 9 & (4)
\end{array}$$

Adding twice equation (2) to equation (3) will also eliminate the variable y. The resulting equation is numbered (5):

$$\begin{array}{lr}
4x - 2y + 2z = 6 & \text{Twice (2)} \\
\underline{x + 2y - 3z = -4} & (3) \\
5x \quad\quad\;\; - z = 2 & (5)
\end{array}$$

Equations (4) and (5) form a linear system in two variables. By multiplying equation (5) by 2 and adding the result to equation (4), we will succeed in eliminating the variable z from the new pair of equations:

$$\begin{array}{lr}
3x + 2z = 9 & (4) \\
\underline{10x - 2z = 4} & \text{Twice (5)} \\
13x \quad\quad\; = 13 & \\
x = 1 &
\end{array}$$

Substituting $x = 1$ into equation (4), we have

$$3(1) + 2z = 9$$
$$2z = 6$$
$$z = 3$$

Using $x = 1$ and $z = 3$ in equation (1) gives us
$$1 + y + 3 = 6$$
$$y + 4 = 6$$
$$y = 2$$

The solution set for the system is the ordered triple $\{(1, 2, 3)\}$.   ▲

▼   **Example 2** Solve the system

$$
\begin{array}{ll}
2x + y - z = 3 & (1) \\
3x + 4y + z = 6 & (2) \\
2x - 3y + z = 1 & (3)
\end{array}
$$

Solution   It is easiest to eliminate z from the equations. The equation produced by adding (1) and (2) is

$$5x + 5y = 9 \qquad (4)$$

The equation that results from adding (1) and (3) is

$$4x - 2y = 4 \qquad (5)$$

Equations (4) and (5) form a linear system in two variables. We can eliminate the variable y from this system as follows:

$$
\begin{array}{l}
5x + 5y = 9 \quad \xrightarrow{\text{Multiply by 2}} \quad 10x + 10y = 18 \\
4x - 2y = 4 \quad \xrightarrow[\text{Multiply by 5}]{} \quad 20x - 10y = 20 \\
\hline
 \quad 30x = 38 \\
 x = \tfrac{38}{30} \\
 x = \tfrac{19}{15}
\end{array}
$$

Substituting $x = \tfrac{19}{15}$ into equation (5) or equation (4) and solving for y gives

$$y = \tfrac{8}{15}$$

Using $x = \tfrac{19}{15}$ and $y = \tfrac{8}{15}$ in equation (1), (2), or (3) and solving for z results in

$$z = \tfrac{1}{15}$$

The ordered triple that satisfies all three equations is $(\tfrac{19}{15}, \tfrac{8}{15}, \tfrac{1}{15})$. ▲

▼ **Example 3**   Solve the system

$$2x + 3y - z = 5 \qquad (1)$$
$$4x + 6y - 2z = 10 \qquad (2)$$
$$x - 4y + 3z = 5 \qquad (3)$$

Solution   Multiplying equation (1) by -2 and adding the result to equation (2) yields

$$\begin{array}{ll} -4x - 6y + 2z = -10 & -2 \text{ times (1)} \\ \underline{4x + 6y - 2z = 10} & (2) \\ 0 = 0 & \end{array}$$

All three variables have been eliminated, and we are left with a true statement. As was the case in Section 8.1, this implies that the two equations are dependent. There is no unique solution to the system.   ▲

▼ **Example 4**   Solve the system

$$x - 5y + 4z = 8 \qquad (1)$$
$$3x + y - 2z = 7 \qquad (2)$$
$$-9x - 3y + 6z = 5 \qquad (3)$$

Solution   Multiplying equation (2) by 3 and adding the result to equation (3) produces

$$\begin{array}{ll} 9x + 3y - 6z = 21 & 3 \text{ times (2)} \\ \underline{-9x - 3y + 6z = 5} & (3) \\ 0 = 26 & \end{array}$$

In this case all three variables have been eliminated, and we are left with a false statement. The two equations are inconsistent, there are no ordered triples that satisfy both equations. The solution set for the system is $\varnothing$. If equations (2) and (3) have no ordered triples in common, then certainly (1), (2), and (3) do not either.   ▲

Note   The graph of a linear equation in three variables is a plane in three-dimensional space. In Example 4 two of the equations are inconsistent, implying their graphs are parallel planes. In Example 3 we found two of the equations to be dependent, which implies that the two planes that represent the graphs of these equations coincide.

The graphs of dependent equations always coincide, while the graphs of inconsistent equations have no points in common.

▼   **Example 5** Solve the system

$$
\begin{aligned}
x + 3y &= 5 & (1) \\
6y + z &= 12 & (2) \\
x - 2z &= -10 & (3)
\end{aligned}
$$

Solution   It may be helpful to rewrite the system as

$$
\begin{aligned}
x + 3y &= 5 & (1) \\
6y + z &= 12 & (2) \\
x - 2z &= -10 & (3)
\end{aligned}
$$

Equation (2) does not contain the variable x. If we multiply equation (3) by -1 and add the result to equation (1), we will be left with another equation that does not contain the variable x:

$$
\begin{array}{ll}
x + 3y = 5 & (1) \\
\underline{-x \qquad\quad + 2z = 10} & -1 \text{ times (3)} \\
3y + 2z = 15 & (4)
\end{array}
$$

Equations (2) and (4) form a linear system in two variables. Multiplying equation (2) by -2 and adding to equation (4) eliminates the variable z:

$$
\begin{array}{ll}
6y + z = 12 & \xrightarrow{\text{Multiply by } -2} \quad -12y - 2z = -24 \\
3y + 2z = 15 & \xrightarrow[\text{No change}]{} \qquad\quad \underline{3y + 2z = \quad 15} \\
& \qquad\qquad\qquad\quad -9y \qquad\; = -9 \\
& \qquad\qquad\qquad\qquad\qquad y = 1
\end{array}
$$

Using $y = 1$ in equation (4) and solving for z, we have

$$ z = 6 $$

Substituting $y = 1$ into equation (1) gives

$$ x = 2 $$

The ordered triple that satisfies all three equations is $(2, 1, 6)$. ▲

Solve the following systems: *Problem Set 8.2*

1. $x + y + z = 4$
    $x - y + 2z = 1$
    $x - y - z = -2$

2. $x - y - 2z = -1$
    $x + y + z = 6$
    $x + y - z = 4$

3. $x + y + z = 6$
$x - y + 2z = 7$
$2x - y - z = 0$

4. $x + y + z = 0$
$x + y - z = 6$
$x - y + z = -4$

5. $x + 2y + z = 3$
$2x - y + 2z = 6$
$3x + y - z = 5$

6. $2x + y - 3z = -14$
$x - 3y + 4z = 22$
$3x + 2y + z = 0$

7. $2x - y - 3z = 1$
$x + 2y + 4z = 3$
$4x - 2y - 6z = 2$

8. $3x + 2y + z = 3$
$x - 3y + z = 4$
$-6x - 4y - 2z = 1$

9. $2x - y + 3z = 4$
$x + 2y - z = 2$
$4x + 3y + 2z = 9$

10. $6x - 2y + z = 5$
$3x + y + 3z = 7$
$x + 4y - z = 4$

11. $x - 4y + 3z = 2$
$2x - 8y + 6z = -1$
$3x - y + z = 8$

12. $5x - 2y + z = 6$
$2x + 3y - 4z = 2$
$4x + 6y - 8z = 4$

13. $x + y = 9$
$y + z = 7$
$x - z = 2$

14. $x - y = -3$
$x + z = 2$
$y - z = 7$

15. $2x + y = 2$
$y + z = 3$
$4x - z = 0$

16. $2x + y = 6$
$3y - 2z = -8$
$x + z = 5$

17. $3x + 4y = 15$
$2x - 5z = -3$
$4y - 3z = 9$

18. $6x - 4y = 2$
$3y + 3z = 9$
$2x - 5z = -8$

19. $2x - y + 2z = -8$
$3x - y - 4z = 3$
$x + 2y - 3z = 9$

20. $2x + 2y + 4z = 2$
$2x - y - z = 0$
$3x + y + 2z = 2$

21. In the following diagram of an electrical circuit, x, y, and z represent the amount of current (in amperes) flowing across the 5-ohm, 20-ohm, and 10-ohm resistors, respectively. (In circuit diagrams resistors are represented by ⏛ and potential differences by ⊣⊢.)

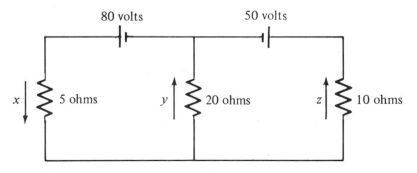

The system of equations used to find the three currents x, y, and z is

$$x - y - z = 0$$
$$5x + 20y = 80$$
$$20y - 10z = 50$$

Solve the system for all variables.

22. If a car rental company charges $10 a day and 8¢ a mile to rent one of its cars, then the cost z, in dollars, to rent a car for x days and drive y miles can be found from the equation

$$z = 10x + .08y$$

   a. How much does it cost to rent a car for 2 days and drive it 200 miles under these conditions?
   b. A second company charges $12 a day and 6¢ a mile for the same car. Write an equation that gives the cost z, in dollars, to rent a car from this company for x days and drive it y miles.
   c. A car is rented from each of the companies mentioned above for 2 days. To find the mileage at which the cost of renting the cars from each of the two companies will be equal, solve the following system for y.

$$z = 10x + .08y$$
$$z = 12x + .06y$$
$$x = 2$$

Review Problems   The problems below review material we covered in Sections 6.1 and 6.3. Reviewing these problems will help you with some of the material in the next section.

Solve by factoring.

23. $x^2 - 2x - 8 = 0$ 24. $x^2 - 5x - 6 = 0$
25. $6x^2 - 13x + 6 = 0$ 26. $9x^2 + 6x - 8 = 0$

Solve using the quadratic formula.

27. $x^2 + x + 1 = 0$ 28. $x^2 + x - 1 = 0$
29. $2x^2 - 6x + 3 = 0$ 30. $3x^2 - 4x + 6 = 0$

In this section we will expand and evaluate determinants. The purpose of this section is simply to be able to find the value of a given determinant. As we will see in the next section, determinants are very useful in solving systems of linear equations. Before we apply determinants to systems of linear equations, however, we must practice calculating the value of some determinants. The problems in this section are rather mechanical in nature. We will have to wait until the next section to see the practical applications of the determinants developed in this section.

**8.3
Introduction to
Determinants**

DEFINITION   The value of the 2 × 2 (2 by 2) determinant

$$\begin{vmatrix} a & c \\ b & d \end{vmatrix}$$

is given by

$$\begin{vmatrix} a & c \\ b & d \end{vmatrix} = ad - bc$$

From the definition above we see that a determinant is simply a square array of numbers with two vertical lines enclosing it. The value of a 2 × 2 determinant is found by cross-multiplying on the diagonals, a diagram of which looks like

$$\begin{vmatrix} a & c \\ b & d \end{vmatrix} = ad - bc$$

▼   Example 1   Find the value of the following 2 × 2 determinants:

a. $\begin{vmatrix} 1 & 2 \\ 3 & 4 \end{vmatrix} = 1(4) - 3(2) = 4 - 6 = -2$

b. $\begin{vmatrix} 3 & 5 \\ -2 & 7 \end{vmatrix} = 3(7) - (-2)5 = 21 + 10 = 31$ ▲

▼   Example 2   Solve for x if

$$\begin{vmatrix} -3 & x \\ 2 & x \end{vmatrix} = 20$$

Solution   Applying the definition of a determinant to expand the left side, we have

$$-3(x) - 2(x) = 20$$
$$-5x = 20$$
$$x = -4$$ ▲

▼   Example 3   Solve for x if

$$\begin{vmatrix} x^2 & 2 \\ x & 1 \end{vmatrix} = 8$$

Solution   We expand the determinant on the left side to get

$$x^2(1) - x(2) = 8$$
$$x^2 - 2x = 8$$
$$x^2 - 2x - 8 = 0$$
$$(x - 4)(x + 2) = 0$$
$$x - 4 = 0 \quad \text{or} \quad x + 2 = 0$$
$$x = 4 \quad \text{or} \quad x = -2$$

The solution set is $\{4, -2\}$.   ▲

We now turn our attention to 3×3 determinants. A 3×3 determinant is also a square array of numbers, the value of which is given by the following definition.

DEFINITION   The value of the 3×3 determinant

$$\begin{vmatrix} a_1 & b_1 & c_1 \\ a_2 & b_2 & c_2 \\ a_3 & b_3 & c_3 \end{vmatrix}$$

is given by

$$\begin{vmatrix} a_1 & b_1 & c_1 \\ a_2 & b_2 & c_2 \\ a_3 & b_3 & c_3 \end{vmatrix} = a_1b_2c_3 + a_3b_1c_2 + a_2b_3c_1 - a_3b_2c_1 - a_1b_3c_2 - a_2b_1c_3$$

At first glance, the expansion of a 3×3 determinant looks a little complicated. There are actually two different methods used to find the six products given above that simplify matters somewhat. The first method involves a process similar to the one used to find the value of a 2×2 determinant. It involves a cross-multiplication scheme.

We begin by writing the determinant with the first two columns   *Method 1*
repeated on the right:

$$\begin{vmatrix} a_1 & b_1 & c_1 \\ a_2 & b_2 & c_2 \\ a_3 & b_3 & c_3 \end{vmatrix} \begin{matrix} a_1 & b_1 \\ a_2 & b_2 \\ a_3 & b_3 \end{matrix}$$

The positive products in the definition come from multiplying down the three full diagonals:

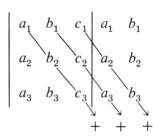

The negative products come from multiplying up the three full diagonals:

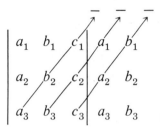

Check the products found by multiplying up and down the diagonals given here with the products given in the definition of a 3×3 determinant to see that they match.

▼   **Example 4**   Find the value of

$$\begin{vmatrix} 1 & 3 & -2 \\ 2 & 0 & 1 \\ 4 & -1 & 1 \end{vmatrix}$$

Solution   Repeating the first two columns and then finding the products up the diagonals and the products down the diagonals as given in Method 1, we have

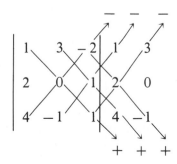

$$= 1(0)(1) + 3(1)(4) + (-2)(2)(-1)$$
$$- 4(0)(-2) - (-1)(1)(1) - 1(2)(3)$$
$$= 0 + 12 + 4 - 0 - (-1) - (6)$$
$$= 11 \qquad \blacktriangle$$

The second method of evaluating a 3×3 determinant is called *expansion by minors*.   *Method 2*

DEFINITION   The *minor* for an element in a 3×3 determinant is the determinant consisting of the elements remaining when the row and column to which the element belongs are deleted. For example, in the determinant

$$\begin{vmatrix} a_1 & b_1 & c_1 \\ a_2 & b_2 & c_2 \\ a_3 & b_3 & c_3 \end{vmatrix}$$

$$\text{Minor for element } a_1 = \begin{vmatrix} b_2 & c_2 \\ b_3 & c_3 \end{vmatrix}$$

$$\text{Minor for element } b_2 = \begin{vmatrix} a_1 & c_1 \\ a_3 & c_3 \end{vmatrix}$$

$$\text{Minor for element } c_3 = \begin{vmatrix} a_1 & b_1 \\ a_2 & b_2 \end{vmatrix}$$

Before we can evaluate a 3×3 determinant by Method 2, we must first define what is known as the sign array for a 3×3 determinant.

DEFINITION   The *sign array* for a 3×3 determinant is a 3×3 array of signs in the following pattern:

$$\begin{vmatrix} + & - & + \\ - & + & - \\ + & - & + \end{vmatrix}$$

The sign array begins with a $+$ sign in the upper left-hand corner. The signs then alternate between $+$ and $-$ across every row and down every column.

Note   If you have read this far and are confused, hang on. After you have done a couple of examples you will find expansion by minors to be a fairly simple process. It just takes a lot of writing to explain it.

We can evaluate a 3×3 determinant by expanding across any row or down any column as follows:

Step 1.   Choose a row or column to expand about.
Step 2.   Write the product of each element in the row or column chosen in step 1 with its minor.
Step 3.   Connect the three products in step 2 with the signs in the corresponding row or column in the sign array.

We will use the same determinant used in Example 4 to illustrate the procedure.

▼   **Example 5**   Expand across the first row:

$$\begin{vmatrix} 1 & 3 & -2 \\ 2 & 0 & 1 \\ 4 & -1 & 1 \end{vmatrix}$$

Solution   The products of the three elements in row 1 with their minors are

$$1\begin{vmatrix} 0 & 1 \\ -1 & 1 \end{vmatrix} \quad 3\begin{vmatrix} 2 & 1 \\ 4 & 1 \end{vmatrix} \quad (-2)\begin{vmatrix} 2 & 0 \\ 4 & -1 \end{vmatrix}$$

Connecting these three products with the signs from the first row of the sign array, we have

$$+1\begin{vmatrix} 0 & 1 \\ -1 & 1 \end{vmatrix} - 3\begin{vmatrix} 2 & 1 \\ 4 & 1 \end{vmatrix} + (-2)\begin{vmatrix} 2 & 0 \\ 4 & -1 \end{vmatrix}$$

We complete the problem by evaluating each of the three 2×2 determinants and then simplifying the resulting expression:

$$\begin{aligned} +1[0 &- (-1)] - 3(2 - 4) + (-2)(-2 - 0) \\ &= 1(1) - 3(-2) + (-2)(-2) \\ &= 1 + 6 + 4 \\ &= 11 \end{aligned}$$

The results of Examples 4 and 5 match. It makes no difference which method we use—the value of a 3×3 determinant is unique.   ▲

▼ **Example 6** Expand down column 2:

$$\begin{vmatrix} 2 & 3 & -2 \\ 1 & 4 & 1 \\ 1 & 5 & -1 \end{vmatrix}$$

Solution We connect the products of elements in column 2 and their minors with the signs from the second column in the sign array:

$$\begin{vmatrix} 2 & 3 & -2 \\ 1 & 4 & 1 \\ 1 & 5 & -1 \end{vmatrix} = -3\begin{vmatrix} 1 & 1 \\ 1 & -1 \end{vmatrix} + 4\begin{vmatrix} 2 & -2 \\ 1 & -1 \end{vmatrix} - 5\begin{vmatrix} 2 & -2 \\ 1 & 1 \end{vmatrix}$$

$$= -3(-1-1) + 4[-2-(-2)] - 5[2-(-2)]$$
$$= -3(-2) + 4(0) - 5(4)$$
$$= 6 + 0 - 20$$
$$= -14 \qquad \blacktriangle$$

Find the value of the following 2×2 determinants:

1. $\begin{vmatrix} 1 & 0 \\ 2 & 3 \end{vmatrix}$ **2.** $\begin{vmatrix} 5 & 4 \\ 3 & 2 \end{vmatrix}$

3. $\begin{vmatrix} 2 & 1 \\ 3 & 4 \end{vmatrix}$ **4.** $\begin{vmatrix} 4 & 1 \\ 5 & 2 \end{vmatrix}$

5. $\begin{vmatrix} 0 & 1 \\ 1 & 0 \end{vmatrix}$ **6.** $\begin{vmatrix} 1 & 0 \\ 0 & 1 \end{vmatrix}$

7. $\begin{vmatrix} -3 & 2 \\ 6 & -4 \end{vmatrix}$ **8.** $\begin{vmatrix} 8 & -3 \\ -2 & 5 \end{vmatrix}$

9. $\begin{vmatrix} -3 & -1 \\ 4 & -2 \end{vmatrix}$ **10.** $\begin{vmatrix} 5 & 3 \\ 7 & -6 \end{vmatrix}$

Solve each of the following for x:

11. $\begin{vmatrix} 2x & 1 \\ x & 3 \end{vmatrix} = 10$ **12.** $\begin{vmatrix} 3x & -2 \\ 2x & 3 \end{vmatrix} = 26$

13. $\begin{vmatrix} 1 & 2x \\ 2 & -3x \end{vmatrix} = 21$ **14.** $\begin{vmatrix} -5 & 4x \\ 1 & -x \end{vmatrix} = 27$

15. $\begin{vmatrix} 2x & -4 \\ 2 & x \end{vmatrix} = -8x$ **16.** $\begin{vmatrix} 3x & 2 \\ 2 & x \end{vmatrix} = -11x$

17. $\begin{vmatrix} x^2 & 3 \\ x & 1 \end{vmatrix} = 10$

18. $\begin{vmatrix} x^2 & -2 \\ x & 1 \end{vmatrix} = 35$

Find the value of each of the following 3×3 determinants by using Method 1 of this section:

19. $\begin{vmatrix} 1 & 2 & 0 \\ 0 & 2 & 1 \\ 1 & 1 & 1 \end{vmatrix}$

20. $\begin{vmatrix} -1 & 0 & 2 \\ 3 & 0 & 1 \\ 0 & 1 & 3 \end{vmatrix}$

21. $\begin{vmatrix} 1 & 2 & 3 \\ 3 & 2 & 1 \\ 1 & 1 & 1 \end{vmatrix}$

22. $\begin{vmatrix} -1 & 2 & 0 \\ 3 & -2 & 1 \\ 0 & 5 & 4 \end{vmatrix}$

Find the value of each determinant by using Method 2 and expanding across the first row:

23. $\begin{vmatrix} 0 & 1 & 2 \\ 1 & 0 & 1 \\ -1 & 2 & 0 \end{vmatrix}$

24. $\begin{vmatrix} 3 & -2 & 1 \\ 0 & -1 & 0 \\ 2 & 0 & 1 \end{vmatrix}$

25. $\begin{vmatrix} 3 & 0 & 2 \\ 0 & -1 & -1 \\ 4 & 0 & 0 \end{vmatrix}$

26. $\begin{vmatrix} 1 & 1 & 1 \\ 1 & -1 & 1 \\ 1 & 1 & -1 \end{vmatrix}$

Find the value of each of the following determinants:

27. $\begin{vmatrix} 2 & -1 & 0 \\ 1 & 0 & -2 \\ 0 & 1 & 2 \end{vmatrix}$

28. $\begin{vmatrix} 5 & 0 & -4 \\ 0 & 1 & 3 \\ -1 & 2 & -1 \end{vmatrix}$

29. $\begin{vmatrix} 1 & 3 & 7 \\ -2 & 6 & 4 \\ 3 & 7 & -1 \end{vmatrix}$

30. $\begin{vmatrix} 2 & 1 & 5 \\ 6 & -3 & 4 \\ 8 & 9 & -2 \end{vmatrix}$

31. Show that the determinant equation below is another way to write the slope-intercept form of the equation of a line.

$$\begin{vmatrix} y & x \\ m & 1 \end{vmatrix} = b$$

32. Show that the determinant equation below is another way to write the equation $F = \frac{9}{5}C + 32$.

$$\begin{vmatrix} C & F & 1 \\ 5 & 41 & 1 \\ -10 & 14 & 1 \end{vmatrix} = 0$$

Review Problems The problems below review material we covered in Section 6.4.

Solve each equation.

33. $x^4 - 2x^2 - 8 = 0$

34. $x^4 - 8x^2 - 9 = 0$

35. $x^{2/3} - 5x^{1/3} + 6 = 0$

36. $x^{2/3} - 3x^{1/3} + 2 = 0$

37. $2x - 5\sqrt{x} + 3 = 0$

38. $3x - 8\sqrt{x} + 4 = 0$

In this section we will use determinants to solve systems of linear equations in two and three variables. We begin by solving a general system of linear equations in two variables by the elimination method. The general system is

$$a_1x + b_1y = c_1$$
$$a_2x + b_2y = c_2$$

8.4
Cramer's Rule

We begin by eliminating the variable y from the system. To do so we multiply both sides of the first equation by b_2, and both sides of the second equation by $-b_1$:

$$a_1b_2x + b_1b_2y = c_1b_2$$
$$-a_2b_1x - b_1b_2y = -c_2b_1$$

We now add the two equations and solve the result for x:

$$(a_1b_2 - a_2b_1)x = c_1b_2 - c_2b_1$$
$$x = \frac{c_1b_2 - c_2b_1}{a_1b_2 - a_2b_1}$$

To solve for y we go back to the original system and eliminate the variable x. To do so we multiply the first equation by $-a_2$ and the second equation by a_1:

$$-a_1a_2x - a_2b_1y = -a_2c_1$$
$$a_1a_2x + a_1b_2y = a_1c_2$$

Adding these two equations and solving for y, we have

$$(a_1 b_2 - a_2 b_1) y = a_1 c_2 - a_2 c_1$$
$$y = \frac{a_1 c_2 - a_2 c_1}{a_1 b_2 - a_2 b_1}$$

The solutions we have obtained for x and y can be written in terms of 2×2 determinants as follows:

$$x = \frac{c_1 b_2 - c_2 b_1}{a_1 b_2 - a_2 b_1} = \frac{\begin{vmatrix} c_1 & b_1 \\ c_2 & b_2 \end{vmatrix}}{\begin{vmatrix} a_1 & b_1 \\ a_2 & b_2 \end{vmatrix}}$$

$$y = \frac{a_1 c_2 - a_2 c_1}{a_1 b_2 - a_2 b_1} = \frac{\begin{vmatrix} a_1 & c_1 \\ a_2 & c_2 \end{vmatrix}}{\begin{vmatrix} a_1 & b_1 \\ a_2 & b_2 \end{vmatrix}}$$

We summarize these results, known as Cramer's rule, into the following theorem:

THEOREM 8.2 (CRAMER'S RULE)   The solution to the system

$$a_1 x + b_1 y = c_1$$
$$a_2 x + b_2 y = c_2$$

is given by

$$x = \frac{D_x}{D}, \qquad y = \frac{D_y}{D}$$

where

$$D = \begin{vmatrix} a_1 & b_1 \\ a_2 & b_2 \end{vmatrix} \qquad (D \neq 0)$$

$$D_x = \begin{vmatrix} c_1 & b_1 \\ c_2 & b_2 \end{vmatrix}$$

$$D_y = \begin{vmatrix} a_1 & c_1 \\ a_2 & c_2 \end{vmatrix}$$

The determinant D is made up of the coefficients of x and y in the original system. The terms D_x and D_y are found by replacing the coefficients of x or y by the constant terms in the original system. Notice also that Cramer's rule does not apply if $D = 0$. In this case the equations are either inconsistent or dependent.

▼ **Example 1** Use Cramer's rule to solve

$$2x - 3y = 4$$
$$4x + 5y = 3$$

Solution We begin by calculating the determinants D, D_x, and D_y:

$$D = \begin{vmatrix} 2 & -3 \\ 4 & 5 \end{vmatrix} = 2(5) - 4(-3) = 22$$

$$D_x = \begin{vmatrix} 4 & -3 \\ 3 & 5 \end{vmatrix} = 4(5) - 3(-3) = 29$$

$$D_y = \begin{vmatrix} 2 & 4 \\ 4 & 3 \end{vmatrix} = 2(3) - 4(4) = -10$$

$$x = \frac{D_x}{D} = \frac{29}{22} \quad \text{and} \quad y = \frac{D_y}{D} = \frac{-10}{22} = -\frac{5}{11}$$

The solution set for the system is $\{(\frac{29}{22}, -\frac{5}{11})\}$.   ▲

Cramer's rule can be applied to systems of linear equations in three variables also.

THEOREM 8.3 (ALSO CRAMER'S RULE) The solution set to the system

$$a_1x + b_1y + c_1z = d_1$$
$$a_2x + b_2y + c_2z = d_2$$
$$a_3x + b_3y + c_3z = d_3$$

is given by

$$x = \frac{D_x}{D}, \quad y = \frac{D_y}{D}, \quad \text{and } z = \frac{D_z}{D}$$

where

$$D = \begin{vmatrix} a_1 & b_1 & c_1 \\ a_2 & b_2 & c_2 \\ a_3 & b_3 & c_3 \end{vmatrix} \quad (D \neq 0)$$

$$D_x = \begin{vmatrix} d_1 & b_1 & c_1 \\ d_2 & b_2 & c_2 \\ d_3 & b_3 & c_3 \end{vmatrix}$$

$$D_y = \begin{vmatrix} a_1 & d_1 & c_1 \\ a_2 & d_2 & c_2 \\ a_3 & d_3 & c_3 \end{vmatrix}$$

$$D_z = \begin{vmatrix} a_1 & b_1 & d_1 \\ a_2 & b_2 & d_2 \\ a_3 & b_3 & d_3 \end{vmatrix}$$

Again the determinant D consists of the coefficients of x, y, and z in the original system. The determinants D_x, D_y, and D_z are found by replacing the coefficients of x, y, and z respectively with the constant terms from the original system. If $D = 0$, there is no unique solution to the system.

▼   **Example 2** Use Cramer's rule to solve

$$\begin{aligned} x + y + z &= 6 \\ 2x - y + z &= 3 \\ x + 2y - 3z &= -4 \end{aligned}$$

Solution   This is the same system used in Example 1 in Section 8.2, so we can compare Cramer's rule with our previous methods of solving a system in three variables. We begin by setting up and evaluating D, D_x, D_y, and D_z. (Recall that there are a number of ways to evaluate a 3×3 determinant. Since we have four of these determinants, we can use both Methods 1 and 2 from the previous section.) We evaluate D using Method 1 from Section 8.3:

$$D = \begin{vmatrix} 1 & 1 & 1 & 1 & 1 \\ 2 & -1 & 1 & 2 & -1 \\ 1 & 2 & -3 & 1 & 2 \end{vmatrix}$$

$$= 3 + 1 + 4 - (-1) - (2) - (-6) = 13$$

We evaluate D_x using Method 2 from Section 8.3 and expanding across row 1:

$$D_x = \begin{vmatrix} 6 & 1 & 1 \\ 3 & -1 & 1 \\ -4 & 2 & -3 \end{vmatrix} = 6 \begin{vmatrix} -1 & 1 \\ 2 & -3 \end{vmatrix} - 1 \begin{vmatrix} 3 & 1 \\ -4 & -3 \end{vmatrix} + 1 \begin{vmatrix} 3 & -1 \\ -4 & 2 \end{vmatrix}$$

$$= 6(1) - 1(-5) + 1(2)$$
$$= 13$$

Find D_y by expanding across row 2:

$$D_y = \begin{vmatrix} 1 & 6 & 1 \\ 2 & 3 & 1 \\ 1 & -4 & -3 \end{vmatrix} = -2 \begin{vmatrix} 6 & 1 \\ -4 & -3 \end{vmatrix} + 3 \begin{vmatrix} 1 & 1 \\ 1 & -3 \end{vmatrix} - 1 \begin{vmatrix} 1 & 6 \\ 1 & -4 \end{vmatrix}$$

$$= -2(-14) + 3(-4) - 1(-10)$$
$$= 26$$

Find D_z by expanding down column 1:

$$D_z = \begin{vmatrix} 1 & 1 & 6 \\ 2 & -1 & 3 \\ 1 & 2 & -4 \end{vmatrix} = 1 \begin{vmatrix} -1 & 3 \\ 2 & -4 \end{vmatrix} - 2 \begin{vmatrix} 1 & 6 \\ 2 & -4 \end{vmatrix} + 1 \begin{vmatrix} 1 & 6 \\ -1 & 3 \end{vmatrix}$$

$$= 1(-2) - 2(-16) + 1(9)$$
$$= 39$$

$$x = \frac{D_x}{D} = \frac{13}{13} = 1, \qquad y = \frac{D_y}{D} = \frac{26}{13} = 2, \qquad z = \frac{D_z}{D} = \frac{39}{13} = 3$$

The solution set is $\{(1, 2, 3)\}$.   ▲

▼ **Example 3** Use Cramer's rule to solve

$$\begin{aligned} x + y &= -1 \\ 2x - z &= 3 \\ y + 2z &= -1 \end{aligned}$$

Solution It is helpful to rewrite the system using zeros for the coefficients of those variables not shown:

$$\begin{aligned} x + y + 0z &= -1 \\ 2x + 0y - z &= 3 \\ 0x + y + 2z &= -1 \end{aligned}$$

The four determinants used in Cramer's rule are

$$D = \begin{vmatrix} 1 & 1 & 0 \\ 2 & 0 & -1 \\ 0 & 1 & 2 \end{vmatrix} = -3$$

$$D_x = \begin{vmatrix} -1 & 1 & 0 \\ 3 & 0 & -1 \\ -1 & 1 & 2 \end{vmatrix} = -6$$

$$D_y = \begin{vmatrix} 1 & -1 & 0 \\ 2 & 3 & -1 \\ 0 & -1 & 2 \end{vmatrix} = 9$$

$$D_z = \begin{vmatrix} 1 & 1 & -1 \\ 2 & 0 & 3 \\ 0 & 1 & -1 \end{vmatrix} = -3$$

$$x = \frac{D_x}{D} = \frac{-6}{-3} = 2, \qquad y = \frac{D_y}{D} = \frac{9}{-3} = -3,$$

$$z = \frac{D_z}{D} = \frac{-3}{-3} = 1$$

The solution set is $\{(2, -3, 1)\}$. ▲

Note When solving a system of linear equations by Cramer's rule, it is best to find the determinant D first. If $D = 0$, Cramer's rule does not apply and there is no sense in calculating D_x, D_y, or D_z.

Solve each of the following systems using Cramer's rule: *Problem Set 8.4*

1. $2x - 3y = 3$
$4x - 2y = 10$

2. $3x + y = -2$
$-3x + 2y = -4$

3. $5x - 2y = 4$
$-10x + 4y = 1$

4. $-4x + 3y = -11$
$5x + 4y = 6$

5. $2x + 5y = 2$
$4x + 10y = 4$

6. $3x + 5y = 2$
$-6x - 10y = 6$

7. $4x - 7y = 3$
$5x + 2y = -3$

8. $3x - 4y = 7$
$6x - 2y = 5$

9. $9x - 8y = 4$
$2x + 3y = 6$

10. $4x - 7y = 10$
$-3x + 2y = -9$

11. $x + y - z = 2$
$-x + y + z = 3$
$x + y + z = 4$

12. $-x - y + z = 1$
$x - y + z = 3$
$x + y - z = 4$

13. $x + y + z = 4$
$x - y - z = 2$
$2x + 2y - z = 2$

14. $-x + y + 3z = 6$
$x + y + 2z = 7$
$2x + 3y + z = 4$

15. $3x - y + 2z = 4$
$6x - 2y + 4z = 8$
$x - 5y + 2z = 1$

16. $2x - 3y + z = 1$
$3x - y - z = 4$
$4x - 6y + 2z = 3$

17. $2x - y + 3z = 4$
$x - 5y - 2z = 1$
$-4x - 2y + z = 3$

18. $4x - y + 5z = 1$
$2x + 3y + 4z = 5$
$x + y + 3z = 2$

19. $-x - 2y = 1$
$x + 2z = 11$
$2y + z = 0$

20. $x + y = 2$
$-x + z = 0$
$2x + z = 3$

21. $x - y = 2$
$3x + z = 11$
$y - 2z = -3$

22. $4x + 5y = -1$
$2y + 3z = -5$
$x + 2z = -1$

23. If a company has fixed costs of $100 per week and each item it produces costs $10 to manufacture, then the total cost (y) per week to produce x items is

$$y = 10x + 100$$

If the company sells each item it manufactures for $12, then the total amount of money (y) the company brings in for selling x items is

$$y = 12x$$

Use Cramer's rule to solve the system

$$y = 10x + 100$$
$$y = 12x$$

for x to find the number of items the company must sell per week in order to break even.

24. Suppose a company has fixed costs of $200 per week, and each item it produces costs $20 to manufacture.

    a. Write an equation that gives the total cost per week, y, to manufacture x items.

    b. If each item sells for $25, write an equation that gives the total amount of money, y, the company brings in for selling x items.

    c. Use Cramers rule to find the number of items the company must sell each week to break even.

Review Problems   The problems below review material we covered in Section 4.7. They are taken from the book *Academic Algebra*, written by William J. Milne and published by the American Book Company in 1901.

25. Find a fraction whose value is $\frac{4}{7}$ and whose denominator is 15 greater than its numerator.

26. Find a fraction whose value is $\frac{2}{3}$ and whose numerator is 3 greater than half of its denominator.

27. Three pipes empty into a cistern. One can fill the cistern in 5 hours, another in 6 hours, and the third in 10 hours. How long will it take three pipes together to fill it?

28. A cistern can be filled by one pipe in 20 minutes, by another in 15 minutes, and it can be emptied by a third in 10 minutes. If the three pipes are running at the same time, how long will it take to fill the cistern?

**8.5
Word Problems**

Many times word problems involve more than one unknown quantity. If a problem is stated in terms of two unknowns and we represent each unknown quantity with a different variable, then we must write the relationship between the variables with two equations. The two equations written in terms of the two variables form a system of linear equations which we solve using the methods developed in this chapter. If we find a problem that relates three unknown quantities, then we need three different equations in order to form a linear system we can solve.

▼   **Example 1**   One number is 2 more than 3 times another. Their sum is 26. Find the two numbers.

Solution   If we let x and y represent the two numbers, then the translation of the first sentence in the problem into an equation would be

$$y = 3x + 2$$

The second sentence gives us a second equation:

$$x + y = 26$$

The linear system that describes the situation is

$$x + y = 26$$
$$y = 3x + 2$$

Substituting the expression for y from the second equation into the first and solving for x yields

$$x + (3x + 2) = 26$$
$$4x + 2 = 26$$
$$4x = 24$$
$$x = 6$$

Using $x = 6$ in $y = 3x + 2$ gives the second number:

$$y = 3(6) + 2$$
$$y = 20$$

The two numbers are 6 and 20. Their sum is 26, and the second is 2 more than 3 times the first.   ▲

▼   **Example 2**   Suppose 850 tickets were sold for the game for a total of $1,100. If adult tickets cost $1.50 and children's tickets cost $1.00, how many of each kind of ticket were sold?

Solution   If we let $x = $ the number of adult tickets and $y = $ the number of children's tickets, then

$$x + y = 850$$

since a total of 850 tickets were sold. Since each adult ticket costs $1.50 and each children's ticket costs $1.00 and the total amount of money paid for tickets was $1,100, a second equation is

$$1.50x + 1.00y = 1,100$$

which is equivalent to

$$15x + 10y = 11,000$$

The system that describes the situation is

$$x + y = 850$$
$$15x + 10y = 11,000$$

Multiplying the first equation by -10 and adding the result to the second equation eliminates the variable y from the two systems:

$$
\begin{aligned}
-10x - 10y &= -8,500 \\
\underline{15x + 10y} &= \underline{11,000} \\
5x &= 2,500 \\
x &= 500
\end{aligned}
$$

The number of adult tickets sold was 500. To find the number of children's tickets, we substitute $x = 500$ into $x + y = 850$ to get

$$
\begin{aligned}
500 + y &= 850 \\
y &= 350
\end{aligned}
$$

The number of children's tickets sold was 350.   ▲

▼   **Example 3**   Suppose a person invests a total of $10,000 in two accounts. One account earns 5% annually and the other earns 6% annually. If the total interest earned from both accounts in a year is $560, how much is invested in each account?

Solution   The form of the solution to this problem is very similar to that of Example 2. We let x equal the amount invested at 6% and y be the amount invested at 5%. Since the total investment is $10,000, one relationship between x and y can be written as

$$x + y = 10,000$$

The total interest earned from both accounts is $560. The amount of interest earned on x dollars at 6% is $.06x$, while the amount of interest earned on y dollars at 5% is $.05y$. This relationship is represented by the equation

$$.06x + .05y = 560$$

which is equivalent to

$$6x + 5y = 56,000$$

The original problem can be stated in terms of x and y with the system

$$
\begin{aligned}
x + \quad y &= 10,000 \\
6x + 5y &= 56,000
\end{aligned}
$$

Eliminating x from the system and solving for y, we have

$$-6x - 6y = -60,000$$
$$\underline{6x + 5y = 56,000}$$
$$- y = -4,000$$
$$y = 4,000$$

The amount of money invested at 5% is $4,000. Since the total investment was $10,000, the amount invested at 6% must be $6,000.

▲

▼   **Example 4**   How much 20% alcohol solution and 50% alcohol solution must be mixed to get 12 gallons of 30% alcohol solution?

Solution   To solve this problem we must first understand that a 20% alcohol solution is 20% alcohol and 80% water.

Let x = the number of gallons of 20% alcohol solution needed, and y = the number of gallons of 50% alcohol solution needed. Since we must end up with a total of 12 gallons of solution, one equation for the system is

$$x + y = 12$$

The amount of alcohol in the x gallons of 20% solution is $.20x$, while the amount of alcohol in the y gallons of 50% solution is $.50y$. Since the total amount of alcohol in the 20% and 50% solutions must add up to the amount of alcohol in the 12 gallons of 30% solution, the second equation in our system can be written as

$$.20x + .50y = .30(12)$$

We eliminate the decimals by multiplying both sides by 10 to get

$$2x + 5y = 36$$

The system of equations that describes the situation is

$$x + y = 12$$
$$2x + 5y = 36$$

Multiplying the top equation by -2 to eliminate the x-variable, we have

$$-2x - 2y = -24$$
$$\underline{2x + 5y = 36}$$
$$3y = 12$$
$$y = 4$$

Substituting $y = 4$ into $x + y = 12$, we solve for x:

$$x + 4 = 12$$
$$x = 8$$

It takes 8 gallons of 20% alcohol solution and 4 gallons of 50% alcohol solution to produce 12 gallons of 30% alcohol solution.

▲

▼   **Example 5**   It takes 2 hours for a boat to travel 28 miles downstream. The same boat can travel 18 miles upstream in 3 hours. What is the speed of the boat in still water and the speed of the current of the river?

Solution   Let $x =$ the speed of the boat in still water and $y =$ the speed of the current. Using a table as we did in Section 4.7, we have

	d	r	t
Upstream	18	$x - y$	3
Downstream	28	$x + y$	2

Since $d = r \cdot t$, the system we need to solve the problem is

$$18 = (x - y) \cdot 3$$
$$28 = (x + y) \cdot 2$$

which is equivalent to

$$6 = x - y$$
$$14 = x + y$$

Adding the two equations, we have

$$20 = 2x$$
$$x = 10$$

Substituting $x = 10$ into $14 = x + y$, we see that

$$y = 4$$

The speed of the boat in still water is 10 mph and the speed of the current is 4 mph.   ▲

Note   We solved each system in this section using either the elimination method or the substitution method. We could just as easily have used Cramer's rule to solve the systems. Cramer's rule applies to any system of linear equations with a unique solution.

---

1. One number is 3 more than twice another. The sum of the numbers is 18. Find the two numbers.

2. The sum of two numbers is 32. One of the numbers is 4 less than 5 times the other. Find the two numbers.

3. The difference of two numbers is 6. Twice the smaller is 4 more than the larger. Find the two numbers.

4. The larger of two numbers is 5 more than twice the smaller. If the smaller is subtracted from the larger, the result is 12. Find the two numbers.

5. The sum of three numbers is 8. Twice the smallest is 2 less than the largest, while the sum of the largest and smallest is 5. Use a linear system in three variables to find the three numbers.

6. The sum of three numbers is 14. The largest is 4 times the smallest, while the sum of the smallest and twice the largest is 18. Use a linear system in three variables to find the three numbers.

7. A total of 925 tickets were sold for the game for a total of $1150. If adult tickets sold for $2.00 and children's tickets sold for $1.00, how many of each kind of ticket were sold?

8. If tickets for the show cost $2.00 for adults and $1.50 for children, how many of each kind of ticket were sold if a total of 300 tickets were sold for $525?

9. Bob has 20 coins totaling $1.40. If he has only dimes and nickels, how many of each coin does he have?

10. If Amy has 15 coins totaling $2.70, and the coins are quarters and dimes, how many of each coin does she have?

11. Mr. Jones has $20,000 to invest. He invests part at 6% and the rest at 7%. If he earns $1280 interest after one year, how much did he invest at each rate?

12. A man invests $17,000 in two accounts. One account earns 5% interest per year and the other 6.5%. If his yearly yield in interest is $970, how much does he invest at each rate?

13. Susan invests twice as much money at 7.5% as she does at 6%. If her total interest after a year is $840, how much does she have invested at each rate?

14. A woman earns $1350 interest from two accounts in a year. If she has three times as much invested at 7% as she does at 6%, how much does she have in each account?

15. How many gallons of 20% alcohol solution and 50% alcohol solution must be mixed to get 9 gallons of 30% alcohol solution?

16. How many ounces of 30% hydrochloric acid solution and 80% hydrochloric

acid solution must be mixed to get 10 ounces of 50% hydrochloric acid solution?

17. A mixture of 16% disinfectant solution is to be made from 20% and 14% disinfectant solutions. How much of each solution should be used if 15 gallons of the 16% solution are needed?

18. How much 25% antifreeze and 50% antifreeze should be combined to give 40 gallons of 30% antifreeze?

19. It takes a boat 2 hours to travel 24 miles downstream and 3 hours to travel 18 miles upstream. What is the speed of the boat in still water and of the current of the river?

20. A boat on a river travels 20 miles downstream in only 2 hours. It takes the same boat 6 hours to travel 12 miles upstream. What are the speed of the boat and the speed of the current?

21. An airplane flying with the wind can cover a certain distance in 2 hours. The return trip against the wind takes $2\frac{1}{2}$ hours. How fast is the plane and what is the speed of the air, if the distance is 600 miles?

22. An airplane covers a distance of 1500 miles in 3 hours when it flies with the wind and $3\frac{1}{3}$ hours when it flies against the wind. What is the speed of the plane in still air?

Review Problems   The problems below review material we covered in Section 6.6.

Solve each inequality and graph the solution set.

23. $x^2 - 2x - 8 \leq 0$ 24. $x^2 - x - 12 < 0$
25. $2x^2 + 5x - 3 > 0$ 26. $3x^2 - 5x - 2 \leq 0$
27. $\dfrac{x-3}{x+2} < 0$ 28. $\dfrac{x+5}{x-4} \geq 0$

Examples

1. The solution to the system

$$x + y = 6$$
$$3x - 2y = -2$$

is the ordered pair (2, 4) since it satisfies both equations.

Chapter 8   Summary and Review

SYSTEMS OF LINEAR EQUATIONS [8.1, 8.2]

A system of linear equations consists of two or more linear equations considered simultaneously. The solution set to a linear system in two variables is the set of ordered pairs that satisfies both equations. The solution set to a linear system in three variables consists of all the ordered triples that satisfy each equation in the system.

INCONSISTENT AND DEPENDENT EQUATIONS [8.1, 8.2]

Two linear equations that have no solutions in common are said to be *inconsistent,* while two linear equations that have all their solutions in common are said to be *dependent.*

2. If the two lines are parallel, then the system will be inconsistent and the solution is ∅. If the two lines coincide, then the system is dependent.

DETERMINANTS [8.3]

The values of a 2×2 and a 3×3 determinant are as follows:

$$\begin{vmatrix} a & c \\ b & d \end{vmatrix} = ad - bc$$

$$\begin{vmatrix} a_1 & b_1 & c_1 \\ a_2 & b_2 & c_2 \\ a_3 & b_3 & c_3 \end{vmatrix} = a_1 b_2 c_3 + a_3 b_1 c_2 + a_2 b_3 c_1 \\ - a_3 b_2 c_1 - a_1 b_3 c_2 - a_2 b_1 c_3.$$

There are two methods of finding the six products in the expansion of a 3×3 determinant. One method involves a cross-multiplication scheme. The other method involves expanding the determinant by minors.

3. $\begin{vmatrix} 3 & 4 \\ -2 & 5 \end{vmatrix} = 15 - (-8) = 23$

$$\begin{vmatrix} 3 & 4 & 5 \\ 2 & 0 & 1 \\ -1 & 1 & 2 \end{vmatrix}$$

$$= 3 \begin{vmatrix} 0 & 1 \\ 1 & 2 \end{vmatrix} - 4 \begin{vmatrix} 2 & 1 \\ -1 & 2 \end{vmatrix} + 5 \begin{vmatrix} 2 & 0 \\ -1 & 1 \end{vmatrix}$$

$$= 3(-1) - 4(5) + 5(2)$$

$$= -13$$

CRAMER'S RULE FOR A LINEAR SYSTEM IN TWO VARIABLES [8.4]

The solution to the system

$$a_1 x + b_1 y = c_1$$
$$a_2 x + b_2 y = c_2$$

is given by

$$x = \frac{D_x}{D} \quad \text{and} \quad y = \frac{D_y}{D} \qquad (D \neq 0)$$

where

$$D = \begin{vmatrix} a_1 & b_1 \\ a_2 & b_2 \end{vmatrix}, \quad D_x = \begin{vmatrix} c_1 & b_1 \\ c_2 & b_2 \end{vmatrix}, \quad \text{and} \quad D_y = \begin{vmatrix} a_1 & c_1 \\ a_2 & c_2 \end{vmatrix}.$$

4. For the system in Example 1:

$$D = \begin{vmatrix} 1 & 1 \\ 3 & -2 \end{vmatrix} = -5$$

$$x = \frac{-10}{-5} = 2$$

$$D_x = \begin{vmatrix} 6 & 1 \\ -2 & -2 \end{vmatrix} = -10$$

$$y = \frac{-20}{-5} = 4$$

$$D_y = \begin{vmatrix} 1 & 6 \\ 3 & -2 \end{vmatrix} = -20$$

5. For the system

$x + y = -1$
$2x - z = 3$
$y + 2z = -1$

$$D = \begin{vmatrix} 1 & 1 & 0 \\ 2 & 0 & -1 \\ 0 & 1 & 2 \end{vmatrix} = -3$$

$$x = \frac{-6}{-3} = 2$$

$$D_x = \begin{vmatrix} -1 & 1 & 0 \\ 3 & 0 & -1 \\ -1 & 1 & 2 \end{vmatrix} = -6$$

$$y = \frac{9}{-3} = -3$$

$$D_y = \begin{vmatrix} 1 & -1 & 0 \\ 2 & 3 & -1 \\ 0 & -1 & 2 \end{vmatrix} = 9$$

$$z = \frac{-3}{-3} = 1$$

$$D_z = \begin{vmatrix} 1 & 1 & -1 \\ 2 & 0 & 3 \\ 0 & 1 & -1 \end{vmatrix} = -3$$

CRAMER'S RULE FOR A LINEAR SYSTEM IN THREE VARIABLES [8.4]

The solution to the system

$$a_1x + b_1y + c_1z = d_1$$
$$a_2x + b_2y + c_2z = d_2$$
$$a_3x + b_3y + c_3z = d_3$$

is given by

$$x = \frac{D_x}{D}, \qquad y = \frac{D_y}{D}, \qquad \text{and } z = \frac{D_z}{D} \qquad (D \neq 0)$$

where

$$D = \begin{vmatrix} a_1 & b_1 & c_1 \\ a_2 & b_2 & c_2 \\ a_3 & b_3 & c_3 \end{vmatrix} \qquad D_y = \begin{vmatrix} a_1 & d_1 & c_1 \\ a_2 & d_2 & c_2 \\ a_3 & d_3 & c_3 \end{vmatrix}$$

$$D_x = \begin{vmatrix} d_1 & b_1 & c_1 \\ d_2 & b_2 & c_2 \\ d_3 & b_3 & c_3 \end{vmatrix} \qquad D_z = \begin{vmatrix} a_1 & b_1 & d_1 \\ a_2 & b_2 & d_2 \\ a_3 & b_3 & d_3 \end{vmatrix}.$$

Chapter 8
Test

Solve the following systems by the elimination method:

1. $2x - 5y = -8$
        $3x + y = 5$

2. $4x - 7y = -2$
        $-5x + 6y = -3$

Solve the following systems by the substitution method:

3. $2x - 5y = 14$
        $y = 3x + 8$

4. $6x - 3y = 0$
        $x + 2y = 5$

5. Solve the system

$$2x - y + z = 9$$
$$x + y - 3z = -2$$
$$3x + y - z = 6$$

Evaluate each determinant:

6. $\begin{vmatrix} 3 & -5 \\ -4 & 2 \end{vmatrix}$

7. $\begin{vmatrix} 1 & 0 & -3 \\ 2 & 1 & 0 \\ 0 & 5 & 4 \end{vmatrix}$

Use Cramer's rule to solve:

8. $5x - 4y = 2$
$-2x + y = 3$

9. $2x + 4y = 3$
$-4x - 8y = -6$

10. $2x - y + 3z = 2$
$x - 4y - z = 6$
$3x - 2y + z = 4$

11. John invests twice as much money at 6% as he does at 5%. If his investments earn a total of $680 in one year, how much does he have invested at each rate?

12. For a woman of average height and weight between the ages of 19 and 22, the Food and Nutrition Board of the National Academy of Sciences has determined the Recommended Daily Allowance (RDA) of ascorbic acid to be 45 mg (milligrams). They also determined the RDA for niacin to be 14 mg for the same woman.

Each ounce of cereal I contains 10 mg of ascorbic acid and 4 mg of niacin, while each ounce of cereal II contains 15 mg of ascorbic acid and 2 mg of niacin. How many ounces of each cereal must the average woman between the ages of 19 and 22 consume in order to have the RDAs for both ascorbic acid and niacin? (The following table is a summary of the information given.)

	Cereal I	Cereal II	Recommended Daily Allowance (RDA)
Ascorbic acid	10 mg	15 mg	45 mg
Niacin	4 mg	2 mg	14 mg

The Conic Sections

To the student:

This chapter is concerned with four special types of graphs and their associated equations. The four types of graphs are called parabolas, circles, ellipses, and hyperbolas. They are called *conic sections* because each can be found by slicing a cone with a plane as is shown in Figure 9-1.

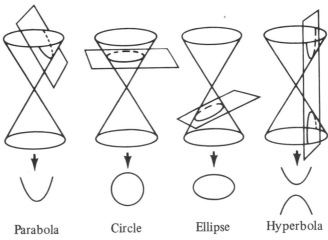

| Parabola | Circle | Ellipse | Hyperbola |

Figure 9-1

There are many applications associated with conic sections. The planets orbit the sun in elliptical orbits. Many of the comets that come

in contact with the gravitational field surrounding the earth travel in parabolic or hyperbolic paths. Flashlight and searchlight mirrors have elliptical or parabolic shapes because of the way surfaces with those shapes reflect light. The arches of many bridges are in the shape of parabolas. Objects fired into the air travel in parabolic paths.

The equations associated with these graphs are all second-degree equations in two variables. The first four sections of this chapter deal with graphing the conic sections, given the appropriate equation. The chapter ends with a section on second-degree inequalities and a section on second-degree systems. A working knowledge of graphing will be useful throughout the chapter. Factoring is needed for the first section of the chapter.

The solution set to the equation

$$y = x^2 - 3$$

9.1

Graphing Parabolas

will consist of ordered pairs. One method of graphing the solution set is to find a number of ordered pairs that satisfy the equation and graph them. We can obtain some ordered pairs that are solutions to $y = x^2 - 3$ by use of a table as follows:

x	$y = x^2 - 3$	y	Solutions
-3	$y = (-3)^2 - 3 = 9 - 3 = 6$	6	$(-3, 6)$
-2	$y = (-2)^2 - 3 = 4 - 3 = 1$	1	$(-2, 1)$
-1	$y = (-1)^2 - 3 = 1 - 3 = -2$	-2	$(-1, -2)$
0	$y = 0^2 - 3 = 0 - 3 = -3$	-3	$(0, -3)$
1	$y = 1^2 - 3 = 1 - 3 = -2$	-2	$(1, -2)$
2	$y = 2^2 - 3 = 4 - 3 = 1$	1	$(2, 1)$
3	$y = 3^2 - 3 = 9 - 3 = 6$	6	$(3, 6)$

Graphing these solutions on a rectangular coordinate system indicates the shape of the graph of $y = x^2 - 3$. (See Figure 9-2.)

If we were to find more solutions by extending the table to include values of x between those we have already, we would see they follow the same pattern. Connecting the points on the graph with a smooth curve we have the graph of the second-degree equation $y = x^2 - 3$. (See Figure 9-3.)

This graph is an example of a parabola. All equations of the form $y = ax^2 + bx + c$, $a \neq 0$, have parabolas for graphs. Although it is always possible to graph the solution set for an equation by using a

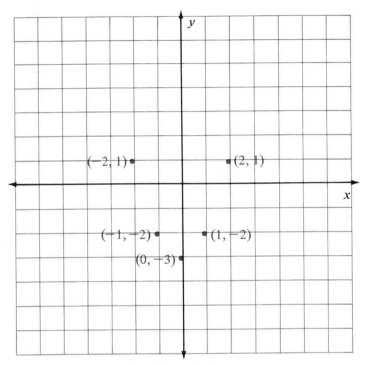

Figure 9-2

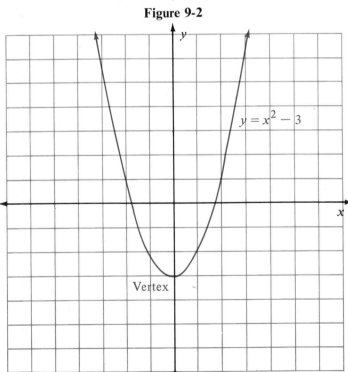

Figure 9-3

table, we would like to find a method that is faster, less tedious, and more accurate.

The important points associated with the graph of a parabola are the highest (or lowest) point on the graph and the x-intercepts. The y-intercepts can also be useful.

The graph of the equation $y = ax^2 + bx + c$ will cross the y-axis at $y = c$, since substituting $x = 0$ into $y = ax^2 + bx + c$ yields $y = c$.   *Intercepts for*
Parabolas

Since the graph will cross the x-axis when $y = 0$, the x-intercepts are those values of x that are solutions to the quadratic equation $0 = ax^2 + bx + c$. Applying the quadratic formula, if necessary, the x-intercepts can be written

$$x = \frac{-b \pm \sqrt{b^2 - 4ac}}{2a}$$

The x-intercepts can also be found by solving the equation $ax^2 + bx + c = 0$ by factoring, if the left side is factorable.

▼ **Example 1** Find the x-intercepts for $y = x^2 - 5x - 6$.

Solution To find the x-intercept, we let $y = 0$ and solve for x:

$$0 = x^2 - 5x - 6$$
$$0 = (x - 6)(x + 1)$$
$$x - 6 = 0 \quad \text{or} \quad x + 1 = 0$$
$$x = 6 \quad \text{or} \quad x = -1$$

The x-intercepts are 6 and -1.   ▲

Not every parabola will cross the x-axis. If the equation $0 = ax^2 + bx + c$ does not have real solutions, there will be no x-intercepts for the graph of $y = ax^2 + bx + c$.

▼ **Example 2** Find the intercepts for $y = 2x^2 + 3x + 5$.

Solution Substituting $y = 0$ into the equation and solving for x, we have

$$0 = 2x^2 + 3x + 5$$

The right side does not factor so we must use the quadratic formula:

$$x = \frac{-3 \pm \sqrt{9 - 4(2)(5)}}{2(2)}$$

$$x = \frac{-3 \pm \sqrt{-31}}{4}$$

$$x = \frac{-3 \pm \sqrt{31}\, i}{4}$$

Since the solutions are complex numbers, the graph does not cross the x-axis. There are no x-intercepts. ▲

The Vertex of a Parabola

The highest or lowest point on a parabola is called the *vertex*. The vertex for the graph of $y = ax^2 + bx + c$ will always occur when

$$x = \frac{-b}{2a}$$

That is, the x-coordinate of the vertex is $-b/2a$. To find the y-coordinate of the vertex, we substitute $x = -b/2a$ into $y = ax^2 + bx + c$ and solve for y.

If we complete the square on the first two terms on the right side of the equation $y = ax^2 + bx + c$, we can transform the equation as follows:

$$y = ax^2 + bx + c$$
$$y = a\left(x^2 + \frac{b}{a}x\right) + c$$
$$y = a\left[x^2 + \frac{b}{a}x + \left(\frac{b}{2a}\right)^2\right] + c - a\left(\frac{b}{2a}\right)^2$$
$$y = a\left(x + \frac{b}{2a}\right)^2 + \frac{4ac - b^2}{4a}$$

It may not look like it, but this last line indicates the vertex of the graph of $y = ax^2 + bx + c$ has an x-coordinate of $-b/2a$. Since a, b, and c are constants, the only quantity that is varying in the last expression is the x in $(x + b/2a)^2$. Since the quantity $(x + b/2a)^2$ is the square of $x + b/2a$, the smallest it will ever be is 0, and that will happen when $x = -b/2a$.

We can use the vertex point along with the x- and y-intercepts to sketch the graph of any equation of the form $y = ax^2 + bx + c$. Here is a summary of the information given above:

The graph of $y = ax^2 + bx + c$ will have

1. a y-intercept at $y = c$
2. x-intercepts (if they exist) at

$$x = \frac{-b \pm \sqrt{b^2 - 4ac}}{2a}$$

3. a vertex when $x = -b/2a$

▼ **Example 3** Sketch the graph of $y = x^2 - 4x - 5$.

Solution To find the x-intercepts, we let $y = 0$ and solve for x:

$$0 = x^2 - 4x - 5$$
$$0 = (x - 5)(x + 1)$$
$$x - 5 = 0 \quad \text{or} \quad x + 1 = 0$$
$$x = 5 \quad \text{or} \quad x = -1$$

The x-coordinate of the vertex will occur when

$$x = \frac{-b}{2a} = \frac{-(-4)}{2(1)}$$
$$= 2$$

To find the y-coordinate of the vertex we substitute 2 for x:

$$y = 2^2 - 4(2) - 5$$
$$= 4 - 8 - 5$$
$$= -9$$

The graph crosses the x-axis at 5 and -1 and has its vertex at $(2, -9)$. Plotting these points and connecting them with a smooth curve, we have the graph of $y = x^2 - 4x - 5$. (See Figure 9-4.) A parabola that opens up as this one does is said to be *concave-up*. The vertex is the lowest point on the graph. ▲

▼ **Example 4** Graph $y = -x^2 - 4x + 12$.

Solution Letting $y = 0$, the x-intercepts come from

$$0 = -x^2 - 4x + 12$$
$$0 = x^2 + 4x - 12$$
$$0 = (x + 6)(x - 2)$$
$$x + 6 = 0 \quad \text{or} \quad x - 2 = 0$$
$$x = -6 \quad \text{or} \quad x = 2$$

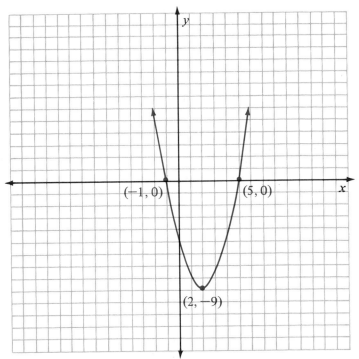

Figure 9-4

The x-coordinate of the vertex is given by

$$x = \frac{-b}{2a} = \frac{-(-4)}{2(-1)} = -2$$

We find the y-coordinate of the vertex by substituting $x = -2$ into the original equation:

$$
\begin{aligned}
y &= -(-2)^2 - 4(-2) + 12 \\
&= -4 + 8 + 12 \\
&= 16
\end{aligned}
$$

The graph crosses the x-axis at -6 and 2, and has its vertex at $(-2, 16)$. (See Figure 9-5.)

    The graph is *concave-down*. The vertex in this case is the highest point on the graph.   ▲

▼ **Example 5** Graph $y = 3x^2 - 6x + 1$.

Solution To find the x-intercepts, we let $y = 0$ and solve for x:

$$0 = 3x^2 - 6x + 1$$

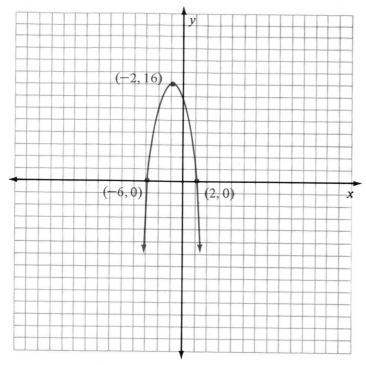

Figure 9-5

Since the right side of this equation will not factor, we must use the quadratic formula:

$$x = \frac{-(-6) \pm \sqrt{36 - 4(3)(1)}}{2(3)}$$

$$= \frac{6 \pm \sqrt{24}}{6}$$

$$= \frac{6 \pm 2\sqrt{6}}{6}$$

$$= \frac{3 \pm \sqrt{6}}{3}$$

The x-intercepts are

$$\frac{3 + \sqrt{6}}{3} \quad \text{and} \quad \frac{3 - \sqrt{6}}{3}$$

Since we are going to graph these points, it is a good idea to get a

decimal approximation for them. Since $\sqrt{6} \approx 2.45$ ($\approx$ is read "approximately equals"), the x-intercepts are

$$\frac{3 + \sqrt{6}}{3} \approx \frac{3 + 2.45}{3} = \frac{5.45}{3} \approx 1.8$$

$$\frac{3 - \sqrt{6}}{3} \approx \frac{3 - 2.45}{3} = \frac{.55}{3} \approx .2$$

The x-coordinate of the vertex is

$$x = \frac{-b}{2a} = \frac{-(-6)}{2(3)} = 1$$

To find the y-coordinate of the vertex, we let $x = 1$:

$$\begin{aligned} y &= 3(1)^2 - 6(1) + 1 \\ &= 3 - 6 + 1 \\ &= -2 \end{aligned}$$

The graph crosses the x-axis at 1.8 and .2 (approximately) and has its vertex at $(1, -2)$. (See Figure 9-6.)

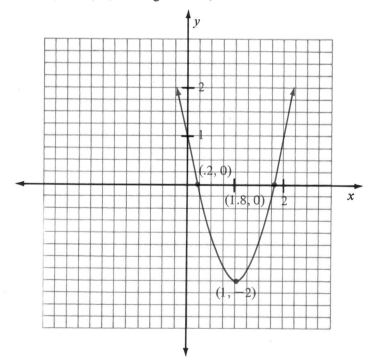

Figure 9-6

The graph is concave-up. The vertex is the lowest point on the graph. ▲

Note It is apparent from these last three examples that the x-coordinate of the vertex will always fall halfway between the x-intercepts—*if* the graph has x-intercepts. If the x-intercepts are x_1 and x_2, then the value halfway between them will always be their average, or

$$\frac{x_1 + x_2}{2}$$

▼ **Example 6** Graph $y = -2x^2 + 6x - 5$.

Solution Letting $y = 0$, we have

$$0 = -2x^2 + 6x - 5$$

Since the right side does not factor, we must use the quadratic formula:

$$x = \frac{-6 \pm \sqrt{36 - 4(-2)(-5)}}{2(-2)}$$

$$= \frac{-6 \pm \sqrt{-4}}{-4}$$

$$= \frac{-6 \pm 2i}{-4}$$

The results are complex numbers, so the x-intercepts do not exist. The graph does not cross the x-axis. The x-coordinate of the vertex is

$$x = \frac{-b}{2a} = \frac{-6}{2(-2)} = \frac{6}{4} = \frac{3}{2}$$

To find the y-coordinate, we let $x = \frac{3}{2}$:

$$y = -2\left(\frac{3}{2}\right)^2 + 6\left(\frac{3}{2}\right) - 5$$

$$= \frac{-18}{4} + \frac{18}{2} - 5$$

$$= \frac{-18 + 36 - 20}{4}$$

$$= \frac{-2}{4}$$

$$= \frac{-1}{2}$$

The vertex is $(\frac{3}{2}, -\frac{1}{2})$. Since this is the only point we have so far, we must find two others. Let's let $x = 3$ and $x = 0$, since each point is the same distance from $x = \frac{3}{2}$ and on either side:

When $x = 3$,
$$y = -2(3)^2 + 6(3) - 5$$
$$= -18 + 18 - 5$$
$$= -5$$

When $x = 0$,
$$y = -2(0)^2 + 6(0) - 5$$
$$= 0 + 0 - 5$$
$$= -5$$

The two additional points on the graph are $(3, -5)$ and $(0, -5)$. Figure 9-7 shows the graph.

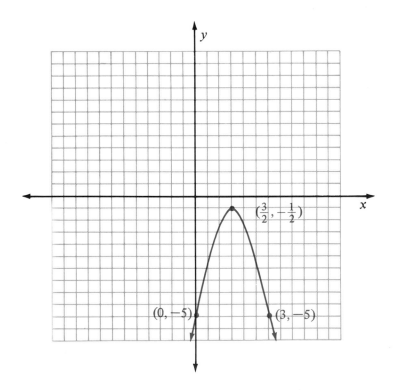

Figure 9-7

The graph is concave-down. The vertex is the highest point on the graph. ▲

We see from the examples of this section that the graph of $y = ax^2 + bx + c$ is concave-up whenever $a > 0$ and concave-down whenever $a < 0$.

For each of the following equations, give the x-intercepts and the coordinates of the vertex and sketch the graph:

Problem Set 9.1

1. $y = x^2 - 6x - 7$
2. $y = x^2 + 6x - 7$
3. $y = x^2 + 2x - 3$
4. $y = x^2 - 2x - 3$
5. $y = -x^2 - 4x + 5$
6. $y = -x^2 + 4x - 5$
7. $y = -x^2 + 2x + 15$
8. $y = -x^2 - 2x + 8$
9. $y = x^2 - 1$
10. $y = x^2 - 4$
11. $y = -x^2 + 9$
12. $y = -x^2 + 1$
13. $y = 3x^2 - 10x - 8$
14. $y = 3x^2 + 10x + 8$
15. $y = x^2 - 2x - 11$
16. $y = x^2 - 4x - 14$
17. $y = 4x^2 + 12x - 11$
18. $y = 2x^2 + 8x - 14$
19. $y = -4x^2 - 4x + 31$
20. $y = -9x^2 - 6x + 19$

Find the vertex and any two convenient points to sketch the graphs of the following:

21. $y = x^2 - 4x - 4$
22. $y = x^2 - 2x + 3$
23. $y = -x^2 + 2x - 5$
24. $y = -x^2 + 4x - 2$
25. $y = x^2 + 1$
26. $y = x^2 + 4$
27. $y = -x^2 - 3$
28. $y = -x^2 - 2$
29. $y = 3x^2 + 4x + 1$
30. $y = 2x^2 + 4x + 3$
31. $y = 2x^2 + 3x + 2$
32. $y = 3x^2 + 3x + 1$
33. $y = -4x^2 + x - 1$
34. $y = -3x^2 + x - 1$

35. An arrow is shot straight up into the air with an initial velocity of 128 feet/second. If h is the height of the arrow at any time t, then the equation that gives h in terms of t is

$$h = 128t - 16t^2$$

Find the maximum height attained by the arrow.

36. A company finds that its weekly profit P, obtained by selling x items is given by the equation

$$P = -2x^2 + 160x + 1{,}000$$

a. How many items must they sell to obtain their maximum profit?
b. What is their maximum profit?

As you know, if the graph of $y = ax^2 + bx + c$ crosses the x-axis, it will cross at

$$x_1 = \frac{-b + \sqrt{b^2 - 4ac}}{2a} \quad \text{and} \quad x_2 = \frac{-b - \sqrt{b^2 - 4ac}}{2a}.$$

There is an interesting relationship between the sum and product of these two numbers and the coefficients a, b, and c in the original equation.

37. Compute the sum $x_1 + x_2$, where x_1 and x_2 are given above.

38. Compute the product $x_1 \cdot x_2$, where x_1 and x_2 are given above.

39. If the perimeter of a rectangle is 40 inches and the width is x, then the length is $20 - x$. The area of the rectangle then is $A = x(20 - x) = 20x - x^2$. Find the value of x that will make A a maximum.

40. The perimeter of a rectangle is 60 feet. Find the dimensions that will make the area a maximum.

Review Problems   The problems below review material we covered in Section 6.2. Reviewing these problems will help you in the next section.

Add a last term to each of the following so that the trinomial that results is a perfect square trinomial. In each case write the binomial square that it is equal to.

41. $x^2 + 6x$ **42.** $x^2 - 6x$

43. $x^2 - 10x$ **44.** $x^2 + 4x$

45. $x^2 + 8x$ **46.** $x^2 - 14x$

47. $x^2 + 3x$ **48.** $x^2 + 5x$

9.2
The Circle

Before we find the general equation of a circle, we must first derive what is known as the *distance formula*.

Suppose (x_1, y_1) and (x_2, y_2) are any two points in the first quadrant. (Actually, we could choose the two points to be anywhere on the coordinate plane. It is just more convenient to have them in the first quadrant.) We can name the points P_1 and P_2, respectively, and draw the diagram shown in Figure 9-8.

Notice the coordinates of point Q. The x-coordinate is x_2 since Q is directly below point P_2. The y-coordinate of Q is y_1 since Q is directly across from point P_1. It is evident from the diagram that the length of P_2Q is $y_2 - y_1$ and the length of P_1Q is $x_2 - x_1$. Using the Pythagorean

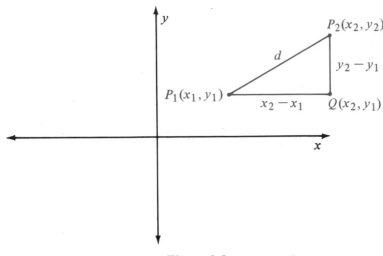

Figure 9-8

theorem, we have

$$(P_1P_2)^2 = (P_1Q)^2 + (P_2Q)^2$$

or

$$d^2 = (x_2 - x_1)^2 + (y_2 - y_1)^2$$

Taking the square root of both sides, we have

$$d = \sqrt{(x_2 - x_1)^2 + (y_2 - y_1)^2}$$

We know this is the positive square root, since d is the distance from P_1 to P_2 and must therefore be positive. This formula is called the *distance formula*.

▼   **Example 1**   Find the distance between $(3, 5)$ and $(2, -1)$.

Solution   If we let $(3, 5)$ be (x_1, y_1) and $(2, -1)$ be (x_2, y_2) and apply the distance formula, we have

$$\begin{aligned} d &= \sqrt{(2 - 3)^2 + (-1 - 5)^2} \\ &= \sqrt{(-1)^2 + (-6)^2} \\ &= \sqrt{1 + 36} \\ &= \sqrt{37} \end{aligned}$$ ▲

Note   The choice of $(3, 5)$ as (x_1, y_1) and $(2, -1)$ as (x_2, y_2) is arbitrary. We could just as easily have reversed them.

▼   **Example 2**   Find x if the distance from $(x, 5)$ to $(3, 4)$ is $\sqrt{2}$.

Solution   Using the distance formula, we have

$$\sqrt{2} = \sqrt{(x - 3)^2 + (5 - 4)^2}$$
$$2 = (x - 3)^2 + 1^2$$
$$2 = x^2 - 6x + 9 + 1$$
$$0 = x^2 - 6x + 8$$
$$0 = (x - 4)(x - 2)$$
$$x = 4 \quad \text{or} \quad x = 2$$

The two solutions are 4 and 2, which indicates there are two points, $(4, 5)$ and $(2, 5)$, which are $\sqrt{2}$ units from $(3, 4)$.   ▲

We can use the distance formula to derive the equation of a circle.

THEOREM 9.1   The equation of the circle with center at (a, b) and radius r is given by

$$(x - a)^2 + (y - b)^2 = r^2$$

Proof   By definition, all points on the circle are a distance r from the center (a, b). If we let (x, y) represent any point on the circle, then (x, y) is r units from (a, b). Applying the distance formula, we have

$$r = \sqrt{(x - a)^2 + (y - b)^2}$$

Squaring both sides of this equation gives the equation of the circle:

$$(x - a)^2 + (y - b)^2 = r^2$$

We can use Theorem 9.1 to find the equation of a circle given its center and radius, or to find its center and radius given the equation.

▼   **Example 3**   Find the equation of the circle with center at $(-3, 2)$ having a radius of 5.

Solution   We have $(a, b) = (-3, 2)$ and $r = 5$. Applying Theorem 9.1 yields

$$[x - (-3)]^2 + (y - 2)^2 = 5^2$$
$$(x + 3)^2 + (y - 2)^2 = 25$$ ▲

▼   **Example 4**   Give the equation of the circle with radius 3 whose center is at the origin.

Solution The coordinates of the center are $(0, 0)$, and the radius is 3. The equation must be

$$(x - 0)^2 + (y - 0)^2 = 3^2$$
$$x^2 + y^2 = 9$$ ▲

We can see from Example 4 that the equation of any circle with its center at the origin and radius r will be

$$x^2 + y^2 = r^2$$

▼ **Example 5** Find the center and radius, and sketch the graph, of the circle whose equation is

$$(x - 1)^2 + (y + 3)^2 = 4$$

Solution Writing the equation in the form

$$(x - a)^2 + (y - b)^2 = r^2$$

we have

$$(x - 1)^2 + [y - (-3)]^2 = 2^2$$

The center is at $(1, -3)$ and the radius is 2. (See Figure 9-9.)

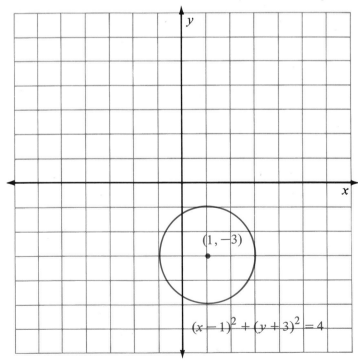

Figure 9-9   ▲

▼   **Example 6**   Sketch the graph of $x^2 + y^2 = 9$.

Solution   Since the equation can be written in the form

$$(x - 0)^2 + (y - 0)^2 = 3^2$$

it must have its center at $(0, 0)$ and a radius of 3. (See Figure 9-10.)

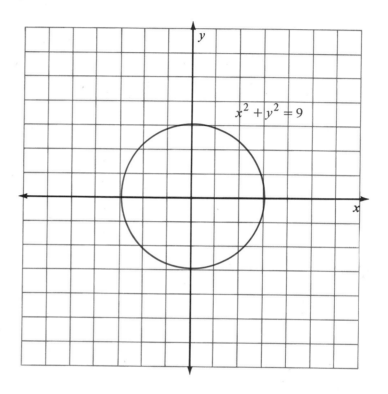

Figure 9-10   ▲

▼   **Example 7**   Sketch the graph of $x^2 + y^2 + 6x - 4y - 12 = 0$.

Solution   To sketch the graph we must find the center and radius. The center and radius can be identified if the equation has the form

$$(x - a)^2 + (y - b)^2 = r^2$$

The original equation can be written in this form by completing the squares on x and y:

$$x^2 + y^2 + 6x - 4y - 12 = 0$$
$$x^2 + 6x + y^2 - 4y = 12$$
$$x^2 + 6x + 9 + y^2 - 4y + 4 = 12 + 9 + 4$$
$$(x + 3)^2 + (y - 2)^2 = 25$$
$$(x + 3)^2 + (y - 2)^2 = 5^2$$

From the last line it is apparent that the center is at $(-3, 2)$ and the radius is 5. (See Figure 9-11.)

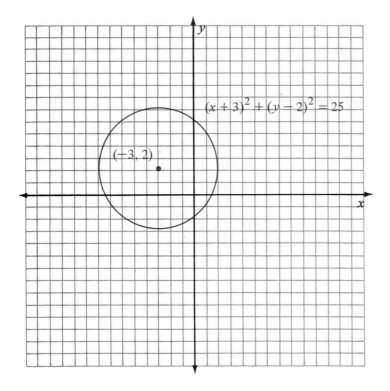

Figure 9-11

Problem Set 9.2   Find the distance between the following points:

1. $(3, 7)$ and $(6, 3)$ 2. $(4, 7)$ and $(8, 1)$
3. $(0, 9)$ and $(5, 0)$ 4. $(-3, 0)$ and $(0, 4)$
5. $(3, -5)$ and $(-2, 1)$ 6. $(-8, 9)$ and $(-3, -2)$
7. $(-1, -2)$ and $(-10, 5)$ 8. $(-3, -8)$ and $(-1, 6)$

9. Find x so the distance between $(x, 2)$ and $(1, 5)$ is $\sqrt{13}$.
10. Find x so the distance between $(-2, 3)$ and $(x, 1)$ is 3.
11. Find y so the distance between $(7, y)$ and $(8, 3)$ is 1.
12. Find y so the distance between $(3, -5)$ and $(3, y)$ is 9.

Write the equation of the circle with the given center and radius:

13. Center $(2, 3)$; $r = 4$ 14. Center $(3, -1)$; $r = 5$
15. Center $(3, -2)$; $r = 3$ 16. Center $(-2, 4)$; $r = 1$
17. Center $(-5, -1)$; $r = \sqrt{5}$ 18. Center $(-7, -6)$; $r = \sqrt{3}$
19. Center $(0, -5)$; $r = 1$ 20. Center $(0, -1)$; $r = 7$
21. Center $(0, 0)$; $r = 2$ 22. Center $(0, 0)$; $r = 5$
23. Center $(-1, 0)$; $r = 2\sqrt{3}$ 24. Center $(3, 0)$; $r = 2\sqrt{2}$

Give the center and radius, and sketch the graph, of each of the following circles:

25. $x^2 + y^2 = 4$ 26. $x^2 + y^2 = 16$
27. $x^2 + y^2 = 5$ 28. $x^2 + y^2 = 3$
29. $(x - 1)^2 + (y - 3)^2 = 25$ 30. $(x - 4)^2 + (y - 1)^2 = 36$
31. $(x + 2)^2 + (y - 4)^2 = 8$ 32. $(x - 3)^2 + (y + 1)^2 = 12$
33. $(x + 1)^2 + (y + 1)^2 = 1$ 34. $(x + 3)^2 + (y + 2)^2 = 9$
35. $x^2 + y^2 - 6y = 7$ 36. $x^2 + y^2 - 4y = 5$
37. $x^2 + y^2 + 2x = 1$ 38. $x^2 + y^2 + 10x = 2$
39. $x^2 + y^2 - 4x - 6y = -12$ 40. $x^2 + y^2 - 4x + 2y = 4$
41. $x^2 + y^2 + 2x + y = 3$ 42. $x^2 + y^2 - 6x - y = 1$
43. $x^2 + y^2 - x - 3y = 2$ 44. $x^2 + y^2 + x - 5y = 1$

45. Find the equation of the circle with center at the origin that contains the point $(3, 4)$.
46. Find the equation of the circle with center at the origin and x-intercepts $(3, 0)$ and $(-3, 0)$.
47. Find the equation of the circle with y-intercepts $(0, 4)$ and $(0, -4)$, and center at the origin.

48. A circle with center at $(-1, 3)$ passes through the point $(4, 3)$. Find the equation.

49. A circle with center at $(2, 5)$ passes through the point $(-1, 4)$. Find the equation.

50. A circle with center at $(-3, -4)$ contains the point $(2, -1)$. Find the equation.

51. If we were to solve the equation $x^2 + y^2 = 9$ for y, we would obtain the equation $y = \pm\sqrt{9 - x^2}$. This last equation is equivalent to the two equations $y = \sqrt{9 - x^2}$, in which y is always positive, and $y = -\sqrt{9 - x^2}$, in which y is always negative. Look at the graph of $x^2 + y^2 = 9$ in Example 6 of this section and indicate what part of the graph each of the two equations corresponds to.

52. Solve the equation $x^2 + y^2 = 9$ for x, and then indicate what part of the graph in Example 6 each of the two resulting equations corresponds to.

53. The formula for the circumference of a circle is $C = 2\pi r$. If the units of the coordinate system used in Problem 29 are in meters, what is the circumference of that circle?

54. The formula for the area of a circle is $A = \pi r^2$. What is the area of the circle mentioned in Problem 53?

Review Problems   The problems below review material we covered in Section 7.2.

Find the slope of the line that contains the following pairs of points.

55. $(-4, -1)$ and $(-2, 5)$ **56.** $(-2, -3)$ and $(-5, 1)$

57. Find y if the slope of the line through $(5, y)$ and $(4, 2)$ is 3.

58. Find x if the slope of the line through $(4, 9)$ and $(x, -2)$ is $-7/3$.

A line has a slope of $2/3$. Find the slope of any line:

59. parallel to it.   **60.** perpendicular to it.

This section is concerned with the graphs of ellipses and hyperbolas. To simplify matters somewhat we will only consider those graphs that are centered about the origin.

9.3
Ellipses and Hyperbolas

Suppose we want to graph the equation

$$\frac{x^2}{25} + \frac{y^2}{9} = 1$$

We can find the y-intercepts by letting $x = 0$, and the x-intercepts by letting $y = 0$:

When $x = 0$,   When $y = 0$,

$$\frac{0^2}{25} + \frac{y^2}{9} = 1 \qquad\qquad \frac{x^2}{25} + \frac{0^2}{9} = 1$$

$$y^2 = 9 \qquad\qquad\qquad x^2 = 25$$

$$y = \pm 3 \qquad\qquad\qquad x = \pm 5$$

The graph crosses the y-axis at $(0, 3)$ and $(0, -3)$ and the x-axis at $(5, 0)$ and $(-5, 0)$. Graphing these points and then connecting them with a smooth curve gives the graph shown in Figure 9-12.

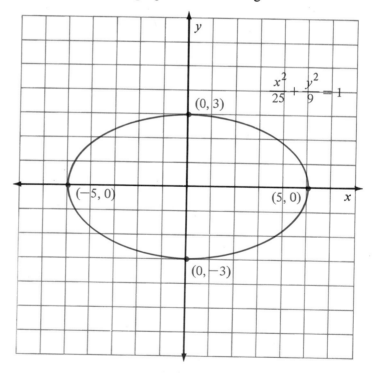

Figure 9-12

A graph of this type is called an *ellipse*. If we were to find some other ordered pairs that satisfy our original equation, we would find that their

graphs lie on the ellipse. Also, the coordinates of any point on the ellipse will satisfy the equation. We can generalize these results as follows.

The graph of any equation of the form

$$\frac{x^2}{a^2} + \frac{y^2}{b^2} = 1$$

will be an ellipse. The ellipse will cross the x-axis at $(a, 0)$ and $(-a, 0)$. It will cross the y-axis at $(0, b)$ and $(0, -b)$. When a and b are equal, the ellipse will be a circle.

The most convenient method for graphing an ellipse is by locating the intercepts.

▼   **Example 1**   Graph the ellipse described by $\dfrac{x^2}{16} + \dfrac{y^2}{36} = 1$.

Solution   The x-intercepts are 4 and -4. The y-intercepts are 6 and -6. (See Figure 9-13.)

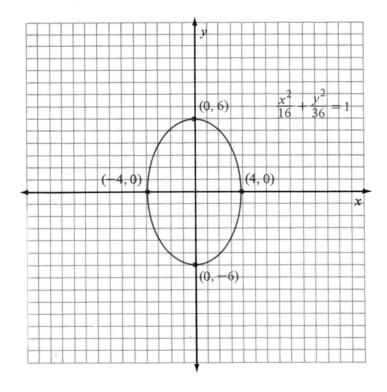

Figure 9-13   ▲

▼   **Example 2**   Sketch the graph of $4x^2 + 9y^2 = 36$.

Solution   To write the equation in the form

$$\frac{x^2}{a^2} + \frac{y^2}{b^2} = 1$$

we must divide both sides by 36:

$$\frac{4x^2}{36} + \frac{9y^2}{36} = \frac{36}{36}$$

$$\frac{x^2}{9} + \frac{y^2}{4} = 1$$

The graph crosses the x-axis at $(3, 0)$, $(-3, 0)$ and the y-axis at $(0, 2)$, $(0, -2)$. (See Figure 9-14.)

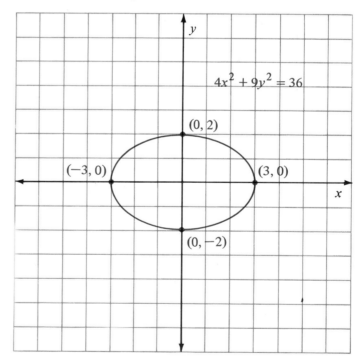

Figure 9-14   ▲

Consider the equation

$$\frac{x^2}{9} - \frac{y^2}{4} = 1$$

If we were to find a number of ordered pairs that are solutions to the equation and connect their graphs with a smooth curve, we would have Figure 9-15.

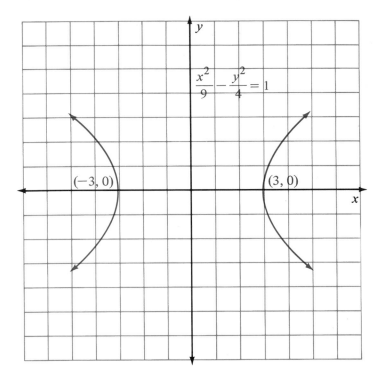

Figure 9-15

This graph is an example of a *hyperbola*. Notice that the graph has x-intercepts at $(3, 0)$ and $(-3, 0)$. The graph has no y-intercepts and hence does not cross the y-axis, since substituting $x = 0$ into the equation yields

$$\frac{0^2}{9} - \frac{y^2}{4} = 1$$
$$-y^2 = 4$$
$$y^2 = -4$$

for which there is no real solution. We can, however, use the number below y^2 to help sketch the graph. If we draw a rectangle that has its sides parallel to the x- and y-axes and that passes through the x-intercepts and the points on the y-axis corresponding to the square roots of the number

below y, $+2$ and -2, it looks like the rectangle in Figure 9-16.

The lines that connect opposite corners of the rectangle are called *asymptotes*. The graph of the hyperbola

$$\frac{x^2}{9} - \frac{y^2}{4} = 1$$

will approach these lines.

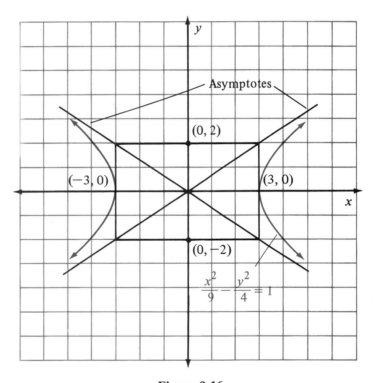

Figure 9-16

Although we won't attempt to prove so, it is true that there are two asymptotes associated with the graph of every hyperbola.

▼   **Example 3**   Graph the equation $\dfrac{y^2}{9} - \dfrac{x^2}{16} = 1$.

Solution   In this case the y-intercepts are 3 and -3, and the x-intercepts do not exist. We can use the square root of the number

below x^2, however, to find the asymptotes associated with the graph. The sides of the rectangle used to draw the asymptotes must pass through 3 and -3 on the y-axis, and 4 and -4 on the x-axis. (See Figure 9-17.)

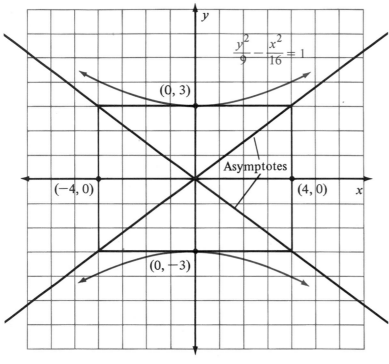

$$\frac{y^2}{9} - \frac{x^2}{16} = 1$$

$(0, 3)$

Asymptotes

$(-4, 0)$ $(4, 0)$

$(0, -3)$

Figure 9-17   ▲

▼   **Example 4**   Sketch the graph of $5x^2 - y^2 = 25$.

Solution   We begin by dividing both sides by 25:

$$\frac{5x^2}{25} - \frac{y^2}{25} = \frac{25}{25}$$

$$\frac{x^2}{5} - \frac{y^2}{25} = 1$$

The graph crosses the x-axis at $(\sqrt{5}, 0)$ and $(-\sqrt{5}, 0)$. The rectangle used to draw the asymptotes passes through these points and the points $(0, 5)$ and $(0, -5)$. (See Figure 9-18.)

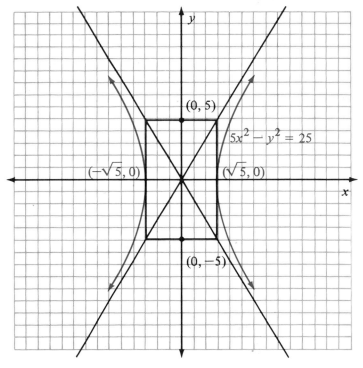

Figure 9-18   ▲

Problem Set 9.3

Graph each of the following. Be sure to label both the x- and y-intercepts.

1. $\dfrac{x^2}{9} + \dfrac{y^2}{16} = 1$

2. $\dfrac{x^2}{25} + \dfrac{y^2}{4} = 1$

3. $\dfrac{x^2}{16} + \dfrac{y^2}{9} = 1$

4. $\dfrac{x^2}{4} + \dfrac{y^2}{25} = 1$

5. $\dfrac{x^2}{3} + \dfrac{y^2}{4} = 1$

6. $\dfrac{x^2}{4} + \dfrac{y^2}{3} = 1$

7. $4x^2 + 25y^2 = 100$

8. $4x^2 + 9y^2 = 36$

9. $x^2 + 8y^2 = 16$

10. $12x^2 + y^2 = 36$

Graph each of the following. Show all intercepts and the asymptotes in each case.

11. $\dfrac{x^2}{9} - \dfrac{y^2}{16} = 1$

12. $\dfrac{x^2}{25} - \dfrac{y^2}{4} = 1$

13. $\dfrac{x^2}{16} - \dfrac{y^2}{9} = 1$

14. $\dfrac{x^2}{4} - \dfrac{y^2}{25} = 1$

15. $\dfrac{y^2}{9} - \dfrac{x^2}{16} = 1$ **16.** $\dfrac{y^2}{25} - \dfrac{x^2}{4} = 1$

17. $\dfrac{y^2}{36} - \dfrac{x^2}{4} = 1$ **18.** $\dfrac{y^2}{4} - \dfrac{x^2}{36} = 1$

19. $x^2 - 4y^2 = 4$ **20.** $y^2 - 4x^2 = 4$

21. $16y^2 - 9x^2 = 144$ **22.** $4y^2 - 25x^2 = 100$

23. Give the equation of the two asymptotes in the graph you found in Problem 15.

24. Give the equation of the two asymptotes in the graph you found in Problem 16.

25. For the ellipses you have graphed in this section, the longer line segment connecting opposite intercepts is called the *major axis* of the ellipse. Give the length of the major axis of the ellipse you graphed in Problem 3.

26. For the ellipses you have graphed in this section, the shorter line segment connecting opposite intercepts is called the *minor axis* of the ellipse. Give the length of the minor axis of the ellipse you graphed in Problem 3.

Review Problems   The problems that follow review material we covered in Section 7.4. Reviewing these problems will help you with the next section.

Graph each inequality.

27. $x + y < 5$ **28.** $x - y < 5$

29. $y \geq 2x - 1$ **30.** $y \leq 2x + 1$

31. $2x - 3y > 6$ **32.** $3x + 2y > 6$

In Section 7.4 we graphed linear inequalities by first graphing the boundary and then choosing a test point not on the boundary to indicate the region used for the solution set. The problems in this section are very similar. We will use the same general methods for graphing the inequalities in this section as we did in Section 7.4.

**9.4
Second-Degree
Inequalities**

▼   **Example 1**   Graph $x^2 + y^2 < 16$.

Solution   The boundary is $x^2 + y^2 = 16$, which is a circle with center at the origin and a radius of 4. Since the inequality sign is $<$, the boundary is not included in the solution set and must therefore

be represented with a broken line. The graph of the boundary is shown in Figure 9-19.

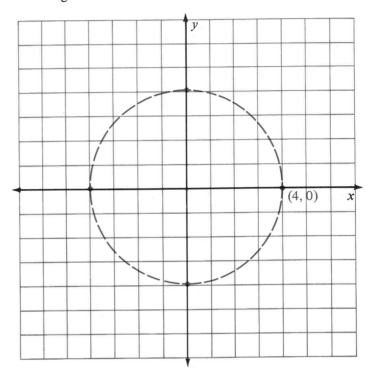

Figure 9-19

The solution set for $x^2 + y^2 < 16$ is either the region inside the circle or the region outside the circle. To see which region represents the solution set, we choose a convenient point not on the boundary and test it in the original inequality. The origin $(0, 0)$ is a convenient point. Since the origin satisfies the inequality $x^2 + y^2 < 16$, all points in the same region will also satisfy the inequality. The graph of the solution set is shown in Figure 9-20. ▲

▼   **Example 2**   Graph the inequality $y \leq x^2 - 2$.

Solution   The parabola $y = x^2 - 2$ is the boundary and is included in the solution set. Using $(0, 0)$ as the test point, we see that $0 \leq 0^2 - 2$ is a false statement, which means that the region containing $(0, 0)$ is not in the solution set. (See Figure 9-21.) ▲

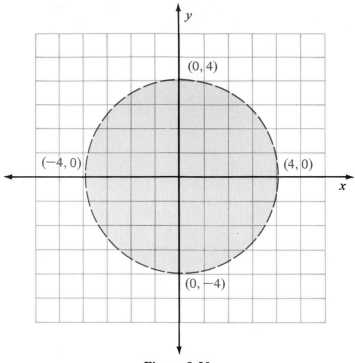

Figure 9-20

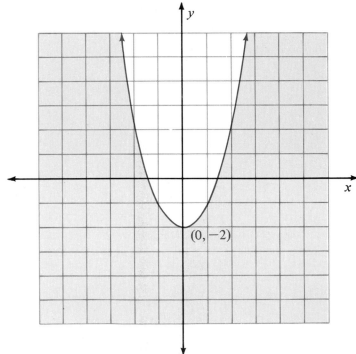

Figure 9-21

▼   **Example 3**   Graph $4y^2 - 9x^2 < 36$.

Solution   The boundary is the hyperbola $4y^2 - 9x^2 = 36$ and is not included in the solution set. Testing $(0, 0)$ in the original inequality yields a true statement, which means that the region containing the origin is the solution set. (See Figure 9-22.)

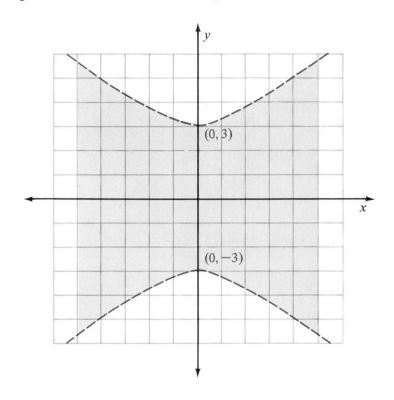

Figure 9-22   ▲

We now turn our attention to systems of inequalities. To solve a system of inequalities by graphing, we simply graph each inequality on the same set of axes. The solution set for the system is the region common to both graphs—the intersection of the individual solution sets.

▼   **Example 4**   Graph the solution set for the system

$$x^2 + y^2 \leq 9$$
$$\frac{x^2}{4} + \frac{y^2}{25} \geq 1$$

Solution The boundary for the top equation is a circle with center at the origin and a radius of 3. The solution set lies inside the boundary. The boundary for the second equation is an ellipse. In this case the solution set lies outside the boundary. (See Figure 9-23.)

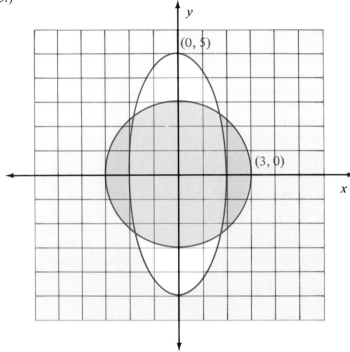

Figure 9-23

The solution set for the system is the intersection of the two individual solution sets. ▲

Graph each of the following inequalities: *Problem Set 9.4*

1. $x^2 + y^2 > 49$ 2. $x^2 + y^2 \geq 49$
3. $x^2 + y^2 \leq 49$ 4. $x^2 + y^2 < 49$
5. $(x - 2)^2 + (y + 3)^2 < 16$ 6. $(x + 3)^2 + (y - 2)^2 \geq 25$
7. $y \leq x^2 - 4$ 8. $y \geq x^2 + 3$
9. $y < x^2 - 6x + 7$ 10. $y \geq x^2 + 2x - 8$

11. $\dfrac{x^2}{9} + \dfrac{y^2}{25} < 1$ 12. $\dfrac{x^2}{9} - \dfrac{y^2}{25} > 1$

13. $\dfrac{x^2}{25} - \dfrac{y^2}{9} \geq 1$ 14. $\dfrac{x^2}{25} - \dfrac{y^2}{9} \leq 1$

15. $4x^2 + 25y^2 \leq 100$ **16.** $25x^2 - 4y^2 > 100$

Graph the solution sets to the following systems:

17. $x^2 + y^2 < 9$
$$y \geq x^2 - 1$$

18. $x^2 + y^2 \leq 16$
$$y < x^2 + 2$$

19. $\dfrac{x^2}{9} + \dfrac{y^2}{25} \leq 1$

$$\dfrac{x^2}{4} - \dfrac{y^2}{9} > 1$$

20. $\dfrac{x^2}{4} + \dfrac{y^2}{16} \geq 1$

$$\dfrac{x^2}{9} - \dfrac{y^2}{25} < 1$$

21. $4x^2 + 9y^2 \leq 36$
$$y > x^2 + 2$$

22. $9x^2 + 4y^2 \geq 36$
$$y < x^2 + 1$$

23. $x^2 + y^2 \geq 25$

$$\dfrac{x^2}{36} - \dfrac{y^2}{36} \geq 1$$

24. $x^2 + y^2 < 4$

$$x^2 - 4y^2 > 4$$

Review Problems   The problems that follow review material we covered in Section 8.1. Reviewing these problems will help you understand the next section.

Solve each system by the elimination method.

25. $4x + 3y = 10$
$$2x + y = 4$$

26. $3x - 5y = -2$
$$2x - 3y = 1$$

Solve each system by the substitution method.

27. $x + y = 3$
$$y = x + 3$$

28. $x + y = 6$
$$y = x - 4$$

29. $2x - 3y = -6$
$$y = 3x - 5$$

30. $7x - y = 24$
$$x = 2y + 9$$

9.5
Nonlinear Systems

Each system of equations in this section contains at least one second-degree equation. The most convenient method of solving a system that contains one or two second-degree equations is by substitution, although the elimination method can be used at times.

▼   **Example 1**   Solve the system

$$x^2 + y^2 = 4$$
$$x - 2y = 4$$

Solution   In this case the substitution method is the most convenient. Solving the second equation for x in terms of y, we have

$$x - 2y = 4$$
$$x = 2y + 4$$

We now substitute $2y + 4$ for x in the first equation in our original system and proceed to solve for y:

$$(2y + 4)^2 + y^2 = 4$$
$$4y^2 + 16y + 16 + y^2 = 4$$
$$5y^2 + 16y + 12 = 0$$
$$(5y + 6)(y + 2) = 0$$
$$5y + 6 = 0 \quad \text{or} \quad y + 2 = 0$$
$$y = -\tfrac{6}{5} \quad \text{or} \quad y = -2$$

These are the y-coordinates of the two solutions to the system. Substituting $y = -\tfrac{6}{5}$ into $x - 2y = 4$ and solving for x gives us $x = \tfrac{8}{5}$. Using $y = -2$ in the same equation yields $x = 0$. The two solutions to our system are $(\tfrac{8}{5}, -\tfrac{6}{5})$ and $(0, -2)$. Although graphing the system is not necessary, it does help us visualize the situation. (See Figure 9-24.)

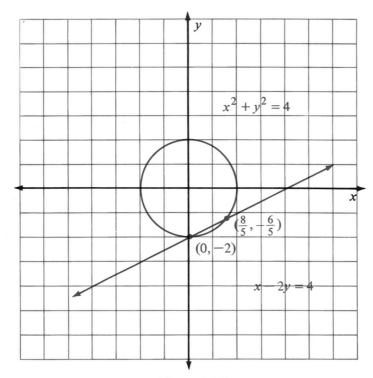

Figure 9-24

▼ **Example 2**   Solve the system

$$16x^2 - 4y^2 = 64$$
$$x^2 + y^2 = 9$$

Solution Since each equation is of the second degree in both x and y, it is easier to solve this system by eliminating one of the variables. To eliminate y we multiply the bottom equation by 4 and add the results to the top equation:

$$
\begin{aligned}
16x^2 - 4y^2 &= 64 \\
4x^2 + 4y^2 &= 36 \\
\hline
20x^2 \qquad &= 100
\end{aligned}
$$

$$x^2 = 5$$
$$x = \pm\sqrt{5}$$

The x-coordinates of the points of intersection are $\sqrt{5}$ and $-\sqrt{5}$. We substitute each back into the second equation in the original system and solve for y:

When $x = \sqrt{5}$,   When $x = -\sqrt{5}$,
$$(\sqrt{5})^2 + y^2 = 9 \qquad\qquad (-\sqrt{5})^2 + y^2 = 9$$
$$5 + y^2 = 9 \qquad\qquad\qquad 5 + y^2 = 9$$
$$y^2 = 4 \qquad\qquad\qquad\qquad y^2 = 4$$
$$y = \pm2 \qquad\qquad\qquad\qquad y = \pm2$$

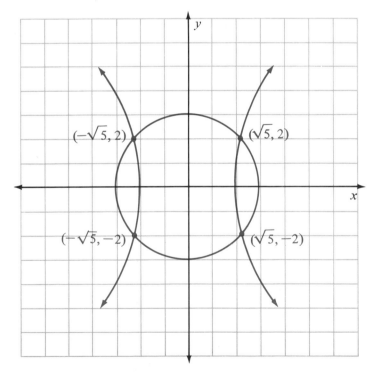

Figure 9-25

The four points of intersection are $(\sqrt{5}, 2)$, $(\sqrt{5}, -2)$, $(-\sqrt{5}, 2)$, and $(-\sqrt{5}, -2)$. Graphically the situation is as shown in Figure 9-25. ▲

▼   **Example 3**   The sum of the squares of two numbers is 34. The difference of their squares is 16. Find the two numbers.

Solution   Let x and y be the two numbers. The sum of their squares is $x^2 + y^2$ and the difference of their squares is $x^2 - y^2$. (We can assume here that x^2 is the larger number.) The system of equations that describes the situation is

$$x^2 + y^2 = 34$$
$$x^2 - y^2 = 16$$

We can eliminate y by simply adding the two equations. The result of doing so is

$$2x^2 = 50$$
$$x^2 = 25$$
$$x = \pm 5$$

Substituting $x = 5$ into either equation in the system gives $y = \pm 3$. Using $x = -5$ gives the same results, $y = \pm 3$. The four pairs of numbers that are solutions to the original problem are

$$\{5, 3\} \qquad \{-5, 3\} \qquad \{5, -3\} \qquad \{-5, -3\} \qquad ▲$$

Solve each of the following systems of equations: *Problem Set 9.5*

1. $x^2 + y^2 = 9$
    $2x + y = 3$
2. $x^2 + y^2 = 9$
    $x + 2y = 3$
3. $x^2 + y^2 = 16$
    $x + 2y = 8$
4. $x^2 + y^2 = 16$
    $x - 2y = 8$
5. $x^2 + y^2 = 25$
    $x^2 - y^2 = 25$
6. $x^2 - y^2 = 4$
    $2x^2 + y^2 = 5$
7. $x^2 + y^2 = 9$
    $y = x^2 - 3$
8. $x^2 + y^2 = 4$
    $y = x^2 - 2$
9. $x^2 + y^2 = 16$
    $y = x^2 - 4$
10. $x^2 + y^2 = 1$
    $y = x^2 - 1$
11. $4x^2 - 3y^2 = 12$
    $2x + 3y = 6$
12. $3x^2 + 4y^2 = 12$
    $3x - 2y = 6$
13. $3x + 2y = 10$
    $y = x^2 - 5$
14. $4x + 2y = 10$
    $y = x^2 - 10$
15. $4x^2 - 9y^2 = 36$
    $4x^2 + 9y^2 = 36$
16. $4x^2 + 25y^2 = 100$
    $4x^2 - 25y^2 = 100$

17. $2x^2 - 4y^2 = 8$
    $x^2 + 2y^2 = 10$

18. $4x^2 + 2y^2 = 8$
    $2x^2 - y^2 = 4$

19. $x - y = 4$
    $x^2 + y^2 = 16$

20. $x + y = 2$
    $x^2 - y^2 = 4$

21. The sum of the squares of two numbers is 89. The difference of the numbers is 3. Find the numbers.

22. The difference of the squares of two numbers is 35. The sum of their squares is 37. Find the numbers.

23. One number is 3 less than the square of another. Their sum is 9. Find the numbers.

24. The square of one number is 2 less than twice the square of another. The sum of the squares of the two numbers is 25. Find the numbers.

Review Problems   The problems below review material we covered in Section 7.3.

25. Give the equation of the line with slope -3 and y-intercept 5.
26. Give the slope and y-intercept of the line $2x - 3y = 6$.
27. Find the equation of the line with slope 5 that contains the point $(3, -2)$.
28. Find the equation of the line through $(1, 3)$ and $(-1, -5)$.
29. Find the equation of the line with x-intercept 3 and y-intercept -2.
30. Find the equation of the line through $(-1, 4)$ whose graph is perpendicular to the graph of $y = 2x + 3$.

Chapter 9 Summary and Review

CONIC SECTIONS [9.1]

Each of the four conic sections can be obtained by slicing a cone with a plane at different angles as shown in Figure 9-26.

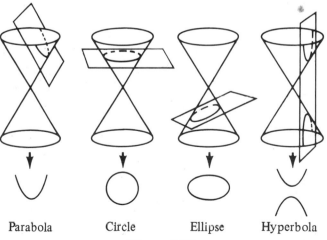

Parabola   Circle   Ellipse   Hyperbola

Figure 9-26

THE PARABOLA [9.1]

The graph of any equation of the form

$$y = ax^2 + bx + c \qquad a \neq 0$$

is a parabola. The graph is concave-up if $a > 0$, and concave-down if $a < 0$. The highest or lowest point on the graph is called the *vertex* and will always occur when $x = -b/2a$.

DISTANCE FORMULA [9.2]

The distance between the two points (x_1, y_1) and (x_2, y_2) is given by the formula

$$d = \sqrt{(x_2 - x_1)^2 + (y_2 - y_1)^2}.$$

THE CIRCLE [9.2]

The graph of any equation of the form

$$(x - a)^2 + (y - b)^2 = r^2$$

will be a circle having its center at (a, b) and a radius of r.

THE ELLIPSE [9.3]

Any equation that can be put in the form

$$\frac{x^2}{a^2} + \frac{y^2}{b^2} = 1$$

will have an ellipse for its graph. The x-intercepts will be at a and $-a$, the y-intercepts at b and $-b$.

THE HYPERBOLA [9.3]

The graph of an equation that can be put in either of the forms

$$\frac{x^2}{a^2} - \frac{y^2}{b^2} = 1 \quad \text{or} \quad \frac{y^2}{b^2} - \frac{x^2}{a^2} = 1$$

will be a hyperbola. The x-intercepts, if they exist, will be a and $-a$. The y-intercepts, if they occur, will be at b and $-b$. There are two straight lines called asymptotes associated with

Examples

1. The graph of $y = x^2 - 4$ will be a parabola. It will cross the x-axis at 2 and -2, and the vertex will be $(0, -4)$.

2. The distance between $(5, 2)$ and $(-1, 1)$ is

$$d = \sqrt{(5 + 1)^2 + (2 - 1)^2}$$
$$= \sqrt{37}.$$

3. The graph of the circle $(x - 3)^2 + (y + 2)^2 = 25$ will have its center at $(3, -2)$ and the radius will be 5.

4. The ellipse $\frac{x^2}{9} + \frac{y^2}{4} = 1$ will cross the x-axis at 3 and -3, and will cross the y-axis at 2 and -2.

5. The hyperbola $\frac{x^2}{4} - \frac{y^2}{9} = 1$ will cross the x-axis at 2 and -2. It will not cross the y-axis.

the graph of every hyperbola. Although the asymptotes are not part of the hyperbola, they are useful in sketching the graph.

**Chapter 9
Test**

1. Sketch the graph of each of the following. Give the coordinates of the vertex in each case.

    a. $y = x^2 - x - 12$ **b.** $y = -x^2 + 2x + 8$

2. Find the distance between the points $(3, -7)$ and $(4, 2)$.
3. Find x so that $(x, 2)$ is $2\sqrt{5}$ units from $(-1, 4)$.
4. Give the equation of the circle with center at $(-2, 4)$ and radius 3.
5. Give the equation of the circle with center at the origin that contains the point $(-3, -4)$.
6. Find the center, radius, and sketch the graph of the circle whose equation is $x^2 + y^2 - 10x + 6y = 5$.

Graph each of the following.

7. $4x^2 - y^2 = 16$ 8. $\dfrac{x^2}{25} + \dfrac{y^2}{4} = 1$

9. $(x - 2)^2 + (y + 1)^2 \leq 9$ 10. $\dfrac{x^2}{4} + \dfrac{y^2}{25} \leq 1$

Solve the following systems.

11. $x^2 + y^2 = 25$ 12. $x^2 + y^2 = 16$
     $2x + y = 5$ $y = x^2 - 4$

10

Relations and Functions

To the student:

In this chapter we will study two main concepts, relations and functions. Relations and functions have many applications in the real world. The idea of a relation is already familiar to us on an intuitive level. When we say "the price of gasoline is increasing because there is more demand for it this year," we are expressing a relationship between the price of gasoline and the demand for it. We are implying the price of gasoline is a function of the demand for it. Mathematics becomes a part of this problem when we express, with an equation, the exact relationship between the two quantities.

Actually, we have been working with functions and relations for some time now, we just haven't said so. Any time we have used an equation containing two variables we have been using a relation. This chapter is really just a formalization and classification of concepts and equations that we have already encountered. The chapter begins with some basic definitions associated with functions and relations. We will develop a new notation associated with functions called function notation; we will consider combinations of functions and exponential functions. The chapter ends with a section on the inverse of a function.

10.1
Relations and
Functions

We begin this section with the definition of a relation. It is apparent from the definition that we have worked with relations many times previously in this book.

DEFINITION   A *relation* is any set of ordered pairs. The set of all first coordinates is called the *domain* of the relation, and the set of all second coordinates is said to be the *range* of the relation.

There are two ways to specify the ordered pairs in a relation. One method is to simply list them. The other method is to give the rule (equation) for obtaining them.

▼   **Example 1**   The set $\{(1, 2), (-3, \frac{1}{2}), (\pi, -4), (0, 1)\}$ is a relation. The domain for this relation is $\{1, -3, \pi, 0\}$ and the range is $\{2, \frac{1}{2}, -4, 1\}$.   ▲

▼   **Example 2**   The set of ordered pairs given by $\{(x, y) \mid x + y = 5\}$ is an example of a relation. In this case we have written the relation in terms of the equation used to obtain the ordered pairs in the relation. This relation is the set of all ordered pairs whose coordinates have a sum of 5. Some members of this relation are $(1, 4)$, $(0, 5)$, $(5, 0)$, $(-1, 6)$, and $(\frac{1}{2}, \frac{9}{2})$. It is impossible to list all the members of this relation since there are an infinite number of ordered pairs that are solutions to the equation

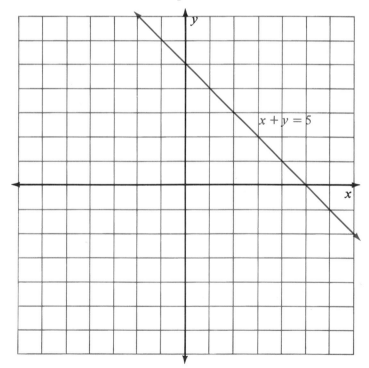

Figure 10-1

$x + y = 5$. The domain, although not given directly, is the set of real numbers; the range is also. The graph of this relation is a straight line, shown in Figure 10-1. ▲

DEFINITION A *function* is a relation in which no two different ordered pairs have the same first coordinates. The *domain* and *range* of a function are the sets of first and second coordinates, respectively.

A function is simply a relation that does not repeat any first coordinates. The restriction on first coordinates may seem arbitrary at first. As it turns out, restricting ordered pairs to those that do not repeat first coordinates allows us to generalize and develop properties we would otherwise be unable to work with.

▼ **Example 3** The relation {(2, 3), (5, 2), (6, 3)} is also a function, since no two ordered pairs have the same first coordinates. The relation {(1, 7), (3, 7), (1, 5)} is not a function since two of its ordered pairs, (1, 7) and (1, 5), have the same first coordinates. ▲

Vertical Line Test

If the ordered pairs of a relation are given in terms of an equation rather than a list, we can use the graph of the relation to determine if the relation is a function or not. Any two ordered pairs with the same first coordinates will lie along a vertical line parallel to the y-axis. Therefore, if a vertical line crosses the graph of a relation in more than one place, the relation cannot be a function. If no vertical line can be found that crosses the graph in more than one place, the relation must be a function. Testing to see if a relation is a function by observing if a vertical line crosses the graph of the relation in more than one place is called the *vertical line test*.

▼ **Example 4** Use the vertical line test to see which of the following are functions: (a) $y = x^2 - 2$; (b) $x^2 + y^2 = 9$.

Solution The graph of each relation is given in Figure 10-2. The equation $y = x^2 - 2$ is a function, since there are no vertical lines that cross its graph in more than one place. The equation $x^2 + y^2 = 9$ does not represent a function, since we can find a vertical line that crosses its graph in more than one place. ▲

The Domain and Range of a Function

When a function (or relation) is given in terms of an equation, the domain is the set of all possible replacements for the variable x. If the domain of a function (or relation) is not specified, it is assumed to be all real numbers that do not give undefined terms in the equation. That is, we cannot use values of x in the domain that will produce 0 in a denominator or the square root of a negative number.

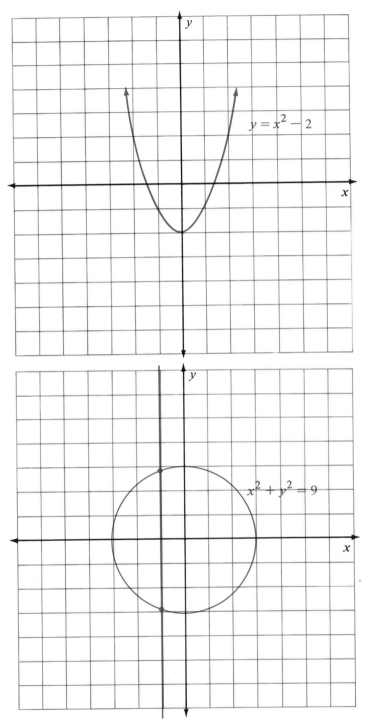

Figure 10-2

▼   **Example 5**   Specify the domain for $y = \dfrac{1}{x - 3}$.

Solution   The domain can be any real number that does not produce an undefined term. If $x = 3$, the denominator on the right side will be 0. Hence, the domain is all real numbers except 3.   ▲

▼   **Example 6**   Give the domain for $y = \sqrt{x - 4}$.

Solution   Since the domain must consist of real numbers, the quantity under the radical will have to be greater than or equal to 0:

$$x - 4 \geq 0$$
$$x \geq 4$$

The domain in this case is $\{x \,|\, x \geq 4\}$.   ▲

The graph of a function (or relation) is sometimes helpful in determining the domain and range.

▼   **Example 7**   Give the domain and range for $9x^2 + 4y^2 = 36$.

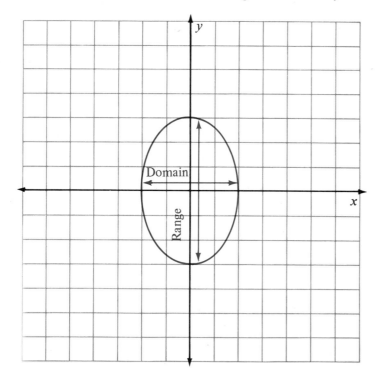

Figure 10-3

Solution   The graph is the ellipse shown in Figure 10-3.

From the graph we have

$$\text{Domain} = \{x \mid -2 \leq x \leq 2\}$$
$$\text{Range} = \{y \mid -3 \leq y \leq 3\}$$

Note also that by applying the vertical line test we see the relation is not a function.   ▲

▼   **Example 8**   Give the domain and range for $y = x^2 - 3$.

Solution   The graph is the parabola shown in Figure 10-4.

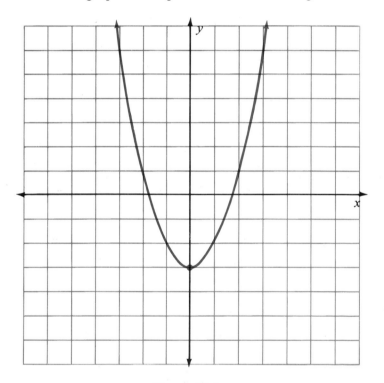

Figure 10-4

The domain is all real numbers, and the range is the set $\{y \mid y \geq -3\}$. Since no vertical line will cross the graph in more than one place, the graph represents a function.   ▲

▼   **Example 9**   Give the domain and range for $y - \sqrt{9 - x^2}$.

Solution   Since y is equal to a positive square root, y must be positive or zero. Keeping this in mind, we square both sides of the equation:

$$\begin{aligned} y^2 &= 9 - x^2 & y &\geq 0 \\ x^2 + y^2 &= 9 & y &\geq 0 \end{aligned}$$

The graph of the last line is shown in Figure 10-5.

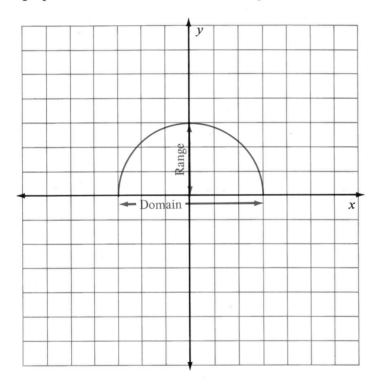

Figure 10-5

The graph is the graph of a function. The domain is $\{x \mid -3 \leq x \leq 3\}$ and the range is $\{y \mid 0 \leq y \leq 3\}$.   ▲

For each of the following relations, give the domain and range and indicate which are also functions:

Problem Set 10.1

1. $\{(1, 3), (2, 5), (4, 1)\}$
2. $\{(3, 1), (5, 7), (2, 3)\}$
3. $\{(-1, 3), (1, 3), (2, -5)\}$
4. $\{(3, -4), (-1, 5), (3, 2)\}$
5. $\{(7, -1), (3, -1), (7, 4)\}$
6. $\{(5, -2), (3, -2), (5, -1)\}$
7. $\{(4, 3), (3, 4), (3, 5)\}$
8. $\{(4, 1), (1, 4), (-1, -4)\}$

9. $\{(5, -3), (-3, 2), (2, -3)\}$ 10. $\{(2, 4), (3, 4), (4, 4)\}$

Which of the following graphs represent functions?

11.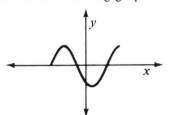

12.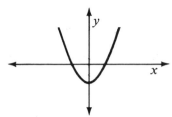

13.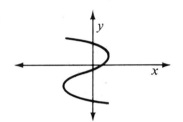

14.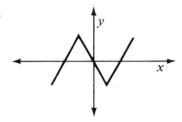

15.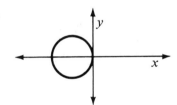

16.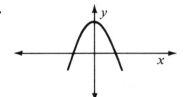

17.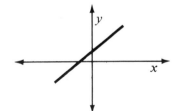

18.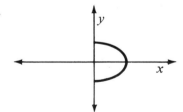

19.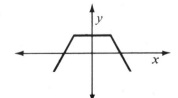

20.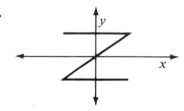

Give the domain for each of the following functions:

21. $y = \sqrt{x + 3}$ **22.** $y = \sqrt{x + 4}$

23. $y = \sqrt{2x - 1}$ **24.** $y = \sqrt{3x + 2}$

25. $y = \sqrt{1 - 4x}$ **26.** $y = \sqrt{2 + 3x}$

27. $y = \dfrac{x + 2}{x - 5}$ **28.** $y = \dfrac{x - 3}{x + 4}$

29. $y = \dfrac{3}{2x^2 + 5x - 3}$ **30.** $y = \dfrac{-1}{3x^2 - 5x + 2}$

31. $y = \dfrac{-3}{x^2 - x - 6}$ **32.** $y = \dfrac{4}{x^2 - 2x - 8}$

33. $y = \dfrac{4}{x^2 - 4}$ **34.** $y = \dfrac{2}{x^2 - 9}$

Graph each of the following relations. Use the graph to find the domain and range, and indicate which relations are also functions.

35. $x^2 + 4y^2 = 16$ **36.** $4x^2 + y^2 = 16$

37. $(x - 2)^2 + (y + 1)^2 = 9$ **38.** $(x + 3)^2 + (y - 4)^2 = 25$

39. $x^2 + y^2 - 4x = 12$ **40.** $x^2 + y^2 + 6x = 16$

41. $y = x^2 - x - 12$ **42.** $y = x^2 + 2x - 8$

43. $y = 2x - 3$ **44.** $y = 3x - 2$

45. $\dfrac{x^2}{4} - \dfrac{y^2}{9} = 1$ **46.** $\dfrac{x^2}{9} - \dfrac{y^2}{4} = 1$

47. $\dfrac{x^2}{9} + \dfrac{y^2}{4} = 1$ **48.** $\dfrac{x^2}{4} + \dfrac{y^2}{9} = 1$

49. $y = \sqrt{16 - x^2}$ **50.** $y = -\sqrt{25 - x^2}$

51. $y = -\sqrt{9 + x^2}$ **52.** $y = \sqrt{16 + x^2}$

53. $2y = \sqrt{36 - 9x^2}$ **54.** $2y = -\sqrt{36 - 9x^2}$

55. A ball is thrown straight up into the air from ground level. The relationship between the height (h) of the ball at any time (t) is illustrated by the following graph:

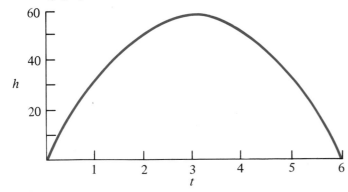

The horizontal axis represents time (t) and the vertical axis represents height (h).

   a. Is this graph the graph of a function?
   b. Identify the domain and range.
   c. At what time does the ball reach its maximum height?
   d. What is the maximum height of the ball?
   e. At what time does the ball hit the ground?

56. The graph below shows the relationship between a company's profits, P, and the number of items it sells, x. (P is in dollars.)

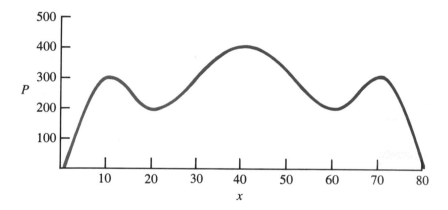

   a. Is this graph the graph of a function?
   b. Identify the domain and range.
   c. How many items must the company sell to make their maximum profit?
   d. What is their maximum profit?

Review Problems The problems below review material we covered in Section 8.2.

Solve each system.

57. $\begin{aligned} x + y + z &= 6 \\ 2x - y + z &= 3 \\ x + 2y - 3z &= -4 \end{aligned}$
        58. $\begin{aligned} x + y + z &= 6 \\ x - y + 2z &= 7 \\ 2x - y - z &= 0 \end{aligned}$

59. $\begin{aligned} 3x + 4y &= 15 \\ 2x - 5z &= -3 \\ 4y - 3z &= 9 \end{aligned}$
        60. $\begin{aligned} x + 3y &= 5 \\ 6y + z &= 12 \\ x - 2z &= -10 \end{aligned}$

10.2
Function Notation

Consider the function

$$y = 3x - 2$$

Solution The expression $f(x + h)$ is given by

$$f(x + h) = 2(x + h) - 3$$
$$= 2x + 2h - 3$$

Using this result gives us

$$\frac{f(x + h) - f(x)}{h} = \frac{(2x + 2h - 3) - (2x - 3)}{h}$$

$$= \frac{2h}{h}$$

$$= 2$$

The expression

$$\frac{f(x + h) - f(x)}{h}$$

is a very important formula used in calculus. We are using it here just for practice. ▲

▼ **Example 6**   If $f(x) = x^2 - 3x$, find $\dfrac{f(x + h) - f(x)}{h}$.

Solution

$$\frac{f(x + h) - f(x)}{h} = \frac{[(x + h)^2 - 3(x + h)] - (x^2 - 3x)}{h}$$

$$= \frac{x^2 + 2xh + h^2 - 3x - 3h - x^2 + 3x}{h}$$

$$= \frac{2xh - 3h + h^2}{h}$$

$$= 2x - 3 + h$$ ▲

▼ **Example 7**   If $f(x) = 2x - 4$, find $\dfrac{f(x) - f(a)}{x - a}$.

Solution

$$\frac{f(x) - f(a)}{x - a} = \frac{(2x - 4) - (2a - 4)}{x - a}$$

$$= \frac{2x - 2a}{x - a}$$

$$= \frac{2(x - a)}{x - a}$$

$$= 2$$ ▲

Problem Set 10.2

Let $f(x) = 2x - 5$ and $g(x) = x^2 + 3x + 4$. Evaluate the following:

1. $f(2)$ 2. $g(3)$
3. $f(-3)$ 4. $g(-2)$
5. $f(0)$ 6. $f(5)$
7. $g(-1)$ 8. $f(-4)$
9. $g(-3)$ 10. $g(2)$
11. $g(4) + f(4)$ 12. $f(2) - g(3)$
13. $f(3) - g(2)$ 14. $g(-1) + f(-1)$

If $f = \{(1, 4), (-2, 0), (3, \frac{1}{2}), (\pi, 0)\}$ and $g = \{(1, 1), (-2, 2), (\frac{1}{2}, 0)\}$, find each of the following values of f and g:

15. $f(1)$ 16. $g(1)$
17. $g(\frac{1}{2})$ 18. $f(3)$
19. $g(-2)$ 20. $f(\pi)$
21. $f(-2) + g(-2)$ 22. $g(1) + f(1)$
23. $f(-2) - g(1) + f(3)$ 24. $g(1) + g(-2) + f(-2)$

Let $f(x) = 3x^2 - 4x + 1$ and $g(x) = 2x - 1$. Evaluate each of the following:

25. $f(0)$ 26. $g(0)$
27. $g(-4)$ 28. $f(1)$
29. $f(a)$ 30. $g(z)$
31. $f(a + 3)$ 32. $g(a - 2)$
33. $f[g(2)]$ 34. $g[f(z)]$
35. $g[f(-1)]$ 36. $f[g(-2)]$
37. $g[f(0)]$ 38. $f[g(0)]$
39. $f[g(x)]$ 40. $g[f(x)]$

For each of the following functions, evaluate the quantity $\dfrac{f(x + h) - f(x)}{h}$:

41. $f(x) = 2x + 3$ 42. $f(x) = 3x - 2$
43. $f(x) = x^2$ 44. $f(x) = x^2 - 3$
45. $y = -4x - 1$ 46. $y = -x + 4$
47. $y = 3x^2 - 2$ 48. $y = 4x^2 + 3$
49. $f(x) = 2x^2 + 3x + 4$ 50. $f(x) = 4x^2 + 3x + 2$

For each of the following functions evaluate the quantity $\dfrac{f(x) - f(a)}{x - a}$:

51. $f(x) = 3x$ 52. $f(x) = -2x$
53. $f(x) = 4x - 5$ 54. $f(x) = 3x + 1$
55. $f(x) = x^2$ 56. $f(x) = 2x^2$
57. $y = 5x - 3$ 58. $y = -2x + 7$
59. $y = x^2 + 1$ 60. $y = x^2 - 1$

61. Suppose the total cost C of manufacturing x items is given by the equation $C(x) = 2x^2 + 5x + 100$. If the notation $C(x)$ is used in the same way we have been using $f(x)$, then $C(10) = 2(10)^2 + 5(10) + 100 = 350$ is the

total cost, in dollars, of manufacturing 10 items. Find the total cost of producing 1 item, 5 items, and 20 items.

62. Suppose the profit P a company makes by selling x items is given by $P(x) = -x^3 + 100x^2$. Find the profit made if 0 items are sold and if 50 items are sold.

63. Suppose $f(x) = x^2 - 5x + 6$. Find the values of x for which $f(x) = 0$.

64. Let $f(x) = x^2 - x - 9$, and find the values of x for which $f(x) = 3$.

65. Let $f(x) = 3x - 5$ and $g(x) = 7x + 1$, and find the value of x for which $f(x) = g(x)$.

66. Use the functions $f(x)$ and $g(x)$ as given in Problem 65 and find the value of x for which $f(x) = -g(x)$.

Review Problems The problems below review material we covered in Section 8.3.

Find the value of each determinant.

67. $\begin{vmatrix} 2 & 3 \\ 5 & 1 \end{vmatrix}$ 68. $\begin{vmatrix} 6 & 1 \\ 2 & 0 \end{vmatrix}$.

69. $\begin{vmatrix} -3 & 5 \\ -2 & -7 \end{vmatrix}$ 70. $\begin{vmatrix} 4 & -3 \\ -1 & -8 \end{vmatrix}$

71. $\begin{vmatrix} 2 & 3 & -2 \\ 1 & 4 & 1 \\ 1 & 5 & -1 \end{vmatrix}$ 72. $\begin{vmatrix} 1 & 2 & 0 \\ 1 & 1 & 1 \\ 0 & 2 & 1 \end{vmatrix}$

If we are given two functions, f and g, with a common domain, we can define four other functions as follows.

**10.3
Algebra with
Functions**

DEFINITION

$$(f + g)(x) = f(x) + g(x)$$ The function $f + g$ is the sum of the functions f and g.

$$(f - g)(x) = f(x) - g(x)$$ The function $f - g$ is the difference of the functions f and g.

$$(fg)(x) = f(x)g(x)$$ The function fg is the product of the functions f and g.

$$\frac{f}{g}(x) = \frac{f(x)}{g(x)}$$ The function f/g is the quotient of the functions f and g, where $g(x) \neq 0$.

▼   **Example 1**   If $f(x) = 4x^2 + 3x + 2$ and $g(x) = 2x^2 - 5x - 6$, write the formula for the functions $f + g, f - g, fg$, and f/g.

Solution   The function $f + g$ is defined by

$$
\begin{aligned}
(f + g)(x) &= f(x) + g(x) \\
&= (4x^2 + 3x + 2) + (2x^2 - 5x - 6) \\
&= 6x^2 - 2x - 4
\end{aligned}
$$

The function $f - g$ is defined by

$$
\begin{aligned}
(f - g)(x) &= f(x) - g(x) \\
&= (4x^2 + 3x + 2) - (2x^2 - 5x - 6) \\
&= 4x^2 + 3x + 2 - 2x^2 + 5x + 6 \\
&= 2x^2 + 8x + 8
\end{aligned}
$$

The function fg is defined by

$$
\begin{aligned}
(fg)(x) &= f(x)g(x) \\
&= (4x^2 + 3x + 2)(2x^2 - 5x - 6) \\
&= 8x^4 - 20x^3 - 24x^2 \\
&\quad\quad + 6x^3 - 15x^2 - 18x \\
&\quad\quad\quad\quad + 4x^2 - 10x - 12 \\
&= 8x^4 - 14x^3 - 35x^2 - 28x - 12
\end{aligned}
$$

The function f/g is defined by

$$
\begin{aligned}
(f/g)(x) &= f(x)/g(x) \\
&= \frac{4x^2 + 3x + 2}{2x^2 - 5x - 6}
\end{aligned}
$$

▲

▼   **Example 2**   Let $f(x) = 4x - 3$, $g(x) = 4x^2 - 7x + 3$, and $h(x) = x - 1$. Find $f + g, fh, fg$, and g/f.

Solution   The function $f + g$, the sum of functions f and g, is defined by

$$
\begin{aligned}
(f + g)(x) &= f(x) + g(x) \\
&= (4x - 3) + (4x^2 - 7x + 3) \\
&= 4x^2 - 3x
\end{aligned}
$$

The function fh, the product of functions f and h, is defined by

$$(fh)(x) = f(x)h(x)$$
$$= (4x - 3)(x - 1)$$
$$= 4x^2 - 7x + 3$$
$$= g(x)$$

The product of the functions f and g, fg, is given by

$$(fg)(x) = f(x)g(x)$$
$$= (4x - 3)(4x^2 - 7x + 3)$$
$$= 16x^3 - 28x^2 + 12x - 12x^2 + 21x - 9$$
$$= 16x^3 - 40x^2 + 33x - 9$$

The quotient of the functions g and f, g/f, is defined as

$$\frac{g}{f}(x) = \frac{g(x)}{f(x)}$$
$$= \frac{4x^2 - 7x + 3}{4x - 3}$$

Factoring the numerator, we can reduce to lowest terms:

$$\frac{g}{f}(x) = \frac{(4x - 3)(x - 1)}{4x - 3}$$
$$= x - 1$$
$$= h(x)$$ ▲

▼   **Example 3**   If f, g, and h are the same functions defined in Example 2, evaluate $(f + g)(2)$, $(fh)(-1)$, $(fg)(0)$, and $(g/f)(5)$.

Solution   We use the formulas for $f + g$, fh, fg, and g/f found in Example 2:

$$(f + g)(2) = 4(2)^2 - 3(2)$$
$$= 16 - 6$$
$$= 10$$

$$(fh)(-1) = 4(-1)^2 - 7(-1) + 3$$
$$= 4 + 7 + 3$$
$$= 14$$
$$(fg)(0) = 16(0)^3 - 40(0)^2 + 33(0) - 9$$
$$= 0 - 0 + 0 - 9$$
$$= -9$$

$$\frac{g}{f}(5) = 5 - 1$$
$$= 4 \qquad \blacktriangle$$

Problem Set 10.3

Let $f(x) = 4x - 3$ and $g(x) = 2x + 5$. Write a formula for each of the following functions:

1. $f + g$ 2. $f - g$
3. $g - f$ 4. $g + f$
5. fg 6. f/g
7. g/f 8. ff

If the functions f, g, and h are defined by $f(x) = 3x - 5$, $g(x) = x - 2$, and $h(x) = 3x^2 - 11x + 10$, write a formula for each of the following functions:

9. $g + f$ 10. $f + h$
11. $g + h$ 12. $f - g$
13. $g - f$ 14. $h - g$
15. fg 16. gf
17. fh 18. gh
19. h/f 20. h/g
21. f/h 22. g/h
23. $f + g + h$ 24. $h - g + f$
25. $h + fg$ 26. $h - fg$

Let $f(x) = 2x + 1$, $g(x) = 4x + 2$, and $h(x) = 4x^2 + 4x + 1$, and find the following:

27. $(f + g)(2)$ 28. $(f - g)(-1)$
29. $(fg)(3)$ 30. $(f/g)(-3)$
31. $(h/g)(1)$ 32. $(hg)(1)$
33. $(fh)(0)$ 34. $(h - g)(-4)$

35. $(f + g + h)(2)$
36. $(h - f + g)(0)$
37. $(h + fg)(3)$
38. $(h - fg)(5)$

39. Suppose a phone company charges 33¢ for the first minute and 24¢ for each additional minute to place a long-distance call out of state between 5 P.M. and 11 P.M. If x is the number of additional minutes and $f(x)$ is the cost of the call, then $f(x) = 24x + 33$.

    a. How much does it cost to talk for 10 minutes?
    b. What does $f(5)$ represent in this problem?
    c. If a call costs $1.29, how long was it?

40. The same phone company mentioned in Problem 39 charges 52¢ for the first minute and 36¢ for each additional minute to place an out-of-state call between 8 A.M. and 5 P.M.

    a. Let $g(x)$ be the total cost of an out-of-state call between 8 A.M. and 5 P.M. and write an equation for $g(x)$.
    b. Find $g(5)$.
    c. Find the difference in price between a 10-minute call made between 8 A.M. and 5 P.M. and the cost of the same call made between 5 P.M. and 11 P.M.

41. If $f(x) = x^3 - 2x$, show that $f(-x) = -f(x)$.
42. If $f(x) = x^4 + 2x^2$, show that $f(-x) = f(x)$.
43. Let $f(x) = x^2$ and $g(x) = x + 6$, and find all values of x for which $f(x) - g(x) = 0$.
44. Let $f(x) = x^2 - 2x$ and $g(x) = x - 2$, and find all values of x for which $f(x) - g(x) = 0$.

Review Problems The problems below review material we covered in Section 8.4.

Solve each system by using Cramer's rule.

45. $x + y = 5$
    $3x - y = 3$
46. $2x + y = 5$
    $5x + 3y = 11$
47. $4x - 7y = 3$
    $5x + 2y = -3$
48. $9x - 8y = 4$
    $2x + 3y = 6$
49. $3x + 4y = 15$
    $2x - 5z = -3$
    $4y - 3z = 9$
50. $x + 3y = 5$
    $6y + z = 12$
    $x - 2z = -10$

10.4
Classification of
Functions

Much of the work we have done in previous chapters has involved functions. All linear equations in two variables, except those with vertical lines for graphs, are functions. The parabolas we worked with in Chapter 9 are graphs of functions. We will begin this section by looking at some of the major classifications of functions. We will then spend the remainder of this section working with exponential functions.

Constant Functions

Any function that can be written in the form

$$f(x) = c$$

where c is a real number, is called a *constant function*. The graph of every constant function is a horizontal line.

▼ **Example 1**   The function $f(x) = 3$ is an example of a constant function. Since all ordered pairs belonging to f have a y-coordinate of 3, the graph is the horizontal line given by $y = 3$. Remember, y and $f(x)$ are equivalent—that is, $y = f(x)$. (See Figure 10-6.)

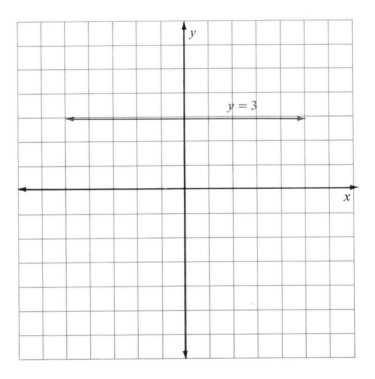

Figure 10-6   ▲

Any function that can be written in the form   *Linear Functions*

$$f(x) = ax + b$$

where a and b are real numbers, $a \neq 0$, is called a *linear function*. The graph of every linear function is a straight line. In the past we have written linear functions in the form

$$y = mx + b$$

▼   **Example 2** The function $f(x) = 2x - 3$ is an example of a linear function. The graph of this function is a straight line with slope 2 and y-intercept -3. (See Figure 10-7.)

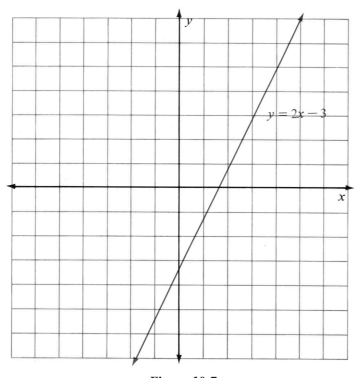

$$y = 2x - 3$$

Figure 10-7   ▲

A *quadratic function* is any function that can be written in the form   *Quadratic Functions*

$$f(x) = ax^2 + bx + c$$

where a, b, and c are real numbers and $a \neq 0$. The graph of every quadratic function is a parabola. We considered parabolic graphs in

Section 9.1. At that time the quadratic functions were written as $y = ax^2 + bx + c$ using y instead of $f(x)$.

▼   **Example 3**   The function $f(x) = x^2 + 3x - 4$ is an example of a quadratic function. Using the methods developed in Section 9.1, we sketch the graph shown in Figure 10-8.

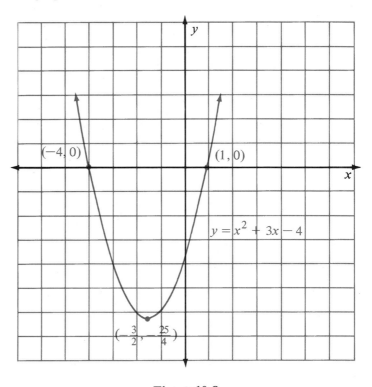

Figure 10-8   ▲

There are many other classifications of functions. One such classification we have not worked with previously is the exponential function. Our introduction to exponential functions, along with the material on inverse functions in the next section, will lay the groundwork necessary to get started in Chapter 11.

Exponential Functions   **DEFINITION**   An *exponential function* is any function that can be written in the form

$$f(x) = b^x$$

where b is a positive real number other than 1.

Each of the following is an exponential function:

$$f(x) = 2^x, \qquad y = 3^x, \qquad f(x) = (\tfrac{1}{4})^x$$

The first step in becoming familiar with exponential functions is to find some values for specific exponential functions.

▼ **Example 4** If the exponential functions f and g are defined by

$$f(x) = 2^x \quad \text{and} \quad g(x) = 3^x$$

then

$$f(0) = 2^0 = 1 \qquad\qquad g(0) = 3^0 = 1$$
$$f(1) = 2^1 = 2 \qquad\qquad g(1) = 3^1 = 3$$
$$f(2) = 2^2 = 4 \qquad\qquad g(2) = 3^2 = 9$$
$$f(3) = 2^3 = 8 \qquad\qquad g(3) = 3^3 = 27$$

$$f(-2) = 2^{-2} = \frac{1}{2^2} = \frac{1}{4} \qquad g(-2) = 3^{-2} = \frac{1}{3^2} = \frac{1}{9}$$

$$f(-3) = 2^{-3} = \frac{1}{2^3} = \frac{1}{8} \qquad g(-3) = 3^{-3} = \frac{1}{3^3} = \frac{1}{27} \qquad ▲$$

We will now turn our attention to the graphs of exponential functions. Since the notation y is easier to use when graphing, and $y = f(x)$, for convenience we will write the exponential functions as

$$y = b^x$$

▼ **Example 5** Sketch the graph of the exponential function

$$y = 2^x$$

Solution Using the results of Example 4 in addition to some other convenient values of x, we have the following table:

x	y
-3	$1/8$
-2	$1/4$
-1	$1/2$
0	1
1	2
2	4
3	8

Graphing the ordered pairs given in the table and connecting them with a smooth curve, we have the graph of $y = 2^x$ shown in Figure 10-9.

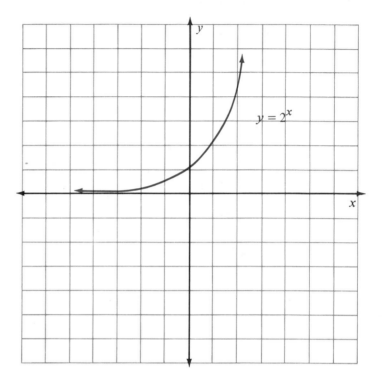

Figure 10-9

Notice that the graph does not cross the x-axis. It *approaches* the x-axis—in fact, we can get it as close to the x-axis as we want without its actually intersecting the x-axis. In order for the graph of $y = 2^x$ to intersect the x-axis, we would have to find a value of x that would make $2^x = 0$. Because no such value of x exists, the graph of $y = 2^x$ cannot intersect the x-axis.   ▲

▼   **Example 6**   Sketch the graph of $y = (\frac{1}{3})^x$.

Solution   We can make a table that will give some ordered pairs that satisfy the equation:

x	$y = \left(\dfrac{1}{3}\right)^x$	y
-3	$y = \left(\dfrac{1}{3}\right)^{-3} = 3^3 = 27$	27
-2	$y = \left(\dfrac{1}{3}\right)^{-2} = 3^2 = 9$	9
-1	$y = \left(\dfrac{1}{3}\right)^{-1} = 3^1 = 3$	3
0	$y = \left(\dfrac{1}{3}\right)^{0} = 1$	1
1	$y = \left(\dfrac{1}{3}\right)^{1} = \dfrac{1}{3}$	$\dfrac{1}{3}$
2	$y = \left(\dfrac{1}{3}\right)^{2} = \dfrac{1}{9}$	$\dfrac{1}{9}$
3	$y = \left(\dfrac{1}{3}\right)^{3} = \dfrac{1}{27}$	$\dfrac{1}{27}$

Using the ordered pairs from the table, we have the graph shown in Figure 10-10.

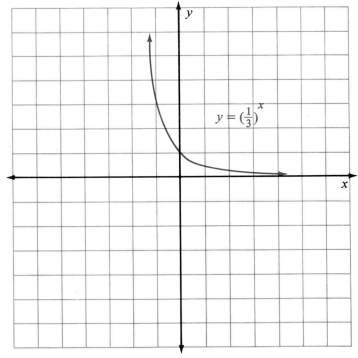

$$y = \left(\tfrac{1}{3}\right)^x$$

Figure 10-10

The graphs of all exponential functions have two things in common: (1) each crosses the y-axis at $(0, 1)$, since $b^0 = 1$; and (2) none can cross the x-axis, since $b^x = 0$ is impossible because of the restrictions on b.

Problem Set 10.4

Sketch the graph of each of the following functions. Identify each as a constant function, linear function, or quadratic function:

1. $f(x) = x^2 - 3$
2. $g(x) = 2x^2$
3. $g(x) = 4x - 1$
4. $f(x) = 3x + 2$
5. $f(x) = 5$
6. $f(x) = -3$
7. $g(x) = 0$
8. $g(x) = 1$
9. $f(x) = x$
10. $g(x) = \frac{1}{2}x + 1$
11. $f(x) = x^2 + 5x - 6$
12. $f(x) = -x^2 + 2x + 8$

Let $f(x) = 3^x$ and $g(x) = (\frac{1}{2})^x$ and evaluate each of the following:

13. $f(-2)$
14. $f(4)$
15. $g(2)$
16. $g(-2)$
17. $g(0)$
18. $f(0)$
19. $g(-1)$
20. $g(-4)$
21. $f(-3)$
22. $f(-1)$
23. $f(2) + g(-2)$
24. $f(2) - g(-2)$
25. $g(-3) + f(2)$
26. $g(2) + f(-1)$
27. $f(0) + g(0)$
28. $g(-3) + f(3)$
29. $f(-2) + g(3)$
30. $g(0) - f(0)$

Graph each of the following functions:

31. $y = 4^x$
32. $y = 2^{-x}$
33. $y = 3^{-x}$
34. $y = (\frac{1}{3})^{-x}$
35. $y = 2^{x+1}$
36. $y = 2^{x-3}$
37. $y = 3^{x-2}$
38. $y = 3^{x+1}$
39. $y = 2^{2x}$
40. $y = 3^{2x}$

The base b in an exponential function is restricted to positive numbers other than 1. To see why this is done, graph each of the following if possible:

41. $y = 1^x$
42. $y = 0^x$
43. $y = (-1)^x$
44. $y = (-2)^x$

45. Suppose it takes 1 day for a certain strain of bacteria to reproduce by dividing in half. If there are 100 bacteria present to begin with, then the total number present after x days will be

$$f(x) = 100 \cdot 2^x$$

Find the total number present after 1 day, 2 days, 3 days, and 4 days. How many days must elapse before there are over 100,000 bacteria present?

46. Suppose it takes 12 hours for a certain strain of bacteria to reproduce by
     dividing in half. If there are 50 bacteria present to begin with, then the
     total number present after x days will be

$$f(x) = 50 \cdot 4^x.$$

     Find the total number present after 1 day, 2 days, and 3 days.

Review Problems   The problems below review material we covered in Section
9.2.

Find the distance between the following points.

47. (0, 4) (3, 0)   48. $(2, -5)$ $(-4, 3)$

49. Find x so the distance between $(x, 3)$ and $(1, 6)$ is $\sqrt{13}$.
50. Find y so the distance between $(2, y)$ and $(1, 2)$ is $\sqrt{2}$.

Give the center and radius of each of the following circles.

51. $x^2 + y^2 = 25$ 52. $x^2 + y^2 = 9$
53. $x^2 + y^2 + 6x - 4y = 3$ 54. $x^2 + y^2 - 8x + 2y = 8$

Suppose the function f is given by

$$f = \{(1, 4), (2, 5), (3, 6), (4, 7)\}$$

10.5
The Inverse of a
Function

The inverse of f, written f^{-1}, is obtained by reversing the order of the
coordinates in each ordered pair in f. The inverse of f is the relation
given by

$$f^{-1} = \{(4, 1), (5, 2), (6, 3), (7, 4)\}$$

It is obvious that the domain of f is now the range of f^{-1}, and the range
of f is now the domain of f^{-1}. Every function (or relation) has an inverse
that is obtained from the original function by interchanging the components of each ordered pair. We can generalize this discussion by
giving the following formal definition for the inverse of a function or
relation.

DEFINITION   The inverse of the function (relation) f is written f^{-1} and is such that

$$(x, y) \in f \quad \text{if and only if} \quad (y, x) \in f^{-1}$$

The $^{-1}$ in the expression f^{-1} serves to denote the inverse of f. It is *not* an exponent—that is, $f^{-1}(x)$ is *not* the reciprocal of f. *In general,*

$$f^{-1}(x) \neq \frac{1}{f(x)}$$

Suppose a function f is defined with an equation instead of a list of ordered pairs. We can obtain the equation of the inverse f^{-1} by applying the above definition to interchange the role of x and y in the equation for f.

▼   **Example 1**   If the function f is defined by $f(x) = 2x - 3$, find the equation that represents the inverse of f.

Solution   Since the inverse of f is obtained by interchanging the components of all the ordered pairs belonging to f, and each ordered pair in f satisfies the equation $y = 2x - 3$, we simply exchange x and y in the equation $y = 2x - 3$ to get the formula for f^{-1}:

$$x = 2y - 3$$

We now solve this equation for y in terms of x:

$$x + 3 = 2y$$

$$\frac{x + 3}{2} = y$$

$$y = \frac{x + 3}{2}$$

The last line gives the equation that defines the function f^{-1}. We can write this equation with function notation as

$$f^{-1}(x) = \frac{x + 3}{2}$$

Let's compare the graphs of f and f^{-1} as given in Example 1. (See Figure 10-11.)

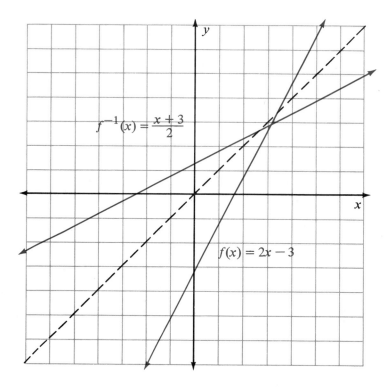

$$f^{-1}(x) = \frac{x + 3}{2}$$

$$f(x) = 2x - 3$$

Figure 10-11

The graphs of f and f^{-1} have symmetry about the line $y = x$. This is a reasonable result, since the one function was obtained from the other by interchanging x and y in the equation. The ordered pairs (a, b) and (b, a) always have symmetry about the line $y = x$. ▲

▼   **Example 2**   Graph the function $y = x^2 - 2$ and its inverse. Give the equation for the inverse.

Solution   We can obtain the graph of the inverse of $y = x^2 - 2$ by graphing $y = x^2 - 2$ by the usual methods, and then reflecting the graph about the line $y = x$. The equation that corresponds to the inverse of $y = x^2 - 2$ is obtained by interchanging x and y to get $x = y^2 - 2$. (See Figure 10-12.)

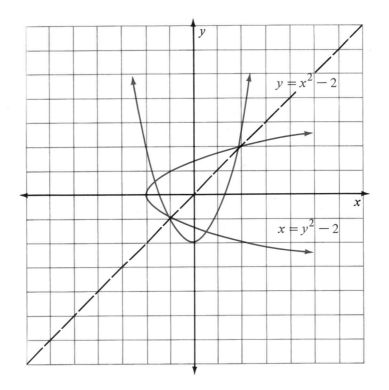

Figure 10-12

We can solve the equation $x = y^2 - 2$ for y in terms of x as follows:

$$x = y^2 - 2$$
$$x + 2 = y^2$$
$$y = \pm \sqrt{x + 2}$$

Using function notation, we can write the function and its inverse as

$$f(x) = x^2 - 2, \qquad f^{-1}(x) = \pm \sqrt{x + 2} \qquad \blacktriangle$$

Comparing the graphs from Examples 1 and 2, we observe that the inverse of a function is not always a function. In Example 1, both f and f^{-1} have graphs that are straight lines and therefore both represent functions. In Example 2, the inverse of function f is not a function, since a vertical line crosses the graph of f^{-1} in more than one place.

▼   **Example 3**   Graph the relation

$$\frac{x^2}{9} + \frac{y^2}{25} = 1$$

and its inverse. Write an equation for the inverse.

Solution   The graph of

$$\frac{x^2}{9} + \frac{y^2}{25} = 1$$

is an ellipse that crosses the x-axis at $(3, 0)$ and $(-3, 0)$ and the y-axis at $(0, 5)$ and $(0, -5)$. Exchanging x and y in the equation, we obtain the equation of the inverse shown in Figure 10-13:

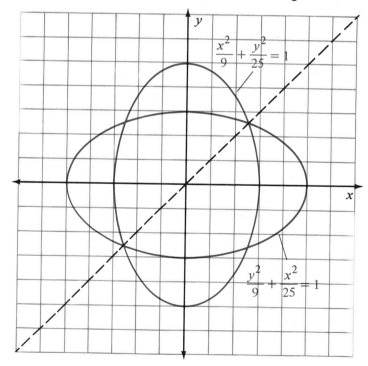

Figure 10-13

$$\frac{x^2}{9} + \frac{y^2}{25} = 1 \text{ and } \frac{y^2}{9} + \frac{x^2}{25} = 1 \text{ are inverse relations.} \quad \blacktriangle$$

▼ **Example 4** Graph the function $y = 2^x$ and its inverse $x = 2^y$.

Solution We graphed $y = 2^x$ in the preceding section. We simply reflect its graph about the line $y = x$ to obtain the graph of its inverse $x = 2^y$. (See Figure 10-14.)

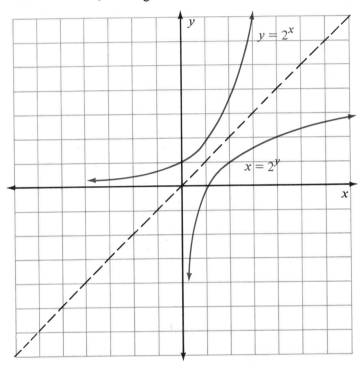

Figure 10-14   ▲

Problem Set 10.5   For each of the following functions, find the equation of the inverse. Write the inverse using the notation $f^{-1}(x)$.

1. $f(x) = 3x - 1$ 2. $f(x) = 2x - 5$
3. $f(x) = 1 - 3x$ 4. $f(x) = 3 - 4x$
5. $f(x) = x^2 + 4$ 6. $f(x) = 3x^2$

7. $f(x) = \dfrac{x - 3}{4}$ 8. $f(x) = \dfrac{x + 7}{2}$

9. $f(x) = \frac{1}{2}x - 3$ 10. $f(x) = \frac{1}{3}x + 1$
11. $f(x) = -x^2 + 3$ 12. $f(x) = 4 - x^2$

For each of the following relations, sketch the graph of the relation and its inverse, and write an equation for the inverse:

13. $y = 2x - 1$
14. $y = 3x + 1$
15. $y = x^2 - 3$
16. $y = x^2 + 1$
17. $y = x^2 - 2x - 8$
18. $y = -x^2 + 2x + 8$
19. $y = 3^x$
20. $y = (\tfrac{1}{2})^x$
21. $y = 4$
22. $y = -2$
23. $4x^2 - 9y^2 = 36$
24. $9x^2 + 4y^2 = 36$
25. $y = \tfrac{1}{2}x + 2$
26. $y = \tfrac{1}{3}x - 1$
27. $x^2 + y^2 = 16$
28. $x^2 - y^2 = 16$
29. $y = \sqrt{9 - x^2}$
30. $y = -\sqrt{25 - x^2}$
31. $\dfrac{x^2}{4} + \dfrac{y^2}{25} = 1$
32. $\dfrac{x^2}{25} + \dfrac{y^2}{9} = 1$

33. If $f(x) = 3x - 2$, then $f^{-1}(x) = \dfrac{x + 2}{3}$. Use these two functions to find

   a. $f(2)$
   c. $f[f^{-1}(2)]$
   b. $f^{-1}(2)$
   d. $f^{-1}[f(2)]$

34. If $f(x) = \tfrac{1}{2}x + 5$, then $f^{-1}(x) = 2x - 10$. Use these two functions to find
   a. $f(-4)$
   c. $f[f^{-1}(-4)]$
   b. $f^{-1}(-4)$
   d. $f^{-1}[f(-4)]$

35. Let $f(x) = \dfrac{1}{x}$, and find $f^{-1}(x)$.

36. Let $f(x) = \dfrac{a}{x}$, and find $f^{-1}(x)$. (a is a real number constant.)

Review Problems   The problems below review material we covered in Section 9.5.

Solve each system.

37. $x^2 + y^2 = 9$
    $x - 2y = 3$
38. $x^2 - y^2 = 9$
    $x + 2y = 3$
39. $x^2 + y^2 = 16$
    $4x^2 + y^2 = 16$
40. $x^2 + y^2 = 25$
    $x^2 - y^2 = 25$
41. $x^2 + y^2 = 4$
    $y = x^2 - 2$
42. $x^2 + y^2 = 49$
    $y = x^2 - 7$

Chapter 10   Summary and Review

Examples

RELATIONS AND FUNCTIONS [10.1]

A *relation* is any set of ordered pairs. The set of all first coordinates is called the *domain* of the relation. The set of all second coordinates is the *range* of the relation.

   A *function* is a relation in which no two different ordered pairs have the same first coordinate.

   If the domain for a relation or a function is not specified, it

1.   The relation $\{(8, 1), (6, 1), (-3, 0)\}$ is also a function since no ordered pairs have the same first coordinates. The domain is $\{8, 6, -3\}$ and the range is $\{1, 0\}$.

is assumed to be all real numbers for which the relation (or function) is defined. Since we are only concerned with real number functions, a function is not defined for those values of x that give 0 in the denominator or the square root of a negative number.

2. The graph of any circle, ellipse, or hyperbola found in this chapter will fail the vertical line test: A vertical line can always be found that crosses the graph in more than one place.

VERTICAL LINE TEST [10.1]

If a vertical line crosses the graph of a relation in more than one place, then the relation is not a function. If no vertical line can cross the graph of a relation in more than one place, then the relation is a function.

3. If $f(x) = 5x - 3$ then

$$f(0) = 5(0) - 3 = -3$$
$$f(1) = 5(1) - 3 = 2$$
$$f(-2) = 5(-2) - 3 = -13$$
$$f(a) = 5a - 3$$

FUNCTION NOTATION [10.2]

The notation $f(x)$ is read "f of x." It is defined to be the value of the function f at x. The value of $f(x)$ is the value of y associated with a given value of x. The expressions $f(x)$ and y are equivalent. That is, $y = f(x)$.

4. If $f(x) = 4x$ and $g(x) = x^2 - 3$, then

$$(f + g)(x) = x^2 + 4x - 3$$

$$(f - g)(x) = -x^2 + 4x + 3$$

$$(fg)(x) = 4x^3 - 12x$$

$$\frac{f}{g}(x) = \frac{4x}{x^2 - 3}$$

ALGEBRA FOR FUNCTIONS [10.3]

If f and g are any two functions with a common domain, then:
The sum of f and g, written $f + g$, is defined by

$$(f + g)(x) = f(x) + g(x).$$

The difference of f and g, written $f - g$, is defined by

$$(f - g)(x) = f(x) - g(x).$$

The product of f and g, written fg, is defined by

$$(fg)(x) = f(x)g(x).$$

The quotient of f and g, written f/g, is defined by

$$\left(\frac{f}{g}\right)(x) = \frac{f(x)}{g(x)} \qquad g(x) \neq 0.$$

5. Functions

Constant: $f(x) = 5$
Linear: $f(x) = 3x - 2$
Quadratic: $f(x) = x^2 - 5x + 6$
Exponential: $f(x) = 2^x$

CLASSIFICATION OF FUNCTIONS [10.4]

Constant function	$f(x) = c$	$(c = \text{Constant})$
Linear function:	$f(x) = ax + b$	$(a \neq 0)$
Quadratic function:	$f(x) = ax^2 + bx + c$	$(a \neq 0)$
Exponential function:	$f(x) = b^x$	$(b > 0, b \neq 1)$

THE INVERSE OF A FUNCTION [10.5]

The inverse of a function is obtained by reversing the order of the coordinates of the ordered pairs belonging to the function.

The inverse of a function is not necessarily a function.

THE GRAPH OF THE INVERSE OF A FUNCTION [10.5]

The graph of f^{-1} can be obtained from the graph of f by simply reflecting the graph of f across the line $y = x$. That is, the graphs of f and f^{-1} are symmetrical about the line $y = x$.

6.   The inverse of $f(x) = 2x - 3$ is

$$f^{-1}(x) = \frac{x + 3}{2}$$

COMMON MISTAKES

1.   The most common mistake made when working with functions is to interpret the notation $f(x)$ as meaning the product of f and x. The notation $f(x)$ does *not* mean f times x. It is the value of the function f at x and is equivalent to y.

2.   Another common mistake occurs when the expression $f^{-1}(x)$ is interpreted as meaning the reciprocal of x. The notation $f^{-1}(x)$ is used to denote the *inverse* of $f(x)$:

$$f^{-1}(x) \neq \frac{1}{f(x)}$$

$$f^{-1}(x) = \text{Inverse of } f$$

Specify the domain and range for the following relations and indicate which relations are also functions:

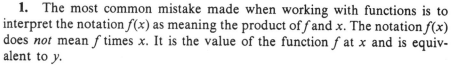

1. $\{(-3, 1)(2, 1)(3, 5)(0, 3)\}$ 2. $\{(-2, 0)(-3, 0)(-2, 1)\}$

3. $y = x^2 - 9$ 4. $4x^2 + 9y^2 = 36$

Indicate any restrictions on the domain of the following:

5. $y = \sqrt{x - 4}$ 6. $y = \dfrac{4}{\sqrt{3 - x}}$

7. $y = \dfrac{6}{x + 1}$ 8. $y = \dfrac{x - 1}{x^2 - 2x - 8}$

Let $f(x) = x - 2$, $g(x) = 3x + 4$, and $h(x) = 3x^2 - 2x - 8$, and find the following:

9. $f(3) + g(2)$ 10. $h(0) + g(0)$
11. $f + g$ 12. h/g
13. $(f + g + h)(-3)$ 14. $(h + fg)(1)$

Graph each of the following exponential functions:

15. $y = 2^x$ 16. $y = 3^{-x}$

Find an equation for the inverse of each of the following functions. Sketch the graph of f and f^{-1} on the same set of axes.

17. $f(x) = 2x - 3$ 18. $f(x) = x^2 - 4$

Logarithms

To the student:

This chapter is mainly concerned with applications of a new notation for exponents. Logarithms are exponents. The properties of logarithms are actually the properties of exponents. As it turns out, writing exponents with the new notation, logarithms, allows us to solve some problems we would otherwise be unable to solve. Logarithms used to be used extensively to simplify tedious calculations. Since hand-held calculators are so common now, logarithms are seldom used in connection with computations. Nevertheless, there are many other applications of logarithms to both science and higher mathematics. For example, the pH of a liquid is defined in terms of logarithms. (That's the same pH that is given on the label of many hair conditioners.) The Richter scale for measuring earthquake intensity is a logarithmic scale, as is the decibel scale used for measuring the intensity of sound.

We will begin this chapter with the definition of logarithms and the three main properties of logarithms. The rest of the chapter involves applications of the definition and properties. Understanding the last two sections in Chapter 10 will be very useful in getting started in this chapter.

11.1
Logarithms Are
Exponents

In Section 10.4 we considered some equations of the form

$$y = b^x \qquad (b > 0, b \neq 1)$$

called exponential functions. In the next section, Section 10.5, we found that the equation of the inverse of a function can be obtained by exchanging x and y in the equation of the original function. We can therefore exchange x and y in the equation of an exponential function to get the equation of its inverse. The equation of the inverse of an exponential function must have the form

$$x = b^y \qquad (b > 0, b \neq 1)$$

The problem with this expression is that y is not written explicitly in terms of x. That is, we would like to be able to write the equation $x = b^y$ as an equivalent equation with just y on the left side. One way to do so is with the following definition.

DEFINITION   The expression $y = \log_b x$ is read "y is the logarithm to the base b of x," and is equivalent to the expression

$$x = b^y \qquad (b > 0, b \neq 1)$$

We say y is the number we raise b to in order to get x.

It may seem rather strange to take an expression we already know something about, like $x = b^y$, and write it in another form, $y = \log_b x$, that we know nothing about. As you will see as we progress through the chapter, there are many advantages to using the notation $y = \log_b x$ in place of $x = b^y$, even though they say exactly the same thing about the relationship between x, y, and b. What we have to do now is realize that $y = \log_b x$ is *defined* to be equivalent to $x = b^y$. There isn't anything to prove about the definition—we can only accept it as it is.

NOTATION   When an expression is in the form $x = b^y$, it is said to be in *exponential form*. On the other hand, if an expression is in the form $y = \log_b x$, it is said to be in *logarithmic form*.

The following table illustrates the two forms:

Exponential form		Logarithmic form
$8 = 2^3$	$\Leftrightarrow$	$\log_2 8 = 3$
$25 = 5^2$	$\Leftrightarrow$	$\log_5 25 = 2$
$.1 = 10^{-1}$	$\Leftrightarrow$	$\log_{10} .1 = -1$
$\frac{1}{8} = 2^{-3}$	$\Leftrightarrow$	$\log_2 \frac{1}{8} = -3$
$r = z^s$	$\Leftrightarrow$	$\log_z r = s$

As the table indicates, logarithms are exponents. That is, $\log_2 8$ is 3 *because* 3 is the exponent to which we raise 2 in order to get 8. *Logarithms are exponents.*

The ability to write expressions in logarithmic form as equivalent expressions in exponential form allows us to solve some equations involving logarithms, as the following examples illustrate.

▼ **Example 1** Solve for x: $\log_3 x = -2$

Solution In exponential form the equation looks like this:

$$x = 3^{-2}$$
$$\text{or} \quad x = \tfrac{1}{9}$$

The solution set is $\{\tfrac{1}{9}\}$. ▲

▼ **Example 2** Solve $\log_x 4 = 3$.

Solution Again, we use the definition of logarithms to write the expression in exponential form:

$$4 = x^3$$

Taking the cube root of both sides, we have

$$\sqrt[3]{4} = \sqrt[3]{x^3}$$
$$x = \sqrt[3]{4}$$

The solution set is $\{\sqrt[3]{4}\}$. ▲

▼ **Example 3** Solve $\log_8 4 = x$.

Solution We write the expression again in exponential form:

$$4 = 8^x$$

Since both 4 and 8 can be written as powers of 2, we write them in terms of powers of 2:

$$2^2 = (2^3)^x$$
$$2^2 = 2^{3x}$$

The only way the left and right sides of this last line can be equal is if the exponents are equal—that is, if

$$2 = 3x$$
$$\text{or} \quad x = \tfrac{2}{3}$$

The solution is $\tfrac{2}{3}$. We check as follows:

$$\log_8 4 = \tfrac{2}{3} \Longleftrightarrow 4 = 8^{2/3}$$
$$4 = (\sqrt[3]{8})^2$$
$$4 = 2^2$$
$$4 = 4$$

The solution checks when used in the original equation.   ▲

Graphing logarithmic functions can be done using the graphs of exponential functions and the fact that the graphs of inverse functions have symmetry about the line $y = x$. Here's an example to illustrate.

▼   **Example 4** Graph the equation $y = \log_2 x$.

Solution   The equation $y = \log_2 x$ is, by definition, equivalent to the exponential equation

$$x = 2^y$$

which is the equation of the inverse of the function

$$y = 2^x$$

The graph of $y = 2^x$ was given in Figure 10-9. We simply reflect the graph of $y = 2^x$ about the line $y = x$ to get the graph of $x = 2^y$, which is also the graph of $y = \log_2 x$. (See Figure 11-1.)

It is apparent from the graph that $y = \log_2 x$ is a function, since no vertical line will cross its graph in more than one place. The same is true for all logarithmic equations of the form

$$y = \log_b x$$

where b is a positive number other than 1. All similar equations can be considered as functions.   ▲

Note   Since $y = \log_b x$ is equivalent to $x = b^y$ where $b > 0$ and $b \neq 1$, the only one of the quantities x, y, and b that can be negative is y. Both b and x in the expression $y = \log_b x$ must be positive.

We end this section with two special identities that arise from the definition of logarithms. Each is useful in proving the properties in the next section or for evaluating some simple logarithmic expressions.

Two Special Identities   If b is a positive real number other than 1, then each of the following is a consequence of the definition of a logarithm:

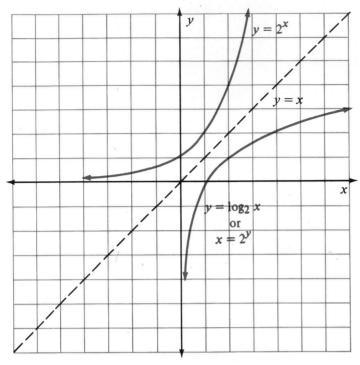

Figure 11-1

$$(1) \quad b^{\log_b x} = x \quad \text{and} \quad (2) \quad \log_b b^x = x$$

The justifications for these identities are similar. Let's consider only the first one. Consider the expression

$$y = \log_b x$$

By definition, it is equivalent to

$$x = b^y$$

Substituting $\log_b x$ for y in the last line gives us

$$x = b^{\log_b x}$$

Here are some examples that illustrate the use of identity 2 given above.

▼ **Example 5**   Simplify $\log_2 8$.

   Solution   Substitute 2^3 for 8:

$$\log_2 8 = \log_2 2^3$$
$$= 3 \qquad \blacktriangle$$

▼ **Example 6**   Simplify $\log_{10} 10{,}000$.

Solution $10{,}000$ can be written as 10^4:

$$\log_{10} 10{,}000 = \log_{10} 10^4$$
$$= 4 \qquad \blacktriangle$$

▼ **Example 7**   Simplify $\log_b b$ $(b > 0, b \neq 1)$.

Solution   Since $b^1 = b$, we have

$$\log_b b = \log_b b^1$$
$$= 1 \qquad \blacktriangle$$

▼ **Example 8**   Simplify $\log_b 1$ $(b > 0, b \neq 1)$.

Solution   Since $1 = b^0$, we have

$$\log_b 1 = \log_b b^0$$
$$= 0 \qquad \blacktriangle$$

▼ **Example 9**   Simplify $\log_4 (\log_5 5)$.

Solution   Since $\log_5 5 = 1$,

$$\log_4 (\log_5 5) = \log_4 1$$
$$= 0 \qquad \blacktriangle$$

Problem Set 11.1

Write each of the following expressions in logarithmic form:

1. $2^4 = 16$

2. $3^2 = 9$

3. $125 = 5^3$

4. $16 = 4^2$

5. $.01 = 10^{-2}$

6. $.001 = 10^{-3}$

7. $2^{-5} = \frac{1}{32}$

8. $4^{-2} = \frac{1}{16}$

9. $(\frac{1}{2})^{-3} = 8$

10. $(\frac{1}{3})^{-2} = 9$

11. $27 = 3^3$

12. $81 = 3^4$

Write each of the following expressions in exponential form:

13. $\log_{10} 100 = 2$

14. $\log_2 8 = 3$

15. $\log_2 64 = 6$

16. $\log_2 32 = 5$

17. $\log_8 1 = 0$

18. $\log_9 9 = 1$

19. $\log_{10} .001 = -3$

20. $\log_{10} .0001 = -4$

21. $\log_6 36 = 2$ **22.** $\log_7 49 = 2$
23. $\log_5 \frac{1}{25} = -2$ **24.** $\log_3 \frac{1}{81} = -4$

Solve each of the following equations for x:

25. $\log_3 x = 2$ **26.** $\log_4 x = 3$
27. $\log_5 x = -3$ **28.** $\log_2 x = -4$
29. $\log_2 16 = x$ **30.** $\log_3 27 = x$
31. $\log_8 2 = x$ **32.** $\log_{25} 5 = x$
33. $\log_x 4 = 2$ **34.** $\log_x 16 = 4$
35. $\log_x 5 = 3$ **36.** $\log_x 8 = 2$

Sketch the graph of each of the following logarithmic equations:

37. $y = \log_3 x$ **38.** $y = \log_{1/2} x$
39. $y = \log_{1/3} x$ **40.** $y = \log_4 x$
41. $y = \log_5 x$ **42.** $y = \log_{1/5} x$
43. $y = \log_{10} x$ **44.** $y = \log_{1/4} x$

Simplify each of the following:

45. $\log_2 16$ **46.** $\log_3 9$
47. $\log_{25} 125$ **48.** $\log_9 27$
49. $\log_{10} 1000$ **50.** $\log_{10} 10,000$
51. $\log_3 3$ **52.** $\log_4 4$
53. $\log_5 1$ **54.** $\log_{10} 1$
55. $\log_3 (\log_6 6)$ **56.** $\log_5 (\log_3 3)$
57. $\log_4 [\log_2 (\log_2 16)]$ **58.** $\log_4 [\log_3 (\log_2 8)]$

59. The formula $M = 0.21(\log_{10} a - \log_{10} b)$ is used in the food processing industry to find the number of minutes M of heat processing a certain food should undergo at 250°F to reduce the probability of survival of *C. botulinum* spores. The letter a represents the number of spores per can before heating, and b represents the total number of spores per can after heating. Find M if $a = 1$ and $b = 10^{-12}$.

60. The formula $N = 10 \log_{10} \dfrac{P_1}{P_2}$ is used in radio electronics to find the ratio of the acoustic powers of two electric circuits in terms of their electric powers. Find N if P_1 is 50 and P_2 is $1/2$.

61. Change the expression $y = \log_2(-4)$ to exponential form. Is there a value of y that will make this true?

62. Write the expression $y = \log_{10}(-10)$ in exponential form. Is there a value of y that will make this true?

Review Problems   The problems below review material we covered in Sections 2.1 and 6.1.

Solve each equation.

63. $5(2x + 1) - 4 = 5x - 9$ **64.** $3(4x - 1) + 5 = 8x + 10$
65. $9 - 2(3x + 7) = 5 - 4x$ **66.** $7 - 3(4x + 9) = 4 - 4x$

67. $x^2 + 2x = 8$ **68.** $x^2 - 2x = 3$

69. $\dfrac{5}{x} - \dfrac{3}{x^2} = -2$ **70.** $2 + \dfrac{1}{x} = \dfrac{15}{x^2}$

11.2
Properties of
Logarithms

We begin this section by stating three properties of logarithms. Since the proofs for all three are very similar in form, we will prove only the first property.

For the following three properties, x, y, and b are all positive real numbers, $b \neq 1$, and r is any real number.

PROPERTY 1 $\log_b (xy) = \log_b x + \log_b y$
In words: The logarithm of a *product* is the *sum* of the logarithms.

PROPERTY 2 $\log_b \left(\dfrac{x}{y}\right) = \log_b x - \log_b y$

In words: The logarithm of a *quotient* is the *difference* of the logarithms.

PROPERTY 3 $\log_b x^r = r \log_b x$
In words: The logarithm of a base raised to a *power* is the *product* of the power and the logarithm of the base.

Proof of Property 1   To prove property 1 we simply apply the first identity for logarithms given at the end of the preceding section:

$$b^{\log_b xy} = xy = (b^{\log_b x})(b^{\log_b y}) = b^{\log_b x + \log_b y}$$

Since the first and last expressions are equal and the bases are the same, the exponents $\log_b xy$ and $\log_b x + \log_b y$ must be equal. Therefore,

$$\log_b xy = \log_b x + \log_b y$$

The proofs of properties 2 and 3 proceed in much the same manner, so we will omit them here. The examples that follow show how the three properties can be used. These types of problems will be useful in applying logarithms to computations which we will do in the next section.

▼   **Example 1**   Expand using the properties of logarithms: $\log_5 \dfrac{3xy}{z}$

Solution Applying property 2, we can write the quotient of $3xy$ and z in terms of a difference:

$$\log_5 \frac{3xy}{z} = \log_5 3xy - \log_5 z$$

Applying property 1 to the product $3xy$, we write it in terms of addition:

$$\log_5 \frac{3xy}{z} = \log_5 3 + \log_5 x + \log_5 y - \log_5 z \qquad \blacktriangle$$

▼ **Example 2** Expand using the properties of logarithms:

$$\log_2 \frac{x^4}{\sqrt{y} \cdot z^3}$$

Solution We write $\sqrt{y}$ as $y^{1/2}$ and apply the properties:

$$\log_2 \frac{x^4}{\sqrt{y} \cdot z^3} = \log_2 \frac{x^4}{y^{1/2} z^3}$$

$$= \log_2 x^4 - \log_2 (y^{1/2} \cdot z^3) \qquad \text{Property 2}$$

$$= \log_2 x^4 - (\log_2 y^{1/2} + \log_2 z^3) \qquad \text{Property 1}$$

$$= \log_2 x^4 - \log_2 y^{1/2} - \log_2 z^3 \qquad \begin{array}{l}\text{Remove} \\ \text{parentheses}\end{array}$$

$$= 4 \log_2 x - \frac{1}{2} \log_2 y - 3 \log_2 z \qquad \text{Property 3} \quad \blacktriangle$$

We can also use the three properties to write an expression in expanded form as just one logarithm.

▼ **Example 3** Write as a single logarithm:

$$2 \log_{10} a + 3 \log_{10} b - \tfrac{1}{3} \log_{10} c$$

Solution We begin by applying property 3:

$$2 \log_{10} a + 3 \log_{10} b - \frac{1}{3} \log_{10} c$$

$$= \log_{10} a^2 + \log_{10} b^3 - \log_{10} c^{1/3} \qquad \text{Property 3}$$
$$= \log_{10} (a^2 \cdot b^3) - \log_{10} c^{1/3} \qquad \text{Property 1}$$
$$= \log_{10} \frac{a^2 b^3}{c^{1/3}} \qquad\qquad\qquad \text{Property 2}$$
$$= \log_{10} \frac{a^2 b^3}{\sqrt[3]{c}} \qquad\qquad c^{1/3} = \sqrt[3]{c} \qquad \blacktriangle$$

The properties of logarithms along with the definition of logarithms are useful in solving equations that involve logarithms.

▼   **Example 4**   Solve for x: $\log_2 (x + 2) + \log_2 x = 3$

Solution   Applying property 1 to the left side of the equation allows us to write it as a single logarithm:

$$\log_2 (x + 2) + \log_2 x = 3$$
$$\log_2 [(x + 2)(x)] = 3$$

The last line can be written in exponential form using the definition of logarithms:

$$(x + 2)(x) = 2^3$$

Solve as usual:

$$x^2 + 2x = 8$$
$$x^2 + 2x - 8 = 0$$
$$(x + 4)(x - 2) = 0$$
$$x + 4 = 0 \quad \text{or} \quad x - 2 = 0$$
$$x = -4 \quad \text{or} \qquad x = 2$$

Earlier in this chapter we noted the fact that x in the expression $y = \log_b x$ cannot be a negative number. Since substitution of $x = -4$ into the original equation gives

$$\log_2 (-2) + \log_2 (-4) = 3$$

which contains logarithms of negative numbers, we cannot use -4 as a solution. The solution set is $\{2\}$.   ▲

Use the three properties of logarithms given in this section to expand each expression as much as possible: *Problem Set 11.2*

1. $\log_3 4x$

2. $\log_2 5x$

3. $\log_6 \dfrac{5}{x}$

4. $\log_3 \dfrac{x}{5}$

5. $\log_2 y^5$

6. $\log_7 y^3$

7. $\log_9 \sqrt[3]{z}$

8. $\log_8 \sqrt{z}$

9. $\log_6 x^2 y^3$

10. $\log_{10} x^2 y^4$

11. $\log_5 \sqrt{x} \cdot y^4$

12. $\log_8 \sqrt[3]{xy^6}$

13. $\log_b \dfrac{xy}{z}$

14. $\log_b \dfrac{3x}{y}$

15. $\log_{10} \dfrac{4}{xy}$

16. $\log_{10} \dfrac{5}{4y}$

17. $\log_{10} \dfrac{x^2 y}{\sqrt{z}}$

18. $\log_{10} \dfrac{\sqrt{x} \cdot y}{z^3}$

19. $\log_{10} \dfrac{x^3 \sqrt{y}}{z^4}$

20. $\log_{10} \dfrac{x^4 \sqrt[3]{y}}{\sqrt{z}}$

21. $\log_b \sqrt[3]{\dfrac{x^2 y}{z^4}}$

22. $\log_b \sqrt[4]{\dfrac{x^4 y^3}{z^5}}$

Write each expression as a single logarithm:

23. $\log_b x + \log_b z$

24. $\log_b x - \log_b z$

25. $2 \log_3 x - 3 \log_3 y$

26. $4 \log_2 x + 5 \log_2 y$

27. $\frac{1}{2} \log_{10} x + \frac{1}{3} \log_{10} y$

28. $\frac{1}{3} \log_{10} x - \frac{1}{4} \log_{10} y$

29. $3 \log_2 x + \frac{1}{2} \log_2 y - \log_2 z$

30. $2 \log_3 x + 3 \log_3 y - \log_3 z$

31. $\frac{1}{2} \log_2 x - 3 \log_2 y - 4 \log_2 z$

32. $3 \log_{10} x - \log_{10} y - \log_{10} z$

33. $\frac{3}{2} \log_{10} x - \frac{3}{4} \log_{10} y - \frac{4}{5} \log_{10} z$

34. $3 \log_{10} x - \frac{4}{3} \log_{10} y - 5 \log_{10} z$

Solve each of the following equations:

35. $\log_2 x + \log_2 3 = 1$

36. $\log_2 x - \log_2 3 = 1$

37. $\log_3 x - \log_3 2 = 2$

38. $\log_3 x + \log_3 2 = 2$

39. $\log_3 x + \log_3 (x - 2) = 1$

40. $\log_6 x + \log_6 (x - 1) = 1$

41. $\log_3 (x + 3) - \log_3 (x - 1) = 1$

42. $\log_4 (x - 2) - \log_4 (x + 1) = 1$

43. $\log_2 x + \log_2 (x - 2) = 3$

44. $\log_4 x + \log_4 (x + 6) = 2$

45. $\log_8 x + \log_8(x - 3) = \frac{2}{3}$
46. $\log_{27} x + \log_{27}(x + 8) = \frac{2}{3}$
47. $\log_5 \sqrt{x} + \log_5 \sqrt{6x + 5} = 1$
48. $\log_2 \sqrt{x} + \log_2 \sqrt{6x + 5} = 1$

Review Problems The problems below review material we covered in Section 3.1. Reviewing these problems will help you with the next section.

Write each number in scientific notation.

49. 394,000,000   **50.** 2760
51. 0.0391   **52.** 0.000276

Write each number in expanded form.

53. 5.23×10^{-3} **54.** 7.48×10^{-2}
55. 5.03×10^4 **56.** 6.89×10^3

11.3
Common Logarithms and Computations

In the past, logarithms have been very useful in simplifying calculations. With the widespread availability of hand-held calculators, the importance of logarithms in calculations has been decreased considerably. Working through arithmetic problems using logarithms is, however, good practice with logarithms. The problems in this section all involve the properties of logarithms developed in the preceding section and indicate how some fairly difficult computation problems can be simplified using logarithms.

DEFINITION A *common logarithm* is a logarithm with a base of 10. Since common logarithms are used so frequently, it is customary, in order to save time, to omit notating the base. That is,

$$\log_{10} x = \log x$$

When the base is not shown, it is assumed to be 10.

The reason logarithms base 10 are so common is that our number system is a base-10 number system.

Common logarithms of powers of 10 are very simple to evaluate. We need only recognize that $\log 10 = \log_{10} 10 = 1$ and apply the third property of logarithms: $\log_b x^r = r \log_b x$.

$$\log 1000 = \log 10^3 \quad = 3 \log 10 \quad = 3(1) \quad = 3$$
$$\log 100 \; = \log 10^2 \quad = 2 \log 10 \quad = 2(1) \quad = 2$$

$$\log 10 \quad = \log 10^1 \quad = 1 \log 10 \quad = 1(1) \quad = 1$$
$$\log 1 \quad\;\, = \log 10^0 \quad = 0 \log 10 \quad = 0(1) \quad = 0$$
$$\log .1 \quad\;\, = \log 10^{-1} = -1 \log 10 = -1(1) = -1$$
$$\log .01 \quad = \log 10^{-2} = -2 \log 10 = -2(1) = -2$$
$$\log .001 = \log 10^{-3} = -3 \log 10 = -3(1) = -3$$

For common logarithms of numbers that are not powers of 10, we have to resort to a table. The table in Appendix D at the back of the book gives common logarithms of numbers between 1.00 and 9.99. To find the common logarithm of, say, 2.76, we read down the left-hand column until we get to 2.7, then across until we are below the 6 in the top row (or above the 6 in the bottom row):

x	0	1	2	3	4	5	6	7	8	9
1.0										
1.1	$\log 2.76 = .4409$									
1.2										
⋮										
2.7							→ .4409			
⋮										
4.1										
4.2										
x	0	1	2	3	4	5	6	7	8	9

The table contains only logarithms of numbers between 1.00 and 9.99. Check the following logarithms in the table to be sure you know how to use the table:

$$\log 7.02 = .8463$$
$$\log 1.39 = .1430$$
$$\log 6.00 = .7782$$
$$\log 9.99 = .9996$$

To find the common logarithm of a number that is not between 1.00 and 9.99, we simply write the number in scientific notation, apply property 2 of logarithms, and use the table. The following examples illustrate the procedure.

Note A number is written in scientific notation when it is written as the product of a number between 1 and 10 and a power of 10. For example, $39,800 = 3.98 \times 10^4$ in scientific notation.

▼ **Example 1** Use the table in Appendix D to find log 2760.

Solution $\log 2760 = \log(2.76 \times 10^3)$
$= \log 2.76 + \log 10^3$
$= .4409 + 3$
$= 3.4409$

The 3 in the answer is called the *characteristic,* and its main function is to keep track of the decimal point. The decimal part of this logarithm is called the *mantissa.* It is found from the table. Here's more of the same: ▲

▼ **Example 2** Find log 843.

Solution $\log 843 = \log(8.43 \times 10^2)$
$= \log 8.43 + \log 10^2$
$= .9258 + 2$
$= 2.9258$ ▲

▼ **Example 3** Find log .0391.

Solution $\log .0391 = \log(3.91 \times 10^{-2})$
$= \log 3.91 + \log 10^{-2}$
$= .5922 + (-2)$

Now there are two ways to proceed from here. We could add .5922 and −2 to get −1.4078. (If you were using a calculator to find log .0391, this is the answer you would see.) The problem with −1.4078 is that the mantissa is negative. Our table contains only positive numbers. If we have to use log .0391 again, and it is written as −1.4078, we will not be able to associate it with an entry in the table. It is most common, when using a table, to write the characteristic −2 as $8 + (-10)$ and proceed as follows:

$$\log .0391 = .5922 + 8 + (-10)$$
$$= 8.5922 - 10$$ ▲

Here is another example:

▼ **Example 4** Find log .00523.

Solution $\log .00523 = \log(5.23 \times 10^{-3})$
$= \log 5.23 + \log 10^{-3}$
$= .7185 + (-3)$
$= .7185 + 7 + (-10)$
$= 7.7185 - 10$ ▲

It is not necessary to show any of the steps we have shown in the examples above. As a matter of fact, it is better if we don't. With a little practice the steps can be eliminated. Once we have become familiar with using the table to find logarithms, we can go in the reverse direction and use the table to solve equations like $\log x = 3.8774$.

▼   **Example 5**   Find x if $\log x = 3.8774$.

Solution   We are looking for the number whose logarithm is 3.8774. The mantissa is .8774, which appears in the table across from 7.5 and under (or above) 4.

x	0	1	2	3	4	5	6	7	8	9
4.3										
4.4										
⋮										
7.5					.8774					
⋮										
7.8										
7.9										
x	0	1	2	3	4	5	6	7	8	9

The characteristic is 3 and came from the exponent of 10. Putting these together, we have

$$\log x = 3.8774$$
$$\log x = .8774 + 3$$
$$x = 7.54 \times 10^3$$
$$x = 7540$$

The number 7540 is called the *antilogarithm* or just *antilog* of 3.8774. That is, 7540 is the number whose logarithm is 3.8774.   ▲

▼   **Example 6**   Find x if $\log x = 7.5821 - 10$.

Solution   The mantissa is .5821, which comes from 3.82 in the table. The characteristic is $7 + (-10)$ or -3, which comes from a power of 10:

$$\log x = 7.5821 - 10$$
$$\log x = .5821 + 7 + (-10)$$
$$\log x = .5821 + (-3)$$
$$x = 3.82 \times 10^{-3}$$
$$x = .00382$$

The antilog of $7.5821 - 10$ is $.00382$. That is, the logarithm of $.00382$ is $7.5821 - 10$.   ▲

The following examples illustrate how logarithms are used as an aid in computations.

▼   **Example 7**   Use logarithms to find $(3780)(45,200)$.

Solution   We will let n represent the answer to this problem:

$$n = (3780)(45,200)$$

Since these two numbers are equal, so are their logarithms. We therefore take the common logarithm of both sides:

$$\log n = \log (3780)(45,200)$$
$$= \log 3780 + \log 45,200$$
$$= 3.5775 + 4.6551$$
$$\log n = 8.2326$$

The number 8.2326 is the logarithm of the answer. The mantissa is $.2326$, which is not in the table. Since $.2326$ is closest to $.2330$, and $.2330$ is the logarithm of 1.71, we write

$$n = 1.71 \times 10^{8}$$ ▲

Note   The answer 1.71×10^{8} is not exactly equal to $(3780)(45,200)$. It is an approximation to it. If we want to be more accurate and have an answer with more significant digits, we would have to use a table with more significant digits.

▼   **Example 8**   Use logarithms to find $\sqrt[3]{875}$.

Solution

$$\text{If} \qquad n = \sqrt[3]{875},$$
$$\text{then} \qquad n = (875)^{1/3}$$

$$\text{and}\quad \log n = \log 875^{1/3}$$
$$= \tfrac{1}{3}\log 875$$
$$= \tfrac{1}{3}(2.9420)$$
$$\log n = .9807$$
$$n = 9.56$$

A good approximation to the cube root of 875 is 9.56.   ▲

▼ **Example 9**   Use logarithms to find $35^{2.7}$.

Solution

$$\text{Let } n = 35^{2.7}.$$
$$\text{Then } \log n = \log 35^{2.7}$$
$$= 2.7 \log 35$$
$$= 2.7(1.5441)$$
$$\log n = 4.1691$$
$$n = 1.48 \times 10^4$$
$$= 14{,}800$$
                                                                        ▲

▼ **Example 10**   Use logarithms to find $\dfrac{(34.5)^2 \sqrt{1080}}{(2.76)^3}$.

Solution

$$\text{If }\quad n = \frac{(34.5)^2 \sqrt{1080}}{(2.76)^3},$$

$$\text{then }\quad \log n = \log\left[\frac{(34.5)^2 \sqrt{1080}}{(2.76)^3}\right]$$

$$= 2 \log 34.5 + \frac{1}{2}\log 1080 - 3 \log 2.76$$

$$= 2(1.5378) + \frac{1}{2}(3.0334) - 3(.4409)$$

$$= 3.0756 + 1.5167 - 1.3227$$
$$\log n = 3.2696$$
$$n = 1.86 \times 10^3$$
$$= 1860$$
                                                                        ▲

Use the table in Appendix D to find the following: *Problem Set 11.3*

1. log 378   **2.** log 426
3. log 37.8   **4.** log 42,600

5. log 3780
7. log .0378
9. log 37,800
11. log 600
13. log 2010
15. log .00971
17. log .0314
19. log .399

6. log .4260
8. log .0426
10. log 4900
12. log 900
14. log 10,200
16. log .0312
18. log .00052
20. log .111

Find x in the following equations:

21. $\log x = 2.8802$
23. $\log x = 7.8802 - 10$
25. $\log x = 3.1553$
27. $\log x = 4.6503 - 10$
29. $\log x = 2.9628 - 10$

22. $\log x = 4.8802$
24. $\log x = 6.8802 - 10$
26. $\log x = 5.5911$
28. $\log x = 8.4330 - 10$
30. $\log x = 5.8000 - 10$

Use logarithms to evaluate the following:

31. $(378)(24.5)$

32. $(921)(2630)$

33. $\dfrac{496}{391}$

34. $\dfrac{512}{216}$

35. $\sqrt{401}$
37. $\sqrt[3]{1030}$

36. $\sqrt[3]{92.3}$
38. $\sqrt{.641}$

39. $\dfrac{(2390)(28.4)}{176}$

40. $\dfrac{(32.9)(5760)}{11.1}$

41. $(296)^2(459)$
43. $\sqrt{526}\,(1080)$
45. $45^{-3.1}$

42. $(3250)(24.2)^3$
44. $(32.7)\sqrt{.580}$
46. $72^{1.8}$

47. $\dfrac{(895)^3\sqrt{41.1}}{17.8}$

48. $\dfrac{(925)^2\sqrt{1.99}}{243}$

49. The weight W of a sphere with diameter d can be found by using the formula

$$W = \frac{\pi}{6}d^3 w$$

where w is the weight of a cubic unit of material. Use logarithms to find W if $D = 2.98$ in, $w = 3.03$ lb and $\pi = 3.14$.

50. The following formula is used in hydraulics:

$$d = 2.57\sqrt[5]{\frac{flQ^2}{h}}$$

Use logarithms to find d, when $f = 0.022$, $l = 2,820$, $h = 133$, and $Q = 12.5$.

51. The area A of a triangle in which all three sides are of equal length l (equilateral) is given by the formula

$$A = \frac{l^2}{4} \sqrt{3}.$$

Find A when $l = 276$. (Use logarithms.)

52. The formula $H = \dfrac{PLAN}{33,000}$ is used to compute the horsepower of a steam or gas engine. In this formula, H is the horsepower; P, the average effective pressure on the piston, in pounds per square inch; L, the distance the piston travels per stroke, in feet; A, the area of cross-section of the cylinder, in square inches; and N, the number of working strokes per minute. Find H, when $P = 62.8$ pounds; $L = 2.63$ feet; $A = 18.4$ square inches; and $N = 92.4$.

Review Problems   The problems below review material we covered in Sections 2.2 and 6.6.

Solve each inequality.

53. $-3x < 21$ **54.** $-2x > -12$

55. $6(3x - 2) \le 4 - (2x - 4)$ **56.** $5(2x - 4) \ge 7 - (3x + 1)$

57. $(2x - 3)(x + 4) > 0$ **58.** $(3x - 5)(x + 2) < 0$

59. $x^2 - 8x + 15 \le 0$ **60.** $x^2 - 8x + 12 \ge 0$

Logarithms are very important in solving equations in which the variable appears as an exponent. The equation

$$5^x = 12$$

is an example of one such equation. Equations of this form are called *exponential equations*. Since the quantities 5^x and 12 are equal, so are their common logarithms. We begin our solution by taking the logarithm of both sides:

$$\log 5^x = \log 12$$

We now apply property 3 for logarithms, $\log x^r = r \log x$, to turn x from an exponent into a coefficient:

$$x \log 5 = \log 12$$

11.4
**Exponential Equations
and Change of Base**

Dividing both sides by log 5 gives us

$$x = \frac{\log 12}{\log 5}$$

If we want a decimal approximation to the solution, we can find log 12 and log 5 in the table in Appendix D and divide:

$$x = \frac{1.0792}{.6990}$$

$$= 1.5439$$

The complete problem looks like this:

$$5^x = 12$$
$$\log 5^x = \log 12$$
$$x \log 5 = \log 12$$
$$x = \frac{\log 12}{\log 5}$$
$$= \frac{1.0792}{.6990}$$
$$= 1.5439$$

Note   A very common mistake can occur in the third-from-the-last step. Many times the expression

$$\frac{\log 12}{\log 5}$$

is mistakenly simplified as log 12 − log 5. There is no property of logarithms that allows us to do this. The only property of logarithms that deals with division is property 2, which is *this:*

$$\log_b \frac{x}{y} = \log_b x - \log_b y \qquad \text{(Right)}$$

not this:

$$\frac{\log_b x}{\log_b y} = \log_b x - \log_b y \qquad \text{(Wrong)}$$

The second statement is simply not a property of logarithms, although it is sometimes mistaken for one.

Here is another example of solving an exponential equation using logarithms.

▼   **Example 1** Solve for x: $25^{2x+1} = 15$.

Solution Taking the logarithm of both sides and then writing the exponent $(2x + 1)$ as a coefficient, we proceed as follows:

$$25^{2x+1} = 15$$
$$\log 25^{2x+1} = \log 15$$
$$(2x + 1) \log 25 = \log 15$$
$$2x + 1 = \frac{\log 15}{\log 25}$$
$$2x = \frac{\log 15}{\log 25} - 1$$
$$x = \frac{1}{2}\left(\frac{\log 15}{\log 25} - 1\right)$$

Using the table of common logarithms (Appendix D), we can write a decimal approximation to the answer:

$$x = \frac{1}{2}\left(\frac{1.1761}{1.3979} - 1\right)$$
$$= (.8413 - 1)$$
$$= \frac{1}{2}(-.1587)$$
$$= -.0793 \qquad\qquad ▲$$

There is a fourth property of logarithms we have not yet considered. This last property allows us to change from one base to another and is therefore called the *change-of-base property*.

PROPERTY 4 (Change of Base) If a and b are both positive numbers other than 1, and if $x > 0$, then

$$\log_a x = \frac{\log_b x}{\log_b a}$$

$$\underset{\text{Base } a}{\uparrow} \qquad \underset{\text{Base } b}{\uparrow}$$

The logarithm on the left side has a base of a, while both logarithms on the right side have a base of b. This allows us to change from base a to any other base b that is a positive number other than 1. Here is a proof of property 4 for logarithms:

Proof   We begin by writing the identity

$$a^{\log_a x} = x$$

Taking the logarithm base b of both sides and setting the exponent $\log_a x$ as a coefficient, we have

$$\log_b a^{\log_a x} = \log_b x$$
$$\log_a x \log_b a = \log_b x$$

Dividing both sides by $\log_b a$, we have the desired result:

$$\frac{\log_a x \log_b a}{\log_b a} = \frac{\log_b x}{\log_b a}$$

$$\log_a x = \frac{\log_b x}{\log_b a}$$

We can use this property to find logarithms we do not have a table for. The next two examples illustrate the use of this property.

▼   **Example 2**   Find $\log_8 24$.

Solution   Since we do not have a table for base-8 logarithms, we can change this expression to an equivalent expression that only contains base-10 logarithms:

$$\log_8 24 = \frac{\log 24}{\log 8}$$

Don't be confused. We did not just drop the base, we changed to base 10. We could have written the last line like this:

$$\log_8 24 = \frac{\log_{10} 24}{\log_{10} 8}$$

Looking up log 24 and log 8 in the table in Appendix D, we write

$$\log_8 24 = \frac{1.3802}{.9031}$$

$$= 1.5283 \qquad\qquad ▲$$

▼ **Example 3** Find $\log_{20} 342$.

Solution Changing to base 10 using property 4, we proceed as follows:

$$\log_{20} 342 = \frac{\log 342}{\log 20}$$

$$= \frac{2.5340}{1.3010}$$

$$= 1.9477$$

The decimal approximation to $\log_{20} 342$ is 1.9477.   ▲

Solve each exponential equation. Use the table in Appendix D to write the   *Problem Set 11.4*
answer in decimal form.

1. $3^x = 5$ 2. $4^x = 3$
3. $5^x = 3$ 4. $3^x = 4$
5. $5^{-x} = 12$ 6. $7^{-x} = 8$
7. $12^{-x} = 5$ 8. $8^{-x} = 7$
9. $8^{x+1} = 4$ 10. $9^{x+1} = 3$
11. $4^{x-1} = 4$ 12. $3^{x-1} = 9$
13. $3^{2x+1} = 2$ 14. $2^{2x+1} = 3$
15. $3^{1-2x} = 2$ 16. $2^{1-2x} = 3$
17. $15^{3x-4} = 10$ 18. $10^{3x-4} = 15$
19. $6^{5-2x} = 4$ 20. $9^{7-3x} = 5$

Use the change-of-base property and the table in Appendix D to find a decimal
approximation to each of the following logarithms:

21. $\log_8 16$ 22. $\log_9 27$
23. $\log_{16} 8$ 24. $\log_{27} 9$
25. $\log_7 15$ 26. $\log_3 12$
27. $\log_{15} 7$ 28. $\log_{12} 3$
29. $\log_{12} 11$ 30. $\log_{14} 15$
31. $\log_{11} 12$ 32. $\log_{15} 14$
33. $\log_8 240$ 34. $\log_6 180$
35. $\log_4 321$ 36. $\log_5 462$

37. The formula $A = P(1 + r)^n$ shows how to find the amount of money A
     that will be in an account if P dollars is invested at interest rate r for n
     years if the interest is compounded yearly. Use logarithms to solve this
     formula for n.
38. If P dollars is invested in an account that pays an annual rate of interest r,
     that is compounded semiannually, the amount of money in the account at

the end of n years is given by the formula

$$A = P\left(1 + \frac{r}{2}\right)^{2n}$$

Use logarithms to solve this formula for n.

Review Problems   The problems below review material we covered in Section 8.5. They are taken from the book *Algebra for the Practical Man,* written by J. E. Thompson and Published by D. Van Nostrand Company in 1931.

39.   A man spent $112.80 for 108 geese and ducks, each goose costing 14 dimes and each duck 6 dimes. How many of each did he buy?

40.   If 15 lb of tea and 10 lb of coffee together cost $15.50, while 25 lb. of tea and 13 lb. of coffee at the same prices cost $24.55, find the price per pound of each.

41.   A number of oranges at the rate of three for ten cents and apples at fifteen cents a dozen cost, together, $6.80. Five times as many oranges and one fourth as many apples at the same rates would have cost $25.45. How many of each were bought?

42.   An estate is divided among three persons, A, B and C. A's share is three times that of B and B's share is twice that of C. If A receives $9000 more than C, how much does each receive?

**11.5
Word Problems**

There are many practical applications of exponential equations and logarithms. Many times an exponential equation will describe a situation arising in nature. Logarithms are then used to solve the exponential equation. Many of the problems in this section deal with expressions and equations that have been derived or defined in some other discipline. We will not attempt to derive them here. We will simply accept them as they are and use them to solve some problems.

In chemistry the pH of a solution is defined in terms of logarithms as

$$\text{pH} = -\log{(H_3O^+)}$$

where (H_3O^+) is the concentration of the hydronium ion, H_3O^+, in solution. An acid solution has a pH lower than 7, and a basic solution has a pH above 7. (This is from chemistry. We don't need to worry here about *why* it's this way. We just want to work some problems involving logarithms.)

▼   **Example 1**   Find the pH of a solution in which (H_3O^+) = 5.1 × 10^{-3}.

Solution   Using the definition given above, we write

$$\begin{aligned} \text{pH} &= -\log{(H_3O^+)} \\ &= -\log{(5.1 \times 10^{-3})} \\ &= -[.7076 + (-3)] \\ &= -.7076 + 3 \\ &\approx 2.29 \end{aligned}$$

According to our previous discussion, this solution would be considered an acid solution, since its pH is less than 7. ▲

▼ **Example 2** What is (H_3O^+) for a solution with a pH of 9.3?

Solution We substitute 9.3 for pH in the definition of pH and proceed as follows:

$$9.3 = -\log{(H_3O^+)}$$
$$\log{(H_3O^+)} = -9.3$$

It is impossible to look up -9.3 in the table in Appendix D because it is negative. We can get around this problem by adding and subtracting 10 to the -9.3:

$$\begin{aligned} \log{(H_3O^+)} &= -9.3 + 10 - 10 \qquad \text{(Actually adding 0)} \\ &= .7 - 10 \end{aligned}$$

The closest thing to .7 in the table is .6998, which is the logarithm of 5.01:

$$(H_3O^+) = 5.01 \times 10^{-10}$$

The concentration of H_3O^+ is 5.01×10^{-10}. (This quantity is usually given in moles/liter. We have left off the units to make things simpler.) ▲

Carbon-14 dating is used extensively in science to find the age of fossils. If at one time a nonliving substance contains an amount A_0 of carbon-14, then t years later it will contain an amount A of carbon-14, where

$$A = A_0 \cdot 2^{-t/5600}$$

▼ **Example 3** If a nonliving substance has 3 grams of carbon-14, how much carbon-14 will be present 500 years later?

Solution The original amount of carbon 14, A_0, is 3 grams. The

number of years, t, is 500. Substituting these quantities into the equation given above, we have the expression

$$A = (3)2^{-500/5600}$$
$$= (3)2^{-5/56}$$

In order to evaluate this expression we must use logarithms. We begin by taking the logarithm of both sides:

$$\log A = \log [(3)2^{-5/56}]$$
$$= \log 3 - \tfrac{5}{56} \log 2$$
$$= .4771 - \tfrac{5}{56}(.3010)$$
$$= .4771 - .0269$$
$$\log A = .4502$$
$$A = 2.82$$

The amount remaining after 500 years is 2.82 grams.   ▲

If an amount of money P (P for principal) is invested in an account that pays a rate r of interest compounded annually, then the total amount of money A (the original amount P plus all the interest) in the account after t years is given by the equation

$$A = P(1 + r)^t$$

▼   **Example 4**   How much money will accumulate over 20 years if a person invests $5000 in an account that pays 6% interest per year?

Solution   The original amount P is $5000, the rate of interest r is .06 (6% = .06), and the length of time t is 20 years. We substitute these values into the equation and solve for A:

$$A = 5000(1 + .06)^{20}$$
$$A = 5000(1.06)^{20}$$
$$\log A = \log [5000(1.06)^{20}]$$
$$\log A = \log 5000 + 20 \log 1.06$$
$$= 3.6990 + 20(.0253)$$
$$= 3.6990 + .5060$$
$$\log A = 4.2050$$
$$A = 1.60 \times 10^4$$
$$= 16,000$$

The original $5000 will more than triple to become $16,000 in 20 years at 6% annual interest.   ▲

▼ **Example 5** How long does it take for $5000 to double if it is deposited in an account that yields 5% per year?

Solution The original amount P is $5000, *the total amount after t years is* $A = \$10,000$, and the interest rate r is .05. Substituting into

$$A = P(1 + r)^t$$

we have

$$10,000 = 5000(1 + .05)^t$$
$$= 5000(1.05)^t$$

This is an exponential equation. We solve by taking the logarithm of both sides:

$$\log 10,000 = \log [(5000)(1.05)^t]$$
$$\log 10,000 = \log 5000 + t \log 1.05$$
$$4 = 3.6990 + t(.0212)$$

Subtract 3.6990 from both sides:

$$.3010 = .0212t$$

Dividing both sides by .0212, we have

$$t = 14.20$$

It takes a little over 14 years for $5000 to double if it earns 5% per year.   ▲

For problems 1–8 find the pH of the solution using the formula $\text{pH} = -\log (H_3O^+)$.

Problem Set 11.5

1. $(H_3O^+) = 4 \times 10^{-3}$
2. $(H_3O^+) = 3 \times 10^{-4}$
3. $(H_3O^+) = 5 \times 10^{-6}$
4. $(H_3O^+) = 6 \times 10^{-5}$
5. $(H_3O^+) = 4.2 \times 10^{-5}$
6. $(H_3O^+) = 2.7 \times 10^{-7}$
7. $(H_3O^+) = 8.6 \times 10^{-2}$
8. $(H_3O^+) = 5.3 \times 10^{-4}$

For problems 9–12 find (H_3O^+) for solutions with the given pH.

9. pH = 3.4
10. pH = 5.7
11. pH = 6.5
12. pH = 2.1

13. A nonliving substance contains 3 micrograms of carbon-14. How much carbon-14 will be left at the end of

   a. 5000 years?
   b. 10,000 years?

   c. 56,000 years?

   d. 112,000 years?

14. A nonliving substance contains 5 micrograms of carbon-14. How much carbon-14 will be left at the end of

   a. 500 years?

   b. 5000 years?

   c. 56,000 years?

   d. 112,000 years?

15. At one time a certain nonliving substance contained 10 micrograms of carbon-14. How many years later did the same substance contain only 5 micrograms of carbon-14?

16. At one time a certain nonliving substance contained 20 micrograms of carbon-14. How many years later did the same substance contain only 5 micrograms of carbon-14?

17. How much money is in an account after 10 years if $5000 was deposited originally at 6% per year?

18. If you put $2000 in an account that earns 5% interest annually, how much will you have in the account at the end of 10 years?

19. If you deposit $2000 in an account that earns 7.5% annually, how much will you have in the account after 10 years?

20. Suppose $15,000 is deposited in an account that yields 10% per year. How much is in the account 5 years later?

21. If $4000 is in an account that earns 5% per year, how long does it take before the account has $8000 in it?

22. If $4000 is deposited in an account that yields 7% annually, how long does it take the account to reach $8000?

23. How long does it take to double $10,000 if it is in an account that earns 6% per year?

24. How long does it take to double $10,000 if it is in an account that earns 10% per year?

Review Problems The problems below review material we covered in Sections 2.3 and 2.4.

Solve each equation.

25. $|3x - 5| = 2$ **26.** $|3x + 1| = 5$

27. $|7x - 8| + 9 = 1$ **28.** $|5x - 3| + 5 = 12$

Solve each inequality.

29. $|x - 4| > 5$ **30.** $|x + 3| < 2$

31. $|2x - 1| \leq 3$ **32.** $|2x + 1| \geq 5$

Chapter 11   Summary and Review

Examples

DEFINITION OF LOGARITHMS [11.1]

If b is a positive number not equal to 1, then the expression

$$y = \log_b x$$

is equivalent to $x = b^y$. That is, in the expression $y = \log_b x$, y is the number to which we raise b in order to get x. Expressions written in the form $y = \log_b x$ are said to be in *logarithmic form*. Expressions like $x = b^y$ are in *exponential form*.

1. The definition allows us to write expressions like

$$y = \log_3 27$$

equivalently in exponential form as

$$3^y = 27$$

which makes it apparent that y is 3.

TWO SPECIAL IDENTITIES [11.1]

For $b > 0$, $b \neq 1$, the following two expressions hold for all positive real numbers x:

1. $b^{\log_b x} = x$
2. $\log_b b^x = x$

2. Examples of the two special properties are

$$5^{\log_5 12} = 12$$

and $\log_8 8^3 = 3$.

PROPERTIES OF LOGARITHMS [11.2]

If x, y, and b are positive real numbers, $b \neq 1$, and r is any real number, then

1. $\log_b (xy) = \log_b x + \log_b y$
2. $\log_b \left(\dfrac{x}{y} \right) = \log_b x - \log_b y$
3. $\log_b x^r = r \log_b x$

3. We can rewrite the expression

$$\log_{10} \frac{45^6}{273}$$

using the properties of logarithms, as

$$6 \log_{10} 45 - \log_{10} 273.$$

COMMON LOGARITHMS [11.3]

Common logarithms are logarithms with a base of 10. To save time in writing, we omit the base when working with common logarithms. That is,

$$\log x = \log_{10} x.$$

4. $\log_{10} 10{,}000 = \log 10{,}000$
$$= \log 10^4$$
$$= 4$$

5. For common logarithms, the characteristic is the power of 10 needed to put the number in scientific notation, and the mantissa is found in Appendix D in the back of the book.

NOTATION [11.3]

In the expression

$$\log 4240 = 3.6274$$

the 3 is called the *characteristic* and the decimal .6274 is called the *mantissa*.

6. $\log_6 475 = \dfrac{\log 475}{\log 6}$

$ = \dfrac{2.6767}{.7782}$

$ = 3.44$

CHANGE OF BASE [11.4]

If x, a, and b are positive real numbers, $a \neq 1$ and $b \neq 1$, then

$$\log_a x = \frac{\log_b x}{\log_b a}.$$

COMMON MISTAKES

The most common mistakes that occur with logarithms come from trying to apply the three properties of logarithms to situations in which they don't apply. For example, a very common mistake looks like this:

$$\frac{\log 3}{\log 2} = \log 3 - \log 2$$

This is not a property of logarithms. In order to write the expression $\log 3 - \log 2$, we would have to start with

$$\log \frac{3}{2} \quad \text{not} \quad \frac{\log 3}{\log 2}$$

There is a difference.

**Chapter 11
Test**

Solve for x:

1. $\log_4 x = 3$ **2.** $\log_x 5 = 2$

Graph each of the following:

3. $y = \log_2 x$ **4.** $y = \log_{1/2} x$

Evaluate each of the following:

5. $\log_8 4$ **6.** $\log_7 21$
7. $\log 23{,}400$ **8.** $\log .0123$

Use logarithms to find an approximate answer for the following computations:

9. (2.34)(6080)

10. $3^{2.5}$

11. $\sqrt[4]{23}$

12. $\dfrac{(352)(41.5)^2}{(2.31)^3}$

Use the properties of logarithms as an aid in solving the following:

13. $5 = 3^x$

14. $4^{2x-1} = 8$

15. $\log_5 x - \log_5 3 = 1$

16. $\log_2 x + \log_2 (x - 7) = 3$

17. Find the total amount of money in an account if $4000 was deposited 5 years ago at 5% annual interest.

18. Give the pH of a solution in which $(H_3O^+) = 3 \times 10^{-4}$.

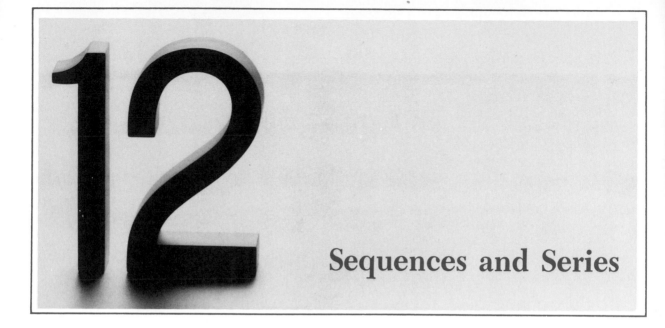

12

Sequences and Series

To the student:

This chapter begins with a look at sequences and series in general. In Section 12.2 we will use a new kind of notation, summation notation, that allows us to write long sums of numbers with few symbols. We will classify two main types of sequences, arithmetic and geometric sequences, and look at what is called the nth partial sum of a sequence. The section on the binomial expansion is very useful as it gives a formula for expanding $(a + b)^n$ without actually having to multiply it out. Each section in this chapter deals in some way with sequences of numbers. Sequences are used frequently in many different branches of mathematics and science. Many real-life situations can be described in terms of arithmetic and geometric sequences. For example, the yearly amount of money in a savings account that compounds interest at regular intervals will form a geometric progression, as will the value of any object that increases or decreases in value by a specified percentage each year. The binomial theorem has some useful applications which are not limited to mathematics. It is used in statistics, biology, and physics as well as other disciplines. Many of the problems we have previously worked in terms of functions can also be worked using sequences and series.

This section serves as a general introduction to sequences and series. We begin with the definition of a sequence.

DEFINITION A *sequence* is a function whose domain is the set of positive integers $\{1, 2, 3, \ldots\}$.

We usually use the letter a with a subscript to denote specific terms in a sequence, instead of the usual function notation. That is, instead of writing $a(3)$ for the third term of a sequence, we use a_3 to denote the third term. For example,

$$a_1 = \text{First term of the sequence}$$
$$a_2 = \text{Second term of the sequence}$$
$$a_3 = \text{Third term of the sequence}$$
$$\vdots$$
$$a_n = n\text{th term of the sequence}$$
$$\vdots$$

The nth term is also called the *general term* of the sequence. The general term is used to define the other terms of the sequence. That is, if we are given the formula for the general term, a_n, we can find any other term in the sequence. The following examples illustrate.

▼   **Example 1**   Find the first four terms of the sequence whose general term is given by $a_n = 2n - 1$.

Solution   The subscript notation a_n works the same way function notation works. To find the first, second, third, and fourth terms of this sequence, we simply substitute 1, 2, 3, and 4 for n in the formula $2n - 1$:

$$\text{If}\quad \text{General term} = a_n = 2n - 1,$$
$$\text{then}\qquad \text{First term} = a_1 = 2(1) - 1 = 1$$
$$\text{Second term} = a_2 = 2(2) - 1 = 3$$
$$\text{Third term} = a_3 = 2(3) - 1 = 5$$
$$\text{Fourth term} = a_4 = 2(4) - 1 = 7$$

The first four terms of this sequence are the odd numbers 1, 3, 5, and 7. The whole sequence can be written as

$$1, 3, 5, \ldots, 2n - 1, \ldots$$

Since each term in this sequence is larger than the preceding term, we say the sequence is an *increasing sequence*.   ▲

▼   **Example 2**   Write the first four terms of the sequence defined by

$$a_n = \frac{1}{n + 1}.$$

Solution   Replacing n with 1, 2, 3, and 4, we have, respectively, the first four terms:

$$\text{First term} = a_1 = \frac{1}{1 + 1} = \frac{1}{2}$$

$$\text{Second term} = a_2 = \frac{1}{2 + 1} = \frac{1}{3}$$

$$\text{Third term} = a_3 = \frac{1}{3 + 1} = \frac{1}{4}$$

$$\text{Fourth term} = a_4 = \frac{1}{4 + 1} = \frac{1}{5}$$

The sequence defined by

$$a_n = \frac{1}{n + 1}$$

can be written as

$$\frac{1}{2}, \frac{1}{3}, \frac{1}{4}, \dots, \frac{1}{n + 1}, \dots$$

Since each term in the sequence is smaller than the term preceding it, the sequence is said to be a *decreasing sequence*.   ▲

▼   **Example 3**   Find the fifth and sixth terms of the sequence whose general term is given by $a_n = \dfrac{(-1)^n}{n^2}.$

Solution   For the fifth term we replace n with 5. For the sixth term we replace n with 6:

$$\text{Fifth term} = a_5 = \frac{(-1)^5}{5^2} = \frac{-1}{25}$$

$$\text{Sixth term} = a_6 = \frac{(-1)^6}{6^2} = \frac{1}{36}$$ ▲

In the first three examples we found some terms of a sequence after being given the general term. In the next two examples we will do the reverse. That is, given some terms of a sequence, we will find the formula for the general term.

▼   **Example 4**   Find a formula for the nth term of the sequence 2, 8, 18, 32, ...

Solution   Solving a problem like this involves some guessing. Looking over the first four terms we see each is twice a perfect square:

$$2 = 2(1)$$
$$8 = 2(4)$$
$$18 = 2(9)$$
$$32 = 2(16)$$

If we write each square with an exponent of 2, the formula for the nth term becomes obvious:

$$a_1 = 2 \ = 2(1)^2$$
$$a_2 = 8 \ = 2(2)^2$$
$$a_3 = 18 = 2(3)^2$$
$$a_4 = 32 = 2(4)^2$$
$$\vdots$$
$$a_n = \qquad 2(n)^2 = 2n^2$$

The general term of the sequence 2, 8, 18, 32, . . . is $a_n = 2n^2$.

▲

▼   **Example 5**   Find the general term for the sequence $2, \frac{3}{8}, \frac{4}{27}, \frac{5}{64}, \ldots$.

Solution   The first term can be written as $\frac{2}{1}$. The denominators are all perfect cubes. The numerators are all 1 more than the base of the cubes in the denominator:

$$a_1 = \frac{2}{1} = \frac{1+1}{1^3}$$

$$a_2 = \frac{3}{8} = \frac{2+1}{2^3}$$

$$a_3 = \frac{4}{27} = \frac{3+1}{3^3}$$

$$a_4 = \frac{5}{64} = \frac{4+1}{4^3}$$

Observing this pattern, we recognize the general term to be

$$a_n = \frac{n+1}{n^3} \qquad\qquad ▲$$

Note Finding the *n*th term of a sequence from the first few terms is not always automatic. That is, it sometimes takes a while to recognize the pattern. Don't be afraid to guess at the formula for the general term. Many times an incorrect guess leads to the correct formula.

Problem Set 12.1

Write the first five terms of the sequences with the following general terms:

1. $a_n = 3n + 1$ **2.** $a_n = 2n + 3$
3. $a_n = 4n - 1$ **4.** $a_n = n + 4$
5. $a_n = n$ **6.** $a_n = -n$
7. $a_n = n^2 + 3$ **8.** $a_n = n^3 + 1$

9. $a_n = \dfrac{n}{n + 3}$ **10.** $a_n = \dfrac{n}{n + 2}$

11. $a_n = \dfrac{n + 1}{n + 2}$ **12.** $a_n = \dfrac{n + 3}{n + 4}$

13. $a_n = \dfrac{1}{n^2}$ **14.** $a_n = \dfrac{1}{n^3}$

15. $a_n = 2^n$ **16.** $a_n = 3^n$
17. $a_n = 3^{-n}$ **18.** $a_n = 2^{-n}$

19. $a_n = 1 + \dfrac{1}{n}$ **20.** $a_n = 1 - \dfrac{1}{n}$

21. $a_n = n - \dfrac{1}{n}$ **22.** $a_n = n + \dfrac{1}{n}$

23. $a_n = (-2)^n$ **24.** $a_n = (-3)^n$

Determine the general term for each of the following sequences:

25. 2, 3, 4, 5, . . . **26.** 3, 6, 9, 12, . . .
27. 4, 8, 12, 16, 20, . . . **28.** 3, 4, 5, 6, . . .
29. 7, 10, 13, 16, . . . **30.** 4, 9, 14, 19, . . .
31. 1, 4, 9, 16, . . . **32.** 1, 8, 27, 64, . . .
33. 3, 12, 27, 48, . . . **34.** 2, 16, 54, 128, . . .
35. 4, 8, 16, 32, . . . **36.** 3, 9, 27, 81, . . .
37. $-2, 4, -8, 16, \ldots$ **38.** $-3, 9, -27, 81, \ldots$
39. $\frac{1}{4}, \frac{1}{8}, \frac{1}{16}, \frac{1}{32}, \ldots$ **40.** $\frac{1}{3}, \frac{1}{9}, \frac{1}{27}, \frac{1}{81}, \ldots$
41. $\frac{1}{4}, \frac{2}{9}, \frac{3}{16}, \frac{4}{25}, \ldots$ **42.** $\frac{1}{4}, \frac{2}{10}, \frac{3}{28}, \frac{4}{82}, \ldots$

43. The value of a home in California is said to increase 1% per month. If a home were currently worth $50,000, then in 1 month it would increase in value 1% of $50,000 or .01(50,000) = $500. It would then be worth $50,500. Write a sequence that gives the value of this home at the end of 1 month, 2 months, 3 months, 4 months, 5 months and 6 months. What is the general term of the sequence? How much should the house be worth in 2 years? (Give answers to nearest dollar.)

44. Suppose the value of a home increases only 0.5% per month. Find the general term of the sequence that gives the value of a home that is now worth $100,000 at the end of *n* months.

Review Problems The problems below review material we covered in Sections 3.1 and 3.2.

Simplify each expression. Write all answers with positive exponents only, and assume all variables are nonzero.

45. $(-2x^2)^3$ **46.** $(2x^2)^{-3}$

47. $3^{-1} + 2^{-2}$ **48.** $3^{-2} + 2^{-3}$

47. $\dfrac{x^{-4}}{x^5}$ **50.** $\dfrac{x^3}{x^{-2}}$

51. $\left(\dfrac{2x^3y^0}{x^{-4}y^3}\right)^{-3}$ **52.** $\left(\dfrac{3x^{-2}y^3}{x^0y^{-1}}\right)^{-2}$

We begin this section with the definition of a series.

**12.2
Series and
Summation Notation**

DEFINITION The sum of a number of terms in a sequence is called a *series*.

A sequence can be finite or infinite depending on whether or not the sequence ends at the *n*th term. For example,

$$1, 3, 5, 7, 9$$

is a finite sequence, while

$$1, 3, 5, \ldots$$

is an infinite sequence. Associated with each of the above sequences is a series found by adding the terms of the sequence:

$$1 + 3 + 5 + 7 + 9 \qquad \text{Finite series}$$
$$1 + 3 + 5 + \ldots \qquad \text{Infinite series}$$

We will consider only finite series in this book. We can introduce a new kind of notation here that is a compact way of indicating a finite series. The notation is called *summation notation,* or *sigma notation* since it is written using the Greek letter sigma. The expression

$$\sum_{i=1}^{4} (3i + 2)$$

is an example of summation notation. The expression is interpreted as

indicating we are to add all the numbers $3i + 2$, where i is replaced by all the consecutive integers from 1 up to and including 4:

$$\sum_{i=1}^{4} (3i + 2) = [3(1) + 2] + [3(2) + 2] + [3(3) + 2] + [3(4) + 2]$$

$$= 5 + 8 + 11 + 14$$
$$= 38$$

The letter i as used here is called the *index of summation*, or just *index* for short.

Here are some examples illustrating the use of summation notation.

▼   **Example 1**   Expand and simplify $\sum_{i=1}^{5} (i^2 - 1)$.

Solution   We replace i in the expression $i^2 - 1$ with all consecutive integers from 1 up to 5, including 1 and 5:

$$\sum_{i=1}^{5} (i^2 - 1) = (1^2 - 1) + (2^2 - 1) + (3^2 - 1) + (4^2 - 1) + (5^2 - 1)$$

$$= 0 + 3 + 8 + 15 + 24$$
$$= 50$$ ▲

▼   **Example 2**   Expand and simplify $\sum_{i=3}^{6} (-2)^i$.

Solution   We replace i in the expression $(-2)^i$ with the consecutive integers beginning at 3 and ending at 6:

$$\sum_{i=3}^{6} (-2)^i = (-2)^3 + (-2)^4 + (-2)^5 + (-2)^6$$

$$= -8 + 16 + (-32) + 64$$
$$= 40$$ ▲

▼   **Example 3**   Expand $\sum_{i=2}^{5} (x^i - 3)$.

Solution   We must be careful here not to confuse the letter x with i.

The index i is the quantity we replace by the consecutive integers from 1 to 5, not x.

$$\sum_{i=2}^{5} (x^i - 3) = (x^2 - 3) + (x^3 - 3) + (x^4 - 3) + (x^5 - 3)$$

▲

In the first three examples we were given an expression with summation notation and asked to expand it. The next examples in this section illustrate how we can write an expression in expanded form as an expression involving summation notation.

▼ **Example 4** Write with summation notation: $1 + 3 + 5 + 7 + 9$.

Solution A formula that gives us the terms of this sum is

$$a_i = 2i - 1$$

where i ranges from 1 up to and including 5. Notice we are using the subscript i here in exactly the same way we used the subscript n in the last section—to indicate the general term. Writing the sum

$$1 + 3 + 5 + 7 + 9$$

with summation notation looks like this:

$$\sum_{i=1}^{5} (2i - 1)$$

▲

▼ **Example 5** Write with summation notation: $3 + 12 + 27 + 48$.

Solution We need a formula, in terms of i, that will give each term in the sum. Writing the sum as

$$3 \cdot 1^2 + 3 \cdot 2^2 + 3 \cdot 3^2 + 3 \cdot 4^2$$

we see the formula

$$a_i = 3 \cdot i^2$$

where i ranges from 1 up to and including 4. Using this formula and summation notation, we can represent the sum

$$3 + 12 + 27 + 48$$

as

$$\sum_{i=1}^{4} 3i^2$$

▲

▼ **Example 6**   Write with summation notation:

$$\frac{x+3}{x^3} + \frac{x+4}{x^4} + \frac{x+5}{x^5} + \frac{x+6}{x^6}.$$

Solution   A formula that gives each of these terms is

$$a_i = \frac{x+i}{x^i}$$

where i assumes all integer values between 3 and 6, including 3 and 6. The sum can be written as

$$\sum_{i=3}^{6} \frac{x+i}{x^i}$$ ▲

Problem Set 12.2   Expand and simplify each of the following:

1. $\displaystyle\sum_{i=1}^{4} (2i + 4)$ **2.** $\displaystyle\sum_{i=1}^{5} (3i - 1)$

3. $\displaystyle\sum_{i=1}^{3} (2i - 1)$ **4.** $\displaystyle\sum_{i=1}^{4} (2i - 1)$

5. $\displaystyle\sum_{i=2}^{3} (i^2 - 1)$ **6.** $\displaystyle\sum_{i=3}^{6} (i^2 + 1)$

7. $\displaystyle\sum_{i=1}^{4} \frac{i}{1 + i}$ **8.** $\displaystyle\sum_{i=1}^{4} \frac{i^2}{1 + i}$

9. $\displaystyle\sum_{i=1}^{3} \frac{i^2}{2i - 1}$ **10.** $\displaystyle\sum_{i=3}^{5} (i^3 + 4)$

11. $\displaystyle\sum_{i=1}^{4} (-3)^i$ **12.** $\displaystyle\sum_{i=1}^{4} \left(-\frac{1}{3}\right)^i$

13. $\displaystyle\sum_{i=3}^{6} (-2)^i$ **14.** $\displaystyle\sum_{i=4}^{6} \left(-\frac{1}{2}\right)^i$

Expand the following:

15. $\displaystyle\sum_{i=1}^{5} (x + i)$ **16.** $\displaystyle\sum_{i=3}^{6} (x - i)$

17. $\displaystyle\sum_{i=2}^{7} (x + 1)^i$ **18.** $\displaystyle\sum_{i=1}^{4} (x + 3)^i$

19. $\displaystyle\sum_{i=1}^{5} \frac{x + i}{x - 1}$ **20.** $\displaystyle\sum_{i=1}^{6} \frac{x - 3i}{x + 3i}$

21. $\displaystyle\sum_{i=3}^{8} (x + i)^i$ **22.** $\displaystyle\sum_{i=4}^{7} (x - 2i)^i$

23. $\displaystyle\sum_{i=1}^{5} (x + i)^{i+1}$ **24.** $\displaystyle\sum_{i=2}^{6} (x + i)^{i-1}$

Write each of the following sums with summation notation:

25. $2 + 4 + 8 + 16$ **26.** $3 + 5 + 7 + 9 + 11$

27. $4 + 8 + 16 + 32 + 64$ **28.** $1 + 3 + 5$

29. $5 + 10 + 17 + 26 + 37$ **30.** $3 + 8 + 15 + 24$

31. $\frac{3}{4} + \frac{4}{5} + \frac{5}{6} + \frac{6}{7} + \frac{7}{8}$ **32.** $\frac{1}{2} + \frac{2}{3} + \frac{3}{4} + \frac{4}{5}$

33. $\frac{1}{3} + \frac{2}{5} + \frac{3}{7} + \frac{4}{9}$ **34.** $\frac{3}{1} + \frac{5}{3} + \frac{7}{5} + \frac{9}{7}$

35. $(x - 3) + (x - 4) + (x - 5) + (x - 6)$

36. $x^2 + x^3 + x^4 + x^5 + x^6$

37. $\dfrac{x}{x + 3} + \dfrac{x}{x + 4} + \dfrac{x}{x + 5}$

38. $\dfrac{x - 3}{x^3} + \dfrac{x - 4}{x^4} + \dfrac{x - 5}{x^5} + \dfrac{x - 6}{x^6}$

39. $x^2(x + 2) + x^3(x + 3) + x^4(x + 4)$

40. $x(x + 2)^2 + x(x + 3)^3 + x(x + 4)^4$

41. A skydiver jumps from a plane and falls 16 feet the first second, 48 feet the second second, and 80 feet the third second. If he continues to fall in the same manner, how far will he fall the seventh second? What is the total distance he falls in 7 seconds?

42. After 1 day a colony of 50 bacteria reproduces to become 200 bacteria. After 2 days they reproduce to become 800 bacteria. If they continue to reproduce at this rate, how many bacteria will be present after 4 days?

Review Problems   The problems below review material we covered in Section 3.7.

Factor completely:

43. $4x^2 + 12x + 9$ **44.** $9x^2 - 12x + 4$

45. $x^4 - 81$ **46.** $x^4 - 16$

47. $x^3 + 8$ **48.** $x^3 - 27$

49. $x^2 - 6x + 9 - y^2$ **50.** $x^2 + 2xy + y^2 - 25$

**12.3
Arithmetic
Progressions**

In this and the following section we will classify two major types of sequences: arithmetic progressions and geometric progressions.

DEFINITION An *arithmetic progression* is a sequence of numbers in which each term is obtained from the preceding term by adding the same amount each time.

The sequence

$$2, 7, 12, 17, \ldots$$

is an example of an arithmetic progression, since each term is obtained from the preceding term by adding 5 each time. The amount we add each time—in this case, 5—is called the *common difference,* since it can be obtained by subtracting any two consecutive terms. (The term with the larger subscript must be written first.) The common difference is denoted by d.

▼ **Example 1** Give the common difference d for the arithmetic progression 4, 10, 16, 22, ...

Solution Since each term can be obtained from the preceding term by adding 6, the common difference is 6. That is, $d = 6$. ▲

▼ **Example 2** Give the common difference for 100, 93, 86, 79, ...

Solution The common difference in this case is $d = -7$, since adding -7 to any term always produces the next consecutive term.
▲

▼ **Example 3** Give the common difference for $\frac{1}{2}$, 1, $\frac{3}{2}$, 2, ...

Solution The common difference is $d = \frac{1}{2}$. ▲

The General Term

The general term, a_n, of an arithmetic progression can always be written in terms of the first term a_1 and the common difference d. Consider the sequence from Example 1:

$$2, 7, 12, 17, \ldots$$

We can write each term in terms of the first term 2 and the common difference 5:

$$\underset{a_1,}{2}, \quad \underset{a_2,}{2 + (1 \cdot 5)}, \quad \underset{a_3,}{2 + (2 \cdot 5)}, \quad \underset{a_4,}{2 + (3 \cdot 5)}, \ldots$$

Observing the relationship between the subscript on the terms in the second line and the coefficients of the 5's in the first line, we write the general term for the sequence as

$$a_n = 2 + (n - 1)5$$

We generalize this result to include the general term of any arithmetic sequence.

The *general term* of an arithmetic progression with the first term a_1 and common difference d is given by

$$a_n = a_1 + (n - 1)d$$

▼   **Example 4**   Find the general term for the sequence

$$7, 10, 13, 16, \ldots .$$

Solution   The first term is $a_1 = 7$, and the common difference is $d = 3$. Substituting these numbers into the formula given above, we have

$$a_n = 7 + (n - 1)3$$

which we can simplify, if we choose, to

$$a_n = 7 + 3n - 3$$
$$= 3n + 4$$ ▲

▼   **Example 5**   Find the general term of the arithmetic progression whose third term, a_3, is 7 and whose eighth term, a_8, is 17.

Solution   According to the formula for the general term, the third term can be written as $a_3 = a_1 + 2d$, and the eighth term can be written as $a_8 = a_1 + 7d$. Since these terms are also equal to 7 and 17, respectively, we can write

$$a_3 = a_1 + 2d = 7$$
$$a_8 = a_1 + 7d = 17$$

To find a_1 and d, we simply solve the system on the right above:

$$a_1 + 2d = 7$$
$$a_1 + 7d = 17$$

We add the opposite of the top equation to the bottom equation. The result is

$$5d = 10$$
$$d = 2$$

To find a_1 we simply substitute 2 for d in either of the original equations and get

$$a_1 = 3$$

The general term for this progression is

$$a_n = 3 + (n - 1)2$$

which we can simplify to

$$a_n = 2n + 1 \qquad\qquad \blacktriangle$$

The sum of the first n terms of an arithmetic progression is denoted by S_n. The following theorem gives the formula for finding S_n, which is sometimes called the nth *partial sum*.

THEOREM 12.1 The sum of the first n terms of an arithmetic progression whose first term is a_1 and whose nth term is a_n is given by

$$S_n = \frac{n}{2}(a_1 + a_n)$$

Proof   We can write S_n in expanded form as

$$S_n = a_1 + [a_1 + d] + [a_1 + 2d] + \cdots + [a_1 + (n - 1)d]$$

This sum can also be written in reverse order as

$$S_n = a_n + [a_n - d] + [a_n - 2d)] + \cdots + [a_n - (n - 1)d]$$

If we add the preceding two expressions term by term, we have

$$2S_n = (a_1 + a_n) + (a_1 + a_n) + (a_1 + a_n) + \cdots + (a_1 + a_n)$$
$$2S_n = n(a_1 + a_n)$$

$$S_n = \frac{n}{2}(a_1 + a_n)$$

▼   **Example 6** Find the sum of the first 10 terms of the arithmetic progression 2, 10, 18, 26, ...

Solution   The first term is 2 and the common difference is 8. The tenth term is

$$a_{10} = 2 + 9(8)$$
$$= 2 + 72$$
$$= 74$$

Substituting $n = 10$, $a_1 = 2$, and $a_{10} = 74$ into the formula

$$S_n = \frac{n}{2}(a_1 + a_n)$$

we have

$$S_{10} = \frac{10}{2}(2 + 74)$$
$$= 5(76)$$
$$= 380$$

The sum of the first 10 terms is 380.   ▲

Determine which of the following sequences are arithmetic progressions. For those that are arithmetic progressions, identify the common difference d.

Problem Set 12.3

1. 1, 2, 3, 4, . . .
2. 4, 6, 8, 10, . . .
3. 1, 2, 4, 7, . . .
4. 1, 2, 4, 8, . . .
5. 50, 45, 40, . . .
6. $1, \frac{1}{2}, \frac{1}{4}, \frac{1}{8}, \ldots$
7. 1, 4, 9, 16, . . .
8. 5, 7, 9, 11, . . .
9. $\frac{1}{3}, 1, \frac{5}{3}, \frac{7}{3}, \ldots$
10. 5, 11, 17, . . .

Each of the following problems gives some information about a specific arithmetic progression. In each case use the given information to find:

   a. the general term a_n
   b. the tenth term a_{10}
   c. the twenty-fourth term a_{24}
   d. the sum of the first 10 terms S_{10}
   e. the sum of the first 24 terms S_{24}

11. $a_1 = 3$, $d = 4$
12. $a_1 = 5$, $d = 10$
13. $a_1 = 6$, $d = -2$
14. $a_1 = 7$, $d = -1$
15. 5, 9, 13, 17, . . .
16. 8, 11, 14, 17, . . .
17. 12, 7, 2, −3, . . .
18. 25, 20, 15, 10, . . .
19. $\frac{1}{2}, 1, \frac{3}{2}, 2, \ldots$
20. $-\frac{1}{3}, 0, \frac{1}{3}, \frac{2}{3}, \ldots$
21. $a_3 = 16$, $a_8 = 46$
22. $a_3 = 16$, $a_8 = 51$
23. $a_6 = 17$, $a_{12} = 29$
24. $a_5 = 23$, $a_{10} = 48$

25. Suppose a woman earns $15,000 a year and gets a raise of $850 per year. Write a sequence that gives her salary for each of the first 5 years she works. What is the general term of this sequence? How much will she be making annually if she stays on the job for 10 years?

26. Suppose a school teacher makes $12,500 the first year he works and then gets a $900 raise every year after that. Write a sequence that gives his

salary for the first 5 years he works. What is the general term of this sequence? How much will he be making annually if he teaches for 20 years?

Review Problems   The problems below review material we covered in Sections 4.3 and 4.5.

Multiply or divide as indicated:

27. $\dfrac{3x - 12}{x^2 - 4} \cdot \dfrac{x^2 - 6x + 8}{x - 4}$

28. $\dfrac{x^2 + 5x + 6}{x + 2} \cdot \dfrac{x - 3}{x^2 - 9}$

29. $\dfrac{4x^2 + 8x + 3}{2x^2 - 5x - 12} \div \dfrac{2x^2 + 7x + 3}{x^2 - 16}$

30. $\dfrac{3x^2 + 7x - 20}{x^2 - 2x + 1} \div \dfrac{3x^2 - 2x - 5}{x^2 + 3x - 4}$

Simplify:

31. $\dfrac{1 + \dfrac{1}{4}}{1 - \dfrac{1}{4}}$

32. $\dfrac{1 - \dfrac{1}{x}}{1 + \dfrac{1}{x}}$

33. $\dfrac{1 - \dfrac{16}{x^2}}{1 - \dfrac{8}{x} + \dfrac{16}{x^2}}$

34. $\dfrac{1 - \dfrac{25}{x^2}}{1 + \dfrac{10}{x} + \dfrac{25}{x^2}}$

12.4 Geometric Progressions

This section is concerned with the second major classification of sequences, called geometric progressions. The problems in this section are very similar to the problems in the preceding section.

DEFINITION   A sequence of numbers in which each term is obtained from the previous term by multiplying by the same amount each time is called a *geometric sequence*.

The sequence

$$3, 6, 12, 24, \ldots$$

is an example of a geometric progression. Each term is obtained from the previous term by multiplying by 2. The amount by which we multiply each time—in this case, 2—is called the *common ratio*. The common ratio is denoted by r, and can be found by taking the ratio of any two consecutive terms. (The term with the larger subscript must be in the numerator.)

▼   **Example 1**   Find the common ratio for the geometric progression

$$\tfrac{1}{2}, \tfrac{1}{4}, \tfrac{1}{8}, \tfrac{1}{16}, \ldots$$

Solution Since each term can be obtained from the term before it by multiplying by $\frac{1}{2}$, the common ratio is $\frac{1}{2}$. That is, $r = \frac{1}{2}$. ▲

▼ **Example 2** Find the common ratio for $\sqrt{3},\ 3,\ 3\sqrt{3},\ 9, \ldots$

Solution If we take the ratio of the second and third terms, we have

$$\frac{3\sqrt{3}}{3} = \sqrt{3}$$

The common ratio is $r = \sqrt{3}$. ▲

The *general term*, a_n, of a geometric progression with a first term a_1 and *The General Term*
common ratio r is given by

$$a_n = a_1 r^{n-1}$$

To see how we arrive at this formula, consider the following geometric progression whose common ratio is 3:

$$2,\ 6,\ 18,\ 54, \ldots$$

We can write each term of the sequence in terms of the first term 2 and the common ratio 3:

$$2 \cdot 3^0, \quad 2 \cdot 3^1, \quad 2 \cdot 3^2, \quad 2 \cdot 3^3, \ldots$$
$$a_1, \qquad a_2, \qquad a_3, \qquad a_4, \quad \ldots$$

Observing the relationship between the two lines written above, we find we can write the general term of this progression as

$$a_n = 2 \cdot 3^{n-1}$$

Since the first term can be designated by a_1 and the common ratio by r, the formula

$$a_n = 2 \cdot 3^{n-1}$$

coincides with the formula

$$a_n = a_1 r^{n-1}$$

▼ **Example 3** Find the general term for the geometric progression

$$5,\ 10,\ 20, \ldots$$

Solution The first term is $a_1 = 5$, and the common ratio is $r = 2$. Using these values in the formula

$$a_n = a_1 r^{n-1}$$

we have

$$a_n = 5 \cdot 2^{n-1}$$ ▲

▼   **Example 4**   Find the tenth term of the sequence $3, \frac{3}{2}, \frac{3}{4}, \frac{3}{8}, \ldots$.

Solution   The sequence is a geometric progression with first term $a_1 = 3$ and common ratio $r = \frac{1}{2}$. The tenth term is

$$a_{10} = 3(\tfrac{1}{2})^9$$ ▲

▼   **Example 5**   Find the general term for the geometric progression whose fourth term is 16 and whose seventh term is 128.

Solution   The fourth term can be written as $a_4 = a_1 r^3$, and the seventh term can be written as $a_7 = a_1 r^6$:

$$a_4 = a_1 r^3 = 16$$
$$a_7 = a_1 r^6 = 128$$

We can solve for r by using the ratio a_7/a_4:

$$\frac{a_7}{a_4} = \frac{a_1 r^6}{a_1 r^3} = \frac{128}{16}$$
$$r^3 = 8$$
$$r = 2$$

The common ratio is 2. To find the first term we substitute $r = 2$ into either of the original two equations. The result is

$$a_1 = 2$$

The general term for this progression is

$$a_n = 2 \cdot 2^{n-1}$$

which we can simplify by adding exponents, since the bases are equal:

$$a_n = 2^n$$ ▲

As was the case in the preceding section, the sum of the first n terms of a geometric progression is denoted by S_n, which is called the nth *partial sum* of the progression.

THEOREM 12.2 The sum of the first n terms of a geometric progression with first term a_1 and common ratio r is given by the formula

$$S_n = \frac{a_1(r^n - 1)}{r - 1}$$

Proof We write the sum of the first n terms in expanded form:

$$S_n = a_1 + a_1 r + a_1 r^2 + \cdots + a_1 r^{n-1} \tag{1}$$

Then multiplying both sides by r, we have

$$r S_n = a_1 r + a_1 r^2 + a_1 r^3 + \cdots + a_1 r^n \tag{2}$$

If we subtract the left side of equation (1) from the left side of equation (2) and do the same for the right sides, we end up with

$$r S_n - S_n = a_1 r^n - a_1$$

We factor S_n from both terms on the left side and a_1 from both terms on the right side of this equation:

$$S_n(r - 1) = a_1(r^n - 1)$$

Dividing both sides by $r - 1$ gives the desired result:

$$S_n = \frac{a_1(r^n - 1)}{r - 1}$$

▼   **Example 6** Find the sum of the first 10 terms of the geometric progression 5, 15, 45, 135, ...

Solution The first term is $a_1 = 5$, and the common ratio is $r = 3$. Substituting these values into the formula for S_{10}, we have the sum of the first 10 terms of the sequence:

$$S_{10} = \frac{5(3^{10} - 1)}{3 - 1}$$

$$= \frac{5(3^{10} - 1)}{2}$$

The answer can be left in this form, since it would be too time-consuming (and boring) to calculate 3^{10} by hand.   ▲

Problem Set 12.4   Identify those sequences that are geometric progressions. For those that are geometric give the common ratio r.

1. 1, 5, 25, 125, . . .   2. 6, 12, 24, 48, . . .
3. $\frac{1}{2}$, $\frac{1}{6}$, $\frac{1}{18}$, $\frac{1}{54}$, . . .   4. 5, 10, 15, 20, . . .
5. 4, 9, 16, 25, . . .   6. -1, $\frac{1}{3}$, $-\frac{1}{9}$, $\frac{1}{27}$, . . .
7. -2, 4, -8, 16, . . .   8. 1, 8, 27, 64, . . .
9. 4, 6, 8, 10, . . .   10. 1, -3, 9, -27, . . .

Each of the following problems gives some information about a specific geometric progression. In each case use the given information to find:

    a. the general term a_n
    b. the sixth term a_6
    c. the twentieth term a_{20}
    d. the sum of the first 10 terms S_{10}
    e. the sum of the first 20 terms S_{20}

11. $a_1 = 4$, $r = 3$ 12. $a_1 = 5$, $r = 2$
13. $a_1 = -2$, $r = -5$ 14. $a_1 = 6$, $r = -2$
15. $a_1 = -4$, $r = \frac{1}{2}$ 16. $a_1 = 6$, $r = \frac{1}{3}$
17. 2, 12, 72, . . .   18. 7, 14, 28, . . .
19. 10, 20, 40, . . .   20. 100, 50, 25, . . .
21. $\sqrt{2}$, 2, $2\sqrt{2}$, . . .   22. 5, $5\sqrt{2}$, 10, . . .
23. $a_4 = 40$, $a_6 = 160$ 24. $a_5 = \frac{1}{8}$, $a_8 = \frac{1}{64}$

25. A savings account earns 6% interest per year. If the account has $10,000 at the beginning of one year, how much will it have at the beginning of the next year? Write a geometric progression that gives the total amount of money in this account every year for 5 years. What is the general term of this sequence?

26. A savings account earns 8% interest per year. If the account has $5,000 at the beginning of one year, how much will it have at the beginning of the next year? Write a geometric progression that gives the total amount of money in this account every year for 4 years. What is the general term of this sequence?

Review Problems   The problems below review material we covered in Section 3.4. Reviewing these problems will help you understand the next section.

Expand and multiply:

27. $(x + 5)^2$ 28. $(x + y)^2$
29. $(x + y)^3$ 30. $(x - 2)^3$
31. $(x + y)^4$ 32. $(x - 1)^4$

The purpose of this section is to write and apply the formula for the expansion of expressions of the form $(x + y)^n$ where n is any positive integer. In order to write the formula, we must generalize the information in the following chart:

$$(x + y)^1 = x + y$$
$$(x + y)^2 = x^2 + 2xy + y^2$$
$$(x + y)^3 = x^3 + 3x^2y + 3xy^2 + y^3$$
$$(x + y)^4 = x^4 + 4x^3y + 6x^2y^2 + 4xy^3 + y^4$$
$$(x + y)^5 = x^5 + 5x^4y + 10x^3y^2 + 10x^2y^3 + 5xy^4 + y^5$$

Note   The polynomials to the right have been found by expanding the binomials on the left—we just haven't shown the work.

There are a number of similarities to notice among the polynomials on the right. Here is a list of them:

1. In each polynomial, the sequence of exponents on the variable x decreases to zero from the exponent on the binomial at the left. (The exponent 0 is not shown, since $x^0 = 1$.)
2. In each polynomial the exponents on the variable y increase from 0 to the exponent on the binomial on the left. (Since $y^0 = 1$, it is not shown in the first term.)
3. The sum of the exponents on the variables in any single term is equal to the exponent on the binomial to the left.

The pattern in the coefficients of the polynomials on the right can best be seen by writing the right side again without the variables. It looks like this:

```
            1    1
         1    2    1
       1    3    3    1
     1    4    6    4    1
   1    5    10   10   5    1
```

This triangular-shaped array of coefficients is called *Pascal's triangle*. Each entry in the triangular array is obtained by adding the two numbers above it. Each row begins and ends with the number 1. If we were to continue Pascal's triangle, the next two rows would be

```
      1    6    15    20    15    6    1
   1    7    21    35    35    21    7    1
```

Pascal's triangle can be used to find coefficients for the expansion of $(x + y)^n$. The coefficients for the terms in the expansion of $(x + y)^n$ are given in the nth row of Pascal's triangle. It wouldn't be too much fun, however, to carry out Pascal's triangle much farther. There is an alternate method of finding the coefficients that does not involve Pascal's triangle. The alternate method of finding the coefficients involves notation we have not seen before.

DEFINITION   The expression $n!$ is read "n factorial" and is the product of all the consecutive integers from n down to 1. For example,

$$1! = 1$$
$$2! = 2 \cdot 1 = 2$$
$$3! = 3 \cdot 2 \cdot 1 = 6$$
$$4! = 4 \cdot 3 \cdot 2 \cdot 1 = 24$$
$$5! = 5 \cdot 4 \cdot 3 \cdot 2 \cdot 1 = 120$$

The expression $0!$ is defined to be 1. We use factorial notation to define binomial coefficients as follows.

DEFINITION   The expression $\binom{n}{r}$ is called a *binomial coefficient* and is defined by

$$\binom{n}{r} = \frac{n!}{r!(n - r)!}$$

(If you are asking yourself, "What is going on here?"—hang on. We will be back to what we were doing shortly.)

▼ **Example 1**   Calculate the following binomial coefficients:

$$\binom{7}{5}, \binom{6}{2}, \binom{3}{0}$$

Solution   We simply apply the definition for binomial coefficients:

$$\binom{7}{5} = \frac{7!}{5!(7 - 5)!}$$
$$= \frac{7!}{5!2!}$$

$$= \frac{7 \cdot 6 \cdot 5 \cdot 4 \cdot 3 \cdot 2 \cdot 1}{(5 \cdot 4 \cdot 3 \cdot 2 \cdot 1)(2 \cdot 1)}$$

$$= \frac{42}{2}$$

$$= 21$$

$$\binom{6}{2} = \frac{6!}{2!(6 - 2)!}$$

$$= \frac{6!}{2! \cdot 4!}$$

$$= \frac{6 \cdot 5 \cdot 4 \cdot 3 \cdot 2 \cdot 1}{(2 \cdot 1)(4 \cdot 3 \cdot 2 \cdot 1)}$$

$$= \frac{30}{2}$$

$$= 15$$

$$\binom{3}{0} = \frac{3!}{0!(3 - 0)!}$$

$$= \frac{3!}{0! \cdot 3!}$$

$$= \frac{3 \cdot 2 \cdot 1}{(1)(3 \cdot 2 \cdot 1)}$$

$$= 1 \qquad\qquad \blacktriangle$$

If we were to calculate all the binomial coefficients in the following array, we would find they match exactly with the numbers in Pascal's triangle. That is why they are called binomial coefficients—because they are the coefficients of the expansion of $(x + y)^n$:

$$\binom{1}{0} \quad \binom{1}{1}$$

$$\binom{2}{0} \quad \binom{2}{1} \quad \binom{2}{2}$$

$$\binom{3}{0} \quad \binom{3}{1} \quad \binom{3}{2} \quad \binom{3}{3}$$

$$\binom{4}{0} \quad \binom{4}{1} \quad \binom{4}{2} \quad \binom{4}{3} \quad \binom{4}{4}$$

$$\binom{5}{0} \quad \binom{5}{1} \quad \binom{5}{2} \quad \binom{5}{3} \quad \binom{5}{4} \quad \binom{5}{5}$$

Using the new notation to represent the entries in Pascal's triangle, we can summarize everything we have noticed about the expansion of binomial powers of the form $(x + y)^n$.

The Binomial Expansion

If x and y represent real numbers and n is a positive integer, then the following formula is known as the *binomial expansion* or *binomial formula:*

$$(x + y)^n = \binom{n}{0}x^n y^0 + \binom{n}{1}x^{n-1}y^1 + \binom{n}{2}x^{n-2}y^2 + \cdots + \binom{n}{n}x^0 y^n$$

It does not make any difference, when expanding binomial powers of the form $(x + y)^n$, whether we use Pascal's triangle or the formula

$$\binom{n}{r} = \frac{n!}{r!(n-r)!}$$

to calculate the coefficients. We will show examples of both methods.

▼   **Example 2**   Expand $(x - 2)^3$.

Solution   The binomial formula for $(x + y)^3$ is

$$(x + y)^3 = \binom{3}{0}x^3 y^0 + \binom{3}{1}x^2 y^1 + \binom{3}{2}x^1 y^2 + \binom{3}{3}x^0 y^3$$

Applying this formula to $(x - 2)^3$, we have

$(x - 2)^3$

$$= \binom{3}{0}x^3(-2)^0 + \binom{3}{1}x^2(-2)^1 + \binom{3}{2}x^1(-2)^2 + \binom{3}{3}x^0(-2)^3$$

The coefficients $\binom{3}{0}$, $\binom{3}{1}$, $\binom{3}{2}$, and $\binom{3}{3}$ can be found in the third row of Pascal's triangle. They are 1, 3, 3, and 1:

$$(x - 2)^3 = 1x^3(-2)^0 + 3x^2(-2)^1 + 3x^1(-2)^2 + 1x^0(-2)^3$$
$$= x^3 - 6x^2 + 12x - 8 \qquad\qquad ▲$$

▼   **Example 3**   Expand $(3x + 2y)^4$.

Solution   The coefficients can be found in the fourth row of Pascal's triangle. They are

$$1, \ 4, \ 6, \ 4, \ 1$$

Here is the expansion of $(3x + 2y)^4$:

$$(3x + 2y)^4$$
$$= 1(3x)^4 + 4(3x)^3(2y) + 6(3x)^2(2y)^2 + 4(3x)(2y)^3 + 1(2y)^4$$
$$= 81x^4 + 216x^3y + 216x^2y^2 + 96xy^3 + 16y^4 \qquad \blacktriangle$$

▼ **Example 4** Write the first three terms in the expansion of $(x + 5)^9$.

Solution The coefficients of the first three terms are $\binom{9}{0}$, $\binom{9}{1}$, and $\binom{9}{2}$, which we calculate as follows:

$$\binom{9}{0} = \frac{9!}{0!9!} = \frac{9 \cdot 8 \cdot 7 \cdot 6 \cdot 5 \cdot 4 \cdot 3 \cdot 2 \cdot 1}{(1)(9 \cdot 8 \cdot 7 \cdot 6 \cdot 5 \cdot 4 \cdot 3 \cdot 2 \cdot 1)} = \frac{1}{1} = 1$$

$$\binom{9}{1} = \frac{9!}{1!8!} = \frac{9 \cdot 8 \cdot 7 \cdot 6 \cdot 5 \cdot 4 \cdot 3 \cdot 2 \cdot 1}{(1)(8 \cdot 7 \cdot 6 \cdot 5 \cdot 4 \cdot 3 \cdot 2 \cdot 1)} = \frac{9}{1} = 9$$

$$\binom{9}{2} = \frac{9!}{2!7!} = \frac{9 \cdot 8 \cdot 7 \cdot 6 \cdot 5 \cdot 4 \cdot 3 \cdot 2 \cdot 1}{(2 \cdot 1)(7 \cdot 6 \cdot 5 \cdot 4 \cdot 3 \cdot 2 \cdot 1)} = \frac{72}{2} = 36$$

From the binomial formula, we write the first three terms:

$$(x + 5)^9 = 1 \cdot x^9 + 9 \cdot x^8(5) + 36x^7(5)^2 + \ldots$$
$$= x^9 + 45x^8 + 900x^7 + \ldots \qquad \blacktriangle$$

Use the binomial formula to expand each of the following:

1. $(x + 2)^4$
2. $(x - 2)^5$
3. $(x + y)^6$
4. $(x - 1)^6$
5. $(2x + 1)^5$
6. $(2x - 1)^4$
7. $(x - 2y)^5$
8. $(2x + y)^5$
9. $(3x - 2)^4$
10. $(2x - 3)^4$
11. $(4x - 3y)^3$
12. $(3x - 4y)^3$
13. $(x^2 + 2)^4$
14. $(x^2 - 3)^3$
15. $(x^2 + y^2)^3$
16. $(x^2 - 3y)^4$

17. $\left(\dfrac{x}{2} - 4\right)^3$
18. $\left(\dfrac{x}{3} + 6\right)^3$

19. $\left(\dfrac{x}{3} + \dfrac{y}{2}\right)^4$
20. $\left(\dfrac{x}{2} - \dfrac{y}{3}\right)^4$

Write the first four terms in the expansion of the following:

21. $(x + 2)^9$
22. $(x - 2)^9$
23. $(x - y)^{10}$
24. $(x + y)^{10}$
25. $(x + 2y)^{10}$
26. $(x - 2y)^{10}$

Write the first three terms in the expansion of each of the following:

27. $(x + 1)^{15}$ 28. $(x - 1)^{15}$
29. $(x - y)^{12}$ 30. $(x + y)^{12}$
31. $(x + 2)^{20}$ 32. $(x - 2)^{20}$

Write the first two terms in the expansion of each of the following:

33. $(x + 2)^{100}$ 34. $(x - 2)^{50}$
35. $(x + y)^{50}$ 36. $(x - y)^{100}$

37. The third term in the expansion of $(\frac{1}{2} + \frac{1}{2})^7$ will give the probability that in a family with 7 children, 5 will be boys and 2 will be girls. Find the third term.

38. The fourth term in the expansion of $(\frac{1}{2} + \frac{1}{2})^8$ will give the probability that in a family with 8 children 3 will be boys and 5 will be girls. Find the fourth term.

Review Problems   The problems below review material we covered in Sections 5.6 and 5.7:

Combine the following complex numbers:

39. $(7 - 6i) - (3 + 4i)$ 40. $(2 - 11i) - (7 + 3i)$

Multiply:

41. $3i(4 - 2i)$ 42. $2i(6 + 3i)$
43. $(2 + 3i)^2$ 44. $(5 + 4i)^2$

Divide:

45. $\dfrac{5i}{3 - 2i}$ 46. $\dfrac{2 - 7i}{3 + 4i}$

Examples

1. In the sequence
$1, 3, 5, \ldots, 2n - 1, \ldots,$
$a_1 = 1, a_2 = 3, a_3 = 5,$
and $a_n = 2n - 1$

Chapter 12   Summary and Review

SEQUENCES [12.1]

A *sequence* is a function whose domain is the set of real numbers. The terms of a sequence are denoted by

$$a_1, a_2, a_3, \ldots, a_n, \ldots$$

where a_1 (read "*a* sub 1") is the first term, a_2 the second term, and a_n the *n*th or general term.

SUMMATION NOTATION [12.2]

The notation

$$\sum_{i=1}^{n} a_i = a_1 + a_2 + a_3 + \cdots + a_n$$

is called *summation notation* or *sigma notation*. The letter i as used here is called the *index of summation* or just *index*.

ARITHMETIC PROGRESSIONS [12.3]

An *arithmetic progression* is a sequence in which each term comes from the preceding term by adding a constant amount each time. If the first term of an arithmetic progression is a_1 and the amount we add each time (called the common difference) is d, then the nth term of the progression is given by

$$a_n = a_1 + (n - 1)d.$$

The sum of the first n terms of an arithmetic progression is

$$S_n = \frac{n}{2}(a_1 + a_n).$$

S_n is called the nth *partial sum*.

GEOMETRIC PROGRESSIONS [12.4]

A *geometric progression* is a sequence of numbers in which each term comes from the previous term by multiplying by a constant amount each time. The constant by which we multiply each term to get the next term is called the *common ratio*. If the first term of a geometric progression is a_1 and the common ratio is r, then the formula that gives the general term, a_n, is

$$a_n = a_1 r^{n-1}.$$

The sum of the first n terms of a geometric progression is given by the formula

$$S_n = \frac{a_1(r^n - 1)}{r - 1}.$$

2. $\displaystyle\sum_{i=3}^{6} (-2)^i$

$$= (-2)^3 + (-2)^4 + (-2)^5 + (-2)^6$$
$$= -8 + 16 + (-32) + 64$$
$$= 40$$

3. For the sequence 3, 7, 11, 15, . . . , $a_1 = 3$ and $d = 4$. The general term is

$$a_n = 3 + (n - 1)4$$
$$= -1 + 4n.$$

4. For the geometric progression 3, 6, 12, 24, . . . , $a_1 = 3$ and $r = 2$. The general term is

$$a_n = 3 \cdot 2^{n-1}.$$

The sum of the first 10 terms is

$$S_{10} = \frac{3(2^{10} - 1)}{2 - 1} = 3069.$$

FACTORIALS [12.5]

The notation $n!$ is called n *factorial,* and is defined to be the product of each consecutive integer from n down to 1. That is,

$$0! = 1 \qquad \text{(By definition)}$$
$$1! = 1$$
$$2! = 2 \cdot 1$$
$$3! = 3 \cdot 2 \cdot 1$$
$$4! = 4 \cdot 3 \cdot 2 \cdot 1$$
$$\text{and so on}$$

5. $\dbinom{7}{3} = \dfrac{7!}{3!(7 - 3)!}$

$= \dfrac{7!}{3!4!}$

$= \dfrac{7 \cdot 6 \cdot 5 \cdot 4 \cdot 3 \cdot 2 \cdot 1}{3 \cdot 2 \cdot 1 \cdot 4 \cdot 3 \cdot 2 \cdot 1}$

$= 35$

BINOMIAL COEFFICIENTS [12.5]

The notation $\dbinom{n}{r}$ is called a *binomial coefficient* and is defined by

$$\binom{n}{r} = \frac{n!}{r!(n - r)!}.$$

Binomial coefficients can be found by using the formula above or by Pascal's triangle, which is

$$
\begin{array}{ccccccccccc}
 & & & & 1 & & 1 & & & & \\
 & & & 1 & & 2 & & 1 & & & \\
 & & 1 & & 3 & & 3 & & 1 & & \\
 & 1 & & 4 & & 6 & & 4 & & 1 & \\
1 & & 5 & & 10 & & 10 & & 5 & & 1 \\
\end{array}
$$

and so on.

6. $(x + 2)^4$

$= x^4 + 4x^3 \cdot 2 + 6x^2 \cdot 2^2$

$\qquad\qquad + 4x \cdot 2^3 + 2^4$

$= x^4 + 8x^3 + 24x^2 + 32x + 16$

BINOMIAL EXPANSION [12.5]

If n is a positive integer, then the formula for expanding $(a + b)^n$ is given by

$$(a + b)^n = \binom{n}{0}a^n b^0 + \binom{n}{1}a^{n-1}b^1 + \binom{n}{2}a^{n-2}b^2 + \cdots$$

$$+ \binom{n}{n}a^0 b^n.$$

Write the first five terms of the sequences with the following general terms:

1. $a_n = 3n - 5$ 2. $a_n = 4n - 1$

3. $a_n = n^2 + 1$ 4. $a_n = 2n^3$

5. $a_n = \dfrac{n+1}{n^2}$ 6. $a_n = (-2)^{n+1}$

Give the general term for each sequence:

7. 6, 10, 14, 18, . . .   8. 1, 2, 4, 8, . . .

9. $\frac{1}{2}, \frac{1}{4}, \frac{1}{8}, \frac{1}{16}, \ldots$ 10. $-3, 9, -27, 81, \ldots$

11. Expand and simplify each of the following:

    a. $\displaystyle\sum_{i=1}^{5} (5i + 3)$

    b. $\displaystyle\sum_{i=3}^{5} (2^i - 1)$

    c. $\displaystyle\sum_{i=2}^{6} (i^2 + 2i)$

12. Find the first term of an arithmetic progression if $a_5 = 11$ and $a_9 = 19$.

13. Find the second term of a geometric progression for which $a_3 = 18$ and $a_5 = 162$.

14. Find the sum of the first 10 terms of the following arithmetic progressions:

    a. 5, 11, 17, . . .

    b. 25, 20, 15, . . .

15. Write a formula for the sum of the first 50 terms of the following geometric progressions:

    a. 3, 6, 12, . . .

    b. 5, 20, 80, . . .

16. Use the binomial formula to expand each of the following:

    a. $(x - 3)^4$

    b. $(2x - 1)^5$

    c. $(3x - 2y)^3$

13

Introduction to Trigonometry

To the student:

Trigonometry means triangle measure. The study of trigonometry arose out of people's natural curiosity about the relationships that exist between the sides and angles in triangles. I have always been fascinated by the extent to which trigonometry can be applied to other fields. Trigonometry has applications in almost every branch of science and mathematics; including physics, chemistry, and biology, as well as engineering, surveying, and navigation.

This chapter and the next two chapters are an introduction to trigonometry. Actually, only a few topics usually found in trigonometry books are not included in these introductory chapters; some proofs have been omitted and some of the more extensive topics have been condensed.

The study of trigonometry can be a rewarding experience. Like all branches of mathematics, it requires lots of practice at working problems. The problems in this chapter and the next two chapters are intended to give you that practice. Remember: As in algebra, the understanding and mastery of trigonometry will come with practice; the more problems you are able to solve, the more you will be able to solve. There is no substitute for problem solving. I hope you enjoy this introduction to trigonometry.

Before we begin our study of trigonometry, there are a few topics from geometry that we should review. Let's begin by looking at some of the terminology associated with angles.

An angle is formed when two half-lines have a common end point. The common end point is called the *vertex* of the angle, and the half lines are called the *sides* of the angle.

In Figure 13-1, the vertex of angle θ (theta) is labeled O, and A and B are points on each side of θ.

We can think of θ as having been formed by rotating side $\overline{OA}$ about the vertex to side $\overline{OB}$. In this case, we call side $\overline{OA}$ the *initial side* of θ and side $\overline{OB}$ the *terminal side* of θ.

When the rotation from the initial side to the terminal side takes place in a counterclockwise direction, the angle formed is considered a *positive angle.* If the rotation is in a clockwise direction, the angle formed is a *negative angle* (Figure 13-2).

13.1
Degrees, Radians, and Special Triangles

Angles in General

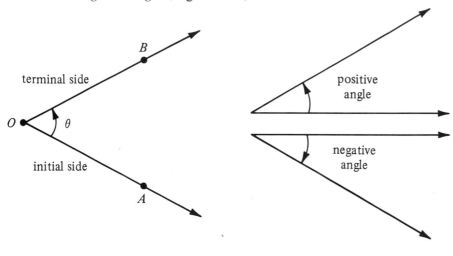

Figure 13-1   **Figure 13-2**

One way to measure the size of an angle is with *degree measure.* The angle formed by rotating a half-line through one complete revolution has a measure of 360 degrees, written 360° (Figure 13-3).

One degree (1°) then, is 1/360 of a full rotation. Likewise, 180° is one-half of a full rotation, and 90° is half of that. As we mentioned in Chapter 6, angles that measure 90° are called *right angles.* Angles that measure between 0° and 90° are called *acute angles,* while angles that measure between 90° and 180° are called *obtuse angles.*

Degree Measure

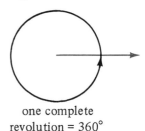

one complete
revolution = 360°

Figure 13-3

If two angles have a sum of 90°, then they are called *complementary angles,* and we say each is the *complement* of the other. Two angles with a sum of 180° are called *supplementary angles.*

▼   **Example 1**   Give the complement and the supplement of each angle.

a. 40°
b. 110°
c. θ

Solution
a. The complement of 40° is 50° since 40° + 50° = 90°.
   The supplement of 40° is 140° since 40° + 140° = 180°.
b. The complement of 110° is −20° since 110° + (−20°) = 90°.
   The supplement of 110° is 70° since 110° + 70° = 180°.
c. The complement of θ is 90° − θ since θ + (90° − θ) = 90°.
   The supplement of θ is 180° − θ since θ + (180° − θ) = 180°.
                                                                                          ▲

Radian Measure

Another way in which to specify the measure of an angle is with *radian measure.* To understand the definition for radian measure, we have to recall from geometry that a *central angle* is an angle with its vertex at the center of a circle.

DEFINITION   In a circle, a central angle that cuts off an arc equal in length to the radius of the circle has a measure of *1 radian.* That is, in a circle of radius *r,* a central angle that measures 1 radian will cut off an arc of length *r* (Figure 13-4).

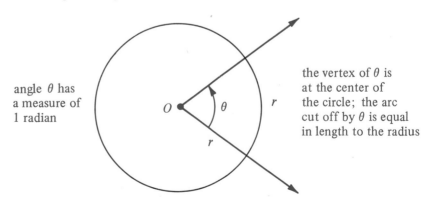

angle θ has a measure of 1 radian

the vertex of θ is at the center of the circle; the arc cut off by θ is equal in length to the radius

Figure 13-4

To find the radian measure of a central angle, we have to find the number of radii in the arc it cuts off. If a central angle θ, in a circle of radium r, cuts off an arc of length s, then the measure of θ, in radians, is given by

$$\theta \text{ (in radians)} = \frac{s}{r}$$

Note It is common practice to omit the word radian when using radian measure. If no units are stated, an angle is understood to be measured in radians; with degree measure the degree symbol, °, must be written.

$\theta = 2$ means the measure of θ is 2 radians

$\theta = 2°$ means the measure of θ is 2 degrees

To see the relationship between degrees and radians, we can compare the number of degrees and the number of radians in one full rotation.   *Comparing Degrees and Radians*

   The angle formed by one full rotation about the center of a circle of radius r will cut off an arc equal to the circumference of the circle (Figure 13-5). Since the circumference of a circle of radius r is $2\pi r$, we have

θ measures one full rotation $\theta = \dfrac{2\pi r}{r} = 2\pi$ The measure of θ in radians is 2π

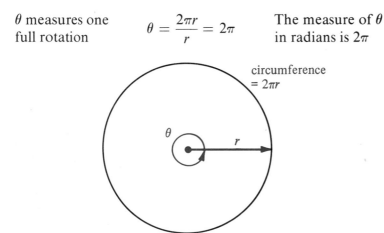

circumference $= 2\pi r$

Figure 13-5

One full rotation is 2π radians and 360°. That is

$$360° = 2\pi \text{ radians}$$

dividing both sides by 2 we have

$$180° = \pi \text{ radians}$$

To obtain conversion factors that will allow us to change back and forth between degrees and radians, we divide both sides of this last equation alternately by 180 and by π.

$$\boxed{\quad} \; 180° = \pi \text{ radians} \; \boxed{\quad}$$

Divide both
sides by 180

Divide both
sides by π.

$$1° = \frac{\pi}{180} \text{ radians} \qquad \left(\frac{180}{\pi}\right)° = 1 \text{ radian}$$

To gain some intuitive insight into the relationship between degrees and radians, we can replace π with 3.14 to obtain the approximate number of degrees in 1 radian.

$$1 \text{ radian} = 1\left(\frac{180}{\pi}\right)°$$

$$= 1\left(\frac{180}{3.14}\right)°$$

$$= 57.32°$$

We see that 1 radian is approximately 57°. A radian is much larger than a degree. Figure 13-6 illustrates the relationship between 20° and 20 radians.

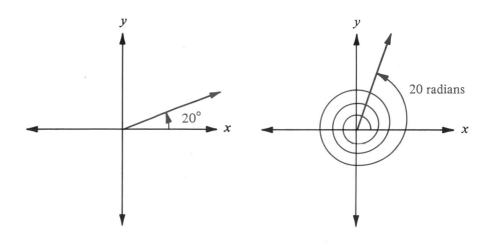

Figure 13-6

Here are some further conversions between degrees and radians.

▼ **Example 2**   Convert 45° to radians.

Solution   Since $1° = \dfrac{\pi}{180}$ radians, and 45° is the same as 45(1°), we have

$$45° = 45\left(\frac{\pi}{180}\right) \text{radians} = \frac{\pi}{4} \text{radians}$$

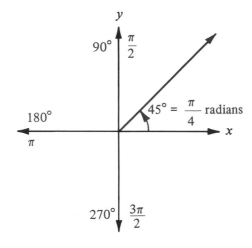

Figure 13-7

When we leave our answer in terms of π, as $\pi/4$, we are writing an exact value. If we want a decimal approximation, we substitute 3.14 for π

$$\begin{matrix} \text{Exact} \\ \text{value} \end{matrix} \qquad \frac{\pi}{4} = \frac{3.14}{4} = 0.785 \qquad \text{Approximate value}$$

Note also that, once we have found the radian equivalent of an angle given in degrees, it is easy to find the radian equivalent of multiples of that angle. For instance, if we wanted the radian equivalent of 90°, we could simply multiply $\pi/4$ by 2, since $90° = 2 \times 45°$.

$$90° = 2 \times \frac{\pi}{4} = \frac{\pi}{2}$$ ▲

▼   **Example 3**   Convert 450° to radians.

Solution   Multiplying by $\pi/180$ we have

$$450° = 450\left(\frac{\pi}{180}\right) = \frac{5\pi}{2} \text{ radians}$$

Again, $5\pi/2$ is the exact value.   ▲

▼   **Example 4**   Convert $\pi/6$ to degrees.

Solution   To convert from radians to degrees, we multiply by $180/\pi$.

$$\frac{\pi}{6}\text{ (radians)} = \frac{\pi}{6}\left(\frac{180}{\pi}\right)°$$

$$= 30°$$

Note that 60° is twice 30°, so $2(\pi/6) = \pi/3$ must be the radian equivalent of 60°.   ▲

▼   **Example 5**   Convert $4\pi/3$ to degrees.

Solution   Multiplying by $180/\pi$ we have

$$\frac{4\pi}{3}\text{ (radians)} = \frac{4\pi}{3}\left(\frac{180}{\pi}\right)°$$

$$= 240°$$ ▲

As is apparent from the preceding examples, changing from degrees to radians and radians to degrees is simply a matter of multiplying by the appropriate conversion factors.

$$
\begin{array}{ccc}
 & \text{Multiply by } \dfrac{180}{\pi} & \\
\text{Degrees} & & \text{Radians} \\
 & \text{Multiply by } \dfrac{\pi}{180} & \\
\end{array}
$$

Figure 13-8 shows the most common angles written in both degrees and radians.

Special Triangles   Recall from Chapter 6 that a *right triangle* is a triangle in which one of the angles is a right angle. Recall also, that the Pythagorean theorem was the important theorem about right triangles.

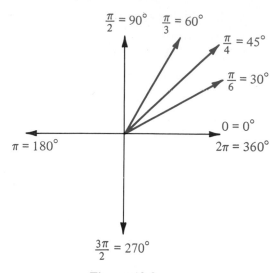

Figure 13-8

PYTHAGOREAN THEOREM   In any right triangle, the square of the length of the longest side (called the hypotenuse) is equal to the sum of the squares of the lengths of the other two sides.

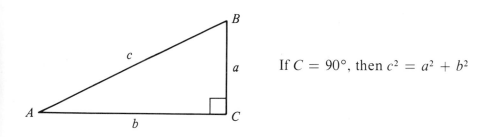

If $C = 90°$, then $c^2 = a^2 + b^2$

Figure 13-9

Note that we denote the lengths of the sides of triangle ABC in Figure 13-9 with lowercase letters and the angles or vertices with uppercase letters. It is standard practice in mathematics to label the sides and angles so that a is opposite A, b is opposite B, and c is opposite C.

We will now turn our attention to the relationship that exists among the sides of two special right triangles.

The 30°–60°–90°
Triangle

In any right triangle in which the two acute angles are 30° and 60°, the longest side (the hypotenuse) is always twice the shortest side, and the medium side is always $\sqrt{3}$ times the shortest side (Figure 13-10).

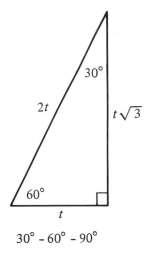

30° – 60° – 90°

Figure 13-10

Note that the shortest side t is opposite the smallest angle 30°. The longest side $2t$ is opposite the largest angle 90°. To verify the relationship between the sides in this triangle we draw an equilateral triangle (one in which all three sides are equal) and label half the base t (Figure 13-11).

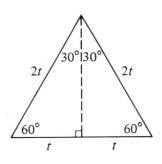

Figure 13-11

The altitude h (the dotted line) bisects the base. We have two 30°–60°–90° triangles. The longest side in each is $2t$. If we were to apply the Pythagorean theorem to solve for the length of the altitude, we would find it to be $t\sqrt{3}$.

$$\text{altitude} = h = \sqrt{4t^2 - t^2}$$
$$= \sqrt{3t^2}$$
$$= t\sqrt{3}$$

▼   **Example 6**   If the shortest side of a 30°–60°–90° triangle is 5, find the other two sides.

Solution   The largest side is 10 (twice the shortest side), and the medium side is $5\sqrt{3}$.

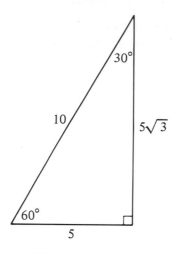

Figure 13-12   ▲

▼   **Example 7**   If the medium side of a 30°–60°–90° triangle is 4, find the other two sides.

Solution   If we let x represent the length of the shortest side, then the largest side is twice this amount or $2x$. The medium side is $x\sqrt{3}$ which must be equal to 4. Therefore,

$$x\sqrt{3} = 4$$

$$x = \frac{4}{\sqrt{3}}$$

$$= \frac{4\sqrt{3}}{3}$$

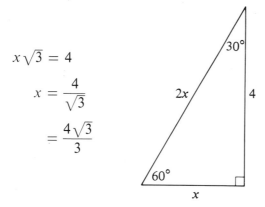

Figure 13-13

The shortest side, x, is $4\sqrt{3}/3$, so the longest side, $2x$, must be $8\sqrt{3}/3$.   ▲

The other special right triangles we want to consider are those in which the two acute angles measure 45° each.

The 45°–45°–90° Triangle

If the two acute angles in a right triangle are both 45°, then the two shorter sides are equal and the longest side (the hypotenuse) is $\sqrt{2}$ times as long as the shorter sides. That is, if the shorter sides are of length t, then the longest side has length $t\sqrt{2}$ (Figure 13-14).

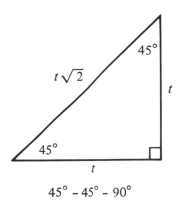

45° – 45° – 90°

Figure 13-14

To verify this relationship, we simply note that if the two acute angles are equal then the sides opposite them are also equal. The hypotenuse is the positive square root of the sum of the squares of these two sides.

$$\text{hypotenuse} = \sqrt{t^2 + t^2}$$
$$= \sqrt{2t^2}$$
$$= t\sqrt{2}$$

▼ **Example 8** If the longest side of a 45°–45°–90° triangle is 10, find the length of the shorter sides.

Solution Since the shorter sides are equal in length, we can let x represent the length of each of them.

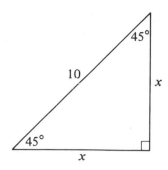

Figure 13-15

The longest side, in terms of x, is $x\sqrt{2}$. Since the longest side is also 10, we have

$$x\sqrt{2} = 10$$

$$x = \frac{10}{\sqrt{2}}$$

$$= 5\sqrt{2}$$

The shorter side is $5\sqrt{2}$. ▲

Indicate which of the angles below are acute angles and which are obtuse angles. Then give the complement and the supplement of each angle.

Problem Set 13.1

1. $10°$ 2. $50°$
3. $45°$ 4. $90°$
5. $120°$ 6. $160°$
7. x 8. y

Convert each of the following from degree measure to radian measure:

9. $30°$ 10. $60°$
11. $90°$ 12. $270°$
13. $120°$ 14. $-240°$
15. $-225°$ 16. $135°$
17. $-150°$ 18. $-210°$
19. $420°$ 20. $390°$

Convert each of the following from radian measure to degree measure:

21. $\dfrac{\pi}{3}$ 22. $\dfrac{\pi}{4}$

23. $\dfrac{2\pi}{3}$ **24.** $\dfrac{3\pi}{4}$

25. $-\dfrac{7\pi}{6}$ **26.** $-\dfrac{5\pi}{6}$

27. $-\dfrac{3\pi}{2}$ **28.** $-\dfrac{5\pi}{3}$

29. $\dfrac{11\pi}{6}$ **30.** $\dfrac{7\pi}{4}$

31. 4π **32.** 3π

33. What positive angle does the line $y = x$ make with the positive x-axis?
34. What positive angle does the line $y = -x$ make with the positive x-axis?
35. Through how many degrees does the hour hand of a clock move in 4 hours?
36. Through how many degrees does the minute hand of a clock move in 40 minutes?
37. It takes the earth 24 hours to make one complete revolution on its axis. Through how many degrees does the earth turn in 12 hours?
38. Through how many degrees does the earth turn in 6 hours?

Each of the following problems refer to right triangle ABC with $C = 90°$:
39. If $a = 4$ and $b = 3$, find c **40.** If $a = 1$ and $b = 2$, find c
41. If $a = 7$ and $c = 25$, find b **42.** If $a = 2$ and $c = 6$, find b
43. If $b = 12$ and $c = 13$, find a **44.** If $b = 6$ and $c = 10$, find a

Find the remaining sides of a 30°–60°–90° triangle if
45. the shortest side is 1.   **46.** the shortest side is 3.
47. the longest side is 8.   **48.** the longest side is 5.
49. the longest side is $\tfrac{2}{3}$. **50.** the longest side is 24.
51. the medium side is $3\sqrt{3}$.   **52.** the medium side is $2\sqrt{3}$.
53. the medium side is 6.   **54.** the medium side is 4.

Find the remaining sides of a 45°–45°–90° triangle if
55. the shorter sides are each 1.   **56.** the shorter sides are each 5.
57. the shorter sides are each $\tfrac{2}{3}$. **58.** the shorter sides are each $\tfrac{1}{2}$.
59. the longest side is $8\sqrt{2}$.   **60.** the longest side is $5\sqrt{2}$.
61. the longest side is 4.   **62.** the longest side is 12.

Review Problems The problems that follow review material we covered in Section 5.2.

Rationalize the denominator.

63. $\dfrac{1}{\sqrt{2}}$ **64.** $\dfrac{1}{\sqrt{3}}$

65. $\dfrac{3}{\sqrt{13}}$ **66.** $\dfrac{2}{\sqrt{5}}$

67. $\dfrac{2}{\sqrt{6}-2}$ **68.** $\dfrac{3}{\sqrt{7}+2}$

69. $\dfrac{\sqrt{3}+\sqrt{2}}{\sqrt{3}-\sqrt{2}}$ **70.** $\dfrac{\sqrt{5}-\sqrt{3}}{\sqrt{5}+\sqrt{3}}$

In this section we will define the six trigonometric functions of an angle. Before we do so, we need one additional definition.

**13.2
Trigonometric
Functions**

DEFINITION An angle is said to be in *standard position* if its vertex is at the origin and its initial side is along the positive x-axis.

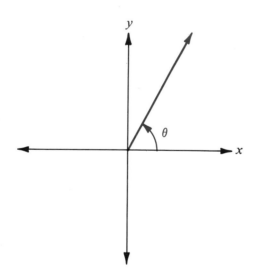

Angle θ in standard
position. θ lies in
quadrant I.

Figure 13-16

We say angle θ lies in quadrant I if θ is in standard position and its terminal side is in quadrant I. Angle θ is said to be in quadrant II if θ is in standard position and its terminal side is in quadrant II. Likewise for quadrants III and IV.

The six trigonometric ratios are defined in terms of angles in standard position.

DEFINITION I   If θ is an angle in standard position, and the point (x, y) is any point on the terminal side of θ other than the origin, then the six trigonometric functions of angle θ are defined as follows:

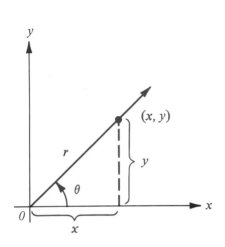

Figure 13-17

Function	Abbreviation	Definition
The sine of θ	$= \sin \theta$	$= \dfrac{y}{r}$
The cosine of θ	$= \cos \theta$	$= \dfrac{x}{r}$
The tangent of θ	$= \tan \theta$	$= \dfrac{y}{x}$
The cotangent of θ	$= \cot \theta$	$= \dfrac{x}{y}$
The secant of θ	$= \sec \theta$	$= \dfrac{r}{x}$
The cosecant of θ	$= \csc \theta$	$= \dfrac{r}{y}$

Where $x^2 + y^2 = r^2$, or $r = \sqrt{x^2 + y^2}$
That is, r is the distance from the origin to (x, y)

As you can see, the six trigonometric functions are simply names given to the six possible ratios that can be made from the numbers $x, y,$ and r as shown in Figure 13-17.

▼   **Example 1**   Find the six trigonometric functions for θ if θ is in standard position and the point $(-2, 3)$ is on the terminal side of θ.

Solution   Figure 13-18 shows θ, the point $(-2, 3)$, and the distance r from the origin to $(-2, 3)$.

Applying the definition for the six trigonometric functions with $x = -2, y = 3$ and $r = \sqrt{13}$ we have

$$\sin \theta = \frac{y}{r} = \frac{3}{\sqrt{13}} \qquad\qquad \csc \theta = \frac{r}{y} = \frac{\sqrt{13}}{3}$$

$$\cos \theta = \frac{x}{r} = -\frac{2}{\sqrt{13}} \qquad\qquad \sec \theta = \frac{r}{x} = -\frac{\sqrt{13}}{2}$$

$$\tan \theta = \frac{y}{x} = -\frac{3}{2} \qquad\qquad \cot \theta = \frac{x}{y} = -\frac{2}{3}$$

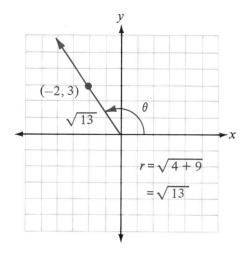

Figure 13-18

Note   In algebra, when we encounter expressions like $3/\sqrt{13}$ that contain a radical in the denominator, we usually rationalize the denominator; in this case by multiplying the numerator and denominator by $\sqrt{13}$.

$$\frac{3}{\sqrt{13}} = \frac{3}{\sqrt{13}} \cdot \frac{\sqrt{13}}{\sqrt{13}} = \frac{3\sqrt{13}}{13}$$

In trigonometry, it is sometimes convenient to use $3\sqrt{13}/13$, and other times it is easier to use $3/\sqrt{13}$. For now, let's agree not to rationalize any denominators unless we are told to do so.

▼   **Example 2**   Find sine, cosine, and tangent of 45°.

Solution   To apply the definition for sine, cosine, and tangent, we need a point on the terminal side of an angle with a measure of 45° when it is drawn in standard position. A convenient point is (1, 1). The distance from the origin to (1, 1) is $\sqrt{2}$.

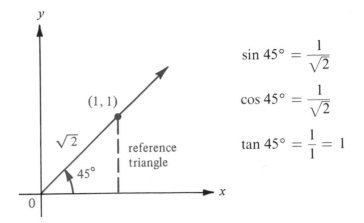

$$\sin 45° = \frac{1}{\sqrt{2}}$$

$$\cos 45° = \frac{1}{\sqrt{2}}$$

$$\tan 45° = \frac{1}{1} = 1$$

Figure 13-19   ▲

Note   The triangle formed by drawing the dotted line from the point on the terminal side to the *x*-axis is called the *reference triangle*.

▼   **Example 3**   Find sine, cosine, and tangent of $\pi/3$.

Solution   It may be easier to visualize this angle if we convert it to degrees. From Section 13.1, we know $\pi/3 = 60°$. If we let the *x*-coordinate of a point on the terminal side be 1 we can find the *y*-coordinate and the distance from the origin to the point by using the fact that the reference triangle is a 30°–60°–90° triangle.

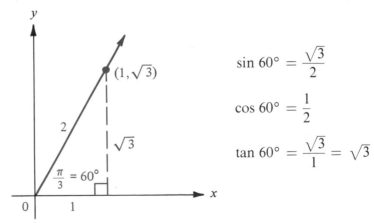

$$\sin 60° = \frac{\sqrt{3}}{2}$$

$$\cos 60° = \frac{1}{2}$$

$$\tan 60° = \frac{\sqrt{3}}{1} = \sqrt{3}$$

Figure 13-20   ▲

▼   **Example 4**   Find sine, cosine, and tangent of 150°.

Solution   To locate a point on the terminal side of 150°, we notice that the reference triangle is a 30°–60°–90° triangle. Since the definitions of sine, cosine, and tangent work for any point on the terminal side of θ, we let the y-coordinate equal 1 and use the relationship between the sides of a 30°–60°–90° triangle to fill in the x-coordinate and r. Since 150° terminates in quadrant II, the x-coordinate of any point on the terminal side will be negative.

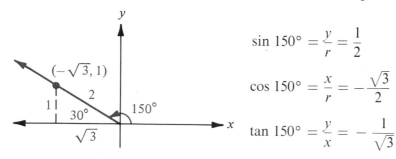

$$\sin 150° = \frac{y}{r} = \frac{1}{2}$$

$$\cos 150° = \frac{x}{r} = -\frac{\sqrt{3}}{2}$$

$$\tan 150° = \frac{y}{x} = -\frac{1}{\sqrt{3}}$$

Figure 13-21   ▲

▼   **Example 5**   Find the six trigonometric ratios of $3\pi/2$.

Solution   From Figure 13-22, we see that the terminal side of $3\pi/2$ is along the negative y-axis. Since we are free to choose any point on the terminal side of $3\pi/2$, we choose the point $(0, -1)$ since it is easy to work with. The distance from the origin down to $(0, -1)$ is $r = 1$.

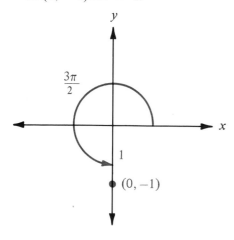

Figure 13-22

$$\sin \frac{3\pi}{2} = \frac{y}{r} = \frac{-1}{1} = -1$$

$$\cos \frac{3\pi}{2} = \frac{x}{r} = \frac{0}{1} = 0$$

$$\tan \frac{3\pi}{2} = \frac{y}{x} = \frac{-1}{0} = \text{undefined}$$

$$\cot \frac{3\pi}{2} = \frac{x}{y} = \frac{0}{-1} = 0$$

$$\sec \frac{3\pi}{2} = \frac{r}{x} = \frac{1}{0} = \text{undefined}$$

$$\csc \frac{3\pi}{2} = \frac{r}{y} = \frac{1}{-1} = -1$$

Note that tangent and secant are undefined since division by 0 is undefined. ▲

Algebraic Signs
of Trigonometric
Functions

The algebraic sign, + or −, of each of the six trigonometric functions will depend on the quadrant in which θ terminates. Since the trigonometric functions are defined in terms of x, y, and r, where (x, y) is a point on the terminal side θ, and we know r is always positive, we can look to the algebraic signs, + or −, of x and y to determine the signs of the trigonometric functions. In quadrant I, x and y are both positive, so all six trigonometric functions will be positive there also. On the other hand, in quadrant II, only $\sin \theta$ and $\csc \theta$ are positive since they are defined in terms of y and y is positive in quadrant II, while x is negative there. Table 13-1 shows the signs of all of the trigonometric functions in each of the four quadrants.

	QI	QII	QIII	QIV
$\sin \theta$ and $\csc \theta$	+	+	−	−
$\cos \theta$ and $\sec \theta$	+	−	−	+
$\tan \theta$ and $\cot \theta$	+	−	+	−

Table 13-1

▼   **Example 6**   If $\sin \theta = -5/13$, and θ terminates in quadrant III, find $\cos \theta$ and $\tan \theta$.

Solution   Since $\sin \theta = -5/13$, we know the ratio of y to r, or y/r is $-5/13$. We can let y be -5 and r be 13 and use these values of y and r to find x. ▲

Note   We are not saying that if $y/r = -5/13$ that y *must* be -5 and r *must* be 13. We know from algebra that there are many pairs of numbers whose ratio is $-5/13$, not just -5 and 13. Our definition for sine and cosine, however, indicates we can choose *any* point on the terminal side of θ to find $\sin \theta$ and $\cos \theta$.

To find x we use the fact that $x^2 + y^2 = r^2$.

$$x^2 + y^2 = r^2$$

$$x^2 + (-5)^2 = 13^2$$

$$x^2 + 25 = 169$$

$$x^2 = 144$$

$$x = \pm 12$$

Is $x - 12$ or $+ 12$?

Since θ terminates in quadrant III, we know any point on its terminal side will have a negative x-coordinate, therefore,

$$x = -12$$

Using $x = -12$, $y = -5$, and $r = 13$ in our original definition we have

$$\cos \theta = \frac{x}{r} = -\frac{12}{13}$$

and

$$\tan \theta = \frac{y}{x} = \frac{-12}{-5} = \frac{12}{5}$$

Figure 13-23 is a diagram of the situation.

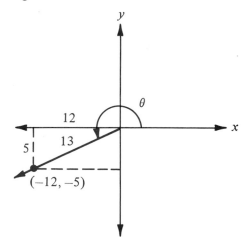

Figure 13-23

Find all six trigonometric functions of θ if the given point is on the terminal side of θ.

Problem Set 13.2

1. $(3, 4)$ 2. $(-3, -4)$

3. $(-5, 12)$ **4.** $(-12, 5)$
5. $(-1, -3)$ **6.** $(1, -3)$
7. (a, b) **8.** (m, n)

Draw each angle below in standard position, and then locate a convenient point on the terminal side of the angle. Use the coordinates of the point to find sine, cosine, and tangent of the angle.

9. $30°$ **10.** $90°$
11. $135°$ **12.** $225°$
13. $5\pi/6$ **14.** $2\pi/3$
15. 2π **16.** π

17. Use the results of Problem 9 and Examples 2 and 3 to complete the table below.

θ	$\sin \theta$	$\cos \theta$	$\tan \theta$
$30°$ or $\pi/6$	$1/2$		
$45°$ or $\pi/4$			1
$60°$ or $\pi/3$		$1/2$	

18. Use the results of Problems 10 and 16, and Example 5, to complete the table below.

θ	$\sin \theta$	$\cos \theta$	$\tan \theta$
$0°$ or 0	0	1	0
$90°$ or $\pi/2$		0	
$180°$ or π	0		
$270°$ or $3\pi/2$			undefined
$360°$ or 2π	0	1	0

Indicate the quadrants in which the terminal side of θ must lie in order that

19. $\cos \theta$ is positive   **20.** $\cos \theta$ is negative
21. $\sin \theta$ is negative   **22.** $\sin \theta$ is positive
23. $\sin \theta$ is negative and $\tan \theta$ is positive
24. $\sin \theta$ is positive and $\cos \theta$ is negative

25. In which quadrant must the terminal side of θ lie if $\sin \theta$ and $\tan \theta$ are to have the same sign?

26. In which quadrant must the terminal side of θ lie if $\cos \theta$ and $\cot \theta$ are to have the same sign?

Find the remaining trigonometric functions of θ if

27. $\sin \theta = \frac{12}{13}$ and θ terminates in QI
28. $\sin \theta = \frac{12}{13}$ and θ terminates in QII
29. $\cos \theta = -\frac{1}{2}$ and θ terminates in QII
30. $\cos \theta = -\sqrt{3}/2$ and θ terminates in QIII
31. $\cos \theta = -\frac{3}{5}$ and θ is not in QII
32. $\tan \theta = \frac{3}{4}$ and θ is not in QIII
33. Find $\sin \theta$ and $\cos \theta$ if the terminal side of θ lies along the line $y = 2x$ in quadrant I.
34. Find $\sin \theta$ and $\cos \theta$ if the terminal side of θ lies along the line $y = 2x$ in quadrant III.
35. Draw $45°$ and $-45°$ in standard position and then show that $\cos(-45°) = \cos 45°$.
36. Draw $45°$ and $-45°$ in standard position and then show that $\sin(-45°) = -\sin 45°$.

Review Problems The problems that follow review material we covered in Section 11.3 on using the table of logarithms in Appendix D. In Section 13.3 we will introduce a table of trigonometric functions, so reviewing our work with tables now will help us with the next section.

Use the table in Appendix D to find the following logarithms:

37. $\log 2.73$ **38.** $\log 273$
39. $\log 0.0273$ **40.** $\log 27{,}300$
41. Find x if $\log x = 1.8299$ **42.** Find x if $\log x = 9.8299 - 10$

Until now, the only angles we have been able to find trigonometric functions for have been angles for which we could find a point on the terminal side, or angles that were part of special triangles. We can find decimal approximations for trigonometric functions of any acute angle by using the table in Appendix E at the back of the book. Appendix E gives the values of sine, cosine, tangent, and cotangent of any acute angle given in decimal degrees—to the nearest tenth of a degree. Angles between $0°$ and $45°$ are listed in the column on the left side of the table, and angles between $45°$ and $90°$ are listed in the column on the right. The angles listed in the column on the left correspond to the trigonometric functions listed at the top of the table. Likewise, the angles in the column on the right correspond to the trigonometric functions at the bottom of the table.

**13.3
Tables
and Calculators**

Tables for Acute Angles

▼   **Example 1**   Use Appendix E to find cos 37.8°.

Solution   We locate 37.8° in the *left column* and then we read across until we are in the column labeled cos θ at the *top*.

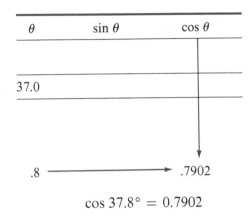

$$\cos 37.8° = 0.7902$$

The number 0.7902 is a four-digit approximation to cos 37.8°. Cos 37.8° is actually an irrational number, as are the trigonometric functions of most of the other angles listed in the table.   ▲

▼   **Example 2**   Use Appendix E to find sin 58.3°.

Solution   We locate 58.3° in the *right column* of the table and then read across until we are in the column labeled sin θ at the *bottom*.

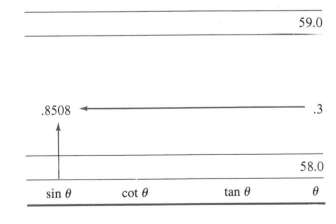

From Appendix E we have sin 58.3° = 0.8508.   ▲

▼   **Example 3**   Use Appendix E to find θ if tan θ = 3.152, and $0° \le \theta \le 90°$.

Solution   In this example, we are given the tangent of an angle and asked to find the angle. Locating 3.152 in one of the tangent columns of the table, we see that it is the tangent of 72.4°.

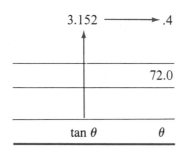

The angle between 0° and 90° whose tangent is 3.152 is 72.4°.   ▲

Most scientific calculators have keys (buttons) for sine, cosine, and tangent. To use these keys you must enter the angle you want sine, cosine, or tangent of in decimal degrees. Check to see that you can obtain the following on your calculator:

Using a Calculator

sin 39.8° = 0.640109699   (Displayed as $6.40109699 \times 10^{-01}$ on some calculators.)

cos 512.3° = −0.885393625

To find cot θ, sec θ, and csc θ on a calculator, we use our reciprocal identities.

$$\sec 73° = \frac{1}{\cos 73°} = \frac{1}{0.2923717} = 3.4203036$$

On a calculator this is found as follows:

73 | cos | | 1/x |

To find the acute angle that corresponds to a given value of a trigonometric function we use the keys labeled

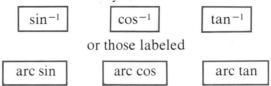

or those labeled

On a calculator, the notation sin $^{-1}$ or arc sin stands for *the angle whose sine is.* Check to see that you can work the problems that follow on your calculator, by entering the sine and tangent and then using the sin^{-1} and tan $^{-1}$ keys, or the keys on your calculator that correspond to these keys, to obtain the given angle.

$$\text{If } \sin \theta = 0.71079947, \text{ then } \theta = 45.3°$$

$$\text{If } \tan \theta = 12.706204, \text{ then } \theta = 85.5°$$

Reference Angle

In order to use the table in Appendix E to find approximate values for trigonometric functions of *any* angle, not just those between 0° and 90°, we need the following definition:

DEFINITION   The *reference angle* (sometimes called related angle) for any angle θ in standard position is the positive acute angle between the terminal side of θ and x-axis.

Note that, for this definition, a reference angle is always positive and always between 0° and 90°. That is, a reference angle is always an acute angle.

▼   **Example 4**   Name the reference angle for each of the following angles:

a. 30°
b. 135°
c. 240°
d. 330°

Solution   We draw each angle in standard position. The reference angle is the positive acute angle formed by the terminal side of the angle in question and the x axis (Figure 13-24).

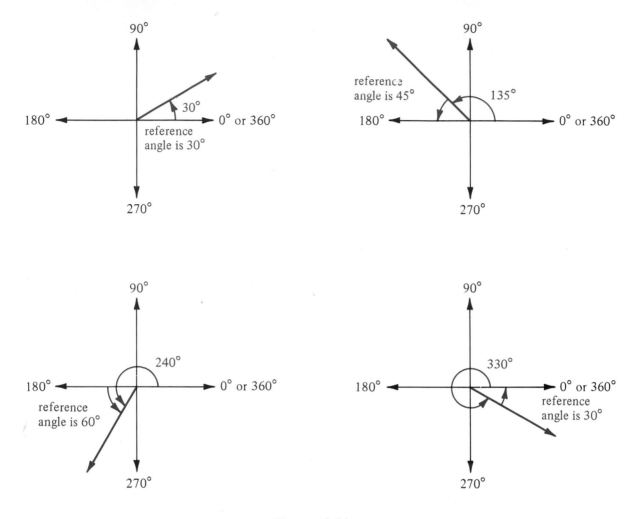

Figure 13-24   ▲

We can use the definition for reference angle, and the signs of the trigonometric functions, to write the following theorem:

REFERENCE ANGLE THEOREM   A trigonometric function of an angle and its reference angle differ at most in sign.

We will not give a detailed proof of this theorem, but rather, justify it by example. Let's look at the sines of all the angles between 0° and 360° that have a reference angle of 30°. These angles are 30°, 150°, 210°, and 330°.

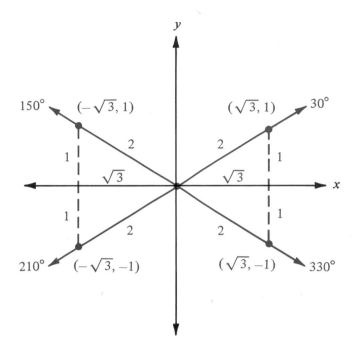

Figure 13-25

$$\sin 150° = \sin 30° = \frac{1}{2}$$

They differ in sign only.

$$\sin 210° = \sin 330° = -\frac{1}{2}$$

As you can see, any angle with a reference angle of 30° will have a sine of $\frac{1}{2}$ or $-\frac{1}{2}$. The sign, + or −, will depend on the quadrant in which the angle terminates.

To summarize, we find trigonometric functions for angles between 0° and 360° by first finding the reference angle. We then find the value of the trigonometric function of the reference angle and then use the quadrant in which the angle terminates to assign the correct sign. Here are some examples.

▼   **Example 5**   Find tan 318.4°.

Solution   Since 318.4° terminates in quadrant IV, tan 318.4° will be negative. The reference angle is 41.6°.

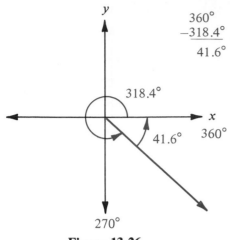

$$\begin{array}{r} 360° \\ -318.4° \\ \hline 41.6° \end{array}$$

$\tan 318.4° = -\tan 41.6°$ Because tangent is negative in QIV

$= -0.8878$ Appendix E

Figure 13-26 ▲

▼ **Example 6** Find $\tan(-135°)$.

Solution The reference angle is 45°. Since $-135°$ terminates in quadrant III, its tangent is positive.

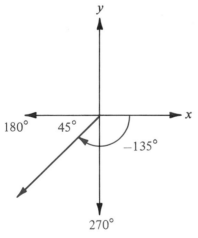

$\tan(-135°) = \tan 45°$ Reference angle
$= 1$ Appendix E

Figure 13-27 ▲

DEFINITION Two angles with the same terminal side are called *co-terminal angles.*

Coterminal angles always differ from each other by multiples of 360°. For example, 10° and 370° are coterminal, as are $-45°$ and 315° (Figure 13-28).

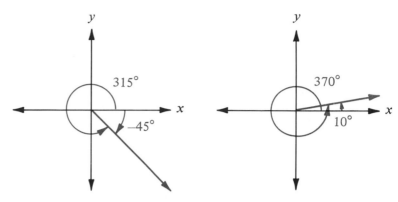

Figure 13-28

The trigonometric functions of an angle and any angle coterminal to it are always equal. For sine and cosine, we can write this in symbols as follows:

$$\sin(\theta + 360°) = \sin\theta \quad \text{and} \quad \cos(\theta + 360°) = \cos\theta$$

To find values of trigonometric functions for an angle larger than 360° or smaller than 0°, we simply find an angle between 0° and 360° that is coterminal to it, and then proceed as we did in Examples 5 and 6.

▼   **Example 7**   Find cos 500°.

Solution   By subtracting 360° from 500°, we obtain 140°, which is coterminal to 500°. The reference angle for 140° is 40°. Since 500° and 140° terminate in quadrant II, cosine is negative.

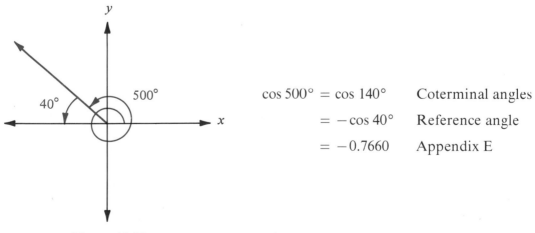

$$\cos 500° = \cos 140° \qquad \text{Coterminal angles}$$
$$= -\cos 40° \qquad \text{Reference angle}$$
$$= -0.7660 \qquad \text{Appendix E}$$

Figure 13-29   ▲

▼   **Example 8** Find θ if $\sin \theta = -0.5592$, with θ in QIII and $0° < \theta < 360°$.

   Solution We look in the sine column until we find 0.5592. Looking to the column on the left, we find that the angle whose sine is 0.5592, is 34°. This is our reference angle. The angle in quadrant III whose reference angle is 34° is $\theta = 180° + 34° = 214°$.

$$\text{If } \sin \theta = -0.5592 \text{ with } \theta \text{ in QIII}$$

$$\text{then } \theta = 180° + 34°$$

$$= 214° \qquad \qquad \blacktriangle$$

Note on Calculators If you were to try Example 7 on your calculator by simply displaying -0.5592 and then pressing the $\sin^{-1}$ key, you would not obtain 214° for your answer. Instead, you would get approximately $-38°$ for the answer, which is wrong. If you want to use a calculator on this kind of problem, use your calculator to find the reference angle and then proceed as we did in Example 8. That is, you would display 0.5592 and then press $\sin^{-1}$ to obtain approximately 34°, to which you would add, in this case, 180°.

▼   **Example 9** Find θ if $\tan \theta = -0.8541$ with θ in QIV and $0° < \theta < 360°$.

   Solution Looking in the tangent column, we find 0.8541 across from 40.5°. This is the reference angle for θ. The angle in quadrant IV with a reference angle of 40.5° is

$$\theta = 360° - 40.5° = 319.5° \qquad \qquad \blacktriangle$$

So far, we have worked through all the examples in this section using the table of approximate values in Appendix E. If we want to find the exact value of a trigonometric function, we can use the information in Table 13-2. The entries in Table 13-2 are the exact values for sine, cosine, and tangent you found in Problem 17 of Problem Set 13.2. Examples 10 through 13 illustrate how we use Table 13-2 and the information we have developed so far in this section to find exact values for trigonometric functions.

Exact Values

Exact Values

θ	$\sin \theta$	$\cos \theta$	$\tan \theta$
30°	$\dfrac{1}{2}$	$\dfrac{\sqrt{3}}{2}$	$\dfrac{1}{\sqrt{3}}$
45°	$\dfrac{1}{\sqrt{2}}$	$\dfrac{1}{\sqrt{2}}$	1
60°	$\dfrac{\sqrt{3}}{2}$	$\dfrac{1}{2}$	$\sqrt{3}$

Table 13-2

▼ **Example 10**   Find the exact value of cos 135°.

Solution   Since 135° terminates in quadrant II, cos 135° will be negative. The reference angle for 135° is 45°.

$$\cos 135° = -\cos 45°$$

$$= -\frac{1}{\sqrt{2}}$$ ▲

▼ **Example 11**   Find the exact value of tan 300°.

Solution   Since 300° terminates in quadrant IV with a reference angle of 60°, we have

$$\tan 300° = -\tan 60°$$

$$= -\sqrt{3}$$ ▲

▼ **Example 12**   Find the exact value of sin 495°.

Solution   If we substract 360° we obtain 495° − 360° = 135°, which terminates in quadrant II with a reference angle of 45°.

$$\sin 495° = \sin 135°$$

$$= \sin 45°$$

$$= \frac{1}{\sqrt{2}}$$ ▲

▼ **Example 13** Find θ if $\sin \theta = -\frac{1}{2}$ with θ in QIII and $0° < \theta < 360°$.

Solution From the table of exact values, we find that the angle whose sine is $\frac{1}{2}$ is 30°. This is the reference angle. The angle in quadrant III with a reference angle of 30° is $180° + 30° = 210°$. ▲

NOTATION The notation $\sin^2 \theta$ means $(\sin \theta)^2 = \sin \theta \cdot \sin \theta$. Raising trigonometric ratios to integer exponents is something that occurs frequently in trigonometry. The expression $\sin^2 \theta$ is the square of the sine of θ. That is, we find the sine of θ first and then square the result.

▼ **Example 14** Use exact values to simplify

$$\sin^2 60° - \cos^2 60° + \cos^3 60°$$

Solution From Table 13-2 we have

$$\sin 60° = \frac{\sqrt{3}}{2} \quad \text{and} \quad \cos 60° = \frac{1}{2}$$

Substituting these values into the original expression we have

$$\sin^2 60° - \cos^2 60° + \cos^3 60°$$

$$= \left(\frac{\sqrt{3}}{2}\right)^2 - \left(\frac{1}{2}\right)^2 + \left(\frac{1}{2}\right)^3$$

$$= \frac{3}{4} - \frac{1}{4} + \frac{1}{8}$$

$$= \frac{6}{8} - \frac{2}{8} + \frac{1}{8}$$

$$= \frac{5}{8} \qquad\qquad ▲$$

Use Appendix E to find each of the following: ***Problem Set 13.3***

1. $\sin 46.4°$ 2. $\cos 24.8°$
3. $\cos 18.3°$ 4. $\sin 41.9°$
5. $\sin 73.7°$ 6. $\cos 59.1°$
7. $\tan 87.1°$ 8. $\cot 78.5°$

Use Appendix E to find θ if θ is between $0°$ and $90°$, and
9. $\cos \theta = 0.8290$ 10. $\sin \theta = 0.8290$
11. $\cos \theta = 0.9348$ 12. $\sin \theta = 0.9348$
13. $\tan \theta = 0.7536$ 14. $\cot \theta = 0.7536$

Use Appendix E to find each of the following:
15. $\cos 101°$ 16. $\sin 166°$
17. $\tan 143.4°$ 18. $\tan 253.8°$
19. $\sin 311.7°$ 20. $\cos 93.2°$
21. $\tan 390°$ 22. $\tan 420°$
23. $\cos 575.4°$ 24. $\sin 590.9°$

Use Appendix E to find θ, if $0° < \theta < 360°$ and
25. $\sin \theta = -0.3090$ with θ in QIII
26. $\sin \theta = -0.3090$ with θ in QIV
27. $\cos \theta = -0.6756$ with θ in QII
28. $\cos \theta = -0.6756$ with θ in QIII
29. $\tan \theta = 0.7729$ with θ in QIII
30. $\tan \theta = 0.7729$ with θ in QI
31. $\cos \theta = 0.5476$ with θ in QIV
32. $\sin \theta = 0.8704$ with θ in QII

Find exact values for each of the following:
33. $\sin 120°$ 34. $\sin 210°$
35. $\tan 135°$ 36. $\tan 315°$
37. $\cos 4\pi/3$ 38. $\cos 5\pi/6$
39. $\cos 300°$ 40. $\sin 300°$
41. $\sin 390°$ 42. $\cos 420°$

Find θ, if $0° < \theta < 360°$ and
43. $\sin \theta = -\dfrac{\sqrt{3}}{2}$ with θ in QIII

44. $\sin \theta = -\dfrac{1}{\sqrt{2}}$ with θ in QIII

45. $\cos \theta = -\dfrac{1}{\sqrt{2}}$ with θ in QII

46. $\cos \theta = \dfrac{\sqrt{3}}{2}$ with θ in QIV

47. $\tan \theta = -\sqrt{3}$ with θ in QIV

48. $\tan \theta - \dfrac{1}{\sqrt{3}}$ with θ in QIII

49. What is the largest value of $\sin \theta$ found in Appendix E?

50. What is the smallest value of sin θ found in Appendix E?
51. If sin θ is between 0 and 1 for θ between 0° and 90°, what are the restrictions on csc θ for θ between 0° and 90°?
52. If sin θ is between 0 and 1 for θ between 0° and 90°, what are the restrictions on sin² θ for θ between 0° and 90°?

Review Problems The problems that follow review material we covered in Section 13.2.
Find sin θ, cos θ, and tan θ for each value of θ. (Do not use tables or calculators.)

53. 90°   54.   135°
55. ·180°   56.   225°

Find the remaining trigonometric functions of θ if
57. cos $\theta = -5/13$ with θ in QIII
58. tan $\theta = -3/4$ with θ in QII

We will begin this section with a look at the graph of $y = \sin x$. As was the case when we first began graphing equations in algebra, we will make a table of values of x and y that satisfy the equation, and then use the information in the table to sketch the graph. To make it easy on ourselves, we will let x take on values that are multiples of $\pi/4$. As an aid in sketching the graphs, we will approximate $1/\sqrt{2}$ with 0.7.

13.4
Graphing I: Amplitude and Period

Graphing y = sin x and y = cos x

x	$y = \sin x$	(x, y)
0	$y = \sin 0 = 0$	$(0, 0)$
$\dfrac{\pi}{4}$	$y = \sin \dfrac{\pi}{4} = \dfrac{1}{\sqrt{2}} = 0.7$	$\left(\dfrac{\pi}{4}, 0.7\right)$
$\dfrac{\pi}{2}$	$y = \sin \dfrac{\pi}{2} = 1$	$\left(\dfrac{\pi}{2}, 1\right)$
$\dfrac{3\pi}{4}$	$y = \sin \dfrac{3\pi}{4} = \dfrac{1}{\sqrt{2}} = 0.7$	$\left(\dfrac{3\pi}{4}, 0.7\right)$
π	$y = \sin \pi = 0$	$(\pi, 0)$
$\dfrac{5\pi}{4}$	$y = \sin \dfrac{5\pi}{4} = -\dfrac{1}{\sqrt{2}} = -0.7$	$\left(\dfrac{5\pi}{4}, -0.7\right)$
$\dfrac{3\pi}{2}$	$y = \sin \dfrac{3\pi}{2} = -1$	$\left(\dfrac{3\pi}{2}, -1\right)$
$\dfrac{7\pi}{4}$	$y = \sin \dfrac{7\pi}{4} = -\dfrac{1}{\sqrt{2}} = -0.7$	$\left(\dfrac{7\pi}{4}, -0.7\right)$
2π	$y = \sin 2\pi = 0$	$(2\pi, 0)$

Table 13-3

Graphing each ordered pair and then connecting them with a smooth curve we obtain the graph shown in Figure 13-30.

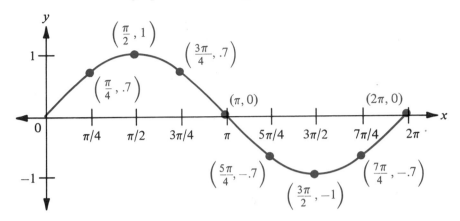

Figure 13-30

To further justify the graph in Figure 13-30, we could find additional ordered pairs that satisfy the equation. For example, we could continue our table by letting x take on multiples of $\pi/6$ between 0 and 2π. If we were to do so, we would find that any new ordered pair that satisfied the equation $y = \sin x$ would be such that its graph would lie on the curve in Figure 13-30. Figure 13-31 shows the curve in Figure 13-30 again but this time with all the ordered pairs with an x-coordinate that is a multiple of $\pi/6$ and y such that $y = \sin x$.

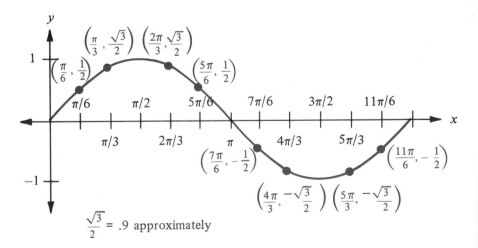

Figure 13-31

Both Figure 13-30 and 13-31 show one complete cycle of $y = \sin x$. We can extend the graph of $y = \sin x$ below $x = 0$ and above $x = 2\pi$ by finding additional valus for x and y. We don't need to go to all that trouble, however, since we know that, once we go past $x = 2\pi$, we will begin to name angles that are coterminal with the angles between 0 and 2π. Because of this, the values of $\sin x$ will start to repeat. Likewise, if we let x take on values below $x = 0$, we will simply get the values of $\sin x$ between 0 and 2π in the reverse order. Figure 13-32 shows the graph of $y = \sin x$ extended beyond the interval from $x = 0$ to $x = 2\pi$.

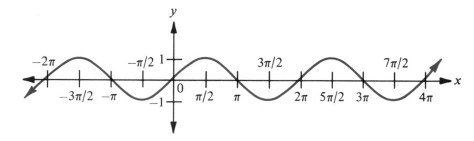

Figure 13-32

Our graph of $y = \sin x$ never goes above 1 or below -1, and it repeats itself every 2π units on the x-axis. This gives rise to the following two definitions:

DEFINITION (PERIOD)   The length of the smallest segment on the x-axis that it takes for a graph to go through one complete cycle is called the *period* of that graph or the equation it came from. In symbols, if p is the smallest positive number for which

$$f(x + p) = f(x) \quad \text{for all } x$$

then we say the *period* of $f(x)$ is p. In the case of $y = \sin x$, the period is 2π since 2π is the smallest positive number for which

$\sin(x + 2\pi) = \sin x$ for all x

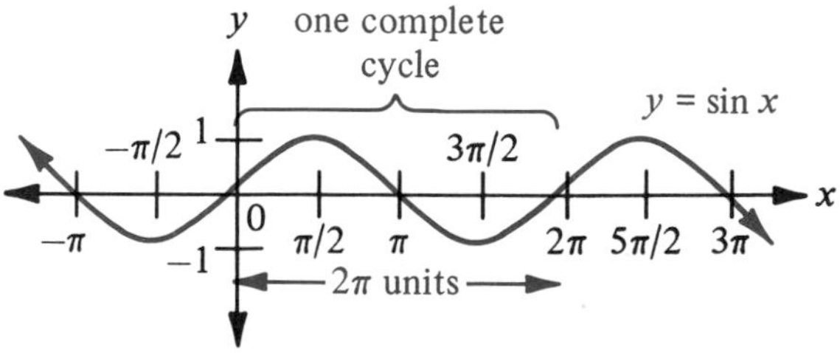

Figure 13-33

DEFINITION (AMPLITUDE)   If the greatest value of y is M and the least value of y is m, then the *amplitude* of the graph of y is defined to be

$$A = \frac{1}{2}|M - m|$$

In the case of $y = \sin x$, the amplitude is 1 because $\frac{1}{2}|1 - (-1)| = \frac{1}{2}(2) = 1$.

The graph of $y = \cos x$ has the same general shape as the graph of $y = \sin x$.

▼   **Example 1**   Sketch the graph of $y = \cos x$ for $0 \le x \le 2\pi$.

Solution   We can arrive at the graph by making a table of convenient values of x and y, as we did for $y = \sin x$.

x	$y = \cos x$	(x, y)
0	$y = \cos 0 = 1$	$(0, 1)$
$\dfrac{\pi}{4}$	$y = \cos \dfrac{\pi}{4} = \dfrac{1}{\sqrt{2}}$	$\left(\dfrac{\pi}{4}, \dfrac{1}{\sqrt{2}}\right)$
$\dfrac{\pi}{2}$	$y = \cos \dfrac{\pi}{2} = 0$	$\left(\dfrac{\pi}{2}, 0\right)$
$\dfrac{3\pi}{4}$	$y = \cos \dfrac{3\pi}{4} = -\dfrac{1}{\sqrt{2}}$	$\left(\dfrac{3\pi}{4}, -\dfrac{1}{\sqrt{2}}\right)$
π	$y = \cos \pi = -1$	$(\pi, -1)$
$\dfrac{5\pi}{4}$	$y = \cos \dfrac{5\pi}{4} = -\dfrac{1}{\sqrt{2}}$	$\left(\dfrac{5\pi}{4}, -\dfrac{1}{\sqrt{2}}\right)$
$\dfrac{3\pi}{2}$	$y = \cos \dfrac{3\pi}{2} = 0$	$\left(\dfrac{3\pi}{2}, 0\right)$
$\dfrac{7\pi}{4}$	$y = \cos \dfrac{7\pi}{4} = \dfrac{1}{\sqrt{2}}$	$\left(\dfrac{7\pi}{4}, \dfrac{1}{\sqrt{2}}\right)$
2π	$y = \cos 2\pi = 1$	$(2\pi, 1)$

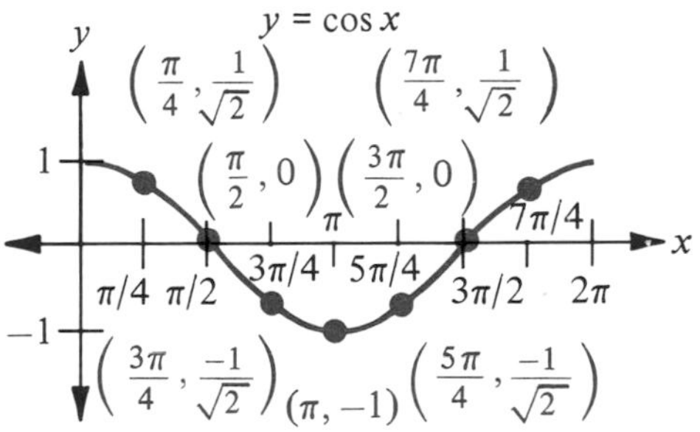

Figure 13-34

Table 13-4   ▲

Now that we have the basic shapes of the curves $y = \sin x$ and $y = \cos x$, we want to investigate the graphs of equations of the form $y = A \sin Bx$ and $y = A \cos Bx$ where A and B are positive numbers. We will begin by making a few tables. We will then use the results obtained by the tables to make some generalizations about the curves in question. Our goal is to be able to sketch the graphs of $y = A \sin Bx$ and $y = A \cos Bx$ without having to make tables.

▼   **Example 2**   Sketch the graph of $y = 2 \sin x$, if $0 \le x \le 2\pi$.

Solution   The coefficient 2 on the right side of the equation will multiply each value of sin x by a factor of 2. Therefore, the values of y in $y = 2 \sin x$ should all be twice the corresponding values of y in $y = \sin x$. Here is a partial table of values for $y = 2 \sin x$, along with the graphs of $y = \sin x$ and $y = 2 \sin x$. (We are including the graph of $y = \sin x$ on the same set of axes simply for reference. With both graphs to look at, it is easier to see what change is brought about by the coefficient 2.)

x	$y = 2 \sin x$	(x, y)
0	$y = 2 \sin 0 = 2(0) = 0$	$(0, 0)$
$\dfrac{\pi}{2}$	$y = 2 \sin \dfrac{\pi}{2} = 2(1) = 2$	$\left(\dfrac{\pi}{2}, 2\right)$
π	$y = 2 \sin \pi = 2(0) = 0$	$(\pi, 0)$
$\dfrac{3\pi}{2}$	$y = 2 \sin \dfrac{3\pi}{2} = 2(-1) = -2$	$\left(\dfrac{3\pi}{2}, -2\right)$
2π	$y = 2 \sin 2\pi = 2(0) = 0$	$(2\pi, 0)$

Table 13-5

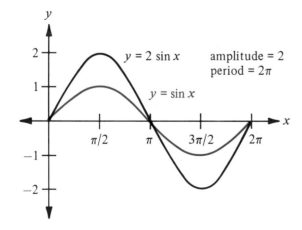

Figure 13-35   ▲

The coefficient 2 in $y = 2 \sin x$ changes the amplitude from 1 to 2. ($A = \frac{1}{2}|2 - (-2)| = \frac{1}{2}(4) = 2$)

▼   **Example 3**   Sketch one complete cycle of the graph of $y = \frac{1}{2}\cos x$.

Solution   Again, we make a partial table and then graph both $y = \frac{1}{2}\cos x$ and $y = \cos x$ on the same set of axes.

x	$y = \dfrac{1}{2}\cos x$	(x, y)
0	$y = \frac{1}{2}\cos 0 = \frac{1}{2}(1) = \frac{1}{2}$	$(0, \frac{1}{2})$
$\dfrac{\pi}{2}$	$y = \frac{1}{2}\cos\dfrac{\pi}{2} = \frac{1}{2}(0) = 0$	$\left(\dfrac{\pi}{2}, 0\right)$
π	$y = \frac{1}{2}\cos\pi = \frac{1}{2}(-1) = -\frac{1}{2}$	$(\pi, -\frac{1}{2})$
$\dfrac{3\pi}{2}$	$y = \frac{1}{2}\cos\dfrac{3\pi}{2} = \frac{1}{2}(0) = 0$	$\left(\dfrac{3\pi}{2}, 0\right)$
2π	$y = \frac{1}{2}\cos 2\pi = \frac{1}{2}(1) = \frac{1}{2}$	$(2\pi, \frac{1}{2})$

Table 13-6

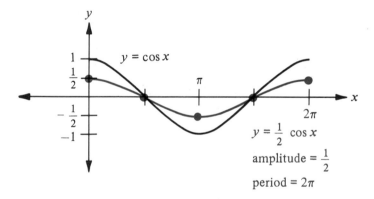

Figure 13-36

The coefficient $\frac{1}{2}$ in $y = \frac{1}{2}\cos x$ is the amplitude of the graph.   ▲

Generalizing the results of these first two examples, we can say

> If A is a positive number, then the graph of $y = A\sin x$ and $y = A\cos x$ will have amplitude $= A$.

▼   **Example 4**   Graph $y = 3\cos x$ and $y = \frac{1}{4}\sin x$, if $0 \le x \le 2\pi$.

Solution   The amplitude for $y = 3\cos x$ is 3, while the amplitude for $y = \frac{1}{4}\sin x$ is $\frac{1}{4}$ (Figure 13-37).   ▲

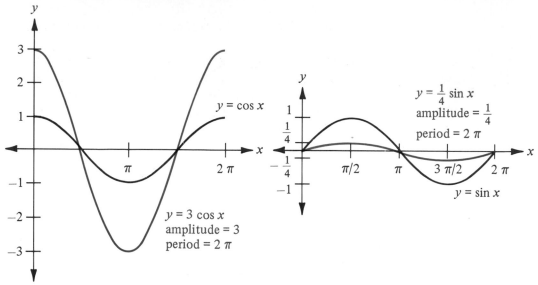

Figure 13-37

▼   **Example 5**   Graph $y = \sin 2x$ for $0 \le x \le 2\pi$.

Solution   To see how the coefficient 2 in $y = \sin 2x$ affects the graph, we will need a table in which the values of x are multiples of $\pi/4$, since the coefficient 2 divides the denominator 4 exactly. Here is the table, along with the graphs of $y = \sin x$ and $y = \sin 2x$.

x	$y = \sin 2x$	(x, y)
0	$y = \sin 2 \cdot 0 = 0$	$(0, 0)$
$\dfrac{\pi}{4}$	$y = \sin 2 \cdot \dfrac{\pi}{4} = \sin \dfrac{\pi}{2} = 1$	$\left(\dfrac{\pi}{4}, 1 \right)$
$\dfrac{\pi}{2}$	$y = \sin 2 \cdot \dfrac{\pi}{2} = \sin \pi = 0$	$\left(\dfrac{\pi}{2}, 0 \right)$
$\dfrac{3\pi}{4}$	$y = \sin 2 \cdot \dfrac{3\pi}{4} = \sin \dfrac{3\pi}{2} = -1$	$\left(\dfrac{3\pi}{4}, -1 \right)$
π	$y = \sin 2\pi = 0$	$(\pi, 0)$
$\dfrac{5\pi}{4}$	$y = \sin 2 \cdot \dfrac{5\pi}{4} = \sin \dfrac{5\pi}{2} = 1$	$\left(\dfrac{5\pi}{4}, 1 \right)$
$\dfrac{3\pi}{2}$	$y = \sin 2 \cdot \dfrac{3\pi}{2} = \sin 3\pi = 0$	$\left(\dfrac{3\pi}{2}, 0 \right)$
$\dfrac{7\pi}{4}$	$y = \sin 2 \cdot \dfrac{7\pi}{4} = \sin \dfrac{7\pi}{2} = -1$	$\left(\dfrac{7\pi}{4}, -1 \right)$
2π	$y = \sin 2 \cdot 2\pi = \sin 4\pi = 0$	$(2\pi, 0)$

Table 13-7

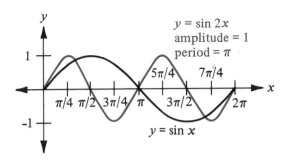

Figure 13-38

The graph of $y = \sin 2x$ has a period of π. It goes through two complete cycles in 2π units on the x-axis.   ▲

▼   **Example 6**   Graph $y = \sin 3x$, if $0 \le x \le 2\pi$.

Solution   To see the effect of the coefficient 3 on the graph, we need a table in which the values of x are multiples of $\pi/6$ since the coefficient 3 divides the denominator 6 exactly.

x	$y = \sin 3x$	(x, y)
0	$y = \sin 3 \cdot 0 = \sin 0 = 0$	$(0, 0)$
$\dfrac{\pi}{6}$	$y = \sin 3 \cdot \dfrac{\pi}{6} = \sin \dfrac{\pi}{2} = 1$	$\left(\dfrac{\pi}{6}, 1\right)$
$\dfrac{\pi}{3}$	$y = \sin 3 \cdot \dfrac{\pi}{3} = \sin \pi = 0$	$\left(\dfrac{\pi}{3}, 0\right)$
$\dfrac{\pi}{2}$	$y = \sin 3 \cdot \dfrac{\pi}{2} = \sin \dfrac{3\pi}{2} = -1$	$\left(\dfrac{\pi}{2}, -1\right)$
$\dfrac{2\pi}{3}$	$y = \sin 3 \cdot \dfrac{2\pi}{3} = \sin 2\pi = 0$	$\left(\dfrac{2\pi}{3}, 0\right)$

Table 13-8

The information in Table 13-8 indicates the period of $y = \sin 3x$ is $2\pi/3$. The graph will go through three complete cycles in 2π units on the x-axis. Figure 13-39 is the graph of $y = \sin 3x$ and the graph of $y = \sin x$.   ▲

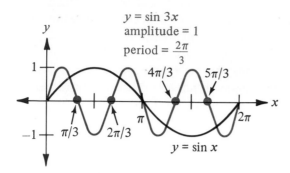

Figure 13-39

Table 13-9 summarizes the information obtained from Examples 5 and 6.

Equation	Number of cycles every 2π units	Period	
$y = \sin x$	1	2π	
$y = \sin 2x$	2	π	
$y = \sin 3x$	3	$2\pi/3$	
$y = \sin Bx$	B	$2\pi/B$	B positive

Table 13-9

▼   **Example 7**   Graph one complete cycle of $y = \cos \tfrac{1}{2}x$.

Solution   The coefficient of x is $\tfrac{1}{2}$. The graph will go through $\tfrac{1}{2}$ a complete cycle every 2π units. The period will be

$$\text{Period} = \frac{2\pi}{1/2} = 4\pi$$

Here is the graph.   ▲

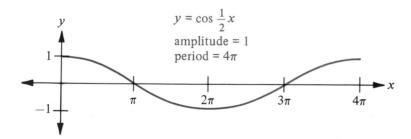

Figure 13-40

The following is a summary of everything we know, so far, about amplitude and period for sine and cosine graphs:

Summary   The graphs of $y = A \sin Bx$ and $y = A \cos Bx$, where A and B are positive numbers, will have amplitude A and period $2\pi/B$.

▼   **Examples**   Graph one complete cycle of each of the following:

$y = 3 \sin 2x$

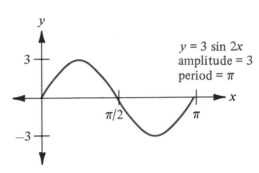

Figure 13-41

9. $y = 4 \cos \frac{1}{2}x$

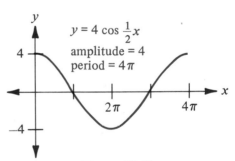

Figure 13-42

10. $y = 2 \sin \pi x$

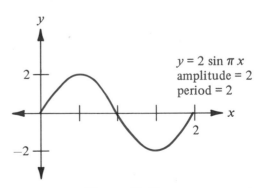

Figure 13-43   ▲

Note that, on the graphs in Figures 13-41, 13-42, and 13-43, the axes have not been labeled proportionally. Instead, they are labeled so that the amplitude and period are easy to read.

Graph each of the following. In each case, label the axes accurately and identify the amplitude and period of each graph.

1. $y = 3\sin x$
2. $y = 2\cos x$
3. $y = \sin 2x$
4. $y = \sin \frac{1}{2}x$
5. $y = \cos \frac{1}{3}x$
6. $y = \cos 3x$
7. $y = \frac{1}{3}\sin x$
8. $y = \frac{1}{2}\cos x$
9. $y = \sin \pi x$
10. $y = \cos \pi x$

11. $y = \sin \frac{\pi}{2}x$
12. $y = \cos \frac{\pi}{2}x$

Graph one complete cycle of each equation. In each case, label the axes accurately and identify the amplitude and period.

13. $y = 4\sin 2x$
14. $y = 2\sin 4x$
15. $y = 2\cos 4x$
16. $y = 3\cos 2x$
17. $y = 3\sin \frac{1}{2}x$
18. $y = 2\sin \frac{1}{3}x$
19. $y = \frac{1}{2}\cos 2x$
20. $y = \frac{1}{2}\sin 2x$

21. $y = \frac{1}{2}\sin \frac{\pi}{2}x$
22. $y = 2\sin \frac{\pi}{2}x$

23. Make a table of values of x and y that satisfy $y = \tan x$. Use multiples of $\pi/4$ and $\pi/6$ starting with $x = 0$ and ending with $x = 2\pi$. At those points for which $\tan x$ is undefined, like $\pi/2$ and $3\pi/2$, the graph of $y = \tan x$ will have a vertical asymptote. See if you can use the values in your table to sketch the graph of $y = \tan x$ between $x = 0$ and $x = 2\pi$.
24. Extend the graph in Problem 23 so that it extends from -2π to 2π.
25. Graph $y = \cot x$ between $x = 0$ and $x = 2\pi$ by making a table as you did for $y = \tan x$ in Problem 23.
26. Extend the graph in Problem 25 so that it extends from -2π to 2π.
27. Graph $y = \csc x$ between $x = 0$ and $x = 2\pi$. You can make a table or use the fact that $\csc x$ and $\sin x$ are reciprocals, or a combination of both, to help find the graph.
28. Graph $y = \sec x$ between $x = 0$ and $x = 2\pi$.

Review Problems The problems that follow review material we covered in Section 13.2.

Find $\sin \theta$, $\cos \theta$, and $\tan \theta$ if the given point is one the terminal side of θ. (θ in standard position.)

29. $(-2, 1)$
30. $(-1, 2)$
31. $(-4, -4)$
32. $(2, -2\sqrt{3})$

13.5
Graphing II: Phase Shift

In this section we will consider equations of the form

$$y = A \sin(Bx + C) \text{ and } y = A \cos(Bx + C) \quad \text{where } A, B > 0$$

We already know how the coefficients A and B affect the graphs of these equations. The only thing we have left to do is discover what effect C has on the graphs. We will start our investigation with a couple of equations in which A and B are 1.

▼ **Example 1**   Graph $y = \sin\left(x + \dfrac{\pi}{2}\right)$ for $0 \le x \le 2\pi$.

Solution   Since we have not graphed an equation of this form before, it is a good idea to begin by making a table. In this case, multiples of $\pi/2$ will be the most convenient replacements for x in the table. Also, if we start with $x = -\pi/2$, our first value of y will be 0.

x	$y = \sin\left(x + \dfrac{\pi}{2}\right)$	(x, y)
$-\dfrac{\pi}{2}$	$y = \sin\left(-\dfrac{\pi}{2} + \dfrac{\pi}{2}\right) = \sin 0 = 0$	$\left(-\dfrac{\pi}{2}, 0\right)$
0	$y = \sin\left(0 + \dfrac{\pi}{2}\right) = \sin\dfrac{\pi}{2} = 1$	$(0, 1)$
$\dfrac{\pi}{2}$	$y = \sin\left(\dfrac{\pi}{2} + \dfrac{\pi}{2}\right) = \sin \pi = 0$	$\left(\dfrac{\pi}{2}, 0\right)$
π	$y = \sin\left(\pi + \dfrac{\pi}{2}\right) = \sin\dfrac{3\pi}{2} = -1$	$(\pi, -1)$
$\dfrac{3\pi}{2}$	$y = \sin\left(\dfrac{3\pi}{2} + \dfrac{\pi}{2}\right) = \sin 2\pi = 0$	$\left(\dfrac{3\pi}{2}, 0\right)$
2π	$y = \sin\left(2\pi + \dfrac{\pi}{2}\right) = \sin\dfrac{\pi}{2} = 1$	$(2\pi, 1)$

Table 13-10

It seems that the graph of $y = \sin(x + \pi/2)$ is shifted $\pi/2$ units to the left of the graph of $y = \sin x$. We say the graph of $y = \sin(x + \pi/2)$ has a *phase shift* of $-\pi/2$; the negative sign indicating the shift is to the left (in the negative direction).

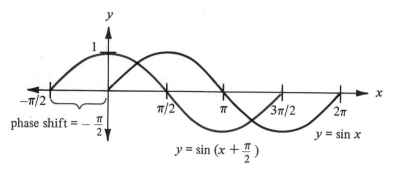

Figure 13-44

From the results in Example 1, we would expect the graph of $y = \sin(x - \pi/2)$ to have a phase shift of $+\pi/2$. That is, we expect the graph of $y = \sin(x - \pi/2)$ to be shifted $\pi/2$ units to the right of the graph of $y = \sin x$.

▼   **Example 2**   Graph one complete cycle of $y = \sin\left(x - \dfrac{\pi}{2}\right)$.

Solution   Proceeding as we did in Example 1, we make a table using multiples of $\pi/2$ for x. In this example, we begin with $x = \pi/2$, since this will give us $y = 0$.

x	$y = \sin\left(x - \dfrac{\pi}{2}\right)$	(x, y)
$\dfrac{\pi}{2}$	$y = \sin\left(\dfrac{\pi}{2} - \dfrac{\pi}{2}\right) = \sin 0 = 0$	$\left(\dfrac{\pi}{2}, 0\right)$
π	$y = \sin\left(\pi - \dfrac{\pi}{2}\right) = \sin\dfrac{\pi}{2} = 1$	$(\pi, 1)$
$\dfrac{3\pi}{2}$	$y = \sin\left(\dfrac{3\pi}{2} - \dfrac{\pi}{2}\right) = \sin \pi = 0$	$\left(\dfrac{3\pi}{2}, 0\right)$
2π	$y = \sin\left(2\pi - \dfrac{\pi}{2}\right) = \sin\dfrac{3\pi}{2} = -1$	$(2\pi, -1)$
$\dfrac{5\pi}{2}$	$y = \sin\left(\dfrac{5\pi}{2} - \dfrac{\pi}{2}\right) = \sin 2\pi = 0$	$\left(\dfrac{5\pi}{2}, 0\right)$

Table 13-11

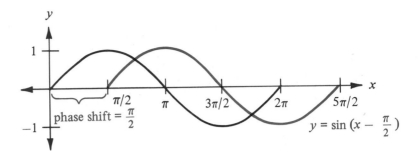

Figure 13-45

The graph of $y = \sin(x - \pi/2)$, as we expected, is a sine curve shifted $\pi/2$ units to the right of the graph of $y = \sin x$. The phase shift, in this case, is $+\pi/2$. ▲

Before we write any conclusions about phase shift, we should look at another example in which A and B are not 1.

▼ **Example 3**   Graph $y = 3\sin\left(2x + \dfrac{\pi}{2}\right)$, if $-\dfrac{\pi}{4} \le x \le \dfrac{3\pi}{4}$.

Solution   We know the coefficient $B = 2$ will change the period from 2π to π. Because the period is smaller, we should use values of x in our table that are closer together, like multiples of $\pi/4$ instead of $\pi/2$.

x	$y = 3\sin\left(2x + \dfrac{\pi}{2}\right)$	(x, y)
$-\dfrac{\pi}{4}$	$y = 3\sin\left[2\left(-\dfrac{\pi}{2}\right) + \dfrac{\pi}{2}\right] = 3\sin 0 = 0$	$\left(-\dfrac{\pi}{4}, 0\right)$
0	$y = 3\sin\left(2 \cdot 0 + \dfrac{\pi}{2}\right) = 3\sin\dfrac{\pi}{2} = 3$	$(0, 3)$
$\dfrac{\pi}{4}$	$y = 3\sin\left(2 \cdot \dfrac{\pi}{4} + \dfrac{\pi}{2}\right) = 3\sin \pi = 0$	$\left(\dfrac{\pi}{4}, 0\right)$
$\dfrac{\pi}{2}$	$y = 3\sin\left(2 \cdot \dfrac{\pi}{2} + \dfrac{\pi}{2}\right) = 3\sin\dfrac{3\pi}{2} = -3$	$\left(\dfrac{\pi}{2}, -3\right)$
$\dfrac{3\pi}{4}$	$y = 3\sin\left(2 \cdot \dfrac{3\pi}{4} + \dfrac{\pi}{2}\right) = 3\sin 2\pi = 0$	$\left(\dfrac{3\pi}{4}, 0\right)$

Table 13-12

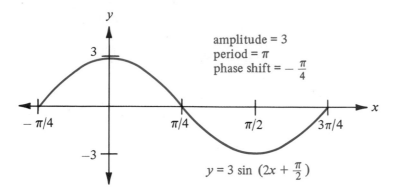

amplitude = 3
period = π
phase shift = $-\dfrac{\pi}{4}$

$y = 3 \sin \left(2x + \dfrac{\pi}{2}\right)$

Figure 13-46

The amplitude and period are as we would expect. The phase shift, however, is half of $-\pi/2$. The phase shift, $-\pi/4$, comes from the ratio $-C/B$ or in this case

$$\frac{-\pi/2}{2} = -\frac{\pi}{4}$$

The phase shift in the equation $y = A \sin(Bx + C)$ depends on both B and C. In this example, the period is half of the period of $y = \sin x$ and the phase shift is half of the phase shift of $y = \sin(x + \pi/2)$ that was found in Example 1. ▲

Although all the examples we have completed so far in this section have been sine equations, the results apply to cosine equations also. Here is a summary.

Summary   The graphs of $y = A \sin(Bx + C)$ and $y = A \cos(Bx + C)$, where $A, B > 0$, will have the following characteristics:

1. Amplitude $= A$

2. Period $= \dfrac{2\pi}{B}$

3. Phase shift $= -\dfrac{C}{B}$

The information on amplitude, period, and phase shift allows us to sketch sine and cosine curves without having to use tables.

▼   **Example 4**   Graph one complete cycle of $y = 2\sin(3x + \pi)$.

Solution   Here is a detailed list of steps to use with this type of graphing.

Step 1.   Use A, B, and C to find the amplitude, period, and phase shift.

$$\text{Amplitude} = A = 2$$
$$\text{Period} = 2\pi/B = 2\pi/3$$
$$\text{Phase shift} = -C/B = -\pi/3$$

Step 2.   On the x-axis, label the starting point, ending point, and the point halfway between them, for each cycle of the curve in question. The starting point is the phase shift. The ending point is the phase shift plus the period.

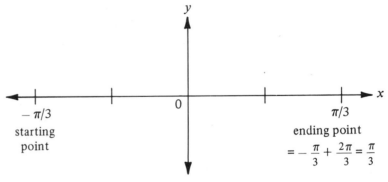

Figure 13-47

Step 3.   Label the y-axis with the amplitude and the opposite of the amplitude. It is okay if the units on the x-axis and the y-axis are not proportional. That is, 1 unit on the y-axis can be a different length than 1 unit on the x-axis. The idea is to make the graph easy to read.

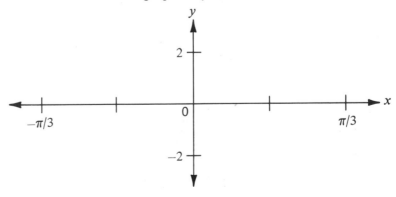

Figure 13-48

Step 4.   Sketch in the curve in question. In this case we want a sine curve, which will be 0 at the starting point, and at the ending point.

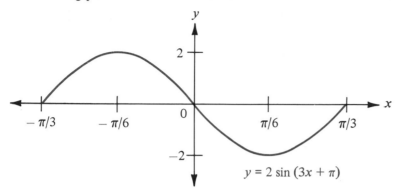

$$y = 2 \sin (3x + \pi)$$

Figure 13-49

The steps listed in Example 4 may seem complicated at first. With a little practice they do not take much time at all. Especially when compared with the time it would take to make a table. Also, once we have graphed one complete cycle of the curve it would be fairly easy to extend the graph in either direction.

▼ **Example 5**   Graph one complete cycle of $y = 2\cos(3x + \pi)$.

Solution A, B, and C are the same here as they were in Example 4. We use the same labeling on the axes as we used in Example 4, but we draw in a cosine curve instead of a sine curve.

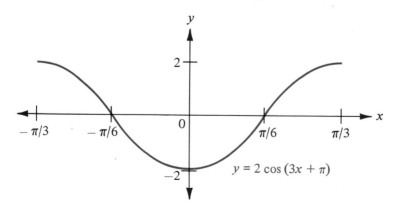

$$y = 2 \cos (3x + \pi)$$

Figure 13-50   ▲

▼ **Example 6**   Graph one complete cycle of $y = 5\sin\left(\pi x + \dfrac{\pi}{4}\right)$.

Solution   In this example, $A = 5$, $B = \pi$, and $C = \pi/4$. Therefore,

$$\text{Amplitude} = 5$$

$$\text{Period} = 2\pi/\pi = 2$$

$$\text{Phase shift} = -\dfrac{\pi/4}{\pi} = -\dfrac{1}{4}$$

The graph is shown in Figure 13-51.

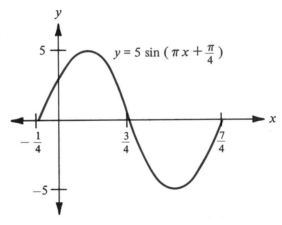

Figure 13-51   ▲

Problem Set 13.5

For each equation, identify the phase shift and then sketch one complete cycle of the graph. In each case, graph one complete cycle of $y = \sin x$ on the same set of axes.

1. $y = \sin\left(x + \dfrac{\pi}{4}\right)$

2. $y = \sin\left(x + \dfrac{\pi}{6}\right)$

3. $y = \sin\left(x - \dfrac{\pi}{4}\right)$

4. $y = \sin\left(x - \dfrac{\pi}{6}\right)$

5. $y = \sin\left(x + \dfrac{\pi}{3}\right)$

6. $y = \sin\left(x - \dfrac{\pi}{3}\right)$

For each equation, identify the phase shift and then sketch one complete cycle of the graph. In each case, graph one complete cycle of $y = \cos x$ on the same set of axes.

7. $y = \cos\left(x - \dfrac{\pi}{2}\right)$

8. $y = \cos\left(x + \dfrac{\pi}{2}\right)$

9. $y = \cos\left(x + \dfrac{\pi}{3}\right)$

10. $y = \cos\left(x - \dfrac{\pi}{4}\right)$

For each equation, identify the amplitude, period, and phase shift. Then label the axes accordingly and sketch one complete cycle of the curve.

11. $y = \sin(2x - \pi)$

12. $y = \sin(2x + \pi)$

13. $y = \cos\left(2x + \dfrac{\pi}{2}\right)$

14. $y = \cos\left(2x - \dfrac{\pi}{2}\right)$

15. $y = 2\sin\left(\dfrac{1}{2}x + \dfrac{\pi}{2}\right)$

16. $y = 3\cos\left(\dfrac{1}{2}x + \dfrac{\pi}{3}\right)$

17. $y = 3\sin\left(2x - \dfrac{\pi}{2}\right)$

18. $y = 3\cos\left(2x - \dfrac{\pi}{3}\right)$

19. $y = 4\cos\left(2x - \dfrac{\pi}{2}\right)$

20. $y = 3\sin\left(2x - \dfrac{\pi}{3}\right)$

21. $y = \dfrac{1}{2}\cos\left(3x - \dfrac{\pi}{2}\right)$

22. $y = \dfrac{4}{3}\cos\left(3x + \dfrac{\pi}{2}\right)$

23. $y = \dfrac{2}{3}\sin\left(3x + \dfrac{\pi}{2}\right)$

24. $y = \dfrac{3}{4}\sin\left(3x - \dfrac{\pi}{2}\right)$

Review Problems   The problems that follow review material we covered in Section 13.3.

Use Appendix E to find the following:

25. $\sin 37.2°$

26. $\cos 37.2°$

27. $\cos 248.1°$

28. $\sin 314.7°$

29. $\tan(-230°)$

30. $\tan 423°$

Chapter 13   Summary and Review   **Examples**

ANGLES [13.1]

1.

An angle is formed by two half-lines with a common end point. The common end point is called the *vertex* of the angle, and the half-lines are called the *sides* of the angle. If we think of an angle as being formed by rotating the initial side about the vertex to the terminal side, then a counter-clockwise rotation gives a positive angle, and a clockwise rotation gives a negative angle.

2.

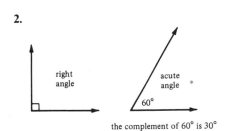

the complement of 60° is 30°

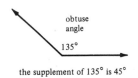

the supplement of 135° is 45°

DEGREE MEASURE [13.1]

There are 360° in a full rotation. This means that 1° is $\frac{1}{360}$ of a full rotation.

An angle that measures 90° is a *right angle*. An angle with measure between 0° and 90° is an *acute angle*, while an angle with measure between 90° and 180° is an *obtuse angle*. If the sum of two angles is 90°, then the two angles are called *complementary angles*. If the sum of two angles is 180°, the angles are called *supplementary angles*.

3.

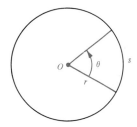

RADIAN MEASURE [13.1]

In a circle, a central angle that cuts off an arc equal in length to the radius of the circle has a measure of 1 radian. Furthermore, if a central angle θ cuts off an arc of length s in a circle of radius r, then the radian measure of θ is $\theta = s/r$.

4. Radians to degrees

$$\frac{4\pi}{3} \text{ rad} = \frac{4\pi}{3}\left(\frac{180}{\pi}\right)^{\circ} = 240°$$

Degrees to radians

$$450° = 450\left(\frac{\pi}{180}\right) = \frac{5\pi}{2} \text{ radians}$$

RADIANS AND DEGREES [13.1]

Changing from degrees to radians and radians to degrees is simply a matter of multiplying by the appropriate conversion factor.

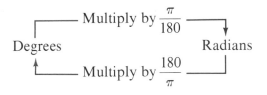

5. 135° in standard position is

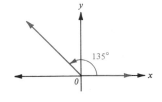

STANDARD POSITION FOR ANGLES [13.2]

An angle is said to be in standard position if its vertex is at the origin and its initial side is along the positive x-axis.

TRIGONOMETRIC FUNCTIONS [13.2]

If θ is an angle in standard position and (x, y) is any point on the terminal side of θ (other than the origin) then

$$\sin \theta = \frac{y}{r} \qquad \csc \theta = \frac{r}{y}$$

$$\cos \theta = \frac{x}{r} \qquad \sec \theta = \frac{r}{x}$$

$$\tan \theta = \frac{y}{x} \qquad \cot \theta = \frac{x}{y}$$

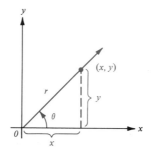

Where $x^2 + y^2 = r^2$, or $r = \sqrt{x^2 + y^2}$. That is, r is the distance from the origin to (x, y).

6. If $(-3, 4)$ is on the terminal side of θ, then

$$r = \sqrt{9 + 16} = 5$$

and

$$\sin \theta = \frac{4}{5} \qquad \csc \theta = \frac{5}{4}$$

$$\cos \theta = -\frac{3}{5} \qquad \sec \theta = \frac{5}{3}$$

$$\tan \theta = -\frac{4}{3} \qquad \cot \theta = \frac{3}{4}$$

SIGNS OF THE TRIGONOMETRIC FUNCTIONS [13.2]

The algebraic signs, $+$ or $-$, of the six trigonometric functions depend on the quadrant in which θ terminates.

	QI	QII	QIII	QIV
$\sin \theta$ and $\csc \theta$	+	+	−	−
$\cos \theta$ and $\sec \theta$	+	−	−	+
$\tan \theta$ and $\cot \theta$	+	−	+	−

7. If $\sin \theta > 0$, and $\cos \theta > 0$, then θ must terminate in QI.
  If $\sin \theta > 0$, and $\cos \theta < 0$, then θ must terminate in QII.
  If $\sin \theta < 0$, and $\cos \theta < 0$, then θ must terminate in QIII.
  If $\sin \theta < 0$, and $\cos \theta > 0$, then θ must terminate in QIV.

APPROXIMATE VALUES FOR TRIGONOMETRIC FUNCTIONS [13.3]

Appendix E gives approximate values for trigonometric functions of angles written in degrees, to the nearest tenth of a degree.

When reading the table in Appendix E notice that the angles in the column on the left correspond to the headings across the top of the table, and the angles in the column on the right correspond to the headings across the bottom of the table.

8.

59.0

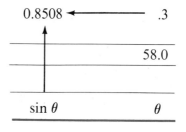

9. 30° is the reference angle for 30°, 150°, 210°, and 330°.

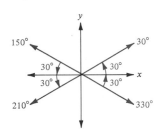

REFERENCE ANGLE [13.3]

The *reference angle* (sometimes called related angle) for any angle θ in standard position, is the positive acute angle between the terminal side of θ and the x-axis.

10. $\sin 150° = \sin 30° = \dfrac{1}{2}$

$\sin 210° = -\sin 30° = -\dfrac{1}{2}$

$\sin 330° = -\sin 30° = -\dfrac{1}{2}$

REFERENCE ANGLE THEOREM [13.3]

A trigonometric function of an angle and its reference angle differ at most in sign.

We find values for trigonometric functions of any angle by first finding the reference angle. Then we find the value of the trigonometric function of the reference angle. Finally, we use the quadrant in which the angle terminates to assign the correct sign.

11.

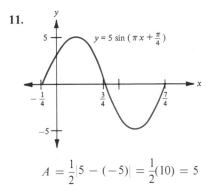

$A = \dfrac{1}{2}|5 - (-5)| = \dfrac{1}{2}(10) = 5$

AMPLITUDE [13.4]

The *amplitude A* of a curve is half the absolute value of the difference between the greatest value of y, denoted by M, and the least value of y, denoted by m.

$$A = \frac{1}{2}|M - m|$$

12.   Since $\sin(x + 2\pi) = \sin x$, the function $y = \sin x$ is periodic with period 2π.

PERIODIC FUNCTIONS [13.4]

A function $y = f(x)$ is said to be periodic with period p if p is the smallest positive number such that $f(x + p) = f(x)$ for all x.

BASIC GRAPHS [13.4]

The graphs of $y = \sin x$ and $y = \cos x$ are both periodic with period 2π. The amplitude of each graph is 1. The sine curve passes through 0 on the y-axis, while the cosine curve passes through 1 on the y-axis.

13. **a.**

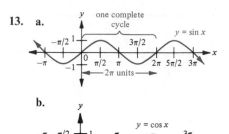

b.

PHASE SHIFT [13.5]

The *phase shift* for a sine or cosine curve, or any of the other trigonometric curves, is the distance the curve has moved right or left from the curve $y = \sin x$ or $y = \cos x$.

14. The phase shift for the graph in Example 11 is $-\frac{1}{4}$.

GRAPHING SINE AND COSINE CURVES [13.5]

The graphs of $y = A\sin(Bx + C)$ and $y = A\cos(Bx + C)$ where $A, B > 0$ will have the following characteristics:

$$\text{Amplitude} = A$$
$$\text{Period} = 2\pi/B$$
$$\text{Phase shift} = -C/B$$

To graph one of these curves we first find the phase shift and label that point on the x-axis (this will be our starting point). We then add the period to the phase shift and mark the result on the x-axis (this is our ending point). We mark the y-axis with the amplitude. Finally, we sketch in one complete cycle of the curve.

15. **a.**

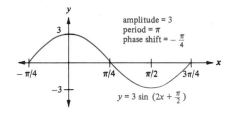

b.

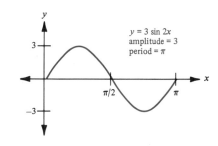

Chapter 13 Test

Convert each angle to radian measure.
1. 75°   **2.** 120°
3. 500°   **4.** 305°

Convert to degree measure.
5. $10\pi/9$ **6.** $11\pi/6$
7. $2\pi/3$ **8.** $9\pi/2$

9. Find the other two sides of a 30°–60°–90° triangle if the shortest side is 5.
10. Find the other two sides of a 45°–45°–90° triangle if the longest side is 12.

Find the six trigonometric functions of θ if the given point is on the terminal side of θ.
11. $(4, -3)$ **12.** $(-2, 1)$

In which quadrant must θ lie if
13. $\sin \theta < 0$ and $\tan \theta > 0$ **14.** $\sin \theta > 0$ and $\sec \theta < 0$

Find the remaining trigonometric functions of θ if
15. $\sin \theta = \frac{1}{2}$ and θ terminates in QII
16. $\tan \theta = -\frac{3}{4}$ and θ terminates in QIV

Use Appendix E to find the following:
17. $\sin 37.3°$ **18.** $\cos 59.8°$
19. $\tan 73.5°$ **20.** $\cot 18.2°$
21. $\cos 195.4°$ **22.** $\tan(-143°)$
23. $\sin 318.1°$ **24.** $\sin 246.8°$
25. $\tan 400°$ **26.** $\sin 525°$

Use Appendix E to find θ if θ is between 0° and 360° and
27. $\sin \theta = 0.3551$ with θ in QI   **28.** $\cos \theta = 0.3551$ with θ in QI
29. $\tan \theta = -2.402$ with θ in QIV   **30.** $\tan \theta = 2.402$ with θ in QIII
31. $\sin \theta = -\sqrt{3}/2$ with θ in QIII **32.** $\cos \theta = -\sqrt{2}/2$ with θ in QII

Find the exact value for each of the following:
33. $\sin 135°$ **34.** $\cos 315°$
35. $\cos 240°$ **36.** $\sin 210°$
37. $\sin 5\pi/6$ **38.** $\cos 5\pi/3$

Graph one complete cycle of each equation. In each case, label your graph appropriately and identify the amplitude, period, and phase shift.

39. $y = 3\sin x$

40. $y = 3\sin 2x$

41. $y = \sin\left(x + \dfrac{\pi}{4}\right)$

42. $y = \cos\left(x - \dfrac{\pi}{4}\right)$

43. $y = 4\sin\left(2x - \dfrac{\pi}{2}\right)$

44. $y = 4\cos\left(3x - \dfrac{\pi}{2}\right)$

14

Trigonometric Identities and Equations

To the student:

In this chapter we will look at some of the consequences of the definition for the six trigonometric functions given in Chapter 13. To begin with, we will list six basic trigonometric identities that are derived directly from the definition for the trigonometric functions given in Section 13.2. A trigonometric identity is simply a statement equating two expressions involving trigonometric functions. For instance, the statement

$$\tan \theta = \frac{\sin \theta}{\cos \theta}$$

is a trigonometric identity. In Section 14.1 we show how the six basic identitites are derived and how they are used in solving problems and proving other identities.

In the second and third sections of Chapter 14, we will use the identities developed in Section 14.1 to derive expansion formulas for expressions of the form $\sin(A + B)$, $\sin 2A$, and $\sin A/2$.

We will end this chapter with a look at trigonometric equations. We solve trigonometric equations by using many of the properties and techniques we developed to solve algebraic equations. With trigonometric equations, however, the variables are associated with trigonometric func-

tions, so we will also use some of the ideas we will develop on identities to help with the solutions to some of the trigonometric equations we will encounter.

All of the material in this chapter is important in the study of trigonometry. Most of the ideas we will encounter in this chapter are used extensively in those subjects for which trigonometry is a prerequisite, such as calculus. To be successful in this chapter, you must understand the definition for the six trigonometric functions given in Section 13.2. You must also know the exact values for sine, cosine, and tangent of 30°, 45°, and 60°, and how to use the table in Appendix E.

In this section, we will turn our attention to identities. In algebra, statements such as $2x = x + x$, $x^3 = x \cdot x \cdot x$, and $x/4x = 1/4$ are called *identities*. They are identities because they are true for all replacements of the variable for which they are defined.

The six basic trigonometric identities are listed in Table 14-1. As we will see, they are all derived from the definition of the trigonometric functions. Since many of the trigonometric identities have more than one form, we will list the basic identity first and then give the most common equivalent forms of that identity.

14.1
Proving Identities

	Basic identities	Common equivalent forms
Reciprocal	$\csc \theta = \dfrac{1}{\sin \theta}$	$\sin \theta = \dfrac{1}{\csc \theta}$
	$\sec \theta = \dfrac{1}{\cos \theta}$	$\cos \theta = \dfrac{1}{\sec \theta}$
	$\cot \theta = \dfrac{1}{\tan \theta}$	$\tan \theta = \dfrac{1}{\cot \theta}$
Ratio	$\tan \theta = \dfrac{\sin \theta}{\cos \theta}$	
	$\cot \theta = \dfrac{\cos \theta}{\sin \theta}$	
Pythagorean	$\cos^2 \theta + \sin^2 \theta = 1$	$\sin^2 \theta = 1 - \cos^2 \theta$ $\sin \theta = \pm \sqrt{1 - \cos^2 \theta}$ $\cos^2 \theta = 1 - \sin^2 \theta$ $\cos \theta = \pm \sqrt{1 - \sin^2 \theta}$

Table 14-1

Note that, in Table 14-1, the six basic identities are grouped in categories. For example, since $\csc \theta = 1/\sin \theta$, cosecant and sine must be reciprocals. It is for this reason that we call the identities in this category *reciprocal identities*.

As we mentioned earlier, the six basic identities are all derived from the definition of the six trigonometric functions. To derive the first reciprocal identity, we use the definition of $\sin \theta$ to write

$$\frac{1}{\sin \theta} = \frac{1}{y/r} = \frac{r}{y} = \csc \theta$$

The other reciprocal identities and their common equivalent forms are derived in a similar manner.

Examples 1, 2, and 3 show how we use the reciprocal identities to find the value of one trigonometric function, given the value of its reciprocal.

▼   **Examples**

1.   If $\sin \theta = \frac{3}{5}$, then $\csc \theta = \frac{5}{3}$, because

$$\csc \theta = \frac{1}{\sin \theta} = \frac{1}{3/5} = \frac{5}{3}$$

2.   If $\cos \theta = -\dfrac{\sqrt{3}}{2}$, then $\sec \theta = -\dfrac{2}{\sqrt{3}}$.

    (*Remember*   Reciprocals always have the same algebraic sign.)

3.   If $\tan \theta = 2$, then $\cot \theta = \frac{1}{2}$.

Unlike the reciprocal identities, the ratio identities do not have any common equivalent forms. Here is how we derive the ratio identity for $\tan \theta$

$$\frac{\sin \theta}{\cos \theta} = \frac{y/r}{x/r} = \frac{y}{x} = \tan \theta \qquad\qquad ▲$$

▼   **Example 4**   If $\sin \theta = -3/5$ and $\cos \theta = 4/5$, find $\tan \theta$ and $\cot \theta$.

    Solution   Using the ratio identities we have

$$\tan \theta = \frac{\sin \theta}{\cos \theta} = \frac{-3/5}{4/5} = -\frac{3}{4}$$

$$\cot \theta = \frac{\cos \theta}{\sin \theta} = \frac{4/5}{3/5} = -\frac{4}{3}$$

Note that, once we found tan θ, we could have used a reciprocal identity to find cot θ.

$$\cot \theta = \frac{1}{\tan \theta} = \frac{1}{-3/4} = -\frac{4}{3} \qquad \blacktriangle$$

Our last basic identity, $\cos^2 \theta + \sin^2 \theta = 1$, is called the *Pythagorean identity* because it is derived from the Pythagorean theorem. Recall from the definition of sin θ and cos θ that if (x, y) is a point on the terminal side of θ and r is the distance to (x, y) from the origin, the relationship between x, y, and r is $x^2 + y^2 = r^2$. This relationship comes from the Pythagorean theorem. Here is how we use it to derive the Pythagorean identity.

$$x^2 + y^2 = r^2$$

$$\frac{x^2}{r^2} + \frac{y^2}{r^2} = 1 \qquad \text{Divide each side by } r^2$$

$$\left(\frac{x}{r}\right)^2 + \left(\frac{y}{r}\right)^2 = 1 \qquad \text{Property of exponents}$$

$$(\cos \theta)^2 + (\sin \theta)^2 = 1 \qquad \text{Definition of sin } \theta \text{ and cos } \theta$$

$$\cos^2 \theta + \sin^2 \theta = 1 \qquad \text{Notation}$$

There are four very useful equivalent forms of the Pythagorean identity. Two of the forms occur when we solve $\cos^2 \theta + \sin^2 \theta = 1$ for $\cos \theta$, while the other two forms are the result of solving for $\sin \theta$.

Solving $\cos^2 \theta + \sin^2 \theta = 1$ for $\cos \theta$ we have

$$\cos^2 \theta + \sin^2 \theta = 1$$

$$\cos^2 \theta = 1 - \sin^2 \theta \qquad \text{Add } -\sin^2 \theta \text{ to each side}$$

$$\cos \theta = \pm \sqrt{1 - \sin^2 \theta} \qquad \text{Take the square root of each side}$$

Similarly, solving for $\sin \theta$ gives us

$$\sin \theta = \pm \sqrt{1 - \cos^2 \theta}$$

▼ **Example 5** If $\sin \theta = \frac{3}{5}$ and θ terminates in quadrant II, find cos θ and tan θ.

Solution We can obtain cos θ from sin θ by using the identity

$$\cos \theta = \pm \sqrt{1 - \sin^2 \theta}$$

$$\text{If} \qquad \sin \theta = \frac{3}{5}$$

$$\text{the identity} \qquad \cos \theta = \pm \sqrt{1 - \sin^2 \theta}$$

$$\text{becomes} \qquad \cos \theta = \pm \sqrt{1 - \left(\frac{3}{5}\right)^2}$$

$$= \pm \sqrt{1 - \frac{9}{25}}$$

$$= \pm \sqrt{\frac{16}{25}}$$

$$= \pm \frac{4}{5}$$

Now we know that $\cos \theta$ is either $+\frac{4}{5}$ or $-\frac{4}{5}$. Looking back to the original statement of the problem, however, we see that θ terminates in quadrant II, therefore, $\cos \theta$ must be negative.

$$\cos \theta = -\frac{4}{5} \qquad\qquad \blacktriangle$$

Proving Identities   Next we want to use our six basic identities and their equivalent forms to verify other trigonometric identities. To prove (or verify) that a trigonometric identity is true, we use trigonometric substitutions and algebraic manipulations to either

1.   Transform the right side into the left side, or
2.   Transform the left side into the right side, or
3.   Transform each side into an equivalent third expression.

The main thing to remember in proving identities is to work on each side of the identity separately. We do not want to use properties from algebra that involve both sides of the identity—like the addition property of equality. We prove identities in order to develop the ability to transform one trigonometric expression into another. When we encounter problems in other courses that require the use of the techniques used to verify identities, we usually find that the solution to these problems hinges upon transforming an expression containing trigonometric functions into a less complicated expression. In these cases, we do not usually have an equal sign to work with.

▼   **Example 6**   Verify the identity $\sin \theta \cot \theta = \cos \theta$.

Proof To help us remember to work with one side or the other, we will separate the two sides with a vertical line.

$$\sin\theta\cot\theta = \cos\theta$$

Ratio identity	$\sin\theta \cdot \dfrac{\cos\theta}{\sin\theta}$
Multiply	$\dfrac{\sin\theta\cos\theta}{\sin\theta}$
Divide out common factor $\sin\theta$	$\cos\theta = \cos\theta$

In this case, we have transformed the left side into the right side. ▲

▼ **Example 7** Prove $\tan x + \cos x = \sin x(\sec x + \cot x)$.

Proof We begin by applying the distributive property to the right side of the identity. Then we change each expression on the right side to an equivalent expression involving only sines and cosines.

$\tan x + \cos x = \sin x(\sec x + \cot x)$

$\sin x \sec x + \sin x \cot x$	Multiply
$\sin x \cdot \dfrac{1}{\cos x} + \sin x \cdot \dfrac{\cos x}{\sin x}$	Reciprocal and ratio identities
$\dfrac{\sin x}{\cos x} + \cos x$	Multiply

$\tan x + \cos x = \tan x + \cos x$ \qquad Ratio identity

In this case, we transformed the right side into the left side.   ▲

Before we go on to the next example, let's list some guidelines that may be useful in learning how to prove identities.

Guidelines for Proving Identities

1. It is usually best to work on the more complicated side first.
2. Look for trigonometric substitutions involving the six basic identities that may help simplify things.
3. Look for algebraic operations, such as adding fractions, the distributive property, or factoring, that may simplify the side you

are working with or that will at least lead to an expression that will be easier to simplify.

4.   If you can't think of anything else to do, change everything to sines and cosines and see if that helps.

5.   Always keep an eye on the side you are not working with, to be sure you are working towards it. There is a certain sense of direction that accompanies a successful proof.

Probably the best advice is to remember that these are simply guidelines. The best way to become proficient at proving trigonometric identities is to practice. The more identities you prove, the more you will be able to prove, and the more confident you will become. Don't be afraid to stop and start over if you don't seem to be getting anywhere. With most identities, there are a number of different proofs that will lead to the same result. Some of the proofs will be longer than others.

▼   **Example 8**   Prove $\dfrac{\cos^4 t - \sin^4 t}{\cos^2 t} = 1 - \tan^2 t$.

Proof   In this example factoring the numerator on the left side will reduce the exponents there from 4 to 2.

$$\frac{\cos^4 t - \sin^4 t}{\cos^2 t} = 1 - \tan^2 t$$

Factor $\qquad\qquad \dfrac{(\cos^2 t + \sin^2 t)(\cos^2 t - \sin^2 t)}{\cos^2 t}$

Pythagorean
identity $\qquad\qquad \dfrac{1(\cos^2 t - \sin^2 t)}{\cos^2 t}$

Separate into
two fractions $\qquad\qquad \dfrac{\cos^2 t}{\cos^2 t} - \dfrac{\sin^2 t}{\cos^2 t}$

Ratio identity $\qquad\qquad 1 - \tan^2 t = 1 - \tan^2 t$

▲

▼   **Example 9**   Prove $1 + \cos \theta = \dfrac{\sin^2 \theta}{1 - \cos \theta}$

Proof   We begin by applying an alternate form of the Pythagorean identity to the right side to write $\sin^2 \theta$ as $1 - \cos^2 \theta$. Then we factor $1 - \sin^2 \theta$ and reduce to lowest terms.

$$1 + \cos \theta = \frac{\sin^2 \theta}{1 - \cos \theta}$$

$$\frac{1 - \cos^2 \theta}{1 - \cos \theta} \qquad \text{Pythagorean identity}$$

$$\frac{(1 - \cos \theta)(1 + \cos \theta)}{1 - \cos \theta} \qquad \text{Factor}$$

$$1 + \cos \theta = 1 + \cos \theta \qquad \text{Reduce} \qquad \blacktriangle$$

▼ **Example 10**   Prove $\tan x + \cot x = \sec x \csc x$.

Proof   We begin this proof by changing everything to sines and cosines. Then we simplify the left side by finding a common denominator, changing to equivalent fractions, and adding, as we did when we combined rational expressions in Chapter 4.

$$\tan x + \cot x = \sec x \csc x$$

Change to sines and cosines $\quad \dfrac{\sin x}{\cos x} + \dfrac{\cos x}{\sin x} \quad \bigg| \quad \dfrac{1}{\cos x} \cdot \dfrac{1}{\sin x} \quad$ Change to sines and cosines

LCD $\quad \dfrac{\sin x}{\cos x} \cdot \dfrac{\mathbf{\sin x}}{\mathbf{\sin x}} + \dfrac{\cos x}{\sin x} \cdot \dfrac{\mathbf{\cos x}}{\mathbf{\cos x}} \quad \bigg| \quad \dfrac{1}{\cos x \sin x} \quad$ Multiply

Add fractions $\quad \dfrac{\sin^2 x + \cos^2 x}{\sin x \cos x}$

Pythagorean identity $\quad \dfrac{1}{\cos x \sin x} = \dfrac{1}{\cos x \sin x}$

This is an example of proving an identity by transforming each side into the same expression. Note that we still work on each side independently $\qquad \blacktriangle$

▼ **Example 11**   Prove $\dfrac{1 + \sin t}{\cos t} = \dfrac{\cos t}{1 - \sin t}$.

Proof   The trick to proving this identity is to multiply the numerator and denominator on the right side by $1 + \sin t$.

$$\frac{1 + \sin t}{\cos t} = \frac{\cos t}{1 - \sin t}$$

$\qquad\qquad \dfrac{\cos t}{1 - \sin t} \cdot \dfrac{1 + \sin t}{1 + \sin t}$ Multiply numerator and denominator by $1 + \sin t$

$\qquad\qquad \dfrac{\cos t(1 + \sin t)}{1 - \sin^2 t}$ Multiply out the denominator

$\qquad\qquad \dfrac{\cos t(1 + \sin t)}{\cos^2 t}$ Pythagorean identity

$$\frac{1 + \sin t}{\cos t} = \frac{1 + \sin t}{\cos t} \qquad\qquad \text{Reduce}$$

Note that it would have been just as easy for us to verify this identity by multiplying the numerator and denominator on the left side by $1 - \sin t$. ▲

Problem Set 14.1

Use the reciprocal identities to work the following problems:

1. If $\sin \theta = \frac{4}{5}$, find $\csc \theta$.

2. If $\cos \theta = \dfrac{\sqrt{3}}{2}$, find $\sec \theta$.

3. If $\sec \theta = -2$, find $\cos \theta$.
4. If $\csc \theta = -\frac{13}{12}$, find $\sin \theta$.
5. If $\tan \theta = a$, find $\cot \theta$.
6. If $\cot \theta = -b$, find $\tan \theta$.

7. If $\cos \theta = \dfrac{1}{\sqrt{2}}$, find $\sec \theta$.

8. If $\tan \theta = -\dfrac{1}{\sqrt{3}}$, find $\cot \theta$.

For Problems 9 through 12, let $\sin \theta = -\frac{12}{13}$, and $\cos \theta = -\frac{5}{13}$, and find
9. $\tan \theta$ 10. $\cot \theta$
11. $\sec \theta$ 12. $\csc \theta$

Use equivalent forms of the Pythagorean identity to work the following problems:
13. If $\sin \theta = -\frac{4}{5}$ and θ terminates in QIII, find $\cos \theta$.
14. If $\sin \theta = -\frac{4}{5}$ and θ terminates in QIV, find $\cos \theta$.

15. If $\cos \theta = \dfrac{\sqrt{3}}{2}$ and θ terminates in QI, find $\sin \theta$.

16. If $\cos \theta = -\frac{1}{2}$ and θ terminates in QII, find $\sin \theta$.

Find the remaining trigonometric ratios of θ if
17. $\sin \theta = \frac{1}{2}$ and θ terminates in quadrant I
18. $\cos \theta = \frac{1}{3}$ and θ terminates in quadrant I

Prove that each of the following identities is true:
19. $\cos \theta \tan \theta = \sin \theta$ 20. $\sec \theta \cot \theta = \csc \theta$
21. $\csc \theta \tan \theta = \sec \theta$ 22. $\tan \theta \cot \theta = 1$
23. $\sec \theta \cot \theta \sin \theta = 1$ 24. $\tan \theta \csc \theta \cos \theta = 1$
25. $\cos x(\csc x + \tan x) = \cot x + \sin x$
26. $\sin x(\sec x + \csc x) = \tan x + 1$
27. $\cos^2 x(1 + \tan^2 x) = 1$ 28. $\sin^2 x(\cot^2 x + 1) = 1$
29. $(1 - \sin x)(1 + \sin x) = \cos^2 x$
30. $(1 - \cos x)(1 + \cos x) = \sin^2 x$

31. $\dfrac{\cos^4 t - \sin^4 t}{\sin^2 t} = \cot^2 t - 1$ 32. $\dfrac{\sin^4 t - \cos^4 t}{\sin^2 t \cos^2 t} = \sec^2 t - \csc^2 t$

33. $1 + \sin \theta = \dfrac{\cos^2 \theta}{1 - \sin \theta}$ 34. $1 - \sin \theta = \dfrac{\cos^2 \theta}{1 + \sin \theta}$

35. $1 + \tan^2 \theta = \sec^2 \theta$ 36. $1 + \cot^2 \theta = \csc^2 \theta$
37. $\cot \theta \cos \theta + \sin \theta = \csc \theta$ 38. $\tan \theta \sin \theta + \cos \theta = \sec \theta$

39. $\dfrac{\sin x}{1 + \cos x} = \dfrac{1 - \cos x}{\sin x}$ 40. $\dfrac{\cos x}{1 + \sin x} = \dfrac{1 - \sin x}{\cos x}$

41. Show that $\sin(A + B)$ is, in general, not equal to $\sin A + \sin B$ by substituting 30° for A and 60° for B in both expressions and simplifying.
42. Show that $\sin 2x \neq 2 \sin x$ by substituting 30° for x and then simplifying both sides.

Review Problems   The problems that follow review material we covered in Sections 13.1 and 13.2. Reviewing these problems will help you with some of the material in the next section.

Give the exact value of each of the following:

43. $\sin \dfrac{\pi}{3}$ 44. $\cos \dfrac{\pi}{3}$

45. $\cos \dfrac{\pi}{6}$ 46. $\sin \dfrac{\pi}{6}$

Convert to degrees.

47. $\dfrac{\pi}{12}$ 48. $\dfrac{5\pi}{12}$

14.2
Sum and
Difference Formulas

The expressions $\sin(A + B)$ and $\cos(A + B)$ occur frequently enough in mathematics that it is necessary to find expressions equivalent to them that involve sines and cosines of single angles. The most obvious question to begin with is

$$\text{Is } \sin(A + B) = \sin A + \sin B?$$

The answer is no. Substituting almost any pair of numbers for A and B in the formula will yield a false statement. For example, if $A = 30°$ and $B = 60°$

$$\sin(30° + 60°) \overset{?}{=} \sin 30° + \sin 60°$$

$$\sin 90° \overset{?}{=} \frac{1}{2} + \frac{\sqrt{3}}{2}$$

$$1 \neq \frac{1 + \sqrt{3}}{2}$$

The formula just doesn't work. The next question is, what are the formulas for $\sin(A + B)$ and $\cos(A + B)$? The answer to that question is what this section is all about. Let's start by deriving the formula for $\cos(A + B)$.

We begin by drawing A in standard position and then adding B and $-B$ to it. These angles are shown in Figure 14-1 in relation to the unit circle. The *unit circle* is the circle with its center at the origin and with a radius of 1. Since the radius of the unit circle is 1, the point through which the terminal side of A passes will have coordinates $(\cos A, \sin A)$. [If P_2 in Figure 14-1 has coordinates (x, y), then by the definition of $\sin A$, $\cos A$, and the unit circle, $\cos A = x/r = x/1 = x$ and $\sin A = y/r = y/1 = y$. Therefore, $(x, y) = (\cos A, \sin A)$.] The points on the unit circle through which the terminal sides of the other angles in Figure 14-1 pass are found in the same manner.

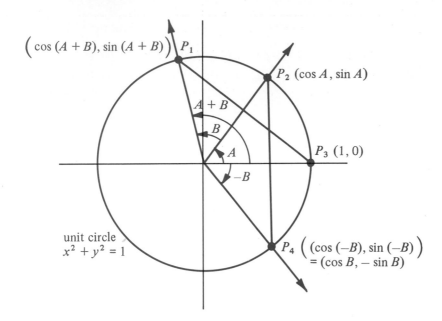

Figure 14-1

To derive the formula for $\cos(A + B)$, we have to see that line segment $\overline{P_1P_3}$ is equal to line segment $\overline{P_2P_4}$. (From geometry, they are chords cut off by equal central angles.)

$$\overline{P_1P_3} = \overline{P_2P_4}$$

Squaring both sides given us

$$(\overline{P_1P_3})^2 = (\overline{P_2P_4})^2$$

Now, applying the distance formula, we have

$$[\cos(A + B) - 1]^2 + [\sin(A + B) - 0]^2$$

$$= (\cos A - \cos B)^2 + (\sin A + \sin B)^2$$

Let's call this Equation 1. Taking the left side of Equation 1, expanding it, and then simplifying by using the Pythagorean identity gives us

Left Side of Equation 1

$$\cos^2(A + B) - 2\cos(A + B) + 1 + \sin^2(A + B) \qquad \text{Expand squares}$$

$$= -2\cos(A + B) + 2 \qquad \begin{array}{l}\text{Pythagorean}\\\text{identity}\end{array}$$

Applying the same two steps to the right side of Equation 1 looks like this

Right Side of Equation 1

$\cos^2 A - 2\cos A \cos B + \cos^2 B + \sin^2 A + 2\sin A \sin B + \sin^2 B$

$$= -2\cos A \cos B + 2\sin A \sin B + 2$$

Equating the simplified versions of the left and right sides of Equation 1 we have

$$-2\cos(A + B) + 2 = -2\cos A \cos B + 2\sin A \sin B + 2$$

Adding -2 to both sides and then dividing both sides by -2 gives us the formula we are after.

$$\cos(A + B) = \cos A \cos B - \sin A \sin B$$

The formula for $\cos(A - B)$ can be derived in a similar manner. It looks like this

$$\cos(A - B) = \cos A \cos B + \sin A \sin B$$

As you can see, the only difference between the two formulas is the signs between the terms. Both formulas are important and should be memorized.

▼   **Example 1**   Find the exact value for $\cos 75°$.

    Solution   We write $75°$ as $45° + 30°$ and then apply the formula for $\cos(A + B)$.

$$\cos 75° = \cos(45° + 30°)$$

$$= \cos 45° \cos 30° - \sin 45° \sin 30°$$

$$= \frac{\sqrt{2}}{2} \cdot \frac{\sqrt{3}}{2} - \frac{\sqrt{2}}{2} \cdot \frac{1}{2}$$

$$= \frac{\sqrt{6} - \sqrt{2}}{4} \qquad\qquad ▲$$

▼   **Example 2**   Prove that $\cos(x + 2\pi) = \cos x$.

    Solution   Applying the formula for $\cos(A + B)$ we have

$$\cos(x + 2\pi) = \cos x \cos 2\pi - \sin x \sin 2\pi$$

$$= \cos x \cdot 1 - \sin x \cdot 0$$

$$= \cos x \qquad\qquad ▲$$

▼ **Example 3** Write $\cos 3x \cos 2x - \sin 3x \sin 2x$ as a single cosine.

Solution We apply the formula for $\cos(A + B)$ in the reverse direction from the way we applied it in the first two examples.

$$\cos 3x \cos 2x - \sin 3x \sin 2x = \cos(3x + 2x)$$

$$= \cos 5x \qquad ▲$$

▼ **Example 4** Show that $\cos(90° - A) = \sin A$.

Solution We will need this formula when we derive the formula for $\sin(A + B)$.

$$\cos(90° - A) = \cos 90° \cos A + \sin 90° \sin A$$

$$= 0 \cdot \cos A + 1 \cdot \sin A$$

$$= \sin A \qquad ▲$$

Note that $90° - A$ and A are complementary angles. The formula we just derived indicates that the sine of an angle is always equal to the cosine of its complement. We could also state it this way

$$\sin(90° - A) = \cos A$$

We can use this information to derive the formula for $\sin(A + B)$. To understand this derivation, you must recognize that $A + B$ and $90° - (A + B)$ are complementary angles.

$\sin(A + B) = \cos[90° - (A + B)]$ The sine of an angle is the cosine of its complement

$\qquad\qquad = \cos[90° - A - B]$ Remove parentheses

$\qquad\qquad = \cos[(90° - A) - B]$ Regroup within brackets

Now we expand using the formula for the cosine of a difference

$$= \cos(90° - A) \cos B + \sin(90° - A) \sin B$$

Since $\cos(90° - A) = \sin A$, and $\sin(90° - A) = \cos A$, we can simplify to

$$= \sin A \cos B + \cos A \sin B$$

This is the formula for the sine of a sum. The formula for $\sin(A - B)$ is similar and looks like this

$$\sin(A - B) = \sin A \cos B - \cos A \sin B$$

Again, these formulas should be memorized.

▼   **Example 5**   Find the exact value of $\sin \dfrac{\pi}{12}$.

Solution   We have to write $\pi/12$ in terms of two numbers the exact values of which are known. The numbers $\pi/3$ and $\pi/4$ will work since their difference is $\pi/12$.

$$\sin \frac{\pi}{12} = \sin\left(\frac{\pi}{3} - \frac{\pi}{4}\right)$$

$$= \sin \frac{\pi}{3} \cos \frac{\pi}{4} - \cos \frac{\pi}{3} \sin \frac{\pi}{4}$$

$$= \frac{\sqrt{3}}{2} \cdot \frac{\sqrt{2}}{2} - \frac{1}{2} \cdot \frac{\sqrt{2}}{2}$$

$$= \frac{\sqrt{6} - \sqrt{2}}{4}$$

This is the same answer we obtained in Example 1 when we found the exact value of $\cos 75°$. It should be, though, because $\pi/12 = 15°$ which is the complement of $75°$, and the cosine of an angle is equal to the sine of its complement.   ▲

▼   **Example 6**   If $\sin A = \frac{3}{5}$ with A in QI and $\cos B = -\frac{5}{13}$ with B in QIII, find $\sin(A + B)$, $\cos(A + B)$, and $\tan(A + B)$.

Solution   We have $\sin A$ and $\cos B$. We need to find $\cos A$ and $\sin B$ before we can apply any of our formulas. Some equivalent forms of our Pythagorean identity with help here.

If $\sin A = \dfrac{3}{5}$ with A in QI, then $\cos A = \sqrt{1 - \sin^2 A}$

$$= \sqrt{1 - \left(\frac{3}{5}\right)^2} = \frac{4}{5}$$

If $\cos B = -\dfrac{5}{13}$ with B in QIII, then $\sin B = -\sqrt{1 - \left(-\frac{5}{13}\right)^2}$

$$= -\frac{12}{13}$$

We have

$$\sin A = \frac{3}{5} \qquad \sin B = -\frac{12}{13}$$

$$\cos A = \frac{4}{5} \qquad \cos B = -\frac{5}{13}$$

Therefore,

$$\sin(A + B) - \sin A \cos B + \cos A \sin B$$

$$= \frac{3}{5}\left(-\frac{5}{13}\right) + \frac{4}{5}\left(-\frac{12}{13}\right)$$

$$= -\frac{63}{65}$$

$$\cos(A + B) = \cos A \cos B - \sin A \sin B$$

$$= \frac{4}{5}\left(-\frac{5}{13}\right) - \frac{3}{5}\left(-\frac{12}{13}\right)$$

$$= \frac{16}{65}$$

$$\tan(A + B) = \frac{\sin(A + B)}{\cos(A + B)}$$

$$= \frac{-63/65}{16/65}$$

$$= -\frac{63}{16} \qquad\qquad \blacktriangle$$

Notice also that $A + B$ must terminate in quadrant IV because $\sin(A + B) < 0$ and $\cos(A + B) > 0$.

While working through the last part of Example 6, you may have wondered if there is a separate formula for $\tan(A + B)$. (More likely, you are hoping there isn't.) There is, and it is derived from the formula we already have.

$$\tan(A + B) = \frac{\sin(A + B)}{\cos(A + B)}$$

$$= \frac{\sin A \cos B + \cos A \sin B}{\cos A \cos B - \sin A \sin B}$$

To be able to write this last line in terms of tangents only, we must divide numerator and denominator by $\cos A \cos B$.

$$= \frac{\dfrac{\sin A \cos B}{\cos A \cos B} + \dfrac{\cos A \sin B}{\cos A \cos B}}{\dfrac{\cos A \cos B}{\cos A \cos B} - \dfrac{\sin A \sin B}{\cos A \cos B}}$$

$$= \frac{\tan A + \tan B}{1 - \tan A \tan B}$$

The formula for $\tan(A - B)$ can be derived in a similar manner.

$$\tan(A - B) = \frac{\tan A - \tan B}{1 + \tan A \tan B}$$

Problem Set 14.2

Find exact values for each of the following:

1. $\sin 15°$ 2. $\sin 75°$
3. $\cos 15°$ 4. $\cos 75°$
5. $\tan 15°$ 6. $\tan 75°$

7. $\sin \dfrac{7\pi}{12}$ $\left(\dfrac{7\pi}{12} = \dfrac{\pi}{3} + \dfrac{\pi}{4}\right)$ 8. $\cos \dfrac{7\pi}{12}$

9. $\cos 195°$ $(195° = 150° + 45°)$
10. $\sin 195°$

Use sum and difference formulas to prove that each statement is true.

11. $\sin(x + 2\pi) = \sin x$ 12. $\cos(x - 2\pi) = \cos x$

13. $\cos\left(x - \dfrac{\pi}{2}\right) = \cos x$ 14. $\sin\left(x - \dfrac{\pi}{2}\right) = -\cos x$

15. $\cos(180° - \theta) = -\cos \theta$ 16. $\sin(180° - \theta) = \sin \theta$
17. $\sin(90° + \theta) = \cos \theta$ 18. $\cos(90° + \theta) = -\sin \theta$

19. $\tan\left(x + \dfrac{\pi}{4}\right) = \dfrac{1 + \tan x}{1 - \tan x}$ 20. $\tan\left(x - \dfrac{\pi}{4}\right) = \dfrac{\tan x - 1}{\tan x + 1}$

Write each expression as a single trigonometric function.

21. $\sin 3x \cos 2x + \cos 3x \sin 2x$ 22. $\cos 3x \cos 2x + \sin 3x \sin 2x$
23. $\cos 5x \cos x - \sin 5x \sin x$ 24. $\sin 8x \cos x - \cos 8x \sin x$
25. $\sin 45° \cos \theta + \cos 45° \sin \theta$ 26. $\cos 60° \cos \theta - \sin 60° \sin \theta$
27. $\cos 30° \sin \theta + \sin 30° \cos \theta$ 28. $\cos 60° \sin \theta + \sin 60° \cos \theta$
29. $\cos 15° \cos 75° - \sin 15° \sin 75°$ 30. $\cos 15° \cos 75° + \sin 15° \sin 75°$

31. Let $\sin A = \frac{3}{5}$ with A in QII and $\sin B = -\frac{5}{13}$ with B in QIII. Find $\sin(A + B)$, $\cos(A + B)$, and $\tan(A + B)$.
32. Let $\cos A = -\frac{5}{13}$ with A in QII and $\sin B = \frac{3}{5}$ with B in QI. Find $\sin(A - B)$, $\cos(A - B)$, and $\tan(A - B)$.
33. If $\sin A = 1/\sqrt{5}$ with A in QI and $\tan B = \frac{3}{4}$ with B in QI, find $\tan(A + B)$ and $\cot(A + B)$.

34. If $\sec A = \sqrt{5}$ with A in QI and $\sec B = \sqrt{10}$ with B in QI, find $\sec(A + B)$. (First find $\cos(A + B)$.)
35. If $\tan(A + B) = 3$ and $\tan B = \frac{1}{2}$, find $\tan A$.
36. If $\tan(A + B) = 2$ and $\tan B = \frac{1}{3}$, find $\tan A$.
37. Write a formula for $\sin 2x$ by writing $\sin 2x$ as $\sin(x + x)$ and using the formula for the sine of a sum.
38. Write a formula for $\cos 2x$ by writing $\cos 2x$ as $\cos(x + x)$ and using the formula for the cosine of a sum.

Review Problems The problems that follow review material we covered in Section 13.3.

Use Appendix E to find each of the following:

39. $\sin 49.3°$ 40. $\cos 49.3°$
41. $\tan 49.3°$ 42. $\cot 49.3°$

Assume θ terminates in QI and
43. Find θ if $\sin \theta = 0.5693$
44. Find θ if $\cos \theta = 0.5693$
45. Find θ if $\tan \theta = 0.4494$
46. Find θ if $\cot \theta = 0.4494$

We will begin this section by deriving the formulas for $\sin 2A$ and $\cos 2A$ using the formulas $\sin(A + B)$ and $\cos(A + B)$. The formulas we derive for $\sin 2A$ and $\cos 2A$ are called *double-angle formulas*. Here is the derivation of the formula for $\sin 2A$.

14.3 Double-Angle and Half-Angle Formulas

$$\sin 2A = \sin(A + A) \qquad \text{Write } 2A \text{ as } A + A$$

$$= \sin A \cos A + \cos A \sin A \qquad \text{Sum formula}$$

$$= \sin A \cos A + \sin A \cos A \qquad \text{Commutative property}$$

$$= 2\sin A \cos A$$

The first thing to notice about this formula is that it indicates that the 2 in $\sin 2A$ *cannot* be factored out and written as a coefficient. That is,

$$\sin 2A \neq 2\sin A$$

Here are some examples of how we can apply the double-angle formula $\sin 2A = 2\sin A \cos A$.

▼ **Example 1** If $\sin A = \frac{3}{5}$ and A terminates in QII, find $\sin 2A$.

Solution In order to apply the formula for $\sin 2A$ we must first find $\cos A$.

$$\cos A = -\sqrt{1 - \sin^2 A} = -\sqrt{1 - \left(\frac{3}{5}\right)^2} = -\sqrt{\frac{16}{25}} = -\frac{4}{5}$$

Now we can apply the formula for $\sin 2A$.

$$\sin 2A = 2\sin A \cos A$$

$$= 2\left(\frac{3}{5}\right)\left(-\frac{4}{5}\right)$$

$$= -\frac{24}{25} \qquad \blacktriangle$$

We can also use our new formula to expand the work we did previously with identities.

▼   **Example 2**   Prove $(\sin \theta + \cos \theta)^2 = 1 + \sin 2\theta$.

Proof	$(\sin \theta + \cos \theta)^2 = 1 + \sin 2\theta$
Expand	$\sin^2 \theta + 2\sin \theta \cos \theta + \cos^2 \theta$
Pythagorean identity	$1 + 2\sin \theta \cos \theta$
Double-angle identity	$1 + \sin 2\theta = 1 + \sin 2\theta$

$\blacktriangle$

There are three forms of the double-angle formula for $\cos 2A$. The first involves both sine and cosine, the second involves only cosine, and the third, just sine. Here is how we obtain the first of these three formulas.

$$\cos 2A = \cos(A + A) \qquad \text{Write } 2A \text{ as } A + A$$

$$= \cos A \cos A - \sin A \sin A \qquad \text{Sum formula}$$

$$= \cos^2 A - \sin^2 A$$

To write this last formula in terms of $\cos A$ only, we substitute $1 - \cos^2 A$ for $\sin^2 A$.

$$\cos 2A = \cos^2 A - (1 - \cos^2 A)$$

$$= \cos^2 A - 1 + \cos^2 A$$

$$= 2\cos^2 A - 1$$

To write the formula for $\cos 2A$ in terms of $\sin A$ only, we substitute $1 - \sin^2 A$ for $\cos^2 A$ in the last line above.

$$\cos 2A = 2\cos^2 A - 1$$
$$= 2(1 - \sin^2 A) - 1$$
$$= 2 - 2\sin^2 A - 1$$
$$= 1 - 2\sin^2 A$$

Here are the three forms of the double-angle formula for $\cos 2A$.

Summary $\cos 2A = \cos^2 A - \sin^2 A$ First form

$= 2\cos^2 A - 1$ Second form

$= 1 - 2\sin^2 A$ Third form

▼   **Example 3**   If $\sin A = 1/\sqrt{5}$, find $\cos 2A$.

Solution   In this example, since we are given $\sin A$, applying the third form of the formula for $\cos 2A$ will give us the answer more quickly than applying either of the other two forms.

$$\cos 2A = 1 - 2\sin^2 A$$
$$= 1 - 2\left(\frac{1}{\sqrt{5}}\right)^2$$
$$= 1 - \frac{2}{5}$$
$$= \frac{3}{5}$$ ▲

▼   **Example 4**   Prove $\tan \theta = \dfrac{1 - \cos 2\theta}{\sin 2\theta}$.

Proof $\tan \theta = \dfrac{1 - \cos 2\theta}{\sin 2\theta}$

$\dfrac{1 - (1 - 2\sin^2 \theta)}{2\sin \theta \cos \theta}$ Double-angle formulas

$\dfrac{2\sin^2 \theta}{2\sin \theta \cos \theta}$ Simplify numerator

$\dfrac{\sin \theta}{\cos \theta}$ Divide out common factor $2\sin \theta$

$\tan \theta = \tan \theta$,   Ratio identity   ▲

Half-Angle Formulas

We are now ready to derive the formulas for $\sin A/2$ and $\cos A/2$. These formulas are called half-angle formulas and are derived from the double-angle formulas for $\cos 2A$.

We can derive the formula for $\sin A/2$ by first solving the double-angle formula $\cos 2x = 1 - 2\sin^2 x$ for $\sin x$, and then applying a simple substitution.

$$1 - 2\sin^2 x = \cos 2x \qquad \text{Double-angle formula}$$

$$-2\sin^2 x = -1 + \cos 2x \qquad \text{Add } -1 \text{ to both sides}$$

$$\sin^2 x = \frac{1 - \cos 2x}{2} \qquad \text{Divide both sides by } -2$$

$$\sin x = \pm\sqrt{\frac{1 - \cos 2x}{2}} \qquad \begin{array}{l}\text{Take the square root}\\ \text{of both sides}\end{array}$$

Since every value of x can be written as $\frac{1}{2}$ of some other number A, we can replace x with $A/2$. This is equivalent to saying $2x = A$.

$$\sin \frac{A}{2} = \pm\sqrt{\frac{1 - \cos A}{2}}$$

This last expression is the half-angle formula for $\sin A/2$. To find the half-angle formula for $\cos A/2$, we solve $\cos 2x = 2\cos^2 x - 1$ for $\cos x$ and then replace x with $A/2$ (and $2x$ with A). Without showing the steps involved in this process, here is the result

$$\cos \frac{A}{2} = \pm\sqrt{\frac{1 + \cos A}{2}}$$

In both half-angle formulas the sign, $+$ or $-$, in front of the radical is determined by the quadrant in which $A/2$ terminates.

▼   **Example 5**   If $\cos A = \frac{3}{5}$ with A in QIV, find $\sin A/2$, $\cos A/2$, and $\tan A/2$.

     Solution   First of all, if A terminates in QIV, then $A/2$ terminates in QII. Here is why.

$$A \in \text{QIV} \Rightarrow 270° < A < 360° \Rightarrow \frac{270°}{2} < \frac{A}{2} < \frac{360°}{2}$$

$$\text{or } 135° < \frac{A}{2} < 180° \Rightarrow \frac{A}{2} \in \text{QII}$$

In quadrant II, sine is positive and cosine is negative.

$$\sin\frac{A}{2} = \sqrt{\frac{1-\cos A}{2}} \qquad \cos\frac{A}{2} = -\sqrt{\frac{1+\cos A}{2}}$$

$$= \sqrt{\frac{1-\frac{3}{5}}{2}} \qquad\qquad = -\sqrt{\frac{1+\frac{3}{5}}{2}}$$

$$= \sqrt{\frac{1}{5}} \qquad\qquad = -\sqrt{\frac{4}{5}}$$

$$= \frac{1}{\sqrt{5}} \qquad\qquad = -\frac{2}{\sqrt{5}}$$

$$\tan\frac{A}{2} = \frac{\sin\dfrac{A}{2}}{\cos\dfrac{A}{2}} = \frac{1/\sqrt{5}}{-2/\sqrt{5}} = -\frac{1}{2}$$

▲

▼ **Example 6**   If $\sin A = -12/13$ and $A \in$ QIII, find $\sin A/2$, $\cos A/2$, and $\tan A/2$.

Solution   To use the half-angle formulas, we need to find $\cos A$.

$$\cos A = -\sqrt{1 - \sin^2 A} = -\sqrt{1 - \left(-\frac{12}{13}\right)^2} = -\sqrt{\frac{25}{169}} = -\frac{5}{13}$$

Also, if $A \in$ QIII, then $\dfrac{A}{2} \in$ QII because

$$A \in \text{QIII} \Rightarrow 180° < A < 270°$$

$$\frac{180°}{2} < \frac{A}{2} < \frac{270°}{2}$$

$$90° < \frac{A}{2} < 135° \Rightarrow \frac{A}{2} \in \text{QII}$$

In quadrant II, sine is positive and cosine is negative.

$$\sin\frac{A}{2} = \sqrt{\frac{1-(-\frac{5}{13})}{2}} \qquad \cos\frac{A}{2} = -\sqrt{\frac{1+(-\frac{5}{13})}{2}}$$

$$= \sqrt{\frac{9}{13}} \qquad\qquad = -\sqrt{\frac{4}{13}}$$

$$= \frac{3}{\sqrt{13}} \qquad\qquad = \frac{2}{\sqrt{13}}$$

$$\tan\frac{A}{2} = \frac{\sin\dfrac{A}{2}}{\cos\dfrac{A}{2}} = \frac{3/\sqrt{13}}{-2/\sqrt{13}} = -\frac{3}{2}$$

▲

▼ **Example 7**   Prove $\sin^2 \dfrac{x}{2} = \dfrac{\tan x - \sin x}{2\tan x}$.

Proof   We can use a half-angle formula on the left side, and write the right side in terms of sine and cosine only.

$$\sin^2 \frac{x}{2} = \frac{\tan x - \sin x}{2\tan x}$$

Half-angle
formula
$$\left[\pm\sqrt{\frac{1 - \cos x}{2}} \right]^2 \quad \middle| \quad \frac{\dfrac{\sin x}{\cos x} - \sin x}{2\dfrac{\sin x}{\cos x}}$$
Ratio
identity

Square the
square root
$$\frac{1 - \cos x}{2} \quad \middle| \quad \frac{\dfrac{1}{\cos x} - 1}{\dfrac{2}{\cos x}}$$
Divide top
and bottom
by $\sin x$

$$\frac{1 - \cos x}{2} = \frac{1 - \cos x}{2}$$
Multiply top
and bottom by
$\cos x$ ▲

Problem Set 14.3

Let $\sin A = -\frac{3}{5}$ with $A \in$ QIII and find

1. $\sin 2A$ 2. $\cos 2A$
3. $\tan 2A$ 4. $\cot 2A$

Let $\cos x = 1/\sqrt{10}$ with $x \in$ QIV and find

5. $\cos 2x$ 6. $\sin 2x$
7. $\cot 2x$ 8. $\tan 2x$

If $\cos A = \frac{1}{2}$ and $A \in$ QIV, find

9. $\sin \dfrac{A}{2}$ 10. $\cos \dfrac{A}{2}$

11. $\csc \dfrac{A}{2}$ 12. $\sec \dfrac{A}{2}$

If $\sin A = -\frac{3}{5}$ and $A \in$ QIII, find

13. $\cos \dfrac{A}{2}$ 14. $\sin \dfrac{A}{2}$

15. $\cot \dfrac{A}{2}$ 16. $\tan \dfrac{A}{2}$

17. $\sec \dfrac{A}{2}$

18. $\csc \dfrac{A}{2}$

If $\sin A = \frac{4}{5}$ with $A \in$ QII, and $\sin B = \frac{3}{5}$ with $B \in$ QI, find

19. $\sin \dfrac{A}{2}$

20. $\cos \dfrac{A}{2}$

21. $\cos 2A$

22. $\sin 2A$

23. $\sec 2A$

24. $\csc 2A$

25. $\cos \dfrac{B}{2}$

26. $\sin \dfrac{B}{2}$

27. $\sin(A + B)$

28. $\cos(A + B)$

29. $\cos(A - B)$

30. $\sin(A - B)$

Use exact values to show that each of the following is true:

31. $\sin 60° = 2\sin 30° \cos 30°$

32. $\cos 60° = 1 - 2\sin^2 30°$

33. $\cos 120° = \cos^2 60° - \sin^2 60°$

34. $\sin 90° = 2\sin 45° \cos 45°$

Prove each of the following identities:

35. $(\sin x - \cos x)^2 = 1 - \sin 2x$

36. $(\cos x - \sin x)(\cos x + \sin x) = \cos 2x$

37. $\cos^2 \theta = \dfrac{1 + \cos 2\theta}{2}$

38. $\sin^2 \theta = \dfrac{1 - \cos 2\theta}{2}$

39. $\cot \theta = \dfrac{\sin 2\theta}{1 - \cos 2\theta}$

40. $\cos 2\theta = \dfrac{1 - \tan^2 \theta}{1 + \tan^2 \theta}$

41. $\cos^4 x - \sin^4 x = \cos 2x$

42. $\sin^4 x - \cos^4 x = -\cos 2x$

43. $\cos^2 \dfrac{\theta}{2} = \dfrac{\tan \theta + \sin \theta}{2\tan \theta}$

44. $2 \sin^2 \dfrac{\theta}{2} = \dfrac{\sin^2 \theta}{1 + \cos \theta}$

Review Problems   The problems that follow review material we covered in Section 13.4

Graph one complete cycle.

45. $y = 3\sin 2x$

46. $y = 2\sin 3x$

47. $y = 2\sin \pi x$

48. $y = 2\sin \dfrac{\pi}{2}x$

The process of solving trigonometric equations is very similar to the process of solving algebraic equations. With trigonometric equations, we look for values of an *angle* that will make the equation into a true statement. We usually begin by solving for a specific trigonometric function of that angle. Then we use some of the concepts we developed in Chapter 13 to find the angle. Let's look at some examples that illustrate this procedure.

**14.4
Solving Trigonometric
Equations**

▼   **Example 1**   Find all values of x for which $2\cos x - \sqrt{3} = 0$, if $0° \leq x \leq 360°$.

Solution   We can solve for $\cos x$ using our methods from algebra. Then we use our knowledge of trigonometry to find x.

$$2\cos x - \sqrt{3} = 0$$

$$2\cos x = \sqrt{3} \qquad \text{Add } \sqrt{3} \text{ to each side}$$

$$\cos x = \frac{\sqrt{3}}{2} \qquad \text{Divide each side by 2}$$

From Section 13.2 we know that if x is between 0° and 360° and $\cos x = \sqrt{3}/2$, then x is either 30° or 330°. (Remember $\cos x$ is positive when x terminates in quadrants I or IV and the reference angle whose cosine is $\sqrt{3}/2$ is 30°.)

$$x = 30° \text{ or } x = 330° \qquad\qquad ▲$$

▼   **Example 2**   Solve $2\sin \theta - 3 = 0$, if θ is between 0° and 360°.

Solution   We begin by solving for $\sin \theta$

$$2\sin \theta - 3 = 0$$
$$2\sin \theta = 3 \qquad \text{Add 3 to each side}$$

$$\sin \theta = \frac{3}{2} \qquad \text{Divide each side by 2}$$

Since $\sin \theta$ is between -1 and 1 for all values of θ, it can never be $\frac{3}{2}$. Therefore, there is no solution to our equation.   ▲

▼   **Example 3**   Solve $3\sin \theta - 2 = 7\sin \theta - 1$ for θ, if $0° \leq \theta \leq 360°$.

Solution   We can solve for $\sin \theta$ by collecting all the variable terms on the left side and all the constant terms on the right side.

$$3\sin \theta - 2 = 7\sin \theta - 1$$

$$-4\sin \theta - 2 = -1 \qquad\qquad \text{Add } -7\sin \theta \text{ to each side}$$

$$-4\sin \theta = 1 \qquad\qquad \text{Add 2 to each side}$$

$$\sin \theta = -\frac{1}{4} \qquad\qquad \text{Divide each side by } -4$$

Since we have not memorized the angle whose sine is $-\frac{1}{4}$, we must convert $-\frac{1}{4}$ to a decimal and use Appendix E to find the reference angle.

$$\sin \theta = -\frac{1}{4} = -0.2500$$

From Appendix E, we find that the angle whose sine is nearest to -0.2500 is $14.5°$. Therefore, the reference angle is $14.5°$. Since $\sin \theta$ is negative, θ will terminate in quadrants III or IV.

In quadrant III we have   In quadrant IV we have
$$\theta = 180° + 14.5°$$ $$\theta = 360° - 14.5°$$
$$= 194.5°$$ $$= 345.5°$$ ▲

The next trigonometric equation we will solve is quadratic in form.

▼ **Example 4** Solve $2\cos^2 t - 9\cos t = 5$, if $0 \le t \le 2\pi$.

Solution The fact that t is between 0 and 2π indicates that we are to write our solutions in radians.

$$2\cos^2 t - 9\cos t = 5$$

$$2\cos^2 t - 9\cos t - 5 = 0 \qquad \text{Standard form}$$

$$(2\cos t + 1)(\cos t - 5) = 0 \qquad \text{Factor}$$

$$2\cos t + 1 = 0 \quad \text{or} \quad \cos t - 5 = 0 \qquad \text{Set each factor to 0}$$

$$\cos t = -\frac{1}{2} \qquad\qquad \cos t = 5$$

The first result, $\cos t = -\frac{1}{2}$, gives us $t = 2\pi/3$ or $t = 4\pi/3$. The second result, $\cos t = 5$, is meaningless. For any value of t, $\cos t$ must be between -1 and 1. It can never be 5.   ▲

▼ **Example 5** Solve $2\sin^2 \theta + 2\sin \theta - 1 = 0$, if $0° \le \theta \le 360°$

Solution The equation is already in standard form. However, if we try to factor the left side, we find it does not factor. We must use the quadratic formula.

The coefficients a, b, and c are

$$a = 2 \quad b = 2 \quad c = -1$$

Using these numbers, we solve for $\sin \theta$ as follows:

$$\sin \theta = \frac{-2 \pm \sqrt{4 - 4(2)(-1)}}{2(2)}$$

$$= \frac{-2 \pm \sqrt{12}}{4}$$

$$= \frac{-2 \pm 2\sqrt{3}}{4}$$

$$= \frac{-1 \pm \sqrt{3}}{2}$$

Using the approximation $\sqrt{3} = 1.7321$, we arrive at the following decimal approximations for $\sin \theta$:

$$\sin \theta = \frac{-1 + 1.7321}{2} \quad \text{or} \quad \sin \theta = \frac{-1 - 1.7321}{2}$$

$$\sin \theta = 0.3661 \qquad \text{or} \quad \sin \theta = -1.3661$$

We will not obtain any solutions from the second expression, $\sin \theta = -1.3661$, since $\sin \theta$ must be between -1 and 1. For $\sin \theta = 0.3661$, we use Appendix E to find the angle whose sine is nearest to 0.3661. That angle is $21.5°$, and it is the reference angle for θ. Since $\sin \theta$ is positive, θ must terminate in quadrants I or II. Therefore,

$$\theta = 21.5° \quad \text{or} \quad \theta = 180° - 21.5° = 158.5° \qquad \blacktriangle$$

Now we will extend the process of solving trigonometric equations to include the use of trigonometric identities. That is, we will use our knowledge of identities to replace some parts of the equations we are solving with equivalent expressions that will make the equations easier to solve. Here are some examples.

▼ **Example 6** Solve $2\cos x - 1 = \sec x$, if $0 \leq x \leq 2\pi$.

Solution To solve this equation as we have solved the equations in the previous examples, we must write each term using the same trigonometric function. To do so, we can use a reciprocal identity to write $\sec x$ in terms of $\cos x$.

$$2\cos x - 1 = \frac{1}{\cos x}$$

To clear the equation of fractions, we multiply both sides by $\cos x$.

$$\cos x(2\cos x - 1) = \frac{1}{\cos x} \cdot \cos x$$

$$2\cos^2 x - \cos x = 1$$

We are left with a quadratic equation, which we write in standard form and then solve.

$$2\cos^2 x - \cos x - 1 = 0 \qquad \text{Standard form}$$

$$(2\cos x + 1)(\cos x - 1) = 0 \qquad \text{Factor}$$

$$2\cos x + 1 = 0 \quad \text{or} \quad \cos x - 1 = 0 \qquad \text{Set each factor to 0}$$

$$\cos x = -\frac{1}{2} \qquad \text{or} \qquad \cos x = 1$$

$$x = 2\pi/3, 4\pi/3 \quad \text{or} \quad x = 0, 2\pi \qquad \blacktriangle$$

▼ **Example 7**   Solve $\sin 2\theta + \sqrt{2}\cos \theta = 0$, if $0° \leq \theta \leq 360°$.

Solution   In order to solve this equation, both trigonometric functions must be functions of the same angle. As the equation stands now, one angle is 2θ, while the other is θ. We can write everything as a function of θ by using the double-angle identity $\sin 2\theta = 2\sin \theta \cos \theta$.

$$\sin 2\theta + \sqrt{2}\cos \theta = 0$$

$$2\sin \theta \cos \theta + \sqrt{2}\cos \theta = 0 \qquad \text{Identity}$$

$$\cos \theta(2\sin \theta + \sqrt{2}) = 0 \qquad \text{Factor out } \cos \theta$$

$$\cos \theta = 0 \quad \text{or} \quad 2\sin \theta + \sqrt{2} = 0 \qquad \text{Set each factor to 0}$$

$$\sin \theta = -\frac{\sqrt{2}}{2}$$

$$\theta = 90°, 270° \quad \text{or} \quad \theta = 225°, 315° \qquad \blacktriangle$$

▼ **Example 8**   Solve $\cos 2\theta + 3\sin \theta - 2 = 0$, if $0° \leq \theta \leq 360°$.

Solution   We have the same problem with this equation as we did with the equation in Example 7. We must rewrite $\cos 2\theta$ in terms of functions of just θ. Recall that there are three forms of the double-angle identity for $\cos 2\theta$. We choose the double-angle identity that involves $\sin \theta$ only, since the middle term of our equation contains $\sin \theta$ and we want all the terms to involve the same trigonometric function.

$$\cos 2\theta + 3\sin \theta - 2 = 0$$

$1 - 2\sin^2 \theta + 3\sin \theta - 2 = 0$	Identity
$-2\sin^2 \theta + 3\sin \theta - 1 = 0$	Simplify
$2\sin^2 \theta - 3\sin \theta + 1 = 0$	Multiply each side by -1
$(2\sin \theta - 1)(\sin \theta - 1) = 0$	Factor
$2\sin \theta - 1 = 0$ or $\sin \theta - 1 = 0$	Set factors to 0

$$\sin \theta = \frac{1}{2} \qquad\qquad \sin \theta = 1$$

$$\theta = 30°, 150° \quad \text{or} \quad \theta = 90° \qquad\qquad \blacktriangle$$

▼   **Example 9**   Solve $4\cos^2 x + 4\sin x - 5 = 0$, if $0 \le x \le 2\pi$.

Solution   We cannot factor and solve this quadratic equation until each term involves the same trigonometric function. If we replace the $\cos^2 x$ in the first term with $1 - \sin^2 x$, we will obtain an equation that involves the sine function only.

$4\cos^2 x + 4\sin x - 5 = 0$	
$4(1 - \sin^2 x) + 4\sin x - 5 = 0$	$\cos^2 x = 1 - \sin^2 x$
$4 - 4\sin^2 x + 4\sin x - 5 = 0$	Distributive property
$-4\sin^2 x + 4\sin x - 1 = 0$	Add 4 and -5
$4\sin^2 x - 4\sin x + 1 = 0$	Multiply each side by -1
$(2\sin x - 1)^2 = 0$	Factor
$2\sin x - 1 = 0$	Set factor to 0

$$\sin x = \frac{1}{2}$$

$$x = \frac{\pi}{6}, \frac{5\pi}{6} \qquad\qquad \blacktriangle$$

Problem Set 14.4   Find all solutions between 0° and 360°. Do not use a calculator or table. (When we ask for solutions between 0° and 360°, we are asking for all solutions θ for which $0° \le \theta \le 360°$.)

1. $2\sin \theta = 1$ 2. $2\cos \theta = 1$

3. $2\cos\theta - \sqrt{3} = 0$ **4.** $2\cos\theta + \sqrt{3} = 0$

Find all solutions between 0 and 2π. Do not use a calculator or table.
5. $4\sin x - \sqrt{3} = 2\sin x$ **6.** $\sqrt{3} + 5\sin x = 3\sin x$
7. $3\sin x - 5 = -2\sin x$ **8.** $3\sin x + 4 = 4$

Find all solutions between 0° and 360°.
9. $(\sin\theta - 1)(2\sin\theta - 1) = 0$ **10.** $(\cos\theta - 1)(2\cos\theta - 1) = 0$
11. $\sin\theta - 2\sin\theta\cos\theta = 0$ **12.** $\cos\theta - 2\sin\theta\cos\theta = 0$

Find all solutions between 0 and 2π.
13. $2\sin^2\theta - \sin\theta - 1 = 0$ **14.** $2\cos^2\theta + \cos\theta - 1 = 0$
15. $2\cos^2 x - 11\cos x = -5$ **16.** $2\sin^2 x - 7\sin x = -3$

Find all solutions between 0° and 360°. Use Appendix E to write your answers to the nearest tenth of a degree.
17. $3\sin\theta - 4 = 0$ **18.** $3\sin\theta + 4 = 0$
19. $2\cos\theta - 5 = -3\cos\theta - 2$ **20.** $-2\sin\theta - 1 = 3\sin\theta + 2$
21. $2\sin^2\theta - 2\sin\theta - 1 = 0$ **22.** $2\cos^2\theta + 2\cos\theta - 1 = 0$

Find all solutions between 0° and 360°.
23. $\sqrt{3}\sec\theta = 2$ **24.** $\sqrt{2}\csc\theta = 2$
25. $4\sin x - 2\csc x = 0$ **26.** $4\cos x - 3\sec x = 0$
27. $\sin 2\theta - \cos\theta = 0$ **28.** $\sqrt{2}\sin\theta + \sin 2\theta = 0$
29. $\cos 2\theta - 3\sin\theta - 2 = 0$ **30.** $\cos 2\theta - \cos\theta - 2 = 0$
31. $2\cos^2\theta + \sin\theta - 1 = 0$ **32.** $2\sin^2\theta - \cos\theta - 1 = 0$

If a projectile (such as a bullet) is fired into the air with an initial velocity of v at an angle of inclination θ, then the height h of the projectile at time t is given by

$$h = -16t^2 + vt\sin\theta$$

33. Give the equation for the height if v is 1500 ft/sec and θ is 30°.
34. Use the equation found in Problem 33 to find the height of the object after 2 seconds.
35. In the human body, the value of θ that makes the expression below zero is the angle at which an artery of radius r will branch off from a larger artery of radius R in order to minimize the energy loss due to friction. Show that the expression below is 0 when $\cos\theta = r^4/R^4$.

$$r^4\csc^2\theta - R^4\csc\theta\cot\theta$$

36. Find the value of θ that makes the equation in Problem 35 equal to 0, if $r = 2$mm and $R = 4$mm. (Give your answer to the nearest tenth of a degree.)

Review Problems   The problems that follow review material we covered in Section 13.4.

Graph one complete cycle.

37. $y = \sin\left(x + \dfrac{\pi}{4}\right)$

38. $y = \sin\left(x - \dfrac{\pi}{4}\right)$

39. $y = 3\sin\left(2x - \dfrac{\pi}{2}\right)$

40. $y = 3\sin\left(2x + \dfrac{\pi}{2}\right)$

Chapter 14 Summary and Review

BASIC IDENTITITES [14.1]

	Basic identities	Common equivalent forms
Reciprocal	$\csc \theta = \dfrac{1}{\sin \theta}$	$\sin \theta = \dfrac{1}{\csc \theta}$
	$\sec \theta = \dfrac{1}{\cos \theta}$	$\cos \theta = \dfrac{1}{\sec \theta}$
	$\cot \theta = \dfrac{1}{\tan \theta}$	$\tan \theta = \dfrac{1}{\cot \theta}$
Ratio	$\tan \theta = \dfrac{\sin \theta}{\cos \theta}$	
	$\cot \theta = \dfrac{\cos \theta}{\sin \theta}$	
Pythagorean	$\cos^2 \theta + \sin^2 \theta = 1$	$\sin^2 \theta = 1 - \cos^2 \theta$ $\sin \theta = \pm \sqrt{1 - \cos^2 \theta}$ $\cos^2 \theta = 1 - \sin^2 \theta$ $\cos \theta = \pm \sqrt{1 - \sin^2 \theta}$

PROVING IDENTITIES [14.1]

An identity in mathematics is a statement that two expressions are equal for all replacements of the variable for which each statement is defined. To prove a trigonometric identity, we use trigonometric substitutions and algebraic manipulations to either

1. Transform the right side into the left side, or

2. Transform the left side into the right side, or

3. Transform each side into an equivalent third expression.

Remember to work on each side separately. We do not want to use properties from algebra that involve both sides of the identity—like the addition property of equality.

Examples

1. To prove $\tan x + \cos x$ $= \sin x(\sec x + \cot x)$, we apply the distributive property to the right side and then change to sines and cosines.

Proof

$$\tan x + \cos x = \sin x(\sec x + \cot x)$$

$$\left|\; \sin x \sec x + \sin x \cot x \right.$$

$$\left|\; \sin x \cdot \frac{1}{\cos x} + \sin x \cdot \frac{\cos x}{\sin x} \right.$$

$$\left|\; \frac{\sin x}{\cos x} + \cos x \right.$$

$$\tan x + \cos x = \tan x + \cos x$$

SUM AND DIFFERENCE FORMULAS [14.2]

$$\sin(A + B) = \sin A \cos B + \cos A \sin B$$

$$\sin(A - B) = \sin A \cos B - \cos A \sin B$$

$$\cos(A + B) = \cos A \cos B - \sin A \sin B$$

$$\cos(A - B) = \cos A \cos B + \sin A \sin B$$

$$\tan(A + B) = \frac{\tan A + \tan B}{1 - \tan A \tan B}$$

$$\tan(A - B) = \frac{\tan A - \tan B}{1 + \tan A \tan B}$$

2. a. Prove $\sin(x + \pi) = -\sin x$
Proof
$$\sin(x + \pi) = \sin x \cos \pi + \cos x \sin \pi$$
$$= \sin x(-1) + \cos x(0)$$
$$= -\sin x$$

b. To find the exact value for $\cos 75°$, we write $75°$ as $45° + 30°$ and then apply the formula for $\cos(A + B)$.

$$\cos 75°$$
$$= \cos(45° + 30°)$$
$$= \cos 45° \cos 30° - \sin 45° \sin 30°$$
$$= \frac{\sqrt{2}}{2} \cdot \frac{\sqrt{3}}{2} - \frac{\sqrt{2}}{2} \cdot \frac{1}{2}$$
$$= \frac{\sqrt{6} - \sqrt{2}}{4}$$

3. a.   If $\sin A = \frac{3}{5}$ with A in QII, then

$\cos 2A = 1 - 2\sin^2 A$

$$= 1 - 2\left(\frac{3}{5}\right)^2 = \frac{7}{25}$$

b.   Prove $\tan \theta = \dfrac{1 - \cos 2\theta}{\sin 2\theta}$

Proof

$\tan \theta = \dfrac{1 - \cos 2\theta}{\sin 2\theta}$

$\quad \left| \quad \dfrac{1 - (1 - 2\sin^2 \theta)}{2\sin \theta \cos \theta} \right.$ Double-angle formulas

$\quad \left| \quad \dfrac{2\sin^2 \theta}{2\sin \theta \cos \theta} \right.$ Simplify numerator

$\quad \left| \quad \dfrac{\sin \theta}{\cos \theta} \right.$ Divide out common factor

$\tan \theta = \tan \theta$ Ratio identity

DOUBLE-ANGLE FORMULAS [14.3]

$$\sin 2A = 2\sin A \cos A$$

$$\cos 2A = \cos^2 A - \sin^2 A \qquad \text{First form}$$

$$= 2\cos^2 A - 1 \qquad \text{Second form}$$

$$= 1 - 2\sin^2 A \qquad \text{Third form}$$

HALF-ANGLE FORMULAS [14.3]

$$\sin \frac{A}{2} = \pm \sqrt{\frac{1 - \cos A}{2}}$$

$$\cos \frac{A}{2} = \pm \sqrt{\frac{1 + \cos A}{2}}$$

4.   Solve for x: $2\cos x - \sqrt{3} = 0$, if $0° \leq x \leq 360°$.

$2\cos x - \sqrt{3} = 0$

$\quad 2\cos x = \sqrt{3}$

$\quad \cos x = \dfrac{\sqrt{3}}{2}$

$\quad x = 30°, 330°$

SOLVING SIMPLE TRIGONOMETRIC EQUATIONS [14.4]

We solve linear equations in trigonometry by applying the properties of equality developed in algebra. The two most important properties from algebra are the addition property of equality and the multiplication property of equality.

To solve a trigonometric equation that is quadratic in $\sin x$ or $\cos x$, we write it in standard form and either factor or use the quadratic formula.

USING IDENTITIES IN TRIGONOMETRIC EQUATIONS [14.4]

Sometimes it is necessary to use identities to make trigono-metric substitutions when solving equations. Identities are usually required if the equation contains more than one trig-onometric function; for instance, if the equation contains both sines and cosines. Identities are also required if there is more than one angle named in the equation; for example, if one term contains $\sin 2x$ and another contains $\sin x$.

5. Solve $\cos 2\theta + 3 \sin \theta - 2 = 0$, if $0° \le \theta \le 360°$.

$$\cos 2\theta + 3\sin \theta - 2 = 0$$
$$1 - 2\sin^2 \theta + 3\sin \theta - 2 = 0$$
$$-2\sin^2 \theta + 3\sin \theta - 1 = 0$$
$$2\sin^2 \theta - 3\sin \theta + 1 = 0$$
$$(2\sin \theta - 1)(\sin \theta - 1) = 0$$
$$2\sin \theta - 1 = 0 \quad \text{or} \quad \sin \theta - 1 = 0$$
$$\sin \theta = \frac{1}{2} \quad \text{or} \quad \sin \theta = 1$$
$$\theta = 30°, 150° \quad \text{or} \quad \theta = 90°$$

Chapter 14 Test

Prove each identity.

1. $\tan \theta = \sin \theta \sec \theta$

2. $\dfrac{\cot \theta}{\csc \theta} = \cos \theta$

3. $(1 - \sin x)(1 + \sin x) = \cos^2 x$
4. $(\sec x - 1)(\sec x + 1) = \tan^2 x$
5. $\sec \theta - \cos \theta = \tan \theta \sin \theta$

6. $\cot \theta \cos \theta + \sin \theta = \csc \theta$

7. $\dfrac{\cos t}{1 - \sin t} = \dfrac{1 + \sin t}{\cos t}$

8. $\dfrac{1}{1 - \sin t} + \dfrac{1}{1 + \sin t} = 2\sec^2 t$

9. $\sin(\theta - 90°) = -\cos \theta$

10. $\cos\left(\dfrac{\pi}{2} + \theta\right) = -\sin \theta$

11. $\cos^4 A - \sin^4 A = \cos 2A$

12. $\cot A = \dfrac{\sin 2A}{1 - \cos 2A}$

Let $\sin A = -\frac{3}{5}$ with A in QIV and $\sin B = \frac{12}{13}$ with B in QII and find
13. $\sin(A + B)$
14. $\cos(A - B)$
15. $\cos 2B$
16. $\sin 2B$
17. $\tan 2B$
18. $\cot 2B$
19. $\sin \dfrac{A}{2}$
20. $\cos \dfrac{A}{2}$

Find exact values for each of the following:
21. $\sin 75°$
22. $\cos 15°$

23. $\tan\dfrac{\pi}{12}$ $\qquad\qquad\qquad\qquad\qquad$ **24.** $\cot\dfrac{\pi}{12}$

Write each expression as a single trigonometric function.
25. $\cos 4x \cos 5x - \sin 4x \sin 5x$
26. $\sin 15° \cos 75° + \cos 15° \sin 75°$

27. If $\sin A = -1/\sqrt{5}$ with A in QIII, find $\cos 2A$ and $\cos\dfrac{A}{2}$.

28. If $\sec A = \sqrt{10}$ with A in QI, find $\sin 2A$ and $\sin\dfrac{A}{2}$.

Find all solutions between $0°$ and $360°$. (*Remember* This includes the solutions of $0°$ and $360°$.)

29. $2\sin\theta - 1 = 0$ $\qquad\qquad\qquad$ **30.** $\sqrt{3}\tan\theta + 1 = 0$
31. $\cos\theta - 2\sin\theta\cos\theta = 0$ $\qquad$ **32.** $\tan\theta - 2\cos\theta\tan\theta = 0$
33. $4\sin^2 t - 4\sin t = -1$ $\qquad\qquad$ **34.** $2\cos^2 t = \cos t + 1$

Find all solutions between 0 and 2π.

35. $4\cos\theta - 2\sec\theta = 0$ $\qquad\qquad$ **36.** $2\sin\theta - \csc\theta = 1$
37. $\sin 2x + \cos x = 0$ $\qquad\qquad\qquad$ **38.** $\cos 2x - \cos x = 0$

Use Appendix E to find all solutions between $0°$ and $360°$ to the nearest tenth of a degree.

39. $\sin^2\theta - 4\sin\theta + 2 = 0$ $\qquad\qquad$ **40.** $\cos^2\theta + 4\cos\theta - 2 = 0$

Triangles

To the student:

We will begin this chapter with a second definition for the six trigonometric functions of an acute angle. This new definition will not conflict with our first definition, but will allow us to apply our knowledge of the trigonometric functions to triangles. We will then use this new definition, along with the Pythagorean theorem and identity, to derive two new formulas that are also important in working with triangles. These two new formulas are called the *law of sines* and the *law of cosines*.

Most of this chapter is concerned with solving triangles, beginning with solutions to right triangles. We solve a triangle by using the information that we are given about it to find the missing sides and angles. There are many application problems with solutions that depend upon your ability to solve triangles.

To be successful in this chapter you should have a good working knowledge of the material we covered in Chapter 13. You need to be proficient at using Appendix E or a calculator to find the value of a trigonometric function or to find an angle. To understand some of the derivations in this chapter you need to know the Pythagorean identity.

15.1
Definition II: Right
Triangle Trigonometry

In Chapter 13, we gave a definition for the six trigonometric functions for any angle in standard position. In this section we will give a second definition for the six trigonometric functions in terms of the sides and angles in a right triangle.

DEFINITION II   If triangle ABC is a right triangle with $C = 90°$, then the six trigonometric functions for A are defined as follows:

$$\sin A = \frac{\text{side opposite } A}{\text{hypotenuse}} = \frac{a}{c}$$

$$\cos A = \frac{\text{side adjacent } A}{\text{hypotenuse}} = \frac{b}{c}$$

$$\tan A = \frac{\text{side opposite } A}{\text{side adjacent } A} = \frac{a}{b}$$

$$\cot A = \frac{\text{side adjacent } A}{\text{side opposite } A} = \frac{b}{a}$$

$$\sec A = \frac{\text{hypotenuse}}{\text{side adjacent } A} = \frac{c}{b}$$

$$\csc A = \frac{\text{hypotenuse}}{\text{side opposite } A} = \frac{c}{a}$$

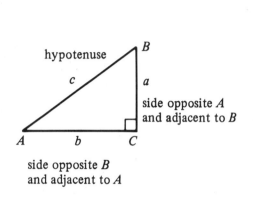

Figure 15-1

▼   **Example 1**   Triangle ABC is a right triangle with $C = 90°$. If $a = 6$ and $c = 10$, find the six trigonometric functions of A.

Solution   We begin by making a diagram of ABC (Figure 15-2), and then use the given information and the Pythagorean theorem to solve for b.

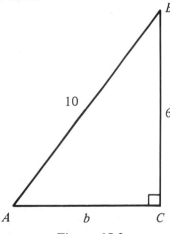

$$b = \sqrt{c^2 - a^2}$$
$$= \sqrt{100 - 36}$$
$$= \sqrt{64}$$
$$= 8$$

Figure 15-2

Now we write the six trigonometric functions of A using $a = 6$, $b = 8$, and $c = 10$.

$$\sin A = \frac{a}{c} = \frac{6}{10} = \frac{3}{5} \qquad \csc A = \frac{c}{a} = \frac{5}{3}$$

$$\cos A = \frac{b}{c} = \frac{8}{10} = \frac{4}{5} \qquad \sec A = \frac{c}{b} = \frac{5}{4}$$

$$\tan A = \frac{a}{b} = \frac{6}{8} = \frac{3}{4} \qquad \cot A = \frac{b}{a} = \frac{4}{3} \qquad \blacktriangle$$

Now that we have done an example using our new definition, let's see how our new definition compares to Definition I from Chapter 13. We can place right triangle ABC on a rectangular coordinate system so that A is in standard position (Figure 15-3). We then note that a point on the terminal side of A is (b, a).

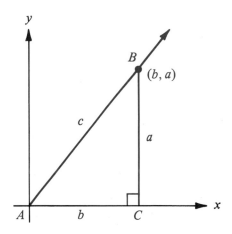

Figure 15-3

From Definition I in Chapter 13, we have

$$\sin A = \frac{a}{c}$$

$$\cos A = \frac{b}{c}$$

$$\tan A = \frac{a}{b}$$

From Definition II in this chapter we have

$$\sin A = \frac{a}{c}$$

$$\cos A = \frac{b}{c}$$

$$\tan A = \frac{a}{b}$$

The two definitions agree as long as A is an acute angle. If A is not an acute angle, then Definition II does not apply since, in right triangle ABC, A must be an acute angle.

Solving Right Triangles   We are now ready to use Definition II, along with the table in Appendix E (or calculators), to solve some right triangles. We solve a right triangle by using the information given about it to find all the missing sides and angles. In Example 2, we are given the values of one side and one of the acute angles and asked to find the remaining two sides and the other acute angle. In all the examples and the problem set that follows, let's assume C is the right angle in all our right triangles.

Note   When we solve for the missing parts of a triangle we will round our answers to the same accuracy as the given information. That is, if the angles we are given are accurate to the nearest tenth of a degree, then if the angles we solve for need to be rounded, we will round them to the nearest tenth of a degree. If the sides we are given all have two digit accuracy, then any sides we solve for should be rounded the same accuracy.

▼   **Example 2**   In right triangle ABC, $A = 40°$ and $c = 12$ cm. Find a, b, and B.

   Solution   We begin by making a diagram of the situation.

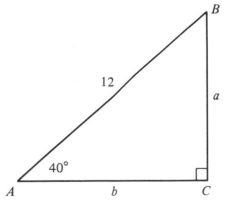

Figure 15-4

To find B, we use the fact that the sum of the two acute angles in any right triangle is $90°$.

$$B = 90° - A$$
$$= 90° - 40°$$
$$= 50°$$

To find a, we can use the formula for $\sin A$.

$$\sin A = \frac{a}{c}$$

Multiplying both sides of this formula by c and then substituting in our given values for A and c we have

$$a = c \sin A$$

$$= 12\sin 40°$$

$$= 12(0.6428) \qquad \sin 40° = 0.6428$$

$$= 7.7 \qquad\qquad \text{Round to two digit accuracy}$$

There is more than one way to find b.

Using $\cos A = \dfrac{b}{c}$ Using the Pythagorean theorem
we have   we have

$$b = c \cos A \qquad\qquad c^2 = a^2 + b^2$$

$$= 12\cos 40° \qquad\qquad b = \sqrt{c^2 - a^2}$$

$$= 12(0.7660) \qquad\qquad = \sqrt{12^2 - (7.7)^2}$$

$$= 9.2 \qquad\qquad\qquad = \sqrt{144 - 59.29}$$

$$= \sqrt{84.71}$$

$$= 9.2 \qquad\qquad\qquad\qquad \blacktriangle$$

In Example 3, we are given two sides and asked to find the remaining parts of a right triangle.

▼   **Example 3**   In right triangle ABC $a = 2.73$ and $b = 3.41$. Find the remaining side and angles.

Solution   Here is a diagram of the triangle.

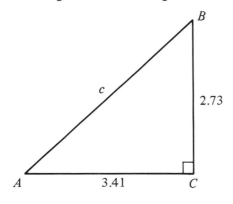

Figure 15-5

We can find A by using the formula for $\tan A$

$$\tan A = \frac{a}{b}$$

$$= \frac{2.73}{3.41}$$

$$= 0.8006$$

Now, to find A, we look for the angle whose tangent is 0.8006 using our table or a calculator.

$$A = 38.7°$$

Next we find B.

$$B = 90° - A$$

$$= 90° - 38.7°$$

$$= 51.3°$$

We can find c using the Pythagorean theorem or one of our trigonometric functions. Let's use a trigonometric function.

$$\text{If } \sin A = \frac{a}{c}$$

$$\text{then} \quad c = \frac{a}{\sin A}$$

$$= \frac{2.73}{\sin 38.7°}$$

$$= \frac{2.73}{0.6252}$$

$$= 4.37 \qquad \blacktriangle$$

We are now ready to put our knowledge of solving right triangles to work to solve some application problems.

▼   **Example 4**   The two equal sides of an isosceles triangle are each 24 cm. If each of the two equal measures 52°, find the length of the base and the length of the altitude.

Solution   An isosceles triangle is any triangle with two equal sides. The angles opposite the two equal sides are called the base angles,

and they are always equal. Figure 15-6 is a picture of our isosceles triangle.

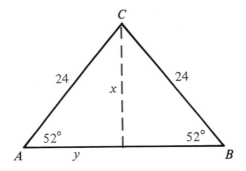

Figure 15-6

We have labeled the altitude x. We can solve for x using a sine ratio.

If
$$\sin 52° = \frac{x}{24}$$

then
$$x = 24 \sin 52°$$
$$= 24(0.7880)$$
$$= 19 \text{ cm}$$

We have labeled half the base with y. To solve for y, we can use a cosine ratio.

If
$$\cos 52° = \frac{y}{24}$$

then
$$y = 24 \cos 52°$$
$$= 24(0.6157)$$
$$= 15 \text{ cm}$$

The base is
$$2y = 2(15) = 30 \text{ cm.}$$
▲

For our next applications, we need the following definition.

DEFINITION   An angle measured from the horizontal up is called an *angle of elevation* (*or inclination*). An angle measured from the horizontal down is called an *angle of depression* (see Figure 15-7).

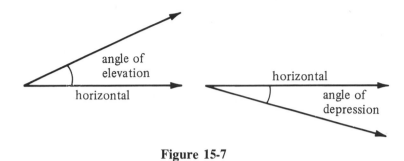

Figure 15-7

These angles of elevation and depression are always considered positive angles.

▼   **Example 5**   If a 75 ft flag pole casts a shadow 43 ft long, what is the angle of elevation of the sun from the tip of the shadow?

Solution   We begin by making a diagram of the situation (Figure 15-8.

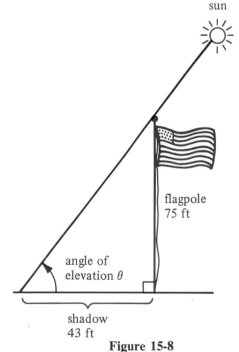

Figure 15-8

If we let θ — the angle of elevation of the sun then

$$\tan \theta = \frac{75}{43}$$

$$\tan \theta = 1.744$$

which means $\theta = 60°$ To the nearest degree   ▲

▼   **Example 6**   A man climbs 213 m up the side of a pyramid and finds the angle of depression to his starting point is 52.6°. How high off the ground is he?

Solution   Again, we begin by making a diagram of the situation (Figure 15-9).

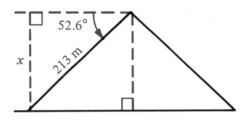

Figure 15-9

If x is his height above the ground, we can solve for x using a sine ratio.

If $\sin 52.6° = \dfrac{x}{213}$

then $x = 213 \sin 52.6°$

$$= 213(0.7944)$$

$$= 169 \text{ m}$$

The man is 169 m above the ground.   ▲

___

Problems 1 through 6 refer to right triangle ABC with $C = 90°$. In each case, use the given information to find the six trigonometric functions of A.

Problem Set 15.1

1. $b = 3, c = 5$ 2. $b = 5, c = 13$
3. $a = 2, b = 1$ 4. $a = 3, b = 2$
5. $a = 2, c = 4$ 6. $a = 3, c = 6$

In each triangle below, find sin A, cos A, sin B, and cos B.

7.

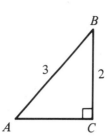

8.

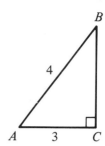

9.

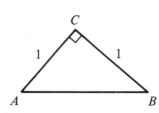

10.

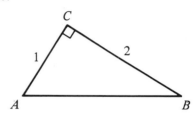

11.

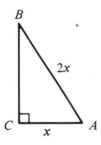

12.

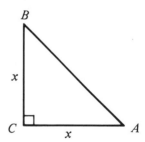

Problems 13 through 20 refer to right triangle ABC with $C = 90°$. Write all answers that are angles to the nearest tenth of a degree.

13. If $A = 40°$ and $c = 16$, find a.   **14.** If $A = 20°$ and $c = 42$, find a.
15. If $B = 20°$ and $b = 12$, find a.   **16.** If $B = 20°$ and $a = 35$, find b.
17. If $a = 5$ and $b = 10$, find A.   **18.** If $a = 6$ and $b = 10$, find B.
19. If $a = 12$ and $c = 20$, find B.   **20.** If $a = 15$ and $c = 25$, find A.

Problems 21 through 28 refer to right triangle ABC with $C = 90°$. In each case, solve for all the missing parts using the given information.

21. $A = 20°, c = 24$ **22.** $A = 40°, c = 36$
23. $B = 76°, c = 5.8$ **24.** $B = 21°, c = 4.2$

25. $a = 35, b = 97$ **26.** $a = 78, b = 83$
27. $b = 150, c = 200$ **28.** $b = 320, c = 650$

In Problems 28 through 31, use the information given in the diagrams to find angle A.

29.

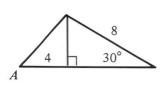

30.

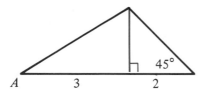

31.

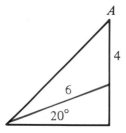

32.

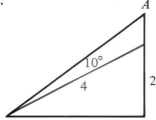

Solve each of the following problems. In each case, be sure to make a diagram of the situation with all the given information labeled.

33. The two equal sides of an isosceles triangle are each 42 cm. If the base measures 30 cm, find the height and the measure of the two equal angles. Round your answers to the nearest tenth.

34. An equilateral triangle (one with all sides the same length) has an altitude of 4.3 in. Find the length of the sides.

35. How long should an escalator be if it is to make an angle of 33° with the floor and carry people a vertical distance of 21 ft between floors? Round to the nearest tenth.

36. A road up a hill makes an angle of 5° with the horizontal. If the road from the bottom of the hill to the top of the hill is 2.5 mi long, how high is the hill?

37. A 72.5 ft rope from the top of a circus tent pole is anchored to the ground 43.2 ft from the bottom of the pole. What angle does the rope make with the pole? (Give your answer to the nearest tenth of a degree.)

38. A ladder is leaning against the top of a 7 ft wall. If the bottom of the ladder is 4.5 ft from the wall, what is the angle between the ladder and the wall? (To the nearest tenth of a degree.)

39. If a 73 ft flag pole casts a shadow 51 ft long, what is the angle of elevation of the sun? (To the nearest tenth of a degree.)

40. If the angle of inclination of the sun is 63.4° when a building casts a shadow of 37.5 ft, what is the height of the building?

Review Problems   The problems below review material we covered in Section 14.1.

Verify each identity.

41. $\tan \theta \cot \theta = 1$

42. $\sec \theta \cot \theta = \csc \theta$

43. $(1 - \sin x)(1 + \sin x) = \cos^2 x$

44. $1 + \tan^2 x = \sec^2 x$

45. $1 + \cos A = \dfrac{\sin^2 A}{1 - \cos A}$

46. $\dfrac{\cos A}{1 + \sin A} = \dfrac{1 - \sin A}{\cos A}$

15.2 The Law of Sines

There are many relationships that exist between the sides and angles in a triangle. One such relationship is called *the law of sines* and it states that the ratio of the sine of an angle to the length of the side opposite that angle is constant in any triangle. Here it is stated in symbols

Law of sines

$$\frac{\sin A}{a} = \frac{\sin B}{b} = \frac{\sin C}{c}$$

or, equivalently

$$\frac{a}{\sin A} = \frac{b}{\sin B} = \frac{c}{\sin C}$$

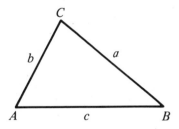

Figure 15-10

Proof   The altitude h of the triangle in Figure 15-11 can be written in terms of $\sin A$ or $\sin B$ depending on which of the two right triangles were are referring to

$$\sin A = \frac{h}{b} \qquad \sin B = \frac{h}{a}$$

$$h = b\sin A \qquad h = a\sin B$$

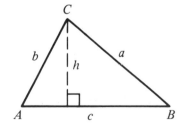

Figure 15-11

since h is equal to itself, we have

$$h = h$$

$$b\sin A = a\sin B$$

$$\frac{b\sin A}{ab} = \frac{a\sin B}{ab}$$ Divide both sides
by ab

$$\frac{\sin A}{a} = \frac{\sin B}{b}$$ Divide out common
factors

If we do the same kind of thing with the altitude that extends from A we will have the third ratio in the law of sines, $\sin C/c$, equal to the two ratios above.

We can use the law of sines to find missing parts of triangles in which we are given two angles and a side.

In our first example, we are given two angles and the side opposite one of them. (You may recall that, in geometry, these were the parts we needed equal in two triangles in order to prove them congruent using the AAS theorem.)   *Two Angles*
and One side

▼   **Example 1**   In triangle ABC, $A = 30°$, $B = 70°$, and $a = 8$ cm. Use the law of sines to find c.

  Solution   We begin by drawing a picture of triangle ABC (it does not have to be accurate) and labeling it so that the information we are given is showing.

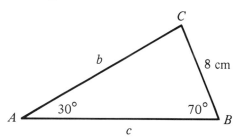

Figure 15-12

When we use the law of sines, we must have one of the ratios given to us. In this case, since we are given a and A, we have the ratio $a/\sin A$. To solve for c we need to first find C. Since the sum of the angles in any triangle is 180°, we have

$$C = 180° - (A + B)$$
$$= 180° - (30° + 70°)$$
$$= 80°$$

To find c, we use the following two ratios given in the law of sines.

$$\frac{c}{\sin C} = \frac{a}{\sin A}$$

To solve for c, we multiply both sides by $\sin C$ and then substitute

$$c = \frac{a\sin C}{\sin A} \qquad \text{Multiply both sides by } \sin C$$

$$= \frac{8\sin 80°}{\sin 30°} \qquad \text{Substitute in given values}$$

$$= \frac{8(0.9848)}{0.5000} \qquad \text{Table or calculator}$$

$$= 16 \text{ cm} \qquad \text{Rounded to the nearest integer} \qquad \blacktriangle$$

In our next example we are given two angles and the side included between them (ASA) and asked to find all the missing parts.

▼   **Example 2**   Solve triangle ABC if $B = 34°$, $C = 82°$, and $a. = 5.6$ cm.

Solution   We begin by finding angle A so that we have one of the ratios in the law of sines completed.

Angle A

$$A = 180° - (B + C)$$
$$= 180° - (34° + 82°)$$
$$= 64°$$

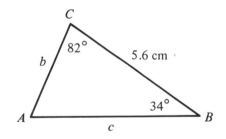

Figure 15-13

Side b

If $\qquad \dfrac{b}{\sin B} = \dfrac{a}{\sin A}$

then $\qquad b = \dfrac{a\sin B}{\sin A} \qquad$ Multiply both sides by $\sin B$

$$= \frac{5.6\sin 34°}{\sin 64°} \qquad \text{Substitute in given values}$$

$$= \frac{5.6(0.5592)}{0.8988} \qquad \text{Table or calculator}$$

$$= 3.5 \text{ cm} \qquad \text{To the nearest tenth}$$

Side c

If $\qquad \dfrac{c}{\sin C} = \dfrac{a}{\sin A}$

then $\qquad c = \dfrac{a\sin C}{\sin A} \qquad$ Multiply both sides by $\sin C$

$$= \frac{5.6\sin 82°}{\sin 64°} \qquad \text{Substitute in given values}$$

$$= \frac{5.6(0.9903)}{0.8988} \qquad \text{Table or calculator}$$

$$= 6.2 \text{ cm} \qquad \text{To the nearest tenth} \qquad \blacktriangle$$

The law of sines, along with some fancy electronic equipment, was used to obtain the results of some of the field events in one of the recent Olympic Games.

Figure 15-14 is a diagram of a shot put ring. The shot is tossed (put) from the left and lands at A. A small electronic device is then placed at A (there is usually a dent in the ground where the shot lands, so it is easy to find where to place the device). The device at A sends a signal to a booth in the stands that gives the measures of angles A and B. The distance a is found ahead of time. To find the distance x, the law of sines is used.

$$\frac{x}{\sin B} = \frac{a}{\sin A}$$

or $\qquad\qquad x = \dfrac{a\sin B}{\sin A}$

▼ **Example 3** Find x in Figure 15-14 if $a = 562$ ft, $B = 5.7°$, and $A = 85.3°$.

Solution
$$x = \frac{a \sin B}{\sin A}$$

$$= \frac{562 \sin 5.7°}{\sin 85.3°}$$

$$= \frac{562(0.0933)}{(0.9966)}$$

$$= 56.0 \text{ ft} \qquad\qquad ▲$$

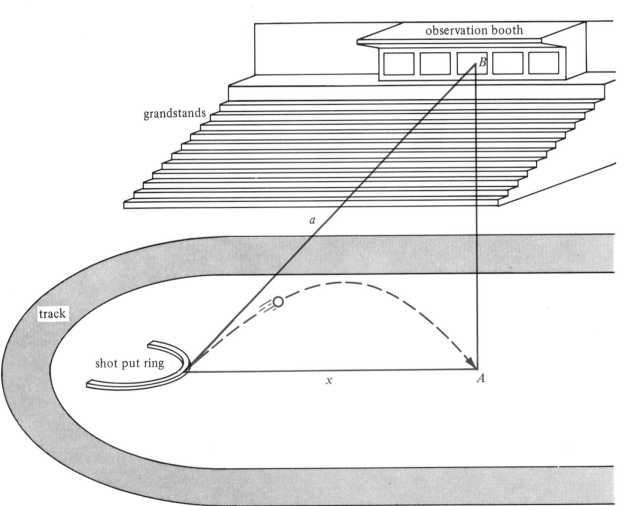

Figure 15-14

Each problem that follows refers to triangle ABC.

1. If $A = 40°$, $B = 60°$, and $a = 12$ cm, find b.
2. If $A = 80°$, $B = 30°$, and $b = 14$ cm, find a.
3. If $B = 120°$, $C = 20°$, and $c = 28$ in, find b.
4. If $B = 110°$, $C = 40°$, and $b = 18$ in, find c.
5. If $A = 10°$, $C = 100°$, and $a = 12$ yd, find c.
6. If $A = 5°$, $C = 125°$, and $c = 51$ yd, find a.
7. If $A = 50°$, $B = 60°$, and $a = 36$ km, find C and then find c.
8. If $B = 40°$, $C = 70°$, and $c = 42$ km, find A and then find a.
9. If $A = 52°$, $B = 48°$, and $c = 14$ cm, find C and then find a.
10. If $A = 33°$, $C = 87°$, and $b = 18$ cm, find B and then find c.

The information below refers to triangle ABC. In each case find all the missing parts.

11. $A = 42.5°$, $B = 71.4°$, $a = 210$ in
12. $A = 110.4°$, $C = 21.8°$, $c = 240$ in
13. $A = 46°$, $B = 95°$, $c = 6.8$ m
14. $B = 57°$, $C = 31°$, $a = 7.3$ m
15. $B = 13.4°$, $C = 24.8°$, $a = 315$ cm
16. $A = 105°$, $B = 45°$, $c = 630$ cm

17. In triangle ABC, $A = 30°$, $b = 20$ ft, and $a = 2$ ft. Show that it is impossible to solve this triangle by using the law of sines to find $\sin B$.

18. In triangle ABC, $A = 40°$, $b = 20$ ft, and $a = 18$ ft. Use the law of sines to find $\sin B$ and then give two possible values for angle B.

19. A man standing near a radio station antenna observes that the angle of inclination to the top of the antenna is 64°. He then walks 100 ft further away and observes the angle of inclination to the top of the antenna to be 46°. Find the height of the antenna to the nearest foot. (*Hint:* Find x first.)

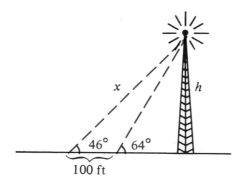

20. A woman standing on the street looks up to the top of a building and finds
the angle of inclination is 38°. She then walks one block further away (440
ft) and finds the angle of inclination to the top of the building is now 28°.
How far away from the building is she when she makes her second
observation?

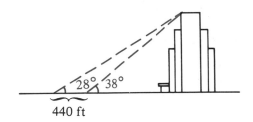

Review Problems   The problems below review material we covered in Section 14.2

21. Use the formula for sin(*A* + *B*) to find the exact value of sin 75°.
22. Use the formula for cos(*A* − *B*) to find the exact value of cos 15°.
23. Prove that sin(θ + 360°) = sin θ.
24. Prove that cos(π − *x*) = −cos *x*.

Let *A* and *B* be in quadrant *I* with sin *A* = 4/5 and sin *B* = 12/13 and find
25. sin(*A* + *B*)   26.   cos(*A* + *B*)
27. tan(*A* + *B*)   28.   cot(*A* + *B*)

**15.3
The Ambiguous Case**

In this section we will extend the law of sines to solve triangles in which
we are given two sides and the angle opposite one of the given sides.

▼   **Example 1**   Find angle *B* in triangle *ABC* if *a* = 2, *b* = 6, and
A = 30°.

Solution   Applying the law of sines we have

$$\sin B = \frac{b\sin A}{a}$$

$$= \frac{6\sin 30°}{2}$$

$$= \frac{6(0.5000)}{2}$$

$$= 1.5$$

Which is impossible. For any value of B, sin B is between -1 and 1. The function sin B can never be larger than 1. No triangle exists for which $a = 2$, $b = 6$, and $A = 30°$. (You may recall from geometry that there was no congruence theorem SSA.) Figure 15-15 illustrates what went wrong here.   ▲

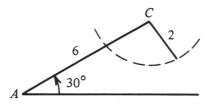

Figure 15-15

When we are given two sides and an angle opposite one of them (SSA), we have several possibilities for the triangle or triangles that result. As was the case in Example 1, one of the possibilities is that no triangle will fit the given information. Another possibility is that two different triangles can be obtained from the given information and a third possibility is that exactly one triangle will fit the given information. Because of these different possibilities, we call the situation where we are solving a triangle in which we are given two sides and the angle opposite one of them the *ambiguous case*.

▼   **Example 2**   Find the missing parts in triangle ABC if $a = 54$ cm, $b = 62$ cm, and $A = 40°$.

Solution   First we solve for sin B with the law of sines

Angle B

$$\sin B = \frac{b\sin A}{a}$$

$$= \frac{62\sin 40°}{54}$$

$$= 0.7380$$

Now, since sin B is positive for any angle in quadrant I or II, we have two possibilities. We will call one of them B and the other B'.

$$B = 48° \quad \text{or} \quad B' = 180° - 48° = 132°$$

We have two different triangles that can be found with $a = 54$, $b = 62$, and $A = 40°$. Figure 15-16 shows both of them. One of them is labeled ABC, while the other is labeled $AB'C$.

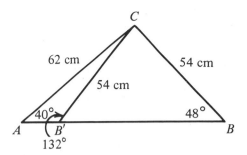

Figure 15-16

Angles C and C'

Since there are two values for B, we have two values for C.

$$C = 180 - (A + B) \qquad \text{and} \qquad C' = 180 - (A + B')$$

$$= 180 - (40° + 48°) \qquad\qquad\qquad = 180 - (40° + 132°)$$

$$= 92° \qquad\qquad\qquad\qquad\qquad = 8°$$

Sides c and c'

$$c = \frac{a\sin C}{\sin A} \qquad \text{and} \qquad c' = \frac{a\sin C'}{\sin A}$$

$$= \frac{54\sin 92°}{\sin 40°} \qquad\qquad\qquad = \frac{54\sin 8°}{\sin 40°}$$

$$= 84 \text{ cm} \qquad\qquad\qquad\quad = 12 \text{ cm}$$

Figure 15-17 shows both triangles.

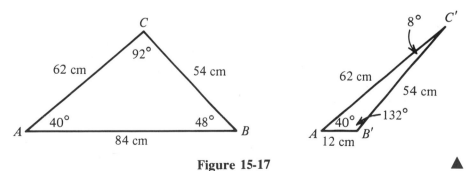

Figure 15-17   ▲

▼   **Example 3**   Find the missing parts of triangle ABC if $C = 35.4°$, $a = 205$ ft, and $c = 314$ ft.

Solution   Applying the law of sines, we find sin A.

Angle A

$$\sin A = \frac{a\sin C}{c}$$

$$= \frac{205\sin 35.4°}{314}$$

$$= 0.3782$$

Since sin A is positive in quadrants I and II we have two possible values for A.

$$A = 22.2° \qquad \text{and} \qquad A' = 180° - 22.2°$$

$$= 157.8°$$

The second possibility, $A' = 157.8°$, will not work, however, since C is already 35.4° and therefore

$$C + A' = 35.4° + 157.8°$$

$$= 193.2°$$

which is larger than 180°. This result indicates that there is exactly one triangle that fits the description given in Example 3. In that triangle

$$A = 22.2°$$

Angle B

$$B = 180° - (35.4° + 22.2°)$$

$$= 122.4°$$

Side b

$$b = \frac{c\sin B}{\sin C}$$

$$= \frac{314\sin 122.4°}{\sin 35.4°}$$

$$= 458 \text{ ft}$$

Figure 15-18 is a diagram of this triangle.

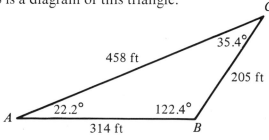

Figure 15-18

Although the different cases that can occur when we solve the kinds of triangles we have been given in Examples 1, 2, and 3 become apparent in the process of solving for the missing parts, we can make a table that shows the set of conditions under which we will have 1, 2, or no triangles in the ambiguous case.

In Table 15-1 we are assuming we are given angle A and sides a and b in triangle ABC, and that h is the altitude from C.

Conditions	Number of triangles	Diagram
$A < 90°$ and $a < h$	0	
$A > 90°$ and $a < b$	0	
$A < 90°$ and $a = h$	1	
$A < 90°$ and $a \geq b$	1	
$A > 90°$ and $a > b$	1	
$A < 90°$ and $h < a < b$	2	

Table 15-1

For each triangle below, solve for B and use the results to explain why the   ***Problem Set 15.3***
triangle has the given number of solutions.
  1. $A = 30°, b = 40$ ft, $a = 10$ ft; no solution
  2. $A = 150°, b = 30$ ft, $a = 10$ ft; no solution
  3. $A = 120°, b = 20$ cm, $a = 30$ cm; one solution
  4. $A = 30°, b = 12$ cm, $a = 6$ cm; one solution
  5. $A = 60°, b = 18$ m, $a = 16$ m; two solutions
  6. $A = 20°, b = 40$ m, $a = 30$ m; two solutions

Find all solutions to each of the following triangles:
  7. $A = 38°, a = 41$ ft, $b = 54$ ft
  8. $A = 43°, a = 31$ ft, $b = 37$ ft
  9. $A = 112.2°, a = 43$ cm, $b = 22$ cm
 10. $B = 30°, b = 42$ cm, $a = 84$ cm
 11. $B = 118°, b = 68$ cm, $a = 92$ cm
 12. $A = 124.3°\ a = 27$ cm, $b = 50$ cm
 13. $A = 142°, b = 2.9$ yd, $a = 1.4$ yd
 14. $A = 65°, b = 7.6$ yd, $a = 7.1$ yd
 15. $C = 26.8°, c = 36.8$ km, $b = 36.8$ km
 16. $C = 73.4°, c = 51.1$ km, $b = 92.4$ km

 17.   A 50 ft wire running from the top of a tent pole to the ground makes an
       angle of 58° with the ground. If the length of the tent pole is 44 ft, how
       far is it from the bottom of the tent pole to the point where the wire is
       fastened to the ground?
 18.   A hot air balloon is held at a constant altitude by two ropes that are
       anchored to the ground. One rope is 120 ft long and makes an angle of
       65° with the ground. The other rope is 115 ft long. What is the distance
       between the points on the ground at which the two ropes are anchored?

Review Problems   The problems below review material we covered in Section 14.3.

Let A terminate in quadrant I with $\sin A = 4/5$ and find
 19. $\sin 2A$ 20. $\cos 2A$

 21. $\sin \dfrac{A}{2}$ 22. $\cos \dfrac{A}{2}$

 23. $\tan 2A$ 24. $\tan \dfrac{A}{2}$

 25.   Use a half-angle formula to find the exact value of $\sin 15°$.
 26.   Use a half-angle formula to find the exact value of $\cos 15°$.

15.4
The Law of Cosines

In this section we will derive another relationship that exists between the sides and angles in any triangle. It is called *the law of cosines* and is stated like this

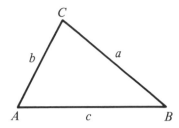

Law of Cosines

$$a^2 = b^2 + c^2 - 2bc\cos A$$

$$b^2 = a^2 + c^2 - 2ac\cos B$$

$$c^2 = a^2 + b^2 - 2ab\cos C$$

Figure 15-19

Derivation

To derive the formulas stated in the law of cosines, we apply the Pythagorean theorem and some of our basic trigonometric identities. Applying the Pythagorean theorem to right triangle BCD in Figure 15-20, we have

$$a^2 = (c - x)^2 + h^2$$

$$= c^2 - 2cx + x^2 + h^2$$

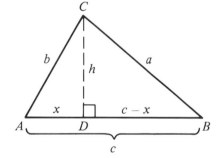

Figure 15-20

But, from right triangle ACD, we have $x^2 + h^2 = b^2$, so

$$a^2 = c^2 - 2cx + b^2$$

$$= b^2 + c^2 - 2cx$$

Now, since $\cos A = x/b$ we have $x = b\cos A$, or

$$a^2 = b^2 + c^2 - 2bc\cos A$$

Applying the same sequence of substitutions and reasoning to the right triangles formed by the altitudes from vertices A and B will give us the other two formulas listed in the law of cosines.

We can use the law of cosines to solve triangles in which we are given two sides and the angle included between them (SAS) or to solve triangles in which we are given all three sides (SSS).

▼   **Example 1**   Find the missing parts of triangle ABC if $A = 60°$,   *Two Sides and the*
$b = 20$ in, and $c = 30$ in.   *Included Angle*

Solution   The solution process will include the use of both the law
of cosines and the law of sines. We begin by using the law of cosines
to find a.

Side a

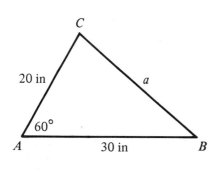

Figure 15-21

$a^2 = b^2 + c^2 - 2bc\cos C$	Law of of cosines
$= 20^2 + 30^2 - 2(20)(30)\cos 60°$	Substitute given values
$= 400 + 900 - 1200(0.5000)$	Table or calculator
$a^2 = 700$	
$a = 26$ in	To the nearest integer

Now that we have a, we can use the law of sines to solve for either
B or C. When we have a choice of angles to solve for, and we are
using the law of sines to do so, it is usually best to solve for the smaller
angle. Since b is smaller than c, B will be smaller than C.

Angle B

$$\sin B = \frac{b\sin A}{a}$$

$$= \frac{20\sin 60°}{26}$$

$$\sin B = 0.6662$$

so $B = 42°$ To the nearest degree

Note that we don't have to check $B' = 180° - 42° = 138°$ because
we know B is an acute angle since it is smaller than either A or C.

Angle C

$$C = 180° - (A + B)$$

$$= 180° - (60° + 42°)$$

$$= 78°$$ ▲

▼   **Example 2**   The diagonals of a parallelogram are 24.2 cm and 35.4 cm and intersect at an angle of 65.5°. Find the length of the shorter side of the paralellogram.

Solution   A diagram of the parallelogram is shown in Figure 15-22. The variable x represents the length of the shorter side. Note also that, since the diagonals bisect each other, we labeled the length of half of each.

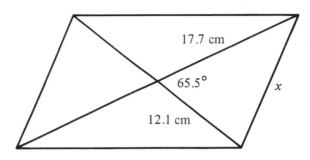

Figure 15-22

$$x^2 = (12.1)^2 + (17.7)^2 - 2(12.1)(17.7)\cos 65.5°$$

$$x^2 = 282.07$$

$$x = 16.8 \text{ cm} \qquad \text{To the nearest tenth} \qquad\qquad \blacktriangle$$

Next we will see how the law of cosines can be used to find the missing parts of a triangle in which all three sides are given.

Three Sides

To use the law of cosines to solve a triangle in which we are given all three sides, it is convenient to rewrite the equations with the cosines isolated on one side. Here is an equivalent form of the law of cosines.

$$\cos A = \frac{b^2 + c^2 - a^2}{2bc}$$

$$\cos B = \frac{a^2 + c^2 - b^2}{2ac}$$

$$\cos C = \frac{a^2 + b^2 - c^2}{2ab}$$

Here is how we arrived at the first of these formulas.

$$a^2 = b^2 + c^2 - 2bc\cos A$$

$$b^2 + c^2 - 2bc\cos A = a^2 \qquad\qquad \text{Exchange sides}$$

$$-2bc\cos A = -b^2 - c^2 + a^2 \qquad \text{Add } -b^2 \text{ and}$$
$$ -c^2 \text{ to both sides}$$

$$\cos A = \frac{b^2 + c^2 - a^2}{2bc} \qquad \text{Divide both sides}$$
$$\phantom{\cos A = \frac{b^2 + c^2 - a^2}{2bc} \qquad} \text{by } -2bc$$

▼ **Example 3**   Solve triangle ABC if $a = 34$ km, $b = 20$ km, and $c = 18$ km.

Solution   We will use the law of cosines to solve for one of the angles, and then the law of sines to find one of the remaining angles. Since there is never any confusion as to whether an angle is acute or obtuse if we have its cosine (the cosine of an obtuse angle is negative) it is best to solve for the largest angle first. Since the longest side is a, we solve for A first.

Angle A

$$\cos A = \frac{b^2 + c^2 - a^2}{2bc}$$

$$= \frac{20^2 + 18^2 - 34^2}{(2)(20)(18)}$$

$$\cos A = -0.6000$$

so $\qquad\qquad A = 126.9° \qquad$ To the nearest tenth

Now we use the law of sines to find angle C.

Angle C

$$\sin C = \frac{c\sin A}{a}$$

$$= \frac{18\sin 126.9°}{34}$$

$$\sin C = 0.4234$$

so $\qquad\qquad C = 25.0° \qquad$ To the nearest tenth

Angle B

$$B = 180° - (A + C)$$

$$= 180° - (126.9° + 25.0°)$$

$$= 28.1° \qquad\qquad\qquad\qquad ▲$$

Problem Set 15.4

Each problem below refers to triangle ABC. Any answers that involve angles should be rounded to the nearest tenth of a degree.

1. If $a = 100$ in, $b = 60$ in, and $C = 60°$, find c.
2. If $a = 100$ in, $b = 60$ in, and $C = 120°$, find c.
3. If $a = 5$ yd, $b = 6$ yd, and $c = 8$ yd, find the largest angle.
4. If $a = 10$ yd, $b = 14$ yd, and $c = 8$ yd, find the largest angle.
5. If $b = 4.2$ m, $c = 6.8$ m, and $A = 116°$, find a.
6. If $a = 3.7$ m, $c = 6.4$ m, and $B = 23°$, find b.
7. If $a = 38$ cm, $b = 10$ cm, and $c = 31$ cm, find the largest angle.
8. If $a = 51$ cm, $b = 24$ cm, and $c = 31$ cm, find the largest angle.

Solve each triangle below.

9. $a = 50$ cm, $b = 70$ cm, $C = 60°$
10. $a = 10$ cm, $b = 12$ cm, $C = 120°$
11. $a = 4$ in, $b = 6$ in, $c = 8$ in (*Remember* Solve for the largest angle first and round to the nearest tenth of a degree.)
12. $a = 5$ in, $b = 10$ in, $c = 12$ in
13. $a = 410$ m, $c = 340$ m, $B = 151.5°$
14. $a = 76.3$ m, $c = 42.8$ m, $B = 16.3°$
15. $a = .048$ yd, $b = .063$ yd, $c = 0.075$ yd
16. $a = 48$ yd, $b = 75$ yd, $c = 63$ yd
17. $a = 4.38$ ft, $b = 3.79$ ft, $c = 5.22$ ft
18. $a = 832$ ft, $b = 623$ ft, $c = 345$ ft

19. Use the law of cosines to show that, if $C = 90°$, then $a^2 = b^2 + c^2$.
20. Use the law of cosines to show that, if $a^2 = b^2 + c^2$, then $C = 90°$.

21. The diagonals of a parallelogram are 56 in and 34 in and intersect at an angle of 120°. Find the length of the shorter side.
22. The diagonals of a parallelogram are 14 m and 16 m and intersect at an angle of 60°. Find the length of the longer side.

Review Problems   The problems below review material covered in Section 14.4.

Find all solutions between $0°$ and $360°$.

23. $2\sin \theta = 1$
24. $2\cos \theta = \sqrt{3}$
25. $2\sin^2 x - \sin x - 1 = 0$
26. $4\cos^2 x + 4\cos x + 1 = 0$
27. $\sin 2A - \sin A = 0$
28. $\cos 2A + \sin A = 0$
29. $4\sin \theta - 2\csc \theta = 0$
30. $4\cos \theta - 3\sec \theta = 0$

Chapter 15 Summary and Review

Examples

DEFINITION II TRIGONOMETRIC FUNCTIONS [15.1]

1.

If triangle ABC is a right triangle with $C = 90°$, then the six trigonometric functions for A are

$$\sin A = \frac{\text{side opposite } A}{\text{hypotenuse}} = \frac{a}{c}$$

$$\cos A = \frac{\text{side adjacent } A}{\text{hypotenuse}} = \frac{b}{c}$$

$$\tan A = \frac{\text{side opposite } A}{\text{side adjacent } A} = \frac{a}{b}$$

$$\cot A = \frac{\text{side adjacent } A}{\text{side opposite } A} = \frac{b}{a}$$

$$\sec A = \frac{\text{hypotenuse}}{\text{side adjacent } A} = \frac{c}{b}$$

$$\csc A = \frac{\text{hypotenuse}}{\text{side opposite } A} = \frac{c}{a}$$

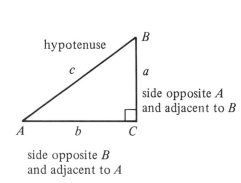

$$\sin A = \frac{3}{5} = \cos B$$

$$\cos A = \frac{4}{5} = \sin B$$

$$\tan A = \frac{3}{4} = \cot B$$

$$\cot A = \frac{4}{3} = \tan B$$

$$\sec A = \frac{5}{4} = \csc B$$

$$\csc A = \frac{5}{3} = \sec B$$

SOLVING RIGHT TRIANGLES [15.1]

We solve a right triangle by using the information given about it to find all the missing sides and angles. In Example 2, we are given the values of one side and one of the acute angles and we use them to find the remaining two sides and the other acute angle.

2.   In right triangle ABC, if $A = 40°$ and $c = 12$ cm, then

$B = 90° - A = 90° - 40° = 50°$

$a = c\sin A = 12\sin 40°$
    $= 12(0.6428) = 7.7$

$b = c\cos A = 12\cos 40°$
    $= 12(0.7660) = 9.2$

ANGLE OF ELEVATION AND ANGLE OF DEPRESSION [15.1]

3.

An angle measured from the horizontal up is called an *angle of elevation*. An angle measured from the horizontal down is called an *angle of depression*.

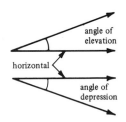

4. If $A = 30°$, $B = 70°$, and $a = 8$ cm in triangle ABC, then by the law of sines

$$b = \frac{a\sin B}{\sin A} = \frac{8\sin 70°}{\sin 30°} = 15 \text{ cm}$$

THE LAW OF SINES [15.2]

For any triangle ABC, the following relationships are always true:

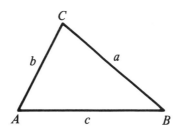

$$\frac{\sin A}{a} = \frac{\sin B}{b} = \frac{\sin C}{c}$$

or, equivalently

$$\frac{a}{\sin A} = \frac{b}{\sin B} = \frac{c}{\sin C}$$

5. In triangle ABC, if $a = 54$ cm, $b = 62$ cm, and $A = 40°$, then

$$\sin B = \frac{b\sin A}{a} = \frac{62\sin 40°}{54} = 0.7380$$

Since $\sin B$ is positive for any angle in quadrant I or II, we have two possibilities for B.

$B = 48°$ or $B' = 180° - 48° = 132°$

This indicates that two triangles exist, both of which fit the given information.

THE AMBIGUOUS CASE [15.3]

When we are given two sides and an angle opposite one of them (SSA), we have several possibilities for the triangle or triangles that result. One of the possibilities is that no triangle will fit the given information. Another possibility is that two different triangles can be obtained from the given information and a third possibility is that exactly one triangle will fit the given information. Because of these different possibilities, we call the situation where we are solving a triangle in which we are given two sides and the angle opposite one of them the *ambiguous case*.

6. In triangle ABC, if $a = 34$ km, $b = 20$ km, and $c = 18$ km, then we can find A using the law of cosines.

$$\cos A = \frac{b^2 + c^2 - a^2}{2bc}$$

$$= \frac{20^2 + 18^2 - 34^2}{(2)(20)(18)}$$

$$\cos A = -0.6000$$

so $A = 126.9°$

THE LAW OF COSINES [15.4]

In any triangle ABC, the following relationships are always true:

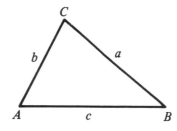

$$a^2 = b^2 + c^2 - 2bc\cos A$$
$$b^2 = a^2 + c^2 - 2ac\cos B$$
$$c^2 = a^2 + b^2 - 2ab\cos C$$

Another form of the law of cosines looks like this

$$\cos A = \frac{b^2 + c^2 - a^2}{2bc}$$

$$\cos B = \frac{a^2 + c^2 - b^2}{2ac}$$

$$\cos C = \frac{a^2 + b^2 - c^2}{2ab}$$

Chapter 15 Test

Problems 1 through 8 refer to right triangle ABC with $C = 90°$.
 1. If $a = 5$ and $b = 12$, find $\sin A$, $\cos A$, $\sin B$, and $\cos B$.
 2. If $a = 4$ and $c = 5$, find $\sin A$, $\tan A$, $\csc A$, and $\cot A$.
 3. If $A = 35°$ and $c = 24$ cm, find a.
 4. If $B = 63°$ and $a = 86$ cm, find b.
 5. If $a = 7.8$ in and $b = 3.6$ in, find A to the nearest tenth of a degree.
 6. If $a = 4.9$ in and $c = 6.2$ in, find B to the nearest tenth of a degree.
 7. If $a = 48$ yd and $A = 26.3°$, find all the missing parts.
 8. If $c = 72$ yd and $B = 18.1°$ find all the missing parts.

Problems 9 through 22 refer to triangle ABC, which is not necessarily a right triangle.
 9. If $A = 32°$, $B = 70°$, and $a = 3.8$ in, use the law of sines to find b.
10. If $B = 118°$, $C = 37°$, and $c = 2.9$ in, use the law of sines to find b.
11. If $A = 38.2°$, $B = 63.4°$, and $c = 42$ cm, find all the missing parts.
12. If $A = 24.7°$, $C = 106.1°$, and $b = 34$ cm, find all the missing parts.
13. Use the law of sines to show that no triangle exists for which $A = 60°$, $a = 12$ in, and $b = 42$ in.
14. Use the law of sines to show that exactly one triangle exists for which $A = 42°$, $a = 29$ in, and $b = 21$ in.
15. Find two triangles for which $A = 51°$, $a = 6.5$ ft, and $b = 7.9$ ft.
16. Find two triangles for which $A = 26°$, $a = 4.8$ ft, and $b = 9.4$ ft.
17. If $C = 60°$, $a = 10$ cm, and $b = 12$ cm, use the law of cosines to find c.
18. If $C = 120°$, $a = 10$ cm, and $b = 12$ cm, use the law of cosines to find c.
19. If $a = 5$ km, $b = 7$ km, and $c = 9$ km, use the law of cosines to find C.
20. If $a = 10$ km, $b = 12$ km, and $c = 11$ km, use the law of cosines to find B.
21. Find all the missing parts if $a = 6.4$ m, $b = 2.8$ m, and $C = 118.9°$.
22. Find all the missing parts if $b = 3.7$ m, $c = 6.2$ m, and $A = 34.7°$.

23. The two equal sides of an isosceles triangle are each 38 cm. If the base measures 48 cm, find the measure of the two equal angles.

24. A lamp pole casts a shadow 53 ft long when the angle of elevation of the sun is 48.1°. Find the height of the lamp pole.

25. A man standing near a building notices the angle of elevation to the top of the building is 64°. He then walks 240 ft farther away from the building and finds the angle of elevation to the top to be 43°. How tall is the building?

26. The diagonals of a parallelogram are 26.8 m and 39.4 m. If they meet at an angle of 134°, find the length of the shorter side of the parallelogram.

Appendix A:
Synthetic Division

Synthetic division is a short form of long division with polynomials. We will consider synthetic division only for those cases in which the divisor is of the form $x + k$, where k is a constant.

Let's begin by looking over an example of long division with polynomials as done in Section 4.2:

$$
\begin{array}{r}
3x^2 - 2x + 4 \\
x + 3\overline{)3x^3 + 7x^2 - 2x - 4} \\
\underline{3x^3 + 9x^2} \\
-2x^2 - 2x \\
\underline{-2x^2 - 6x} \\
4x - 4 \\
\underline{4x + 12} \\
-16
\end{array}
$$

We can rewrite the problem without showing the variable, since the variable is written in descending powers and similar terms are in alignment. It looks like this:

$$
\begin{array}{r}
3 \;-2 \;+4 \\
1 + 3\overline{)\;3 \quad 7 \;-2 \;-4} \\
\underline{(3) + 9} \\
-2 \,(-2) \\
\underline{(-2) -6} \\
4 \,(-4) \\
\underline{(4) \;\;12} \\
-16
\end{array}
$$

We have used parentheses to enclose the numbers which are repetitions of the numbers above them. We can compress the problem by eliminating all repetitions:

$$
\begin{array}{r}
3 \;-2 \quad 4 \\
1 + 3\overline{)3 \quad 7 \;-2 \quad -4} \\
\underline{9 \;-6 \;+12} \\
3 \;-2 \quad 4 \;-16
\end{array}
$$

The top line is the same as the first three terms of the bottom line, so we eliminate the top line. Also, the 1 which was the coefficient of x in the original

problem can be eliminated, since we will only consider division problems where the divisor is of the form $x + k$. The following is the most compact form of the original division problem:

$$+3\overline{)3 \quad 7 \quad -2 \quad -4}$$
$$ 9 \quad -6 \quad 12$$
$$\overline{3 \quad -2 \quad 4 \quad -16}$$

If we check over the problem, we find that the first term in the bottom row is exactly the same as the first term in the top row—and it always will be in problems of this type. Also, the last three terms in the bottom row come from multiplication by $+3$ and then subtraction. We can get an equivalent result by multiplying by -3 and adding. The problem would then look like this:

$$-3 \,\big|\, 3 \quad 7 \quad -2 \quad -4$$
$$ \downarrow \quad -9 \quad 6 \quad -12$$
$$\overline{ 3 \quad -2 \quad 4 \,\big|\, -16}$$

We have used the brackets ⌋ ⌊ to separate the divisor and the remainder. This last expression is synthetic division. It is an easy process to remember. Simply change the sign of the constant term in the divisor, then bring down the first term of the dividend. The process is then just a series of multiplications and additions, as indicated in the following diagram by the arrows:

$$-3 \,\big|\, 3 \quad 7 \quad -2 \quad -4$$
$$ \downarrow \quad -9 \quad 6 \quad -12$$
$$\overline{ 3 \quad -2 \quad 4 \,\big|\, -16}$$

The last term on the bottom row is always the remainder.

Here are some additional examples of synthetic division with polynomials.

▼ **Example 1**   Divide $x^4 - 2x^3 + 4x^2 - 6x + 2$ by $x - 2$.

Solution   We change the sign of the constant term in the divisor to get $+2$ and then use the procedure given above:

$$+2 \,\big|\, 1 \quad -2 \quad 4 \quad -6 \quad 2$$
$$ \downarrow \quad 2 \quad 0 \quad 8 \quad 4$$
$$\overline{ 1 \quad 0 \quad 4 \quad 2 \,\big|\, 6}$$

From the last line we have the answer:

$$1x^3 + 0x^2 + 4x + 2 + \frac{6}{x-2}$$

or

$$\frac{x^4 - 2x^3 + 4x^2 - 6x + 2}{x-2} = x^3 + 4x + 2 + \frac{6}{x-2}$$ ▲

▼ **Example 2**   Divide $\dfrac{3x^3 - 4x + 5}{x + 4}$.

Solution   Since we cannot skip any powers of the variable in the polynomial $3x^3 - 4x + 5$, we rewrite it as $3x^3 + 0x^2 - 4x + 5$ and proceed as we did in Example 1:

$$
\begin{array}{r|rrrr}
-4 & 3 & 0 & -4 & 5 \\
   & \downarrow & -12 & 48 & -176 \\
\hline
   & 3 & -12 & 44 & \boxed{-171}
\end{array}
$$

From the synthetic division, we have

$$\frac{3x^3 - 4x + 5}{x + 4} = 3x^2 - 12x + 44 - \frac{171}{x + 4}$$ ▲

▼ **Example 3**   Divide $\dfrac{x^3 - 1}{x - 1}$.

Solution   Writing the numerator as $x^3 + 0x^2 + 0x - 1$ and using synthetic division, we have

$$
\begin{array}{r|rrrr}
+1 & 1 & 0 & 0 & -1 \\
   & \downarrow & 1 & 1 & 1 \\
\hline
   & 1 & 1 & 1 & \boxed{0}
\end{array}
$$

which indicates

$$\frac{x^3 - 1}{x - 1} = x^2 + x + 1$$ ▲

Problem Set A

Use synthetic division to find the following quotients:

1. $\dfrac{x^2 - 5x + 6}{x + 2}$

2. $\dfrac{x^2 + 8x - 12}{x - 3}$

3. $\dfrac{3x^2 - 4x + 1}{x - 1}$

4. $\dfrac{4x^2 - 2x - 6}{x + 1}$

5. $\dfrac{x^3 + 2x^2 + 3x + 4}{x - 2}$

6. $\dfrac{x^3 - 2x^2 - 3x - 4}{x - 2}$

7. $\dfrac{3x^3 - x^2 + 2x + 5}{x - 3}$

8. $\dfrac{2x^3 - 5x^2 + x + 2}{x - 2}$

9. $\dfrac{2x^3 + x - 3}{x - 1}$

10. $\dfrac{3x^3 - 2x + 1}{x - 5}$

11. $\dfrac{x^4 + 2x^2 + 1}{x + 4}$

12. $\dfrac{x^4 - 3x^2 + 1}{x - 4}$

13. $\dfrac{x^5 - 2x^4 + x^3 - 3x^2 + x + 1}{x - 2}$

14. $\dfrac{2x^5 - 3x^4 + x^3 - x^2 + 2x + 1}{x + 2}$

15. $\dfrac{x^2 + x + 1}{x - 1}$

16. $\dfrac{x^2 + x + 1}{x + 1}$

17. $\dfrac{x^4 - 1}{x + 1}$

18. $\dfrac{x^4 + 1}{x - 1}$

19. $\dfrac{x^3 - 1}{x - 1}$

20. $\dfrac{x^3 - 1}{x + 1}$

Appendix B:
The Discriminant

The quadratic formula

$$x = \frac{-b \pm \sqrt{b^2 - 4ac}}{2a}$$

gives the solutions to any quadratic equation in standard form. There are times, when working with quadratic equations, when it is only important to know what kind of solutions the equation has.

DEFINITION The expression under the radical in the quadratic formula is called the *discriminant:*

$$\text{Discriminant} = D = b^2 - 4ac$$

The discriminant gives the number and type of solutions to a quadratic equation, when the original equation has integer coefficients. For example, when the discriminant is negative, the quadratic formula will contain the square root of a negative number. Hence, the equation will have complex solutions. If the discriminant were 0, the formula would be

$$x = \frac{-b \pm 0}{2a} = \frac{-b}{2a}$$

and the equation would have one rational solution: the number $-b/2a$.

The following table gives the relationship between the discriminant and the type of solutions to the equation.

For the equation $ax^2 + bx + c = 0$ where a, b, and c are integers and $a \neq 0$:

If the discriminant $b^2 - 4ac$ is	Then the equation will have
Negative	Two complex solutions
Zero	One rational solution
A positive number that is also a perfect square	Two rational solutions
A positive number that is not a perfect square	Two irrational solutions

In the second and third cases, when the discriminant is 0 or a positive perfect square, the solutions are rational numbers. The quadratic equations in these two cases are the ones that can be factored.

▼ **Example 1** For each equation give the number and kind of solutions.

a. $x^2 - 3x - 40 = 0$

Solution   Using $a = 1$, $b = -3$, and $c = -40$ in $b^2 - 4ac$, we have
$9 - 4(1)(-40) = 9 + 160 = 169$.

The discriminant is a perfect square. Therefore, the equation has two rational solutions

b. $2x^2 - 3x + 4 = 0$

Solution   Using $a = 2$, $b = -3$, and $c = 4$, we have $b^2 - 4ac = 9 - 4(2)(4) = 9 - 32 = -23$.

The discriminant is negative, implying the equation has two complex solutions.   ▲

▼   **Example 2**   Find an appropriate k so that the equation $4x^2 - kx = -9$ has exactly one rational solution.

Solution   We begin by writing the equation in standard form:

$$4x^2 - kx + 9 = 0$$

Using $a = 4$, $b = -k$, and $c = 9$, we have

$$b^2 - 4ac = (-k)^2 - 4(4)(9)$$
$$= k^2 - 144$$

An equation has exactly one rational solution when the discriminant is 0. We set the discriminant equal to 0 and solve:

$$k^2 - 144 = 0$$
$$k^2 = 144$$
$$k = \pm 12$$

Choosing k to be 12 or -12 will result in an equation with one rational solution.   ▲

Problem Set B

Use the discriminant to find the number and kind of solution for each of the following equations. (See Example 1.)

1. $x^2 - 6x + 5 = 0$	**2.** $x^2 - x - 12 = 0$	**3.** $4x^2 - 4x = -1$
4. $9x^2 + 12x = -4$	**5.** $x^2 + x - 1 = 0$	**6.** $x^2 - 2x + 3 = 0$
7. $2y^2 = 3y + 1$	**8.** $3y^2 = 4y - 2$	**9.** $x^2 - 9 = 0$
10. $4x^2 - 81 = 0$	**11.** $5a^2 - 4a = 5$	**12.** $3a = 4a^2 - 5$

Determine k so that each of the following has exactly one real solution. (See Example 2.)

13. $x^2 - kx + 25 = 0$	**14.** $x^2 + kx + 25 = 0$	**15.** $x^2 = kx - 36$
16. $x^2 = kx - 49$	**17.** $4x^2 - 12x + k = 0$	**18.** $9x^2 + 30x + k = 0$
19. $kx^2 - 40x = 25$	**20.** $kx^2 - 2x = -1$	**21.** $3x^2 - kx + 2 = 0$
22. $5x^2 + kx + 1 = 0$		

Appendix C

Powers, Roots, and Prime Factors

n	n^2	$\sqrt{n}$	n^3	$\sqrt[3]{n}$	Prime factors
1	1	1.000	1	1.000	—
2	4	1.414	8	1.260	prime
3	9	1.732	27	1.442	prime
4	16	2.000	64	1.587	$2 \cdot 2$
5	25	2.236	125	1.710	prime
6	36	2.449	216	1.817	$2 \cdot 3$
7	49	2.646	343	1.913	prime
8	64	2.828	512	2.000	$2 \cdot 2 \cdot 2$
9	81	3.000	729	2.080	$3 \cdot 3$
10	100	3.162	1,000	2.154	$2 \cdot 5$
11	121	3.317	1,331	2.224	prime
12	144	3.464	1,728	2.289	$2 \cdot 2 \cdot 3$
13	169	3.606	2,197	2.351	prime
14	196	3.742	2,744	2.410	$2 \cdot 7$
15	225	3.873	3,375	2.466	$3 \cdot 5$
16	256	4.000	4,096	2.520	$2 \cdot 2 \cdot 2 \cdot 2$
17	289	4.123	4,913	2.571	prime
18	324	4.243	5,832	2.621	$2 \cdot 3 \cdot 3$
19	361	4.359	6,859	2.668	prime
20	400	4.472	8,000	2.714	$2 \cdot 2 \cdot 5$
21	441	4.583	9,261	2.759	$3 \cdot 7$
22	484	4.690	10,648	2.802	$2 \cdot 11$
23	529	4.796	12,167	2.844	prime
24	576	4.899	13,824	2.884	$2 \cdot 2 \cdot 2 \cdot 3$
25	625	5.000	15,625	2.924	$5 \cdot 5$
26	676	5.099	17,576	2.962	$2 \cdot 13$
27	729	5.196	19,683	3.000	$3 \cdot 3 \cdot 3$
28	784	5.292	21,952	3.037	$2 \cdot 2 \cdot 7$
29	841	5.385	24,389	3.072	prime
30	900	5.477	27,000	3.107	$2 \cdot 3 \cdot 5$
31	961	5.568	29,791	3.141	prime
32	1,024	5.657	32,768	3.175	$2 \cdot 2 \cdot 2 \cdot 2 \cdot 2$
33	1,089	5.745	35,937	3.208	$3 \cdot 11$
34	1,156	5.831	39,304	3.240	$2 \cdot 17$
35	1,225	5.916	42,875	3.271	$5 \cdot 7$

Powers, Roots, and Prime Factors (Continued)

n	n^2	$\sqrt{n}$	n^3	$\sqrt[3]{n}$	Prime factors
36	1,296	6.000	46,656	3.302	$2 \cdot 2 \cdot 3 \cdot 3$
37	1,369	6.083	50,653	3.332	prime
38	1,444	6.164	54,872	3.362	$2 \cdot 19$
39	1,521	6.245	59,319	3.391	$3 \cdot 13$
40	1,600	6.325	64,000	3.420	$2 \cdot 2 \cdot 2 \cdot 5$
41	1,681	6.403	68,921	3.448	prime
42	1,764	6.481	74,088	3.476	$2 \cdot 3 \cdot 7$
43	1,849	6.557	79,507	3.503	prime
44	1,936	6.633	85,184	3.530	$2 \cdot 2 \cdot 11$
45	2,025	6.708	91,125	3.557	$3 \cdot 3 \cdot 5$
46	2,116	6.782	97,336	3.583	$2 \cdot 23$
47	2,209	6.856	103,823	3.609	prime
48	2,304	6.928	110,592	3.634	$2 \cdot 2 \cdot 2 \cdot 2 \cdot 3$
49	2,401	7.000	117,649	3.659	$7 \cdot 7$
50	2,500	7.071	125,000	3.684	$2 \cdot 5 \cdot 5$
51	2,601	7.141	132,651	3.708	$3 \cdot 17$
52	2,704	7.211	140,608	3.733	$2 \cdot 2 \cdot 13$
53	2,809	7.280	148,877	3.756	prime
54	2,916	7.348	157,464	3.780	$2 \cdot 3 \cdot 3 \cdot 3$
55	3,025	7.416	166,375	3.803	$5 \cdot 11$
56	3,136	7.483	175,616	3.826	$2 \cdot 2 \cdot 2 \cdot 7$
57	3,249	7.550	185,193	3.849	$3 \cdot 19$
58	3,364	7.616	195,112	3.871	$2 \cdot 29$
59	3,481	7.681	205,379	3.893	prime
60	3,600	7.746	216,000	3.915	$2 \cdot 2 \cdot 3 \cdot 5$
61	3,721	7.810	226,981	3.936	prime
62	3,844	7.874	238,328	3.958	$2 \cdot 31$
63	3,969	7.937	250,047	3.979	$3 \cdot 3 \cdot 7$
64	4,096	8.000	262,144	4.000	$2 \cdot 2 \cdot 2 \cdot 2 \cdot 2 \cdot 2$
65	4,225	8.062	274,625	4.021	$5 \cdot 13$
66	4,356	8.124	287,496	4.041	$2 \cdot 3 \cdot 11$
67	4,489	8.185	300,763	4.062	prime
68	4,624	8.246	314,432	4.082	$2 \cdot 2 \cdot 17$
69	4,761	8.307	328,509	4.102	$3 \cdot 23$
70	4,900	8.367	343,000	4.121	$2 \cdot 5 \cdot 7$
71	5,041	8.426	357,911	4.141	prime
72	5,184	8.485	373,248	4.160	$2 \cdot 2 \cdot 2 \cdot 3 \cdot 3$
73	5,329	8.544	389,017	4.179	prime
74	5,476	8.602	405,224	4.198	$2 \cdot 37$
75	5,625	8.660	421,875	4.217	$3 \cdot 5 \cdot 5$

Powers, Roots, and Prime Factors (Continued)

n	n^2	$\sqrt{n}$	n^3	$\sqrt[3]{n}$	Prime factors
76	5,776	8.718	438,976	4.236	$2 \cdot 2 \cdot 19$
77	5,929	8.775	456,533	4.254	$7 \cdot 11$
78	6,084	8.832	474,552	4.273	$2 \cdot 3 \cdot 13$
79	6,241	8.888	493,039	4.291	prime
80	6,400	8.944	512,000	4.309	$2 \cdot 2 \cdot 2 \cdot 2 \cdot 5$
81	6,561	9.000	531,441	4.327	$3 \cdot 3 \cdot 3 \cdot 3$
82	6,724	9.055	551,368	4.344	$2 \cdot 41$
83	6,889	9.110	571,787	4.362	prime
84	7,056	9.165	592,704	4.380	$2 \cdot 2 \cdot 3 \cdot 7$
85	7,225	9.220	614,125	4.397	$5 \cdot 17$
86	7,396	9.274	636,056	4.414	$2 \cdot 43$
87	7,569	9.327	658,503	4.431	$3 \cdot 29$
88	7,744	9.381	681,472	4.448	$2 \cdot 2 \cdot 2 \cdot 11$
89	7,921	9.434	704,969	4.465	prime
90	8,100	9.487	729,000	4.481	$2 \cdot 3 \cdot 3 \cdot 5$
91	8,281	9.539	753,571	4.498	$7 \cdot 13$
92	8,464	9.592	778,688	4.514	$2 \cdot 2 \cdot 23$
93	8,649	9.644	804,357	4.531	$3 \cdot 31$
94	8,836	9.695	830,584	4.547	$2 \cdot 47$
95	9,025	9.747	857,375	4.563	$5 \cdot 19$
96	9,216	9.798	884,736	4.579	$2 \cdot 2 \cdot 2 \cdot 2 \cdot 2 \cdot 3$
97	9,409	9.849	912,673	4.595	prime
98	9,604	9.899	941,192	4.610	$2 \cdot 7 \cdot 7$
99	9,801	9.950	970,299	4.626	$3 \cdot 3 \cdot 11$
100	10,000	10.000	1,000,000	4.642	$2 \cdot 2 \cdot 5 \cdot 5$

Appendix D

Common Logarithms

x	0	1	2	3	4	5	6	7	8	9
1.0	.0000	.0043	.0086	.0128	.0170	.0212	.0253	.0294	.0334	.0374
1.1	.0414	.0453	.0492	.0531	.0569	.0607	.0645	.0682	.0719	.0755
1.2	.0792	.0828	.0864	.0899	.0934	.0969	.1004	.1038	.1072	.1106
1.3	.1139	.1173	.1206	.1239	.1271	.1303	.1335	.1367	.1399	.1430
1.4	.1461	.1492	.1523	.1553	.1584	.1614	.1644	.1673	.1703	.1732
1.5	.1761	.1790	.1818	.1847	.1875	.1903	.1931	.1959	.1987	.2014
1.6	.2041	.2068	.2095	.2122	.2148	.2175	.2201	.2227	.2253	.2279
1.7	.2304	.2330	.2355	.2380	.2405	.2430	.2455	.2480	.2504	.2529
1.8	.2553	.2577	.2601	.2625	.2648	.2672	.2695	.2718	.2742	.2765
1.9	.2788	.2810	.2833	.2856	.2878	.2900	.2923	.2945	.2967	.2989
2.0	.3010	.3032	.3054	.3075	.3096	.3118	.3139	.3160	.3181	.3201
2.1	.3222	.3243	.3263	.3284	.3304	.3324	.3345	.3365	.3385	.3404
2.2	.3424	.3444	.3464	.3483	.3502	.3522	.3541	.3560	.3579	.3598
2.3	.3617	.3636	.3655	.3674	.3692	.3711	.3729	.3747	.3766	.3784
2.4	.3802	.3820	.3838	.3856	.3874	.3892	.3909	.3927	.3945	.3962
2.5	.3979	.3997	.4014	.4031	.4048	.4065	.4082	.4099	.4116	.4133
2.6	.4150	.4166	.4183	.4200	.4216	.4232	.4249	.4265	.4281	.4298
2.7	.4314	.4330	.4346	.4362	.4378	.4393	.4409	.4425	.4440	.4456
2.8	.4472	.4487	.4502	.4518	.4533	.4548	.4564	.4579	.4594	.4609
2.9	.4624	.4639	.4654	.4669	.4683	.4698	.4713	.4728	.4742	.4757
3.0	.4771	.4786	.4800	.4814	.4829	.4843	.4857	.4871	.4886	.4900
3.1	.4914	.4928	.4942	.4955	.4969	.4983	.4997	.5011	.5024	.5038
3.2	.5051	.5065	.5079	.5092	.5105	.5119	.5132	.5145	.5159	.5172
3.3	.5185	.5198	.5211	.5224	.5237	.5250	.5263	.5276	.5289	.5302
3.4	.5315	.5328	.5340	.5353	.5366	.5378	.5391	.5403	.5416	.5428
3.5	.5441	.5453	.5465	.5478	.5490	.5502	.5514	.5527	.5539	.5551
3.6	.5563	.5575	.5587	.5599	.5611	.5623	.5635	.5647	.5658	.5670
3.7	.5682	.5694	.5705	.5717	.5729	.5740	.5752	.5763	.5775	.5786
3.8	.5798	.5809	.5821	.5832	.5843	.5855	.5866	.5877	.5888	.5899
3.9	.5911	.5922	.5933	.5944	.5955	.5966	.5977	.5988	.5999	.6010
4.0	.6021	.6031	.6042	.6053	.6064	.6075	.6085	.6096	.6107	.6117
4.1	.6128	.6138	.6149	.6160	.6170	.6180	.6191	.6201	.6212	.6222
4.2	.6232	.6243	.6253	.6263	.6274	.6284	.6294	.6304	.6314	.6325
x	0	1	2	3	4	5	6	7	8	9

Common Logarithms (Continued)

x	0	1	2	3	4	5	6	7	8	9
4.3	.6335	.6345	.6355	.6365	.6375	.6385	.6395	.6405	.6415	.6425
4.4	.6435	.6444	.6454	.6464	.6474	.6484	.6493	.6503	.6513	.6522
4.5	.6532	.6542	.6551	.6561	.6571	.6580	.6590	.6599	.6609	.6618
4.6	.6628	.6637	.6646	.6656	.6665	.6675	.6684	.6693	.6702	.6712
4.7	.6721	.6730	.6739	.6749	.6758	.6767	.6776	.6785	.6794	.6803
4.8	.6812	.6821	.6830	.6839	.6848	.6857	.6866	.6875	.6884	.6893
4.9	.6902	.6911	.6920	.6928	.6937	.6946	.6955	.6964	.6972	.6981
5.0	.6990	.6998	.7007	.7016	.7024	.7033	.7042	.7050	.7059	.7067
5.1	.7076	.7084	.7093	.7101	.7110	.7118	.7126	.7135	.7143	.7152
5.2	.7160	.7168	.7177	.7185	.7193	.7202	.7210	.7218	.7226	.7235
5.3	.7243	.7251	.7259	.7267	.7275	.7284	.7292	.7300	.7308	.7316
5.4	.7324	.7332	.7340	.7348	.7356	.7364	.7372	.7380	.7388	.7396
5.5	.7404	.7412	.7419	.7427	.7435	.7443	.7451	.7459	.7466	.7474
5.6	.7482	.7490	.7497	.7505	.7513	.7520	.7528	.7536	.7543	.7551
5.7	.7559	.7566	.7574	.7582	.7589	.7597	.7604	.7612	.7619	.7627
5.8	.7634	.7642	.7649	.7657	.7664	.7672	.7679	.7686	.7694	.7701
5.9	.7709	.7716	.7723	.7731	.7738	.7745	.7752	.7760	.7767	.7774
6.0	.7782	.7789	.7796	.7803	.7810	.7818	.7825	.7832	.7839	.7846
6.1	.7853	.7860	.7868	.7875	.7882	.7889	.7896	.7903	.7910	.7917
6.2	.7924	.7931	.7938	.7945	.7952	.7959	.7966	.7973	.7980	.7987
6.3	.7993	.8000	.8007	.8014	.8021	.8028	.8035	.8041	.8048	.8055
6.4	.8062	.8069	.8075	.8082	.8089	.8096	.8102	.8109	.8116	.8122
6.5	.8129	.8136	.8142	.8149	.8156	.8162	.8169	.8176	.8182	.8189
6.6	.8195	.8202	.8209	.8215	.8222	.8228	.8235	.8241	.8248	.8254
6.7	.8261	.8267	.8274	.8280	.8287	.8293	.8299	.8306	.8312	.8319
6.8	.8325	.8331	.8338	.8344	.8351	.8357	.8363	.8370	.8376	.8382
6.9	.8388	.8395	.8401	.8407	.8414	.8420	.8426	.8432	.8439	.8445
7.0	.8451	.8457	.8463	.8470	.8476	.8482	.8488	.8494	.8500	.8506
7.1	.8513	.8519	.8525	.8531	.8537	.8543	.8549	.8555	.8561	.8567
7.2	.8573	.8579	.8585	.8591	.8597	.8603	.8609	.8615	.8621	.8627
7.3	.8633	.8639	.8645	.8651	.8657	.8663	.8669	.8675	.8681	.8686
7.4	.8692	.8698	.8704	.8710	.8716	.8722	.8727	.8733	.8739	.8745
7.5	.8751	.8756	.8762	.8768	.8774	.8779	.8785	.8791	.8797	.8802
7.6	.8808	.8814	.8820	.8825	.8831	.8837	.8842	.8848	.8854	.8859
7.7	.8865	.8871	.8876	.8882	.8887	.8893	.8899	.8904	.8910	.8915
7.8	.8921	.8927	.8932	.8938	.8943	.8949	.8954	.8960	.8965	.8971
7.9	.8976	.8982	.8987	.8993	.8998	.9004	.9009	.9015	.9020	.9025
x	0	1	2	3	4	5	6	7	8	9

Common Logarithms (Continued)

x	0	1	2	3	4	5	6	7	8	9
8.0	.9031	.9036	.9042	.9047	.9053	.9058	.9063	.9069	.9074	.9079
8.1	.9085	.9090	.9096	.9101	.9106	.9112	.9117	.9122	.9128	.9133
8.2	.9138	.9143	.9149	.9154	.9159	.9165	.9170	.9175	.9180	.9186
8.3	.9191	.9196	.9201	.9206	.9212	.9217	.9222	.9227	.9232	.9238
8.4	.9243	.9248	.9253	.9258	.9263	.9269	.9274	.9279	.9284	.9289
8.5	.9294	.9299	.9304	.9309	.9315	.9320	.9325	.9330	.9335	.9340
8.6	.9345	.9350	.9355	.9360	.9365	.9370	.9375	.9380	.9385	.9390
8.7	.9395	.9400	.9405	.9410	.9415	.9420	.9425	.9430	.9435	.9440
8.8	.9445	.9450	.9455	.9460	.9465	.9469	.9474	.9479	.9484	.9489
8.9	.9494	.9499	.9504	.9509	.9513	.9518	.9523	.9528	.9533	.9538
9.0	.9542	.9547	.9552	.9557	.9562	.9566	.9571	.9576	.9581	.9586
9.1	.9590	.9595	.9600	.9605	.9609	.9614	.9619	.9624	.9628	.9633
9.2	.9638	.9643	.9647	.9652	.9657	.9661	.9666	.9671	.9675	.9680
9.3	.9685	.9689	.9694	.9699	.9703	.9708	.9713	.9717	.9722	.9727
9.4	.9731	.9736	.9741	.9745	.9750	.9754	.9759	.9763	.9768	.9773
9.5	.9777	.9782	.9786	.9791	.9795	.9800	.9805	.9809	.9814	.9818
9.6	.9823	.9827	.9832	.9836	.9841	.9845	.9850	.9854	.9859	.9863
9.7	.9868	.9872	.9877	.9881	.9886	.9890	.9894	.9899	.9903	.9908
9.8	.9912	.9917	.9921	.9926	.9930	.9934	.9939	.9943	.9948	.9952
9.9	.9956	.9961	.9965	.9969	.9974	.9978	.9983	.9987	.9991	.9996
x	0	1	2	3	4	5	6	7	8	9

Appendix E

Table of Trigonometric Functions—Decimal Degrees
$(\csc \theta = 1/\sin \theta; \sec \theta = 1/\cos \theta)$

θ	$\sin \theta$	$\cos \theta$	$\tan \theta$	$\cot \theta$	
0.0	0.00000	1.0000	0.00000	∞	90.0
.1	.00175	1.0000	.00175	573.0	89.9
.2	.00349	1.0000	.00349	286.5	.8
.3	.00524	1.0000	.00524	191.0	.7
.4	.00698	1.0000	.00698	143.24	.6
.5	.00873	1.0000	.00873	114.59	.5
.6	.01047	0.9999	.01047	95.49	.4
.7	.01222	.9999	.01222	81.85	.3
.8	.01396	.9999	.01396	71.62	.2
.9	.01571	.9999	.01571	63.66	89.1
1.0	0.01745	0.9998	0.01746	57.29	89.0
.1	.01920	.9998	.01920	52.08	88.9
.2	.02094	.9998	.02095	47.74	.8
.3	.02269	.9997	.02269	44.07	.7
.4	.02443	.9997	.02444	40.92	.6
.5	.02618	.9997	.02619	38.19	.5
.6	.02792	.9996	.02793	35.80	.4
.7	.02967	.9996	.02968	33.69	.3
.8	.03141	.9995	.03143	31.82	.2
.9	.03316	.9995	.03317	30.14	88.1
2.0	0.03490	0.9994	0.03492	28.64	88.0
.1	.03664	.9993	.03667	27.27	87.9
.2	.03839	.9993	.03842	26.03	.8
.3	.04013	.9992	.04016	24.90	.7
.4	.04188	.9991	.04191	23.86	.6
.5	.04362	.9990	.04366	22.90	.5
.6	.04536	.9990	.04541	22.02	.4
.7	.04711	.9989	.04716	21.20	.3
.8	.04885	.9988	.04891	20.45	.2
.9	.05059	.9987	.05066	19.74	87.1
3.0	0.05234	0.9986	0.05241	19.081	87.0
.1	.05408	.9985	.05416	18.464	86.9
3.2	.05582	.9984	.05591	17.886	86.8
	$\cos \theta$	$\sin \theta$	$\cot \theta$	$\tan \theta$	θ

Table of Trigonometric Functions—Decimal Degrees
(csc $\theta = 1/\sin \theta$; sec $\theta = 1/\cos \theta$) (Continued)

θ	$\sin \theta$	$\cos \theta$	$\tan \theta$	$\cot \theta$	
3.3	.05756	.9983	.05766	17.343	86.7
.4	.05931	.9982	.05941	16.832	.6
.5	.06105	.9981	.06116	16.350	.5
.6	.06279	.9980	.06291	15.895	.4
.7	.06453	.9979	.06467	15.464	.3
.8	.06627	.9978	.06642	15.056	.2
.9	.06802	.9977	.06817	14.669	86.1
4.0	0.06976	0.9976	0.06993	14.301	86.0
.1	.07150	.9974	.07168	13.951	85.9
.2	.07324	.9973	.07344	13.617	.8
.3	.07498	.9972	.07519	13.300	.7
.4	.07672	.9971	.07695	12.996	.6
.5	.07846	.9969	.07870	12.706	.5
.6	.08020	.9968	.08046	12.429	.4
.7	.08194	.9966	.08221	12.163	.3
.8	.08368	.9965	.08397	11.909	.2
.9	.08542	.9963	.08573	11.664	85.1
5.0	0.08716	0.9962	0.08749	11.430	85.0
.1	.08889	.9960	.08925	11.205	84.9
.2	.09063	.9959	.09101	10.988	.8
.3	.09237	.9957	.09277	10.780	.7
.4	.09411	.9956	.09453	10.579	.6
.5	.09585	.9954	.09629	10.385	.5
.6	.09758	.9952	.09805	10.199	.4
.7	.09932	.9951	.09981	10.019	.3
.8	.10106	.9949	.10158	9.845	.2
.9	.10279	.9947	.10334	9.677	84.1
6.0	0.10453	0.9945	0.10510	9.514	84.0
.1	.10626	.9943	.10687	9.357	83.9
.2	.10800	.9942	.10863	9.205	.8
.3	.10973	.9940	.11040	9.058	.7
.4	.11147	.9938	.11217	8.915	.6
.5	.11320	.9936	.11394	8.777	.5
.6	.11494	.9934	.11570	8.643	.4
.7	.11667	.9932	.11747	8.513	.3
.8	.11840	.9930	.11924	8.386	.2
6.9	.12014	.9928	.12101	8.264	83.1
	$\cos \theta$	$\sin \theta$	$\cot \theta$	$\tan \theta$	θ

Table (Continued)

θ	$\sin \theta$	$\cos \theta$	$\tan \theta$	$\cot \theta$	
7.0	0.12187	0.9925	0.12278	8.144	83.0
.1	.12360	.9923	.12456	8.028	82.9
.2	.12533	.9921	.12633	7.916	.8
.3	.12706	.9919	.12810	7.806	.7
.4	.12880	.9917	.12988	7.700	.6
.5	.13053	.9914	.13165	7.596	.5
.6	.13226	.9912	.13343	7.495	.4
.7	.13399	.9910	.13521	7.396	.3
.8	.13572	.9907	.13698	7.300	.2
.9	.13744	.9905	.13876	7.207	82.1
8.0	0.13917	0.9903	0.14054	7.115	82.0
.1	.14090	.9900	.14232	7.026	81.9
.2	.14263	.9898	.14410	6.940	.8
.3	.14436	.9895	.14588	6.855	.7
.4	.14608	.9893	.14767	6.772	.6
.5	.14781	.9890	.14945	6.691	.5
.6	.14954	.9888	.15124	6.612	.4
.7	.15126	.9885	.15302	6.535	.3
.8	.15299	.9882	.15481	6.460	.2
.9	.15471	.9880	.15660	6.386	81.1
9.0	0.15643	0.9877	0.15838	6.314	81.0
.1	.15816	.9874	.16017	6.243	80.9
.2	.15988	.9871	.16196	6.174	.8
.3	.16160	.9869	.16376	6.107	.7
.4	.16333	.9866	.16555	6.041	.6
.5	.16505	.9863	.16734	5.976	.5
.6	.16677	.9860	.16914	5.912	.4
.7	.16849	.9857	.17093	5.850	.3
.8	.17021	.9854	.17273	5.789	.2
.9	.17193	.9851	.17453	5.730	80.1
10.0	0.1736	0.9848	0.1763	5.671	80.0
.1	.1754	.9845	.1781	5.614	79.9
.2	.1771	.9842	.1799	5.558	.8
.3	.1788	.9839	.1817	5.503	.7
.4	.1805	.9836	.1835	5.449	.6
.5	.1822	.9833	.1853	5.396	.5
10.6	.1840	.9829	.1871	5.343	79.4
	$\cos \theta$	$\sin \theta$	$\cot \theta$	$\tan \theta$	θ

Table of Trigonometric Functions—Decimal Degrees
(csc θ = 1/sin θ; sec θ = 1/cos θ) (Continued)

θ	sin θ	cos θ	tan θ	cot θ	
10.7	.1857	.9826	.1890	5.292	79.3
.8	.1874	.9823	.1908	5.242	.2
.9	.1891	.9820	.1926	5.193	79.1
11.0	0.1908	0.9816	0.1944	5.145	79.0
.1	.1925	.9813	.1962	5.097	78.9
.2	.1942	.9810	.1980	5.050	.8
.3	.1959	.9806	.1998	5.005	.7
.4	.1977	.9803	.2016	4.959	.6
.5	.1994	.9799	.2035	4.915	.5
.6	.2011	.9796	.2053	4.872	.4
.7	.2028	.9792	.2071	4.829	.3
.8	.2045	.9789	.2089	4.787	.2
.9	.2062	.9785	.2107	4.745	78.1
12.0	0.2079	0.9781	0.2126	4.705	78.0
.1	.2096	.9778	.2144	4.665	77.9
.2	.2113	.9774	.2162	4.625	.8
.3	.2130	.9770	.2180	4.586	.7
.4	.2147	.9767	.2199	4.548	.6
.5	.2164	.9763	.2217	4.511	.5
.6	.2181	.9759	.2235	4.474	.4
.7	.2198	.9755	.2254	4.437	.3
.8	.2215	.9751	.2272	4.402	.2
.9	.2233	.9748	.2290	4.366	77.1
13.0	0.2250	0.9744	0.2309	4.331	77.0
.1	.2267	.9740	.2327	4.297	76.9
.2	.2284	.9736	.2345	4.264	.8
.3	.2300	.9732	.2364	4.230	.7
.4	.2317	.9728	.2382	4.198	.6
.5	.2334	.9724	.2401	4.165	.5
.6	.2351	.9720	.2419	4.134	.4
.7	.2368	.9715	.2438	4.102	.3
.8	.2385	.9711	.2456	4.071	.2
.9	.2402	.9707	.2475	4.041	76.1
14.0	0.2419	0.9703	0.2493	4.011	76.0
.1	.2436	.9699	.2512	3.981	75.9
14.2	.2453	.9694	.2530	3.952	75.8
	cos θ	sin θ	cot θ	tan θ	θ

Table (Continued)

θ	$\sin \theta$	$\cos \theta$	$\tan \theta$	$\cot \theta$	
14.3	.2470	.9690	.2549	3.923	75.7
.4	.2487	.9686	.2568	3.895	.6
.5	.2504	.9681	.2586	3.867	.5
.6	.2521	.9677	.2605	3.839	.4
.7	.2538	.9673	.2623	3.812	.3
.8	.2554	.9668	.2642	3.785	.2
.9	.2571	.9664	.2661	3.758	75.1
15.0	0.2588	0.9659	0.2679	3.732	75.0
.1	.2605	.9655	.2698	3.706	74.9
.2	.2622	.9650	.2717	3.681	.8
.3	.2639	.9646	.2736	3.655	.7
.4	.2656	.9641	.2754	3.630	.6
.5	.2672	.9636	.2773	3.606	.5
.6	.2689	.9632	.2792	3.582	.4
.7	.2706	.9627	.2811	3.558	.3
.8	.2723	.9622	.2830	3.534	.2
.9	.2740	.9617	.2849	3.511	74.1
16.0	0.2756	0.9613	0.2867	3.487	74.0
.1	.2773	.9608	.2886	3.465	73.9
.2	.2790	.9603	.2905	3.442	.8
.3	.2807	.9598	.2924	3.420	.7
.4	.2823	.9593	.2943	3.398	.6
.5	.2840	.9588	.2962	3.376	.5
.6	.2857	.9583	.2981	3.354	.4
.7	.2874	.9578	.3000	3.333	.3
.8	.2890	.9573	.3019	3.312	.2
.9	.2907	.9568	.3038	3.291	73.1
17.0	0.2924	0.9563	0.3057	3.271	73.0
.1	.2940	.9558	.3076	3.251	72.9
.2	.2957	.9553	.3096	3.230	.8
.3	.2974	.9548	.3115	3.211	.7
.4	.2990	.9542	.3134	3.191	.6
.5	.3007	.9537	.3153	3.172	.5
.6	.3024	.9532	.3172	3.152	.4
.7	.3040	.9527	.3191	3.133	.3
.8	.3057	.9521	.3211	3.115	.2
.9	.3074	.9516	.3230	3.096	72.1
18.0	0.3090	0.9511	0.3249	3.078	72.0
	$\cos \theta$	$\sin \theta$	$\cot \theta$	$\tan \theta$	θ

Table of Trigonometric Functions—Decimal Degrees
($\csc \theta = 1/\sin \theta$; $\sec \theta = 1/\cos \theta$) (Continued)

θ	$\sin \theta$	$\cos \theta$	$\tan \theta$	$\cot \theta$	
18.1	.3107	.9505	.3269	3.060	71.9
.2	.3123	.9500	.3288	3.042	.8
.3	.3140	.9494	.3307	3.024	.7
.4	.3156	.9489	.3327	3.006	.6
.5	.3173	.9483	.3346	2.989	.5
.6	.3190	.9478	.3365	2.971	.4
.7	.3206	.9472	.3385	2.954	.3
.8	.3223	.9466	.3404	2.937	.2
.9	.3239	.9461	.3424	2.921	71.1
19.0	0.3256	0.9455	0.3443	2.904	71.0
.1	.3272	.9449	.3463	2.888	70.9
.2	.3289	.9444	.3482	2.872	.8
.3	.3305	.9438	.3502	2.856	.7
.4	.3322	.9432	.3522	2.840	.6
.5	.3338	.9426	.3541	2.824	.5
.6	.3355	.9421	.3561	2.808	.4
.7	.3371	.9415	.3581	2.793	.3
.8	.3387	.9409	.3600	2.778	.2
.9	.3404	.9403	.3620	2.762	70.1
20.0	0.3420	0.9397	0.3640	2.747	70.0
.1	.3437	.9391	.3659	2.733	69.9
.2	.3453	.9385	.3679	2.718	.8
.3	.3469	.9379	.3699	2.703	.7
.4	.3486	.9373	.3719	2.689	.6
.5	.3502	.9367	.3739	2.675	.5
.6	.3518	.9361	.3759	2.660	.4
.7	.3535	.9354	.3779	2.646	.3
.8	.3551	.9348	.3799	2.633	.2
.9	.3567	.9342	.3819	2.619	69.1
21.0	0.3584	0.9336	0.3839	2.605	69.0
.1	.3600	.9330	.3859	2.592	68.9
.2	.3616	.9323	.3879	2.578	.8
.3	.3633	.9317	.3899	2.565	.7
.4	.3649	.9311	.3919	2.552	.6
.5	.3665	.9304	.3939	2.539	.5
.6	.3681	.9298	.3959	2.526	.4
.7	.3697	.9291	.3979	2.513	.3
21.8	.3714	.9285	.4000	2.500	68.2
	$\cos \theta$	$\sin \theta$	$\cot \theta$	$\tan \theta$	θ

Table (Continued)

θ	$\sin \theta$	$\cos \theta$	$\tan \theta$	$\cot \theta$	
21.9	.3730	.9278	.4020	2.488	68.1
22.0	0.3746	0.9272	0.4040	2.475	68.0
.1	.3762	.9265	.4061	2.463	67.9
.2	.3778	.9259	.4081	2.450	.8
.3	.3795	.9252	.4101	2.438	.7
.4	.3811	.9245	.4122	2.426	.6
.5	.3827	.9239	.4142	2.414	.5
.6	.3843	.9232	.4163	2.402	.4
.7	.3859	.9225	.4183	2.391	.3
.8	.3875	.9219	.4204	2.379	.2
.9	.3891	.9212	.4224	2.367	67.1
23.0	0.3907	0.9205	0.4245	2.356	67.0
.1	.3923	.9198	.4265	2.344	66.9
.2	.3939	.9191	.4286	2.333	.8
.3	.3955	.9184	.4307	2.322	.7
.4	.3971	.9178	.4327	2.311	.6
.5	.3987	.9171	.4348	2.300	.5
.6	.4003	.9164	.4369	2.289	.4
.7	.4019	.9157	.4390	2.278	.3
.8	.4035	.9150	.4411	2.267	.2
.9	.4051	.9143	.4431	2.257	66.1
24.0	0.4067	0.9135	0.4452	2.246	66.0
.1	.4083	.9128	.4473	2.236	65.9
.2	.4099	.9121	.4494	2.225	.8
.3	.4115	.9114	.4515	2.215	.7
.4	.4131	.9107	.4536	2.204	.6
.5	.4147	.9100	.4557	2.194	.5
.6	.4163	.9092	.4578	2.184	.4
.7	.4179	.9085	.4599	2.174	.3
.8	.4195	.9078	.4621	2.164	.2
.9	.4210	.9070	.4642	2.154	65.1
25.0	0.4226	0.9063	0.4663	2.145	65.0
.1	.4242	.9056	.4684	2.135	64.9
.2	.4258	.9048	.4706	2.125	.8
.3	.4274	.9041	.4727	2.116	.7
.4	.4289	.9033	.4748	2.106	.6
.5	.4305	.9026	.4770	2.097	.5
25.6	.4321	.9018	.4791	2.087	64.4
	$\cos \theta$	$\sin \theta$	$\cot \theta$	$\tan \theta$	θ

Table of Trigonometric Functions—Decimal Degrees
(csc θ = 1/sin θ; sec θ = 1/cos θ) (Continued)

θ	sin θ	cos θ	tan θ	cot θ	
25.7	.4337	.9011	.4813	2.078	64.3
.8	.4352	.9003	.4834	2.069	.2
.9	.4368	.8996	.4856	2.059	64.1
26.0	0.4384	0.8988	0.4877	2.050	64.0
.1	.4399	.8980	.4899	2.041	63.9
.2	.4415	.8973	.4921	2.032	.8
.3	.4431	.8965	.4942	2.023	.7
.4	.4446	.8957	.4964	2.014	.6
.5	.4462	.8949	.4986	2.006	.5
.6	.4478	.8942	.5008	1.997	.4
.7	.4493	.8934	.5029	1.988	.3
.8	.4509	.8926	.5051	1.980	.2
.9	.4524	.8918	.5073	1.971	63.1
27.0	0.4540	0.8910	0.5095	1.963	63.0
.1	.4555	.8902	.5117	1.954	62.9
.2	.4571	.8894	.5139	1.946	.8
.3	.4586	.8886	.5161	1.937	.7
.4	.4602	.8878	.5184	1.929	.6
.5	.4617	.8870	.5206	1.921	.5
.6	.4633	.8862	.5228	1.913	.4
.7	.4648	.8854	.5250	1.905	.3
.8	.4664	.8846	.5272	1.897	.2
.9	.4679	.8838	.5295	1.889	62.1
28.0	0.4695	0.8829	0.5317	1.881	62.0
.1	.4710	.8821	.5340	1.873	61.9
.2	.4726	.8813	.5362	1.865	.8
.3	.4741	.8805	.5384	1.857	.7
.4	.4756	.8796	.5407	1.849	.6
.5	.4772	.8788	.5430	1.842	.5
.6	.4787	.8780	.5452	1.834	.4
.7	.4802	.8771	.5475	1.827	.3
.8	.4818	.8763	.5498	1.819	.2
.9	.4833	.8755	.5520	1.811	61.1
29.0	0.4848	0.8746	0.5543	1.804	61.0
.1	.4863	.8738	.5566	1.797	60.9
29.2	.4879	.8729	.5589	1.789	60.8
	cos θ	sin θ	cot θ	tan θ	θ

Table (Continued)

θ	$\sin \theta$	$\cos \theta$	$\tan \theta$	$\cot \theta$	
29.3	.4894	.8721	.5612	1.782	60.7
.4	.4909	.8712	.5635	1.775	.6
.5	.4924	.8704	.5658	1.767	.5
.6	.4939	.8695	.5681	1.760	.4
.7	.4955	.8686	.5704	1.753	.3
.8	.4970	.8678	.5727	1.746	.2
.9	.4985	.8669	.5750	1.739	60.1
30.0	0.5000	0.8660	0.5774	1.7321	60.0
.1	.5015	.8652	.5797	1.7251	59.9
.2	.5030	.8643	.5820	1.7182	.8
.3	.5045	.8634	.5844	1.7113	.7
.4	.5060	.8625	.5867	1.7045	.6
.5	.5075	.8616	.5890	1.6977	.5
.6	.5090	.8607	.5914	1.6909	.4
.7	.5105	.8599	.5938	1.6842	.3
.8	.5120	.8590	.5961	1.6775	.2
.9	.5135	.8581	.5985	1.6709	59.1
31.0	0.5150	0.8572	0.6009	1.6643	59.0
.1	.5165	.8563	.6032	1.6577	58.9
.2	.5180	.8554	.6056	1.6512	.8
.3	.5195	.8545	.6080	1.6447	.7
.4	.5210	.8536	.6104	1.6383	.6
.5	.5225	.8526	.6128	1.6319	.5
.6	.5240	.8517	.6152	1.6255	.4
.7	.5255	.8508	.6176	1.6191	.3
.8	.5270	.8499	.6200	1.6128	.2
.9	.5284	.8490	.6224	1.6066	58.1
32.0	0.5299	0.8480	0.6249	1.6003	58.0
.1	.5314	.8471	.6273	1.5941	57.9
.2	.5329	.8462	.6297	1.5880	.8
.3	.5344	.8453	.6322	1.5818	.7
.4	.5358	.8443	.6346	1.5757	.6
.5	.5373	.8434	.6371	1.5697	.5
.6	.5388	.8425	.6395	1.5637	.4
.7	.5402	.8415	.6420	1.5577	.3
.8	.5417	.8406	.6445	1.5517	.2
32.9	.5432	.8396	.6469	1.5458	57.1
	$\cos \theta$	$\sin \theta$	$\cot \theta$	$\tan \theta$	θ

Table of Trigonometric Functions—Decimal Degrees
(csc θ = 1/sin θ; sec θ = 1/cos θ) (Continued)

θ	sin θ	cos θ	tan θ	cot θ	
33.0	0.5446	0.8387	0.6494	1.5399	57.0
.1	.5461	.8377	.6519	1.5340	56.9
.2	.5476	.8368	.6544	1.5282	.8
.3	.5490	.8358	.6569	1.5224	.7
.4	.5505	.8348	.6594	1.5166	.6
.5	.5519	.8339	.6619	1.5108	.5
.6	.5534	.8329	.6644	1.5051	.4
.7	.5548	.8320	.6669	1.4994	.3
.8	.5563	.8310	.6694	1.4938	.2
.9	.5577	.8300	.6720	1.4882	56.1
34.0	0.5592	0.8290	0.6745	1.4826	56.0
.1	.5606	.8281	.6771	1.4770	55.9
.2	.5621	.8271	.6796	1.4715	.8
.3	.5635	.8261	.6822	1.4659	.7
.4	.5650	.8251	.6847	1.4605	.6
.5	.5664	.8241	.6873	1.4550	.5
.6	.5678	.8231	.6899	1.4496	.4
.7	.5693	.8221	.6924	1.4442	.3
.8	.5707	.8211	.6950	1.4388	.2
.9	.5721	.8202	.6976	1.4335	55.1
35.0	0.5736	0.8192	0.7002	1.4281	55.0
.1	.5750	.8181	.7028	1.4229	54.9
.2	.5764	.8171	.7054	1.4176	.8
.3	.5779	.8161	.7080	1.4124	.7
.4	.5793	.8151	.7107	1.4071	.6
.5	.5807	.8141	.7133	1.4019	.5
.6	.5821	.8131	.7159	1.3968	.4
.7	.5835	.8121	.7186	1.3916	.3
.8	.5850	.8111	.7212	1.3865	.2
.9	.5864	.8100	.7239	1.3814	54.1
36.0	0.5878	0.8090	0.7265	1.3764	54.0
.1	.5892	.8080	.7292	1.3713	53.9
.2	.5906	.8070	.7319	1.3663	.8
.3	.5920	.8059	.7346	1.3613	.7
.4	.5934	.8049	.7373	1.3564	.6
36.5	.5948	.8039	.7400	1.3514	53.5
	cos θ	sin θ	cot θ	tan θ	θ

Table (Continued)

θ	sin θ	cos θ	tan θ	cot θ	
36.6	.5962	.8028	.7427	1.3465	53.4
.7	.5976	.8018	.7454	1.3416	.3
.8	.5990	.8007	.7481	1.3367	.2
.9	.6004	.7997	.7508	1.3319	53.1
37.0	0.6018	0.7986	0.7536	1.3270	53.0
.1	.6032	.7976	.7563	1.3222	52.9
.2	.6046	.7965	.7590	1.3175	.8
.3	.6060	.7955	.7618	1.3127	.7
.4	.6074	.7944	.7646	1.3079	.6
.5	.6088	.7934	.7673	1.3032	.5
.6	.6101	.7923	.7701	1.2985	.4
.7	.6115	.7912	.7729	1.2938	.3
.8	.6129	.7902	.7757	1.2892	.2
.9	.6143	.7891	.7785	1.2846	52.1
38.0	0.6157	0.7880	0.7813	1.2799	52.0
.1	.6170	.7869	.7841	1.2753	51.9
.2	.6184	.7859	.7869	1.2708	.8
.3	.6198	.7848	.7898	1.2662	.7
.4	.6211	.7837	.7926	1.2617	.6
.5	.6225	.7826	.7954	1.2572	.5
.6	.6239	.7815	.7983	1.2527	.4
.7	.6252	.7804	.8012	1.2482	.3
.8	.6266	.7793	.8040	1.2437	.2
.9	.6280	.7782	.8069	1.2393	51.1
39.0	0.6293	0.7771	0.8098	1.2349	51.0
.1	.6307	.7760	.8127	1.2305	50.9
.2	.6320	.7749	.8156	1.2261	.8
.3	.6334	.7738	.8185	1.2218	.7
.4	.6347	.7727	.8214	1.2174	.6
.5	.6361	.7716	.8243	1.2131	.5
.6	.6374	.7705	.8273	1.2088	.4
.7	.6388	.7694	.8302	1.2045	.3
.8	.6401	.7683	.8332	1.2002	.2
.9	.6414	.7672	.8361	1.1960	50.1
40.0	0.6428	0.7660	0.8391	1.1918	50.0
40.1	.6441	.7649	.8421	1.1875	49.9
	cos θ	sin θ	cot θ	tan θ	θ

Table of Trigonometric Functions—Decimal Degrees
(csc θ = 1/sin θ; sec θ = 1/cos θ) (Continued)

θ	sin θ	cos θ	tan θ	cot θ	
40.2	.6455	.7638	.8451	1.1833	49.8
.3	.6468	.7627	.8481	1.1792	.7
.4	.6481	.7615	.8511	1.1750	.6
.5	.6494	.7604	.8541	1.1708	.5
.6	.6508	.7593	.8571	1.1667	.4
.7	.6521	.7581	.8601	1.1626	.3
.8	.6534	.7570	.8632	1.1585	.2
.9	.6547	.7559	.8662	1.1544	49.1
41.0	0.6561	0.7547	0.8693	1.1504	49.0
.1	.6574	.7536	.8724	1.1463	48.9
.2	.6587	.7524	.8754	1.1423	.8
.3	.6600	.7513	.8785	1.1383	.7
.4	.6613	.7501	.8816	1.1343	.6
.5	.6626	.7490	.8847	1.1303	.5
.6	.6639	.7478	.8878	1.1263	.4
.7	.6652	.7466	.8910	1.1224	.3
.8	.6665	.7455	.8941	1.1184	.2
.9	.6678	.7443	.8972	1.1145	48.1
42.0	0.6691	0.7431	0.9004	1.1106	48.0
.1	.6704	.7420	.9036	1.1067	47.9
.2	.6717	.7408	.9067	1.1028	.8
.3	.6730	.7396	.9099	1.0990	.7
.4	.6743	.7385	.9131	1.0951	.6
.5	.6756	.7373	.9163	1.0913	.5
.6	.6769	.7361	.9195	1.0875	.4
.7	.6782	.7349	.9228	1.0837	.3
.8	.6794	.7337	.9260	1.0799	.2
.9	.6807	.7325	.9293	1.0761	47.1
43.0	0.6820	0.7314	0.9325	1.0724	47.0
.1	.6833	.7302	.9358	1.0686	46.9
.2	.6845	.7290	.9391	1.0649	.8
.3	.6858	.7278	.9424	1.0612	.7
.4	.6871	.7266	.9457	1.0575	.6
.5	.6884	.7254	.9490	1.0538	.5
.6	.6896	.7242	.9523	1.0501	.4
43.7	.6909	.7230	.9556	1.0464	46.3
	cos θ	sin θ	cot θ	tan θ	θ

Table (Continued)

θ	$\sin \theta$	$\cos \theta$	$\tan \theta$	$\cot \theta$	
43.8	.6921	.7218	.9590	1.0428	46.2
.9	.6934	.7206	.9623	1.0392	46.1
44.0	0.6947	0.7193	0.9657	1.0355	46.0
.1	.6959	.7181	.9691	1.0319	45.9
.2	.6972	.7169	.9725	1.0283	.8
.3	.6984	.7157	.9759	1.0247	.7
.4	.6997	.7145	.9793	1.0212	.6
.5	.7009	.7133	.9827	1.0176	.5
.6	.7022	.7120	.9861	1.0141	.4
.7	.7034	.7108	.9896	1.0105	.3
.8	.7046	.7096	.9930	1.0070	.2
.9	.7059	.7083	.9965	1.0035	45.1
45.0	0.7071	0.7071	1.0000	1.0000	45.0
	$\cos \theta$	$\sin \theta$	$\cot \theta$	$\tan \theta$	θ

Answers to
Odd-Numbered Exercises
and Chapter Tests

CHAPTER 1

Problem Set 1.1

1. $x + 5$ **3.** $6 - x$ **5.** $2t < y$ **7.** $\dfrac{3x}{2y} > 6$ **9.** $x + y < x - y$

11. $3(x - 5) > y$ **13.** $s - t \neq s + t$ **15.** $2(t + 3) \not> t - 6$ **17.** $\{0, 1, 2, 3, 4, 5, 6\}$
19. $\{2, 4\}$ **21.** $\{-2, -1, 0, 1, 2, 3, 5, 7\}$ **23.** $\{1\}$ **25.** $\{-2, -1, 0, 1, 2, 4, 6\}$
27. $\{0, 2\}$ **29.** $\{0, 1, 2, 3, 4, 5, 6\}$ **31.** $\{0, 1, 2, 3, 4, 5, 6\}$ **33.** $\{0, 2, 4\}$
35. $\{a\}\ \{b\}\ \{c\}\ \{a, b\}\ \{a, c\}\ \{b, c\}\ \{a, b, c\}\ \varnothing$ **37.** Any two sets that do not have any
elements in common will do.

39.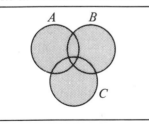

41.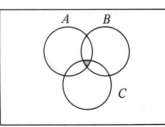

43.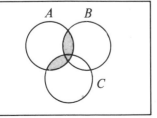

45.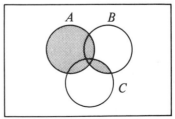

47.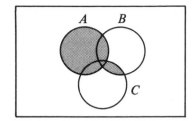

49.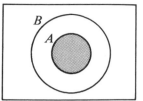

51.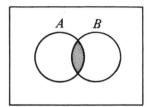

53.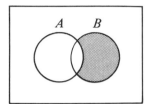

55.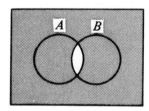

57.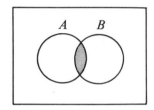

59. $\{1, 2, 3\}$ **61.** $\{1, 3, 5\}$ **63.** $\{2, 4, 8\}$ **65.** $\{2, 4, 6\}$ **67.** 9   **69.** 13

Problem Set 1.2

1.

$-1.75 \quad -\frac{1}{2} \quad \frac{1}{3} \quad 1.3$

$-3 \qquad\qquad\qquad 0 \quad 1 \qquad\qquad 4.5$

$-4 \; -3 \; -2 \; -1 \quad 0 \quad 1 \quad 2 \quad 3 \quad 4$

3. $4, -4, \frac{1}{4}$ **5.** $-\frac{1}{2}, \frac{1}{2}, -2$ **7.** $5, -5, \frac{1}{5}$ **9.** $\frac{3}{8}, -\frac{3}{8}, \frac{8}{3}$ **11.** $-\frac{1}{6}, \frac{1}{6}, -6$
13. $3, -3, \frac{1}{3}$ **15.** $\sqrt{3}, -\sqrt{3}, 1/\sqrt{3}$ **17.** $-1/\sqrt{2}, 1/\sqrt{2}, -\sqrt{2}$ **19.** $x, -x, 1/x$
21. $1, -1$ **23.** 0 **25.** 2 **27.** $\frac{3}{4}$ **29.** π **31.** -4 **33.** -2 **35.** $-\frac{3}{4}$
37. 2 **39.** $-\frac{1}{3}$ **41.** 5 **43.** 8 **45.** 5 **47.** 6 **49.** 0 **51.** $\{1, 2\}$
53. $\{-6, -5.2, 0, 1, 2, 2.3, \frac{9}{2}\}$ **55.** $\{-\sqrt{7}, -\pi, \sqrt{17}\}$ **57.** $\{0, 1, 2\}$
59. False; $1/0$ undefined   **61.** True   **63.** True   **65.** False; $0/1 = 0$, 0 is rational
67. True   **69.** True   **71.** $\frac{21}{40}$ **73.** $\frac{12}{5}$ **75.** $\frac{120}{17}$ **77.** $\frac{280}{9}$ **79.** $\frac{72}{385}$ **81.** 1
83. 1 **85.** 2 **87.** 1 **89.** -3 and 7 **91.** $1 = 0.999\ldots$ **93.** $0.282828\ldots$
95. $-\$15$ **97.** -6 and 6

Problem Set 1.3

1.

$\xleftarrow{\hspace{2cm}} \; | \; \oplus \xrightarrow{\hspace{2cm}}$

$0 \quad 1$

3.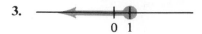
0 1

5.
0   4

7.
0   4

9.
0

11.
−2   0

13.
0   4

15.
−3   0 1

17.
−3   0 1

19.
−3   0 1

21.
−3   0 1

23.
0   2

25.
−1 0

27.
−1 0   3

29. ∅

31.
−4   0   2

33.
−1 0   2

35.
−1 0   2

37.
$-3 \quad 0 \ 1$

39.
$-3 \quad 0$

41.
$-4 \quad 0 \ 1$

43.
$-3 \quad 0 \ 2 \quad 4$

45.
$-5 \quad 0 \quad 3$

47.
$-5 \quad -2 \quad 2 \quad 5$

49. $\$180 < x < \$270$
51. $-3 \leq x \leq 7$
53. $x < -4$ or $x > 4$

Problem Set 1.4

1. $6 + x$ **3.** $a + 8$ **5.** $15y$ **7.** x **9.** a **11.** x **13.** $3x + 18$
15. $12x + 8$ **17.** $15a + 10b$ **19.** $28 + 12y$ **21.** $40x + 8$ **23.** $18x + 12 + 24y$
25. $13x$ **27.** $16y$ **29.** $9a$ **31.** $\frac{6}{7}$ **33.** $\dfrac{9}{\sqrt{3}}$ **35.** $\dfrac{11}{x}$ **37.** $\dfrac{9}{a}$
39. $7x + 7$ **41.** $5x + 12$ **43.** $14a + 7$ **45.** $6y + 6$ **47.** $3x + 6$
49. $10x + 20$ **51.** $7x$ **53.** $\frac{4}{5}$ **55.** Commutative **57.** Commutative
59. Additive inverse **61.** Commutative **63.** Associative and commutative
65. Associative and commutative **67.** Distributive **69.** $y \cdot 5$ **71.** $a + 3$ **73.** 7
75. $(2 + x) + 6$ **77.** $xy + 2y$ **79.** 11 **85.** Not commutative

Problem Set 1.5

1. 4 **3.** -4 **5.** -8 **7.** -5 **9.** 5 **11.** -13 **13.** -14 **15.** 4
17. -10 **19.** -4 **21.** 10 **23.** -35 **25.** 16 **27.** 3 **29.** -5 **31.** 1
33. 19 **35.** -14 **37.** 10 **39.** -1 **41.** 11 **43.** -4 **45.** -3 **47.** 8
49. 16 **51.** $2x - 8$ **53.** $5y - 15$ **55.** $7x - 28$ **57.** $4a - 8 + 4b$
59. $5x - 5y - 20$ **61.** $3x$ **63.** $-13x$ **65.** $4y$ **67.** a **69.** $-6x$
71. $-11a$ **73.** $2x$ **75.** $7x - 14$ **77.** $x + 7$ **79.** $-3a + 2$ **81.** 13
83. $2t - 1$ **85.** -8 **87.** $-7x$ **89.** 13 **91.** $6a$ **93.** $15x - 20$ **95.** $3y + 8$

Problem Set 1.6

1. -15 **3.** 15 **5.** -24 **7.** 20 **9.** -12 **11.** -24 **13.** -24
15. 24 **17.** $-10x$ **19.** $-21a$ **21.** $-32y$ **23.** $15x$ **25.** $-5x - 40$
27. $-8x - 6$ **29.** $-12x + 30$ **31.** -2 **33.** 2 **35.** $-\frac{2}{3}$ **37.** 32 **39.** $\frac{320}{3}$

41. $-\frac{1}{18}$ **43.** $\frac{4}{3}$ **45.** -14 **47.** -7 **49.** 18 **51.** -44 **53.** -30
55. 18 **57.** 4 **59.** $\frac{5}{3}$ **61.** 11 **63.** $\frac{81}{8}$ **65.** $14x + 12$ **67.** $-a - 12$
69. $3x + 13$ **71.** $-7m + 27$ **73.** $-2x + 9$ **75.** $y + 6$ **77.** $7y + 10$
79. $-11x + 10$ **81.** -3 **83.** -11 **85.** $2x$

Chapter 1   Test

1. $2(3x + 4y)$ **2.** $(2a - 3b) < (2a + 3b)$ **3.** $\{1, 3\}$ **4.** $\{1, 2, 3, 4\}$ **5.** $\emptyset$
6. $3, -\frac{1}{3}$ **7.** $-\frac{4}{3}, \frac{3}{4}$ **8.** $\sqrt{5}, -1/\sqrt{5}$ **9.** 3 **10.** 7 **11.** -2
12. $\{-5, 0, 1, 4\}$ **13.** $\{-5, -4.1, -3.75, -\frac{5}{6}, 0, 1, 1.8, 4\}$ **14.** $\{-\sqrt{2}, \sqrt{3}\}$

15.

16.

17.

18. Commutative property of addition
19. Multiplicative identity property
20. Associative and commutative property of multiplication
21. Associative and commutative property of addition   **22.** -19 **23.** 14
24. -19 **25.** 14 **26.** -26 **27.** $\frac{5}{2}$ **28.** 2 **29.** 1 **30.** $\frac{2}{5}$ **31.** 1
32. $24x$ **33.** $-10x$ **34.** $5x + 15$ **35.** $-2x - 18$ **36.** $-6x - 12$
37. $-12x - 8$ **38.** $-6x + 10y - 8$ **39.** $7x$ **40.** $4x$ **41.** $-4x$ **42.** $2x$
43. $-6x - 4$ **44.** $4y - 10$ **45.** $3x - 17$ **46.** $11a - 10$ **47.** $-\frac{7}{3}$ **48.** $-\frac{5}{2}$
49. 16 **50.** -5

CHAPTER 2

Problem Set 2.1

1. 8 **3.** 5 **5.** 2 **7.** -7 **9.** $-\frac{9}{2}$ **11.** -4 **13.** $-\frac{4}{3}$ **15.** -4
17. $\frac{7}{2}$ **19.** $-\frac{11}{5}$ **21.** 12 **23.** -10 **25.** 7 **27.** -4 **29.** -7 **31.** 3
33. -4 **35.** 2 **37.** 1 **39.** 6 **41.** 2 **43.** 0 **45.** -3 **47.** 4
49. 0 **51.** 17 **53.** 2 **55.** -3 **57.** 3 **59.** No

For problems 61 through 66, if the equation
results in a true statement, then the
solution set is all real numbers. If
the equation results in a false statement,
then the solution set is the empty set.

61. $\emptyset$ **63.** All real numbers   **65.** $\emptyset$

67.

69.

71.

73.

Problem Set 2.2

1. $x \leq \frac{3}{2}$

3. $x \geq -5$

5. $a > 2$

7. $x \geq 1$

9. $x < -3$

11. $m \geq -1$

13. $x \geq -3$

15. $y \leq \frac{7}{2}$

17. $y \leq -2$

19. $x \leq 2$

21. $a \geq 1$

23. $x \leq 2$

25. $t < 3$

27. $a \leq -1$

29. $x \geq -17$

31. $y < -5$

33. $x \leq -7$

35. $7 \leq m \leq 12$

37. $-\frac{1}{5} < x < 2$

39. $-4 < a < 2$

41. $-6 \leq a \leq -4$

43. $x \leq -7$ or $x > 3$

45. $y \leq -1$ or $y \geq 5$

47. all real numbers

49. $a < -13$ or $a \geq 4$

51. The opposite of x lies between -5 and -2. $-5 < -x < -2$

53. No, replacing x with -5 yields a false statement.

55.
$$\left. \begin{array}{l} -3(-2) + 2 = 8 \\ 2(-2) + 12 = 8 \end{array} \right\}$$
Both sides simplify to the same number, so -2 is the correct end point.

57. 4   **59.** $-\frac{1}{2}$ **61.** 4   **63.** 1   **65.** $|x| = $ the distance between x and 0 on the number line.

Problem Set 2.3

1. 4, -4 **3.** 2, -2 **5.** $\varnothing$ **7.** 1, -1 **9.** $\varnothing$ **11.** 6, -6
13. 7, -3 **15.** 5, 3   **17.** 2, 4   **19.** $\varnothing$ **21.** $\frac{4}{3}$, -2 **23.** $\frac{2}{3}$, $-\frac{10}{3}$ **25.** $\varnothing$
27. 4, -3 **29.** -2, -3 **31.** $-5, \frac{3}{5}$ **33.** 1, $\frac{1}{9}$ **35.** $-\frac{1}{2}$ **37.** 0   **39.** $-\frac{1}{2}$
41. $-\frac{1}{6}$, $-\frac{7}{4}$ **43.** All real numbers   **45.** All real numbers
47. $\begin{array}{l} |a - b| = |4 - (-7)| = |11| = 11 \\ |b - a| = |-7 - 4| = |-11| = 11 \end{array}$ and $\begin{array}{l} |a - b| = |-5 - (-8)| = |3| = 3 \\ |b - a| = |-8 - (-5)| = |-3| = 3 \end{array}$
49. -2, -1, 0, 1, 2, 3, 4   **51.** $x \geq 2$

53.

55.

$-\frac{11}{3}$ -1

57. $t \leq -\frac{3}{2}$ **59.** $x < -5$

Problem Set 2.4

1. $-3 < x < 3$

3. $x \leq -2$ or $x \geq 2$

5. $-1 < a < 1$

7. $-3 < x < 3$

9. $t < -7$ or $t > 7$

11. $\varnothing$

13. all real numbers

15. $-4 < x < 10$

17. $a \leq -9$ or $a \geq -1$

19. $\varnothing$

21. $-1 \leq x \leq 4$

23. $-1 < x < 5$

25. $y \leq -5$ or $y \geq -1$

27. $x < \frac{2}{3}$ or $x > \frac{5}{3}$

29. $k \leq -5$ or $k \geq 2$

31. $-1 < x < 7$

33. $a \leq -2$ or $a \geq 1$

35. $-6 < x < \frac{8}{3}$

37. $\varnothing$

39. $-\frac{3}{2} \leq a \leq \frac{1}{2}$

41. $x < 2$ or $x > 8$

43. $x \leq -3$ or $x \geq 12$

45. $2 < x < 6$

47. $|x| \leq 4$ **49.** $|x - 2| \leq 5$ **51.** Commutative   **53.** Associative
55. Associative and commutative   **57.** Commutative

Problem Set 2.5

1. 6   **3.** 2   **5.** 4   **7.** 5   **9.** 18.84   **11.** 0   **13.** 8   **15.** 21
17. $l = A/w$ **19.** $t = I/pr$ **21.** $m = E/c^2$ **23.** $T = PV/nR$ **25.** $b^2 = c^2 - a^2$
27. $b = y - mx$ **29.** $h = \dfrac{2A}{b + B}$ **31.** $c = 2s - a - b$ **33.** $a = \dfrac{2(d - vt)}{t^2}$
35. $r = \dfrac{A - P}{Pt}$ **37.** $F = \dfrac{9}{5}C + 32$ **39.** $d = \dfrac{A - a}{n - 1}$ **41.** $y = 3x - 2$
43. $x = zs + \mu$ **45.** $\mu - 2.5s < x < \mu + 2.5s$ **47.** $30° \leq C \leq 40°$ **49.** $2(x + 3)$
51. $5(x - 3)$ **53.** $3x + 2 = x - 4$

Problem Set 2.6

Along with the answers to the odd-numbered problems in this problem set we are including the equations used to solve each problem. Be sure that you try the problems on your own before looking here to see what the correct equations are.

1. $x + 2 = 2x - 5$; 7   **3.** $3(x + 4) = 3$; -3 **5.** $2(2x + 1) = 3(x - 5)$; -17
7. $5x + 2 = 3x + 8$; 3   **9.** $x + (x + 2) = (x + 2) - x + 16$; 8 and 10
11. $x + (x + 1) = 3x - 1$; 2 and 3   **13.** $2x + (x + 1) = 7$; 2
15. $2x - (x + 2) = 5$; 7 and 9
17. The width is x, the length is $2x$; $2x + 4x = 60$; 10 ft by 20 ft
19. The length of a side is x; $4x = 28$; 7 ft
21. The shortest side is x, the medium side is $x + 3$, the longest side is $2x$;
$x + (x + 3) + 2x = 23$; 5 inches

23. The width is x, the length is $2x - 3$; $2(2x - 3) + 2x = 18$; 4 meters
25. Amy's age is x, Patrick's age is $x + 4$; $x + 10 + x + 14 = 36$; Amy is 6, Patrick is 10
27. Bill's age is x, John's age is $2x$; $2x + 5 = \frac{3}{2}(x + 5)$; Bill is 5, John is 10
29. Kate's age is x, Jane's age is $3x$; $3x + 5 = 2(x + 5) - 2$; Kate is 3, Jane is 9
31. 18   **33.** 24   **35.** 10   **37.** $-15x + 21$ **39.** -2

Chapter 2 Test

1. 12   **2.** $-\frac{4}{3}$ **3.** $\frac{17}{3}$ **4.** $-\frac{7}{4}$ **5.** $t \geq -6$ **6.** $x \leq 1$ **7.** $x < 6$
8. $y \geq -52$ **9.** $x = 6$ or $x = 2$ **10.** $a = -1$ or $a = -6$

11. $x < -1$ or $x > \frac{4}{3}$

12. $-\frac{2}{3} \leq x \leq 4$

13. $w = \dfrac{A - 2l}{2}$ **14.** $B = \dfrac{2A}{h} - b$ **15.** $x + (x + 2) = 18$; 8 and 10
16. $x + 9 = 2x$; Amy is 9, Patrick is 13.

CHAPTER 3

Problem Set 3.1

1. 16   **3.** -16 **5.** -27 **7.** 32   **9.** $\frac{1}{8}$ **11.** $\frac{25}{36}$ **13.** x^9 **15.** 64
17. $-8x^6$ **19.** $-6a^6$ **21.** $-36x^{11}$ **23.** $324n^{34}$ **25.** $\frac{1}{9}$ **27.** $-\frac{1}{32}$ **29.** $\frac{1}{9}$
31. $\frac{16}{9}$ **33.** $\frac{7}{12}$ **35.** $-\frac{1}{72}$ **37.** 17   **39.** -4 **41.** x^3 **43.** $\dfrac{a^6}{b^{15}}$ **45.** $\dfrac{48}{x^5}$
47. $\dfrac{8}{125y^{18}}$ **49.** $\dfrac{1}{x^3}$ **51.** y^3 **53.** 3.78×10^5 **55.** 4.9×10^3 **57.** 3.7×10^{-4}
59. 4.95×10^{-3} **61.** 5.62×10^{-1} **63.** 5,340   **65.** 7,800,000   **67.** 0.00344
69. 0.49   **71.** 8   **73.** 25   **75.** $(2^2)^3 = 4^3 = 64,\ 2^{2^3} = 2^8 = 256$ **77.** $g = 32$
79. $(2 + 4)^{-1} = 6^{-1} = \frac{1}{6},\ 2^{-1} + 4^{-1} = \frac{1}{2} + \frac{1}{4} = \frac{3}{4}$ **81.** 22   **83.** 14   **85.** 14
87. -1 **89.** 5

Problem Set 3.2

1. $\dfrac{x^6}{y^4}$ **3.** $\dfrac{4b^2}{a^4}$ **5.** $\frac{1}{9}$ **7.** 8   **9.** $\dfrac{1}{x^{10}}$ **11.** $2a$ **13.** $\frac{1}{81}$ **15.** x^6
17. $\dfrac{b^7}{a}$ **19.** x^4y^6 **21.** $\dfrac{4n^5}{m^2}$ **23.** $\dfrac{x^5}{3}$ **25.** $\dfrac{4x^{18}}{y^{10}}$ **27.** $\dfrac{b^3}{a^4c^3}$ **29.** $\dfrac{x^{10}}{y^{15}}$
31. x^5 **33.** a **35.** 1   **37.** 8×10^4 **39.** 9×10^{10} **41.** 2×10^9
43. 2×10^{-2} **45.** 2.5×10^{-6} **47.** 1.8×10^{-7} **49.** 4.98×10^1 **51.** 2×10^8
53. 2×10^4 **55.** 2.78×10^7 **57.** 6   **59.** 9   **61.** -7 **63.** 1.003×10^{19}
65. $2x + 12$ **67.** 3   **69.** $6y - 9$ **71.** $-4x + 38$

Problem Set 3.3

1. Trinomial 2 5   **3.** Binomial 1 3   **5.** Trinomial 2 8
7. Polynomial 3 4   **9.** Monomial 0 $-\frac{3}{4}$ **11.** Trinomial 3 6   **13.** $7x + 1$
15. $2x^2 + 7x - 15$ **17.** $12a^2 - 7ab - 10b^2$ **19.** $x^2 - 13x + 3$ **21.** $3x^2 - 7x - 3$
23. $-y^3 - y^2 - 4y + 7$ **25.** $2x^3 + x^2 - 3x - 17$ **27.** $x^2 + 2xy + 10y^2$
29. $-3a^3 + 6a^2b - 5ab^2$ **31.** $-3x$ **33.** $3x^2 - 12xy$ **35.** $17x^5 - 12$
37. $14a^2 - 2ab + 8b^2$ **39.** $2 - x$ **41.** $10x - 5$ **43.** $9x - 35$ **45.** $9y - 4x$
47. $9a + 2$ **49.** -2 **51.** 5 **53.** -15
55. $(3 + 4)^2 = 49; \; 3^2 + 16 = 25; \; 3^2 + 8 \cdot 3 + 16 = 49$ **57.** 240 feet
59. The sum of the first n odd numbers is n^2.   **61.** $20x^5$ **63.** $2a^3b^3$ **65.** $6x^3$
67. $-2x^2y$

Problem Set 3.4

1. $12x^3 - 10x^2 + 8x$ **3.** $-3a^5 + 18a^4 - 21a^2$ **5.** $2a^5b - 2a^3b^2 + 2a^2b^4$
7. $-28x^4y^3 + 12x^3y^4 - 24x^2y^5$ **9.** $3r^6s^2 - 6r^5s^3 + 9r^4s^4 + 3r^3s^5$ **11.** $x^2 - 2x - 15$
13. $6x^2 - 19x + 15$ **15.** $x^3 + 9x^2 + 23x + 15$ **17.** $6a^4 + a^3 - 12a^2 + 5a$
19 $a^3 - b^3$ **21.** $8x^3 + y^3$ **23.** $2a^3 - a^2b - ab^2 - 3b^3$ **25.** $18x^3 - 15x^2y - 16y^3$
27. $6x^4 - 44x^3 + 70x^2$ **29.** $6x^3 - 11x^2 - 14x + 24$ **31.** $x^2 + x - 6$
33. $x^2 - 5x + 6$ **35.** $6a^2 + 13a + 6$ **37.** $6x^2 + 2x - 20$ **39.** $5x^2 - 29x + 20$
41. $20a^2 + 9a + 1$ **43.** $20x^2 - 9xy - 18y^2$ **45.** $8x^2 - 22xy + 15y^2$
47. $28a^2 - ab - 2b^2$ **49.** $x^2 + 4x + 4$ **51.** $4a^2 - 12a + 9$
53. $25x^2 + 20xy + 4y^2$ **55.** $x^2 - 16$ **57.** $4a^2 - 9b^2$ **59.** $9r^2 - 49s^2$
61. $25x^2 - 16y^2$ **63.** $x^3 - 6x^2 + 12x - 8$ **65.** $3x^3 - 18x^2 + 33x - 18$
67. $x^{2N} + x^N - 6$ **69.** $x^{4N} - 9$ **71.** $xy + 5x + 3y + 15$ **73.** $a^2b^2 + b^2 + 8a^2 + 8$
75. $3xy^2 + 4x - 6y^2 - 8$ **77.** $(x + y)^2 + (x + y) - 20 = x^2 + 2xy + y^2 + x + y - 20$
79. $(x + y)^2 + 2z(x + y) + z^2 = x^2 + 2xy + y^2 + 2xz + 2yz + z^2$ **81.** $\frac{3}{8}$
83. $2^4 - 3^4 = -65; \; (2 - 3)^4 = 1; \; (2^2 + 3^2)(2 + 3)(2 - 3) = -65$ **85.** 4 **87.** $\frac{5}{3}$
89. $-\frac{1}{2}, \frac{1}{2}$ **91.** $1, -5$

Problem Set 3.5

1. $5x^2(2x - 3)$ **3.** $9y^3(y^3 + 2)$ **5.** $3ab(3a - 2b)$ **7.** $7xy^2(3y^2 + x)$
9. $3(a^2 - 7a + 10)$ **11.** $4x(x^2 - 4x - 5)$ **13.** $10x^2y^2(x^2 + 2xy - 3y^2)$
15. $xy(-x + y - xy)$ **17.** $2xy^2z(2x^2 - 4xz + 3z^2)$ **19.** $5abc(4abc - 6b + 5ac)$
21. $(a - 2b)(5x - 3y)$ **23.** $3(x + y)^2(x^2 - 2y^2)$ **25.** $(x + 5)(2x^2 + 7x + 6)$
27. $(x + 1)(3y + 2a)$ **29.** $(x + 3)(xy + 1)$ **31.** $(a + b)(1 + 5x)$ **33.** $3(3x - 2y)$
35. $(x - 2)(3y^2 + 4)$ **37.** $(x - 4)(2y^3 + 1)$ **39.** $(x - a)(x - b)$
41. $(b + 5)(a - 1)$ **43.** $(b^2 + 1)(a^4 - 5)$ **45.** $(y - 2)(4x^2 + 5)$ **47.** 6
49. $P(1 + r) + P(1 + r)r = (1 + r)(P + Pr) = (1 + r)P(1 + r) = P(1 + r)^2$
51. $x^2 + 5x + 6$ **53.** $x^2 - x - 6$ **55.** $x^2 - 7x + 6$ **57.** $x^2 - 5x - 6$

Problem Set 3.6

1. $(x + 3)(x + 4)$ **3.** $(x + 3)(x - 4)$ **5.** $(y + 3)(y - 2)$ **7.** $(x + 2)(x - 8)$
9. $(x + 2)(x + 6)$ **11.** $3(a - 2)(a - 5)$ **13.** $4x(x - 5)(x + 1)$
15. $(x + 2y)(x + y)$ **17.** $(a + 6b)(a - 3b)$ **19.** $(x - 8a)(x + 6a)$ **21.** $(x - 6b)^2$

23. $3(x - 3y)(x + y)$ **25.** $2a^3(a^2 + 2ab + 2b^2)$ **27.** $10x^2y^2(x + 3y)(x - y)$
29. $(2x - 3)(x + 5)$ **31.** $(2x - 5)(x + 3)$ **33.** $(2x + 3)(x + 5)$
35. $(2x - 5)(x - 3)$ **37.** prime   **39.** $(2a + 1)(3a + 2)$ **41.** $(4y + 3)(y - 1)$
43. $(3x - 2)(2x + 1)$ **45.** $(2r - 3)^2$ **47.** $(4x + y)(x - 3y)$
49. $(2x - 3a)(5x + 6a)$ **51.** $(3a + 4b)(6a - 7b)$ **53.** $2(2x - 1)(2x + 3)$
55. $y^2(3y - 2)(3y + 5)$ **57.** $2a^2(2a - 3)(3a + 4)$ **59.** $2x^2y^2(4x + 3y)(x - y)$
61. $(3x^2 + 1)(x^2 + 3)$ **63.** $(5a^2 + 3)(4a^2 + 5)$ **65.** $3(4r^2 - 3)(r^2 + 1)$
67. $(x + 5)(2x + 3)(x + 2)$ **69.** $(2x + 3)(x + 5)(x + 2)$ **71.** $9x^2 - 25y^2$
73. $a + 250$ **75.** $x^2 + 6x + 9$ **77.** $4x^2 - 20x + 25$ **79.** $x^3 + 8$

Problem Set 3.7

1. $(x - 3)^2$ **3.** $(a - 6)^2$ **5.** $(3x + 4)^2$ **7.** $(4a + 5b)^2$ **9.** $4(2x - 3)^2$
11. $3a(5a + 1)^2$ **13.** $(x + 3)(x - 3)$ **15.** $(2a + 1)(2a - 1)$
17. $(3x + 4y)(3x - 4y)$ **19.** $(x - 3)(x + 3)(x^2 + 9)$ **21.** $(2a - 3)(2a + 3)(4a^2 + 9)$
23. $(x - y)(x + y)(x^2 + xy + y^2)(x^2 - xy + y^2)$
25. $(a - 2)(a + 2)(a^2 + 2a + 4)(a^2 - 2a + 4)$ **27.** $5(x - 5)(x + 5)$
29. $3(a^2 + 4)(a - 2)(a + 2)$ **31.** $(x - 5)(x + 1)$ **33.** $y(y + 8)$ **35.** $(2 - a)(8 + a)$
37. $(x - 5 + y)(x - 5 - y)$ **39.** $(a + 4 + b)(a + 4 - b)$ **41.** $(x + y + a)(x + y - a)$
43. $(x - y)(x^2 + xy + y^2)$ **45.** $(a + 2)(a^2 - 2a + 4)$ **47.** $(y - 1)(y^2 + y + 1)$
49. $(r - 5)(r^2 + 5r + 25)$ **51.** $(2x - 3y)(4x^2 + 6xy + 9y^2)$
53. $3(x - 3)(x^2 + 3x + 9)$ **55.** $(x - 2)(x^2 - 10x + 28)$
57. $(a + 1 - b)[(a + 1)^2 + (a + 1)b + b^2]$ **59.** $(4x + 2)^2 = 4(x - 1)^2$ **61.** $30, -30$
63. $x \geq -7$ **65.** $a < 6$ **67.** $x > 2$ **69.** $y \geq -2$

Chapter 3   Test

1. x^8 **2.** $\frac{1}{32}$ **3.** $\frac{16}{9}$ **4.** $32x^{12}y^{11}$ **5.** a^2 **6.** x^6 **7.** $\dfrac{2a^{12}}{b^{15}}$
8. $3x^3 - 5x^2 - 8x - 4$ **9.** $4x + 75$ **10.** 6.53×10^6 **11.** 8.7×10^{-4}
12. 8.7×10^7 **13.** 3×10^8 **14.** $6y^2 + y - 35$ **15.** $2x^3 + 3x^2 - 26x + 15$
16. $16a^2 - 24ab + 9b^2$ **17.** $36y^2 - 1$ **18.** $4x^3 - 2x^2 - 30x$ **19.** $(x + 4)(x - 3)$
20. $2(3x - 1)(2x + 5)$ **21.** $(4a^2 + 9y^2)(2a + 3y)(2a - 3y)$ **22.** $(7a - b^2)(x^2 - 2y)$
23. $(x + 3)(x^2 - 3x + 9)$ **24.** $4a^3b(a - 8b)(a + 2b)$ **25.** $(x - 5 + b)(x - 5 - b)$

CHAPTER 4

Problem Set 4.1

1. $\frac{1}{3}$ **3.** $-\frac{1}{3}$ **5.** $3x^2$ **7.** $\dfrac{b^3}{2}$ **9.** $\dfrac{-3y^3}{2x}$ **11.** $\dfrac{18c^2}{7a^2}$ **13.** $\dfrac{x - 4}{6}$

15. $\dfrac{4x - 3y}{x(x + y)}$ **17.** $(a^2 + 9)(a + 3)$ **19.** $y + 3$ **21.** $\dfrac{a - 6}{a + 6}$ **23.** $\dfrac{2y + 3}{y + 1}$

25. $\dfrac{x - 2}{x - 1}$ **27.** $\dfrac{x - 3}{x + 2}$ **29.** $\dfrac{a^2 - ab + b^2}{a - b}$ **31.** $\dfrac{2x + 3y}{2x + y}$ **33.** $\dfrac{x + 3}{y - 4}$

35. $\dfrac{x + b}{x - 2b}$ **37.** -1 **39.** $-(y + 6)$ **41.** $\dfrac{-(3a + 1)}{3a - 1}$ **43.** -1 **49.** 13

51. $3x^2 - 7x + 4$ **53.** 9   **55.** $8x^2$

Problem Set 4.2

1. $2x^2 - 4x + 3$ **3.** $-2x^2 - 3x + 4$ **5.** $2y^2 + \frac{5}{2} - \dfrac{3}{2y^2}$ **7.** $-\frac{5}{2}x + 4 + \dfrac{3}{x}$

9. $4ab^3 + 6a^2b$ **11.** $-xy + 2y^2 + 3xy^2$ **13.** $x + 2$ **15.** $a - 3$ **17.** $5x + 6y$

19. $x^2 + xy + y^2$ **21.** $(y^2 + 4)(y + 2)$ **23.** $x - 7 + \dfrac{7}{x + 2}$ **25.** $2x + 5 + \dfrac{2}{3x - 4}$

27. $2x^2 - 5x + 1 + \dfrac{4}{x + 1}$ **29.** $y^2 - 3y - 13$ **31.** $x - 3$

33. $3y^2 + 6y + 8 + \dfrac{37}{2y - 4}$ **35.** $a^3 + 2a^2 + 4a + 6 + \dfrac{17}{a - 2}$ **37.** $y^3 + 2y^2 + 4y + 8$

39. $x^2 - 2x + 1$ **41.** Yes   **43.** 7   **45.** $\frac{21}{10}$ **47.** $\frac{11}{8}$ **49.** $\frac{1}{18}$ **51.** 32

Problem Set 4.3

1. $\frac{1}{6}$ **3.** $\frac{9}{4}$ **5.** $\frac{1}{2}$ **7.** $\dfrac{15y}{x^2}$ **9.** $\dfrac{b}{a}$ **11.** $\dfrac{2y^5}{z^3}$ **13.** $\dfrac{x + 3}{x + 2}$ **15.** $y + 1$

17. $\dfrac{3(x + 4)}{x - 2}$ **19.** $\dfrac{(a - 2)(a + 2)}{a - 5}$ **21.** $\dfrac{x + 3}{x + 4}$ **23.** 1   **25.** $\dfrac{x - 1}{x^2 + 1}$

27. $\dfrac{(a + 4)(a - 3)}{(a - 4)(a + 5)}$ **29.** $\dfrac{(y - 2)(y + 1)}{(y + 2)(y - 1)}$ **31.** $\dfrac{x - 1}{x + 1}$ **33.** $3x$ **35.** $2(x + 5)$

37. $x - 2$ **39.** $-(y - 4)$ **41.** $(a - 5)(a + 1)$ **43.** $x < -2$ or $x > 8$

45. $1 < x < 4$ **47.** $-2 \leq x \leq 1$

Problem Set 4.4

1. $\frac{5}{4}$ **3.** $\frac{1}{3}$ **5.** $\frac{41}{24}$ **7.** $\frac{19}{144}$ **9.** $\frac{31}{24}$ **11.** 1   **13.** -1 **15.** $\dfrac{1}{x + y}$

17. $\dfrac{11}{3(x + 2)}$ **19.** $\dfrac{a^2 + 2a + 3}{a^3}$ **21.** $\dfrac{6 - 5x}{3x^2}$ **23.** $\dfrac{6x + 5}{x^2 - 1}$ **25.** $\dfrac{2(2x - 3)}{(x - 3)(x - 2)}$

27. $\dfrac{a + 9}{(a - 2)(a + 1)(a - 3)}$ **29.** $\dfrac{1}{(y + 4)(y + 3)}$ **31.** $\dfrac{a}{(a + 4)(a + 5)}$

33. $\dfrac{x^3 - 10x - 2}{x^2 - 9}$ **35.** $\dfrac{2(3x - 5)}{(x + 2)(x - 2)(x - 3)}$ **37.** $\dfrac{(x + y)^2}{x^3 - y^3}$ **39.** $\dfrac{2x - 3}{2x}$

41. $\frac{1}{2}$ **43.** $\frac{51}{10}$ **45.** $\frac{1}{5}$ **47.** $(3 + 4)^{-1} = 7^{-1} = \frac{1}{7};\ 3^{-1} + 4^{-1} = \frac{1}{3} + \frac{1}{4} = \frac{7}{12}$ **49.** y

51. $2x - 1$ **53.** $\dfrac{1}{(x - 1)(x - 2)}$

Problem Set 4.5

1. $\frac{9}{8}$ **3.** $\frac{2}{15}$ **5.** $\frac{119}{20}$ **7.** $\dfrac{1}{x + 1}$ **9.** $\dfrac{a + 1}{a - 1}$ **11.** $\dfrac{y - x}{y + x}$

13. $\dfrac{1}{(x + 5)(x - 2)}$ **15.** $\dfrac{1}{a^2 - a + 1}$ **17.** $\dfrac{x + 3}{x + 2}$ **19.** $\dfrac{a + 3}{a - 2}$ **21.** $\dfrac{a - 1}{a + 1}$

23. 1   **25.** $\dfrac{-x^2 + x - 1}{x - 1}$ **27.** $\frac{5}{3}$ **29.** $\dfrac{(2x - 3)(2x + 1)}{(2x - 1)(2x + 3)}$

31. $(a^{-1} + b^{-1})^{-1} = \left(\dfrac{1}{a} + \dfrac{1}{b}\right)^{-1} = \left(\dfrac{a + b}{ab}\right)^{-1} = \dfrac{ab}{a + b}$

33. $\dfrac{1 - x^{-1}}{1 + x^{-1}} = \dfrac{1 - \dfrac{1}{x}}{1 + \dfrac{1}{x}} = \dfrac{\dfrac{x - 1}{x}}{\dfrac{x + 1}{x}} = \dfrac{x - 1}{x + 1}$ **35.** -15 **37.** 20   **39.** 5   **41.** 1

Problem Set 4.6

1. $-\frac{35}{3}$ **3.** $-\frac{18}{5}$ **5.** $\frac{36}{11}$ **7.** 2   **9.** 5   **11.** 2
13. Possible solution -1 which does not check; $\varnothing$ **15.** 5   **17.** $\frac{2}{3}$ **19.** 18
21. Possible solution 4 which does not check; $\varnothing$ **23.** -6 **25.** -5 **27.** $\frac{53}{17}$
29. Possible solution 3 which does not check; $\varnothing$ **31.** $\frac{22}{3}$ **33.** $\frac{15}{8}$ ohms   **35.** 2
37. $2(x + 3) = 16$; 5   **39.** $2x + 2(2x - 3) = 42$; width is 8 meters, length is 13 meters.

Problem Set 4.7

As you can see, in addition to the answers to the problems we have included some of the equations used to solve the problems. Remember, you should attempt the problems on your own before looking here to check your answers or equations.

1. $\dfrac{1}{x} + \dfrac{1}{3x} = \dfrac{20}{3}$; $\frac{1}{5}$ and $\frac{3}{5}$.   **3.** $\dfrac{7 + x}{9 + x} = \dfrac{5}{6}$; 3

5. Let x = speed of current; $\dfrac{1.5}{5 - x} = \dfrac{3}{5 + x}$; $\frac{5}{3}$ mi/hr

7. Train A—75 mi/hr, Train B—60 mi/hr

9. Let x = time it takes them together; $\dfrac{1}{3} + \dfrac{1}{6} = \dfrac{1}{x}$; 2 days

11. 9 hours   **13.** Let x = time to fill with both open; $\dfrac{1}{8} - \dfrac{1}{16} = \dfrac{1}{x}$; 16 hours

15. 15 hours   **17.** $l = \dfrac{A - 2w}{2}$ **19.** $m = \dfrac{y - b}{x}$ **21.** $n = \dfrac{A - a + d}{d}$

Chapter 4   Test

1. $x + y$ **2.** $\dfrac{x - 1}{x + 1}$ **3.** $6x^2 + 3xy - 4y^2$ **4.** $x^2 - 4x - 2 + \dfrac{8}{2x - 1}$

5. $2(a + 4)$ **6.** $4(a + 3)$ **7.** $x + 3$ **8.** $\frac{38}{105}$ **9.** $\frac{7}{8}$ **10.** $\dfrac{1}{a - 3}$

11. $\dfrac{3(x - 1)}{x(x - 3)}$ **12.** $\dfrac{x}{(x + 4)(x + 5)}$ **13.** $\frac{5}{2}$ **14.** $\dfrac{3a + 8}{3a + 10}$ **15.** $\dfrac{x - 3}{x - 2}$

16. $-\frac{3}{5}$ **17.** Possible solution 3, which does not check; $\varnothing$ **18.** $\frac{3}{13}$ **19.** -7
20. 15 hours

CHAPTER 5

Problem Set 5.1

1. 6   **3.** -3 **5.** 2   **7.** -2 **9.** 2   **11.** $\frac{9}{5}$ **13.** $\frac{4}{5}$ **15.** 9
17. 125   **19.** 8   **21.** $\frac{1}{3}$ **23.** $\frac{1}{27}$ **25.** $\frac{6}{5}$ **27.** $\frac{8}{27}$ **29.** 7   **31.** $\frac{3}{4}$
33. $x^{4/5}$ **35.** a **37.** $\dfrac{1}{x^{2/5}}$ **39.** $a^{3/2}b^{2/3}$ **41.** $x^{16/15}$ **43.** $x^{1/6}$

45. $x^{9/25}y^{1/2}z^{1/5}$ **47.** $x^{1/2}y^{1/3}$ **49.** $\dfrac{b^{7/4}}{a^{1/8}}$ **51.** $y^{3/10}$ **53.** $5x^2$ **55.** $6a^4$

57. x^2 **59.** $3a^4$ **61.** xy^2 **63.** $2x^2y$ **65.** $(9^{1/2} + 4^{1/2})^2 = (3 + 2)^2 = 5^2 = 25$
67. $(x^{1/2} + y^{1/2})(x^{1/2} - y^{1/2}) = (x^{1/2})^2 - (y^{1/2})^2 = x - y$
69. $\sqrt{\sqrt{a}} = (a^{1/2})^{1/2} = a^{1/4} = \sqrt[4]{a}$ **71.** 25 mi/hr **73.** x^2 **75.** $54a^6b^2c^4$
77. $-36x^{11}$ **79.** $\dfrac{8}{125y^{18}}$

Problem Set 5.2

1. $2\sqrt{2}$ **3.** $3\sqrt{2}$ **5.** $5\sqrt{3}$ **7.** $12\sqrt{2}$ **9.** $4\sqrt{5}$ **11.** $4\sqrt{3}$
13. $3\sqrt{5}$ **15.** $3\sqrt[3]{2}$ **17.** $4\sqrt[3]{2}$ **19.** $2\sqrt[5]{2}$ **21.** $3\sqrt{6}$ **23.** $2\sqrt[3]{5}$ **25.** $3\sqrt{11}$
27. $3x\sqrt{2x}$ **29.** $4y^3\sqrt{2y}$ **31.** $2xy^2\sqrt[3]{5xy}$ **33.** $4abc^2\sqrt{3b}$ **35.** $2bc\sqrt[3]{6a^2c}$
37. $2xy^2\sqrt[5]{2x^3y^2}$ **39.** $2x^2yz\sqrt{3xz}$ **41.** $\dfrac{2\sqrt{3}}{3}$ **43.** $\dfrac{5\sqrt{6}}{6}$ **45.** $\dfrac{\sqrt{2}}{2}$ **47.** $\dfrac{\sqrt{5}}{5}$
49. $2\sqrt[3]{4}$ **51.** $\dfrac{2\sqrt[3]{3}}{3}$ **53.** $\dfrac{\sqrt{6x}}{2x}$ **55.** $\dfrac{2\sqrt{2y}}{y}$ **57.** $\dfrac{\sqrt[3]{36xy^2}}{3y}$ **59.** $\dfrac{\sqrt[3]{6xy^2}}{3y}$
61. $\dfrac{\sqrt[4]{2x^3}}{2x}$ **63.** $\dfrac{3x\sqrt{15xy}}{5y}$ **65.** $\dfrac{5xy\sqrt{6xz}}{2z}$ **67.** $\dfrac{2ab\sqrt[3]{6ac^2}}{3c}$ **69.** $\dfrac{2xy^2\sqrt[3]{3z^2}}{3z}$
71. $x + 3$ **73.** $\sqrt{9 + 16} = \sqrt{25} = 5;\ \sqrt{9} + \sqrt{16} = 3 + 4 = 7$ **75. a.** 5   **b.** 3
c. 5   **d.** 3   **77.** $5\sqrt{13}\text{ ft}$ **79.** $7x^2$ **81.** $7a^3$ **83.** $x^2 - 2x - 4$ **85.** $3y + 2$

Problem Set 5.3

1. $7\sqrt{5}$ **3.** $16\sqrt{6}$ **5.** $-x\sqrt{7}$ **7.** $\sqrt[3]{10}$ **9.** $9\sqrt{6}$ **11.** 0   **13.** $4\sqrt{2}$
15. $\sqrt{5}$ **17.** $-32\sqrt{2}$ **19.** $-3x\sqrt{2}$ **21.** $-2\sqrt[3]{2}$ **23.** $8x\sqrt[3]{xy^2}$
25. $3a^2b\sqrt{3ab}$ **27.** $11ab\sqrt[3]{3a^2b}$ **29.** $\sqrt{2}$ **31.** $\dfrac{8\sqrt{5}}{15}$ **33.** $\dfrac{2\sqrt{3}}{3}$ **35.** $\dfrac{3\sqrt{2}}{2}$
37. $\dfrac{2\sqrt{6}}{3}$ **39.** $\sqrt{12} = 3.464;\ 2\sqrt{3} = 2(1.732) = 3.464$
41. $\sqrt{8} + \sqrt{18} = 2.828 + 4.243 = 7.071;\ \sqrt{50} = 7.071;\ \sqrt{26} = 5.099$ **43.** $8\sqrt{2x}$
45. 5   **47.** $6x^2 - 10x$ **49.** $2a^2 + 5a - 25$ **51.** $9x^2 - 12xy + 4y^2$ **53.** $x^2 - 4$

Problem Set 5.4

1. $\sqrt{6} - 9$ **3.** $24\sqrt{3} + 6\sqrt{6}$ **5.** $7 + 2\sqrt{6}$ **7.** $x + 2\sqrt{x} - 15$
9. $34 + 20\sqrt{3}$ **11.** $19 + 8\sqrt{3}$ **13.** $x - 6\sqrt{x} + 9$ **15.** $30 + 12\sqrt{6}$
17. $4a - 12\sqrt{ab} + 9b$ **19.** 1   **21.** 15   **23.** $a - 49$ **25.** 92

27. $2\sqrt{3} - 2\sqrt{2}$ **29.** $\dfrac{\sqrt{3} + 1}{2}$ **31.** $\dfrac{2\sqrt{3} + 1}{11}$ **33.** $\dfrac{x + 3\sqrt{x}}{x - 9}$ **35.** $\dfrac{10 + 3\sqrt{5}}{11}$

37. $\dfrac{3\sqrt{x} + 3\sqrt{y}}{x - y}$ **39.** $-2 - \sqrt{3}$ **41.** $\dfrac{a + 2\sqrt{ab} + b}{a - b}$ **43.** $\dfrac{5 - \sqrt{21}}{4}$

45. $\dfrac{\sqrt{x} - 3x + 2}{1 - x}$ **47.** $\dfrac{16 + 14\sqrt{6}}{-23}$ **51.** $10\sqrt{3}$ **53.** $x + 6\sqrt{x} + 9$ **55.** 75

57. a. $\dfrac{5\sqrt{2}}{4}\,\sec$ **b.** $\frac{5}{2}\sec$ **59.** 7   **61.** 5   **63.** 3

Problem Set 5.5

1. 4   **3.** $\varnothing$ **5.** 5   **7.** $\varnothing$ **9.** $\frac{39}{2}$ **11.** $\varnothing$ **13.** 5   **15.** 3
17. $-\frac{32}{3}$ **19.** -1 **21.** $\varnothing$ **23.** 7   **25.** 8   **27.** 0   **29.** -1 **31.** $\frac{19}{2}$
33. 4   **35.** $\frac{9}{4}$ **37.** 1   **39.** $\frac{392}{121} \approx 3.24$ ft.   **41.** $h = 100 - 16t^2$ **43.** x^5
45. x^{40} **47.** x^{30} **49.** $\frac{1}{32}$

Problem Set 5.6

1. $6i$ **3.** $-5i$ **5.** $6i\sqrt{2}$ **7.** $-2i\sqrt{3}$ **9.** $9i\sqrt{2}$ **11.** i **13.** $-i$
15. 1   **17.** -1 **19.** $-i$ **21.** $x = 3, y = -1$ **23.** $x = -2, y = -\frac{1}{2}$
25. $x = -8, y = -5$ **27.** $x = -\frac{6}{5}, y = \frac{3}{4}$ **29.** $x = 7, y = \frac{1}{2}$ **31.** $x = \frac{3}{7}, y = \frac{2}{5}$
33. $5 + 9i$ **35.** $5 - i$ **37.** $2 - 4i$ **39.** $1 - 6i$ **41.** $9 - 2i$ **43.** $3 - 3i$
45. $2 + 2i$ **47.** $-1 - 7i$ **49.** $6 + 8i$ **51.** $2 - 24i$ **53.** $\sqrt{6} - 2$
55. $x + 10\sqrt{x} + 25$ **57.** $\dfrac{x - 3\sqrt{x}}{x - 9}$

Problem Set 5.7

1. $-15 + 12i$ **3.** $7 - 7i$ **5.** $18 + 24i$ **7.** $10 + 11i$ **9.** $21 + 23i$
11. $-26 + 7i$ **13.** $-21 + 20i$ **15.** $-2i$ **17.** $-7 - 24i$ **19.** 5   **21.** 40
23. 13   **25.** 164   **27.** $-3 - 2i$ **29.** $-2 + 5i$ **31.** $\frac{8}{13} + \frac{12}{13}i$ **33.** $-\frac{18}{13} - \frac{12}{13}i$
35. $-\frac{5}{13} + \frac{12}{13}i$ **37.** $\frac{13}{15} - \frac{2}{5}i$ **39.** $\frac{31}{53} - \frac{24}{53}i$ **41.** $x + 3$ **43.** $\dfrac{x - 6}{x + 6}$
45. $\dfrac{x^2 - xy + y^2}{x - y}$

Chapter 5   Test

1. $\sqrt[5]{8^3}$ **2.** $\sqrt[3]{17^4}$ **3.** $3^{1/4}x^{3/4}$ **4.** $7^{1/3}x^{2/3}$ **5.** $\frac{1}{9}$ **6.** $\frac{7}{5}$ **7.** $a^{5/12}$
8. $\dfrac{x^{13/12}}{y}$ **9.** $5xy^2\sqrt{5xy}$ **10.** $2x^2y^2\sqrt[3]{5xy^2}$ **11.** $\dfrac{\sqrt{6}}{3}$ **12.** $\dfrac{2a^2b\sqrt{15bc}}{5c}$
13. $-6\sqrt{3}$ **14.** $-ab\sqrt[3]{3}$ **15.** $17 - \sqrt{14}$ **16.** $21 - 6\sqrt{6}$ **17.** $\dfrac{5 + 5\sqrt{3}}{2}$
18. $\dfrac{7 - 2\sqrt{10}}{3}$ **19.** 10   **20.** $\varnothing$ **21.** -4 **22.** $\varnothing$ **23.** $x = \frac{2}{3}, y = \frac{1}{2}$
24. $x = \frac{1}{2}, y = 7$ **25.** $-1 - 9i$ **26.** $6i$ **27.** $17 - 6i$ **28.** $9 - 40i$

29. $2 - 3i$ **30.** $\frac{10}{13} + \frac{15}{13}i$ **31.** $-\frac{5}{13} - \frac{12}{13}i$

32. $i^{38} = (i^4)^9 \cdot i^2$
$$= 1(-1)$$
$$= -1$$

CHAPTER 6

Problem Set 6.1

1. $6, -1$ **3.** $2, 3$ **5.** $\frac{1}{3}, -4$ **7.** $\frac{2}{3}, \frac{3}{2}$ **9.** $5, -5$ **11.** $-3, 7$
13. $-4, \frac{5}{2}$ **15.** $0, \frac{4}{3}$ **17.** $-\frac{1}{5}, \frac{1}{3}$ **19.** $-\frac{4}{3}, \frac{4}{3}$ **21.** $-10, 0$ **23.** $-5, 1$
25. $1, 2$ **27.** $-2, 3$ **29.** $-2, \frac{1}{4}$ **31.** $-3, 4$ **33.** $-\frac{4}{3}, 1$ **35.** $-3, 3$
37. $-5, 3$ **39.** $-\frac{4}{3}, \frac{1}{2}$ **41.** Possible solutions 2 and 7; only 7 checks
43. Possible solutions 3 and $\frac{8}{5}$; only $\frac{8}{5}$ checks   **45.** $-2, \frac{2}{3}$ **47.** $t = 2$ seconds
49. $x^2 + 6x + 9$ **51.** $x^2 - 8x + 16$ **53.** $(x - 3)^2$ **55.** $(x + 2)^2$

Problem Set 6.2

1. ± 5 **3.** $\pm 3i$ **5.** $\pm \dfrac{\sqrt{3}}{2}$ **7.** $\pm \sqrt{5}$ **9.** $\pm 2i\sqrt{3}$ **11.** $\pm \dfrac{3\sqrt{5}}{2}$

13. $2, 8$ **15.** $-2, 3$ **17.** $\dfrac{-3 \pm 3i}{2}$ **19.** $\dfrac{-2 \pm 2i\sqrt{2}}{5}$

21. $x^2 + 12x + 36 = (x + 6)^2$ **23.** $x^2 - 4x + 4 = (x - 2)^2$
25. $a^2 - 10a + 25 = (a - 5)^2$ **27.** $x^2 + 5x + \frac{25}{4} = (x + \frac{5}{2})^2$
29. $y^2 - 7y + \frac{49}{4} = (y - \frac{7}{2})^2$ **31.** $-6, 2$ **33.** $-3, -9$ **35.** $1 \pm 2i$

37. $4 \pm \sqrt{15}$ **39.** $\dfrac{5 \pm \sqrt{37}}{2}$ **41.** $1 \pm \sqrt{5}$ **43.** $\dfrac{4 \pm \sqrt{13}}{3}$ **45.** $\dfrac{3 \pm i\sqrt{71}}{8}$

47. $\dfrac{-1 \pm \sqrt{13}}{3}$ **49.** $-1, \frac{5}{2}$ **51.** $1 \pm \sqrt{2}$ **53.** $-\frac{10}{3}$

59. $\dfrac{2 + \sqrt{5}}{2} = 2.118$; $\dfrac{2 - \sqrt{5}}{2} = -0.118$ **61.** 9 **63.** 17 **65.** $3 + \sqrt{17}$

Problem Set 6.3

1. $-2, -3$ **3.** $2 \pm \sqrt{3}$ **5.** $1, 2$ **7.** $\dfrac{2 \pm i\sqrt{14}}{3}$ **9.** $0, 5$ **11.** $\dfrac{3 \pm \sqrt{5}}{4}$

13. $-3 \pm \sqrt{17}$ **15.** $-\frac{3}{2}, -1$ **17.** $\dfrac{-1 \pm i\sqrt{5}}{2}$ **19.** 1 **21.** $\dfrac{1 \pm \sqrt{11}}{2}$

23. $\dfrac{1 \pm i\sqrt{47}}{6}$ **25.** $\dfrac{4 \pm \sqrt{2}}{8}$ **27.** $\dfrac{-11 \pm \sqrt{33}}{4}$ **29.** $-\frac{1}{2}, 3$ **31.** 4

33. Possible solutions 0 and 32; only 0 checks   **35.** 4

37. Possible solution 6 and -2; only 6 checks   **39.** 7 **41.** $0, \dfrac{-1 \pm i\sqrt{5}}{2}$

43. $\dfrac{-3 - 2i}{5}$ **45.** 0 or 29   **47.** 5 **49.** 27 **51.** $\frac{1}{4}$ **53.** $\frac{6}{5}$

Problem Set 6.4

1. 1, 2   **3.** $-\frac{5}{2}, -8$ **5.** $\pm 3, \pm i\sqrt{3}$ **7.** $\pm 2i, \pm i\sqrt{5}$ **9.** $\frac{7}{2}, 4$ **11.** $-\frac{9}{8}, \frac{1}{2}$

13. $\pm\dfrac{\sqrt{30}}{6}, \pm i$ **15.** $\pm\dfrac{\sqrt{21}}{3}, \pm\dfrac{i\sqrt{21}}{3}$ **17.** $-27, 8$ **19.** $-\frac{8}{27}$ **21.** 4, 25

23. Possible solutions 25 and 9; only 25 checks   **25.** $-\frac{1}{32}, 32$

27. Possible solutions $\frac{25}{9}$ and $\frac{49}{4}$; only $\frac{25}{9}$ checks   **29.** 27, 38   **31.** 3, 10   **33.** 4, 12

35. $2, -1 \pm i\sqrt{3}$ **37.** $x = -y \pm 2$ **39.** Boots are \$5; a suit is \$18

41. \$18 and \$26

Problem Set 6.5

1. $x^2 + (x + 2)^2 = 34$; 3, 5 or $-5, -3$ **3.** 4, 5 or $-5, -4$ **5.** 5, 13

7. $x + \dfrac{1}{x} = \dfrac{10}{3}$; 3 or $\frac{1}{3}$ **9.** $\dfrac{1}{x} + \dfrac{1}{x + 1} = \dfrac{7}{12}$; 3, 4   **11.** $x + \sqrt{x} = 6$; 4

13. 6, 8, 10   **15.** $x^2 + (3x)^2 = (2\sqrt{10})^2$; 2 cm, 6 cm   **17.** $74 = 5t + 16t^2$; 2 sec

19. $\frac{1}{4}$ sec and 1 sec   **21.** $\dfrac{8}{x - 2} + \dfrac{8}{x + 2} = 3$; 6 mi/hr   **23.** 15 mi/hr

25. $x < -\frac{1}{2}$ or $x > \frac{2}{3}$

27. $x \geq -2$ and $x \leq 4$

29. $\dfrac{x - 1}{x - 3}$ **31.** $\dfrac{x - 5}{(x - 2)(x - 3)}$

Problem Set 6.6

1.

3.

5.

7.

9.

11.

13.

15. all real numbers

17. no solution, ∅

19.

21.

23.

25.

27.

29.

31.

33. Because we don't know if x is positive or negative.   **35.** $4\sqrt{3}$ **37.** $3x\sqrt{2x}$

39. $2xy\sqrt[3]{y}$ **41.** $\dfrac{\sqrt{6}}{3}$

Chapter 6 Test

1. $7, -3$ **2.** $\pm\frac{9}{5}$ **3.** $\frac{3}{4}, -2$ **4.** $2 \pm \sqrt{2}$ **5.** $\dfrac{4 \pm \sqrt{26}}{2}$ **6.** $-\frac{5}{3}, -1$

7. $\frac{3}{2}, -1$ **8.** $\dfrac{7 \pm \sqrt{77}}{2}$ **9.** $\pm\dfrac{i}{2}, \pm\sqrt{2}$

10. Possible solutions 2 and -2; only 2 checks   **11.** $0, 4$ **12.** $5, \frac{8}{5}$ **13.** $1, 8$

14. $\frac{1}{2}, 1$ **15.** $\frac{1}{4}, 9$ **16.** Yes   **17.** $3, 5$ **18.** $2\,\text{mi/hr}$

19.

20.

CHAPTER 7

1-15 (odd).

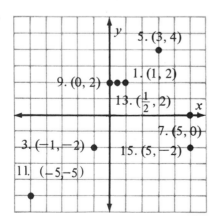

5. $(3, 4)$
1. $(1, 2)$
9. $(0, 2)$
13. $(\frac{1}{2}, 2)$
7. $(5, 0)$
3. $(-1, -2)$
15. $(5, -2)$
11. $(-5, -5)$

17. $\left(-\frac{5}{2}, \frac{9}{2}\right)$ **19.** $\left(-3, \frac{5}{2}\right)$ **21.** $(-2, 0)$ **23.** $(-3, -2)$ **25.** $(-3, -3)$

27. $(3, -4)$

29.

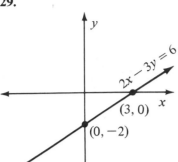

31.

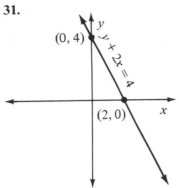

33.

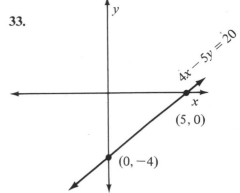

35.

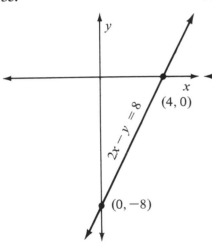

37.

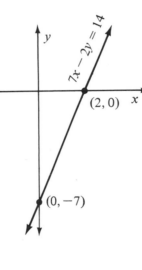

39.

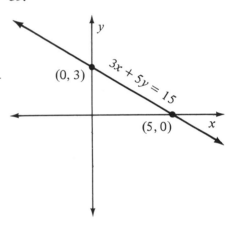

41.

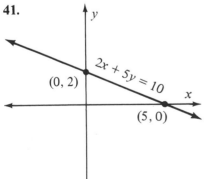

43.

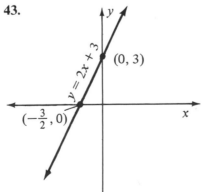

45.

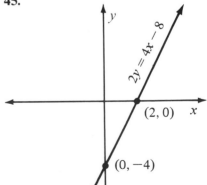

47.

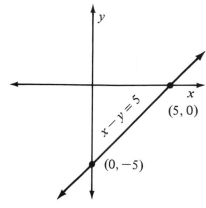

x − y = 5

(5, 0)

(0, −5)

49.

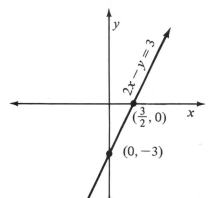

2x − y = 3

$(\frac{3}{2}, 0)$

(0, −3)

51.

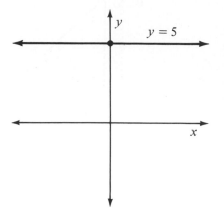

y = 5

53.

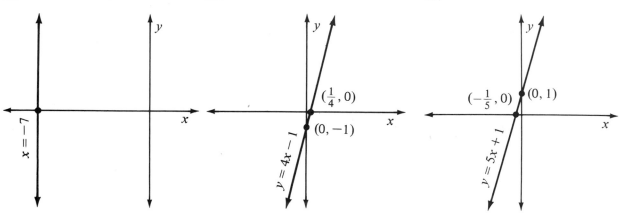

x = −7

55.

$(\frac{1}{4}, 0)$

(0, −1)

y = 4x − 1

57.

$(-\frac{1}{5}, 0)$ (0, 1)

y = 5x + 1

59.

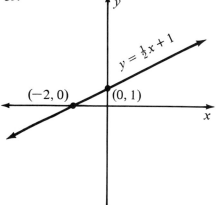

$y = \frac{1}{2}x + 1$

(−2, 0)   (0, 1)

61.

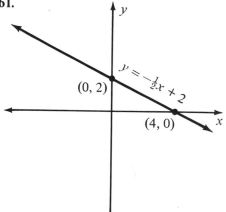

(0, 2)

$y = -\frac{1}{2}x + 2$

(4, 0)

63.

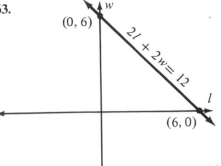

65.

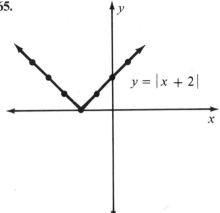

67. $-\frac{7}{4}$ **69.** -5 **71.** 2, 4

Problem Set 7.2

1. 1   **3.** no slope   **5.** -1 **7.** -2 **9.** -4 **11.** $-\frac{3}{2}$ **13.** no slope
15. 0   **17.** $-\frac{39}{2}$ **19.** 5   **21.** -1 **23.** -5 **25.** -8 **27.** $\sqrt{2}/3$ **29.** 0
31. 8   **33.** $-2, 3$ **35.** 24 feet

37.

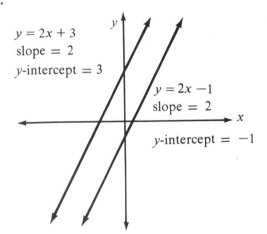

$y = 2x + 3$
slope $= 2$
y-intercept $= 3$

$y = 2x - 1$
slope $= 2$
y-intercept $= -1$

39. $y = \frac{3}{2}x - 3$ **41.** $y = -\frac{2}{3}x + \frac{5}{3}$ **43.** $y = 5x - 17$ **45.** $y = \frac{4}{5}x + 1$

Problem Set 7.3

1. $y = 2x + 3$ **3.** $y = x - 5$ **5.** $y = \dfrac{x}{2} + \frac{3}{2}$ **7.** $y = 4$
9. $y = -\sqrt{2}x + 3\sqrt{2}$

11. $m = 3$,
      y-intercept $= -2$,
      perpendicular slope $= -\frac{1}{3}$

13. $m = 2$,
      y-intercept $= -4$,
      perpendicular slope $= -\frac{1}{2}$

15. $m = \frac{2}{3}$,
      y-intercept $= -4$,
      perpendicular slope $= -\frac{3}{2}$

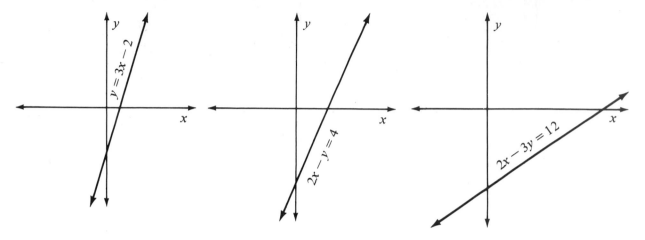

17. $m = -\frac{4}{5}$,
      y-intercept $= 4$,
      perpendicular slope $= \frac{5}{4}$

19. $m = \frac{3}{5}$,
      y-intercept $= -2$,
      perpendicular slope $= -\frac{5}{3}$

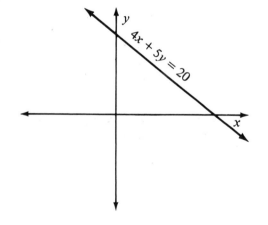

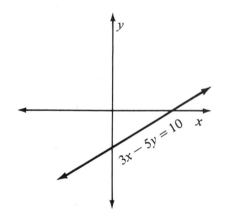

21. $y = 2x + 1$ **23.** $y = -3x + 14$ **25.** $y = -\frac{1}{2}x + 2$ **27.** $y = \frac{2}{3}x - \frac{11}{6}$
29. $y = -2x + 7$ **31.** $y = 2x + 2$ **33.** $y = -\frac{7}{2}x + \frac{17}{2}$ **35.** $y = \frac{5}{3}x + 5$
37. $y = 5$ **39.** $x = 5$

41. slope $= 0$,
y-intercept $= -2$

43. no slope,
x-intercept $= \sqrt{5}$

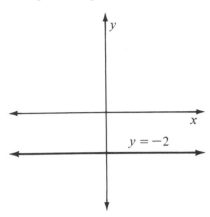

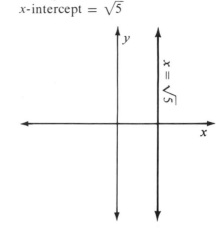

45. $y = 3x + 7$ **47.** $y = -\frac{5}{2}x - 13$ **49.** $y = \frac{1}{4}x + \frac{1}{4}$ **51.** $y = \frac{2}{3}x - 2$

53. **a.** $d = 1100t$
b. 4400 ft
c. 4.8 sec

55. slope $= -\dfrac{a}{b}$; y-intercept $= \dfrac{c}{b}$ **57.** $x + 2\sqrt{x} - 15$ **59.** $9 - 4\sqrt{5}$ **61.** $\dfrac{x - 2\sqrt{x}}{x - 4}$

63. $\dfrac{7 + 2\sqrt{10}}{3}$

Problem Set 7.4

1.

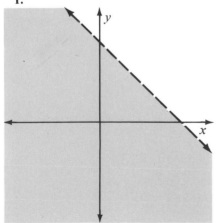

3.

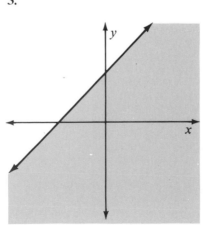

5.

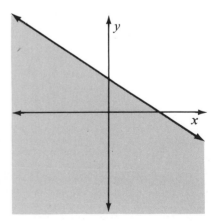

7.

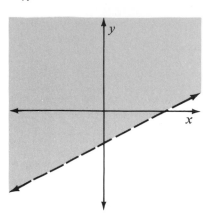

9.

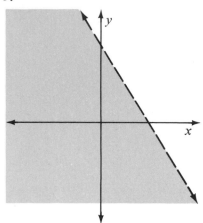

11.

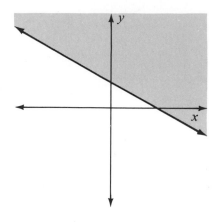

13.

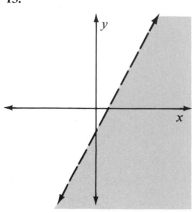

15.

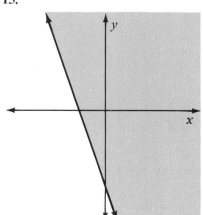

17.

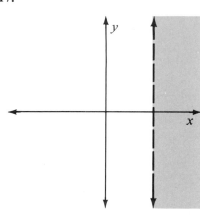

19.

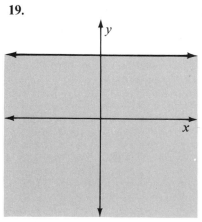

21.

a.

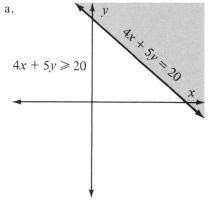

b.

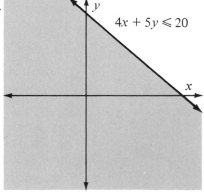

c.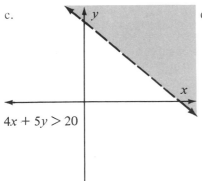

$4x + 5y > 20$

d.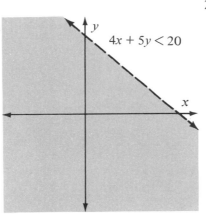

$4x + 5y < 20$

23.

a.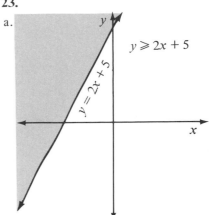

$y \geqslant 2x + 5$

$y = 2x + 5$

b.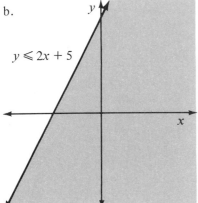

$y \leqslant 2x + 5$

c.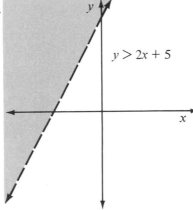

$y > 2x + 5$

d.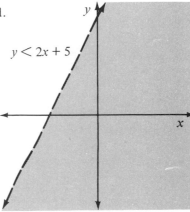

$y < 2x + 5$

25. $y \geq -2x + 4$　　　**27.** $y < \frac{3}{2}x - 3$

29.

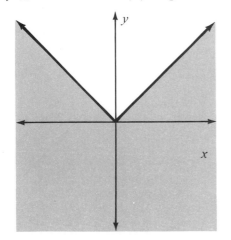

31. $-2 < x < 3$
33. 7
35. 13
37. Possible solutions -6 and 1; only 1 checks.
39. -4

Problem Set 7.5

1. 30   **3.** 5   **5.** $\frac{363}{8}$ **7.** -6 **9.** $\frac{1}{2}$ **11.** 12   **13.** 8   **15.** $4\sqrt{2}$

17. $\frac{81}{5}$ **19.** $\pm 3\sqrt{2}$ **21.** 64   **23.** 8   **25.** $\frac{50}{3}$ lb   **27.** 12 lb   **29.** $\frac{1504}{15}$ sq. in.

31. 1.5 ohms   **33.** $9 - 2i$ **35.** $11 + 10i$ **37.** $5 - 12i$ **39.** $\dfrac{-15 + 12i}{13}$

Chapter 7   Test

1. x-intercept $= 3$,
   y-intercept $= 6$,
   slope $= -2$

2. x-intercept $= \frac{5}{3}$,
   y-intercept $= -\frac{5}{2}$,
   slope $= \frac{3}{2}$

3. x-intercept $= -\frac{3}{2}$,
   y-intercept $= -3$,
   slope $= -2$

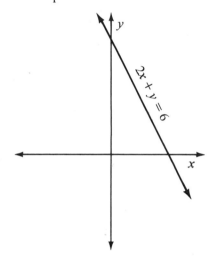

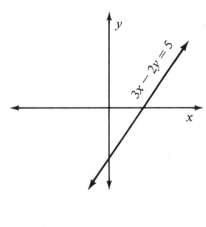

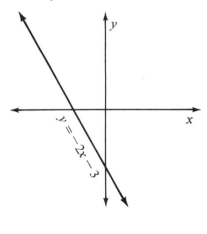

4. x-intercept $= -\frac{8}{3}$,
   y-intercept $= 4$,
   slope $= \frac{3}{2}$

5. x-intercept $= -2$,
   no y-intercept,
   no slope

6. no x-intercept,
   y-intercept $= 3$,
   slope $= 0$

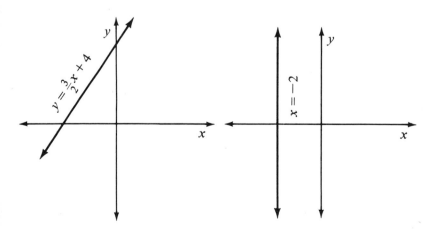

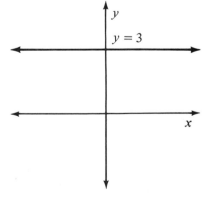

7.

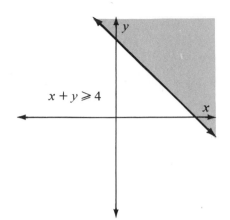

$x + y \geqslant 4$

8.

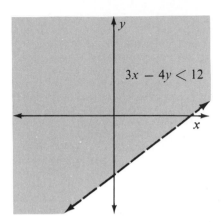

$3x - 4y < 12$

9.

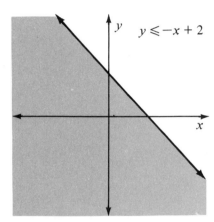

$y \leqslant -x + 2$

10.

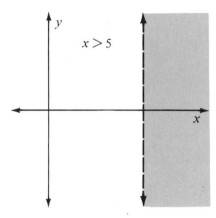

$x > 5$

11. $y = 2x + 5$ **12.** $y = -\frac{3}{7}x + \frac{5}{7}$ **13.** $y = \frac{2}{5}x - 5$ **14.** $y = -\frac{1}{3}x - \frac{7}{3}$

15. $x = 4$ **16.** 18 **17.** 5 **18.** $\frac{2000}{3}$ lb

CHAPTER 8

Problem Set 8.1

1.

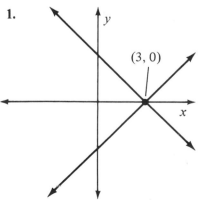

$(3, 0)$

3.

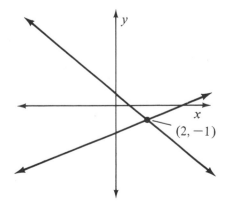

$(2, -1)$

5.   **7.**

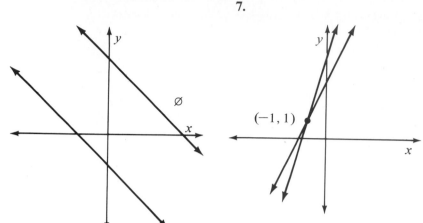

9. $(2, 3)$ **11.** $(1, 1)$ **13.** lines coincide $\{(x, y) \mid 3x - 2y = 6\}$ **15.** $(-1, -2)$
17. $(0, 0)$ **19.** $(4, -3)$ **21.** $(2, 2)$ **23.** parallel lines; $\varnothing$ **25.** $(1, 2)$
27. $(10, 24)$ **29.** $(4, \frac{10}{3})$ **31.** $(0, 3)$ **33.** $(6, 2)$ **35.** $(2, 4)$ **37.** $(3, -3)$
39. lines coincide $\{(x, y) \mid 2x - y = 5\}$ **41.** $(1, 1)$ **43.** $(\frac{1}{3}, 1)$ **45.** $(6{,}000, 4{,}000)$
47. 2   **49. a.** $y = 16x + 22$ **b.** 3 min   **51.** $7\sqrt{3}$ **53.** $-5x\sqrt{5}$ **55.** $21\sqrt{2}$
57. $-2\sqrt[3]{2}$

Problem Set 8.2
1. $(1, 2, 1)$ **3.** $(2, 1, 3)$ **5.** $(2, 0, 1)$ **7.** no unique solution   **9.** $(1, 1, 1)$
11. no unique solution   **13.** no unique solution   **15.** $(\frac{1}{2}, 1, 2)$ **17.** $(1, 3, 1)$
19. $(-1, 2, -2)$ **21.** 4 amp, 3 amp, 1 amp   **23.** $-2, 4$ **25.** $\frac{2}{3}, \frac{3}{2}$ **27.** $\dfrac{-1 \pm i\sqrt{3}}{2}$
29. $\dfrac{3 \pm \sqrt{3}}{2}$

Problem Set 8.3
1. 3   **3.** 5   **5.** -1 **7.** 0   **9.** 10   **11.** 2   **13.** -3 **15.** -2
17. $-2, 5$ **19.** 3   **21.** 0   **23.** 3   **25.** 8   **27.** 6   **29.** -228
31. $\begin{vmatrix} y & x \\ m & 1 \end{vmatrix} = y - mx = b; \ y = mx + b$ **33.** $\pm 2, \pm i\sqrt{2}$ **35.** 27, 8   **37.** $\frac{9}{4}, 1$

Problem Set 8.4
1. $(3, 1)$ **3.** Cramer's Rule does not apply.   **5.** Cramer's Rule does not apply.
7. $(-\frac{15}{43}, -\frac{27}{43})$ **9.** $(\frac{60}{43}, \frac{46}{43})$ **11.** $(\frac{1}{2}, \frac{5}{2}, 1)$ **13.** $(3, -1, 2)$
15. Cramer's Rule does not apply.   **17.** $(-\frac{10}{91}, -\frac{63}{91}, \frac{107}{92})$ **19.** $(3, -2, 4)$
21. $(3, 1, 2)$ **23.** $x = 50$ items   **25.** $\dfrac{x}{x + 15} = \dfrac{4}{7}; \ \frac{20}{35}$ **27.** $\dfrac{1}{5} + \dfrac{1}{6} + \dfrac{1}{10} = \dfrac{1}{x}; \ \frac{15}{7}$ hours

Problem Set 8.5
1. $y = 2x + 3, \ x + y = 18$ The two numbers are 5 and 13   **3.** 10, 16   **5.** 1, 3, 4

7. Let x = the number of adult tickets and y = the number of children's tickets.
   $x + y = 925$ 225 adult and
   $2x + y = 1150$ 700 children's tickets
9. 12 nickels, 8 dimes
11. Let x = the amount invested at 6% and y = the amount invested at 7%.
   $x + y = 20,000$ He has \$12,000 at 6%
   $.06x + .07y = 1280$ and \$8,000 at 7%
13. \$4000 at 6%, \$8,000 at 7.5%   **15.** 3 gal of 50%, 6 gal of 20%
17. 5 gal of 20%, 10 gal of 14%
19. Let x = the speed of the boat and y = the speed of the current.
$3(x - y) = 18$ The speed of the boat is 9 mph
$2(x + y) = 24$ The speed of the current is 3 mph
21. 270 mph airplane, 30 mph wind
23. $-2 \le x \le 4$
25. $x < -3$ or $x > \frac{1}{2}$
27. $-2 < x < 3$

Chapter 8   Test
1. $(1, 2)$ **2.** $(3, 2)$ **3.** $\left(-\frac{54}{13}, -\frac{58}{13}\right)$ **4.** $(1, 2)$ **5.** $(3, -2, 1)$ **6.** -14
7. -26 **8.** $\left(-\frac{14}{3}, -\frac{19}{3}\right)$ **9.** lines coincide $\{(x, y) | 2x + 4y = 3\}$
10. $\left(\frac{5}{11}, -\frac{15}{11}, -\frac{1}{11}\right)$ **11.** \$4000 at 5%, \$8000 at 6%
12. 3 oz of cereal I, 1 oz of cereal II

CHAPTER 9

Problem Set 9.1
1. x-intercepts $= -1, 7$;
   vertex $= (3, -16)$

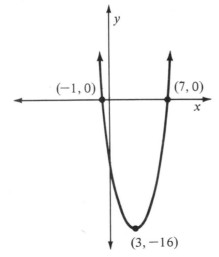

3. x-intercepts $= -3, 1$;
vertex $= (-1, -4)$

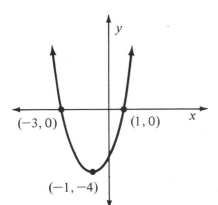

5. x-intercepts $= -5, 1$;
vertex $= (-2, 9)$

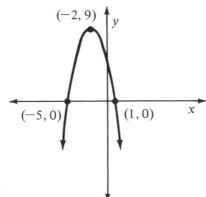

7. x-intercepts $= 5, -3$;
vertex $= (1, 16)$

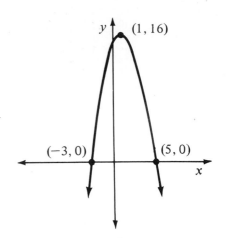

9. x-intercepts $= -1, 1$;
vertex $= (0, -1)$

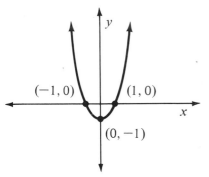

11. x-intercepts $= 3, -3$;
vertex $= (0, 9)$

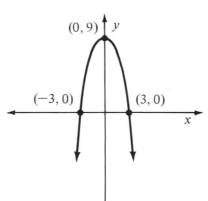

13. x-intercepts $= -\frac{2}{3}, 4$;
vertex $= (\frac{5}{3}, -\frac{49}{3})$

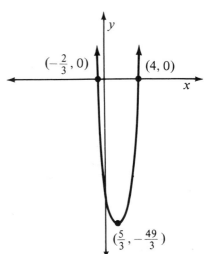

15. x-intercepts $= (1 + 2\sqrt{3}, 0), (1 - 2\sqrt{3}, 0)$;
vertex $= (1, -12)$

17. x-intercepts $= \left(\dfrac{-3 + 2\sqrt{5}}{2}, 0\right), \left(\dfrac{-3 - 2\sqrt{5}}{2}, 0\right)$;
vertex $= (-\frac{3}{2}, -20)$

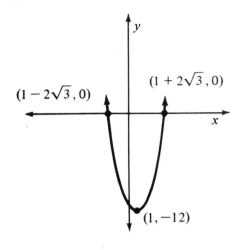

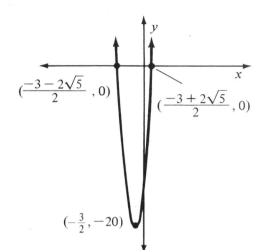

19. x-intercepts $= \left(\dfrac{-1 + 4\sqrt{2}}{2}, 0\right), \left(\dfrac{-1 - 4\sqrt{2}}{2}, 0\right)$;
vertex $= (-\frac{1}{2}, 32)$

21. vertex $= (2, -8)$

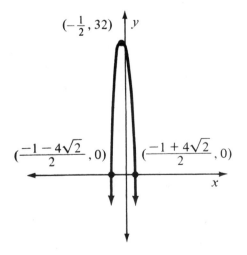

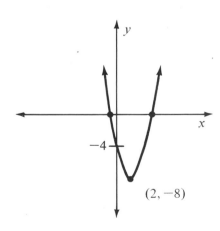

23. vertex = $(1, -4)$

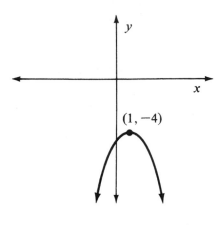

25. vertex = $(0, 1)$

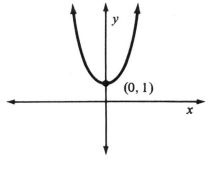

27. vertex = $(0, -3)$

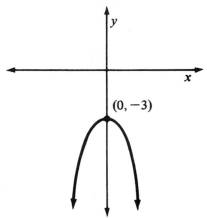

29. vertex = $(-\frac{2}{3}, -\frac{1}{3})$

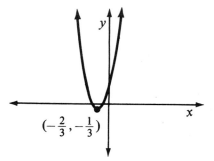

31. vertex = $(-\frac{3}{4}, \frac{7}{8})$

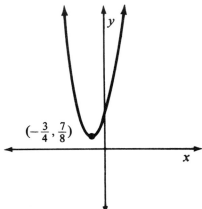

33. vertex = $(\frac{1}{8}, -\frac{15}{16})$

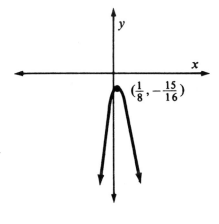

35. 256 ft   **37.** $x_1 + x_2 = -\dfrac{b}{a}$

39. When $x = 10$; that is, a square of side 10 inches gives the maximum area.

41. $x^2 + 6x + 9 = (x + 3)^2$ **43.** $x^2 - 10x + 25 = (x - 5)^2$

45. $x^2 + 8x + 16 = (x + 4)^2$ **47.** $x^2 + 3x + \frac{9}{4} = (x + \frac{3}{2})^2$

Problem Set 9.2

1. 5   **3.** $\sqrt{106}$ **5.** $\sqrt{61}$ **7.** $\sqrt{130}$ **9.** 3 or -1 **11.** 3

13. $(x - 2)^2 + (y - 3)^2 = 16$ **15.** $(x - 3)^2 + (y + 2)^2 = 9$

17. $(x + 5)^2 + (y + 1)^2 = 5$ **19.** $x^2 + (y + 5)^2 = 1$ **21.** $x^2 + y^2 = 4$

23. $(x + 1)^2 + y^2 = 12$

25. center = (0, 0),
radius = 2

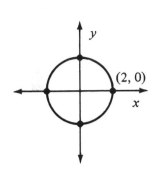

27. center = (0, 0),
radius = $\sqrt{5}$

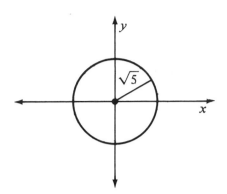

29. center = (1, 3)
radius = 5

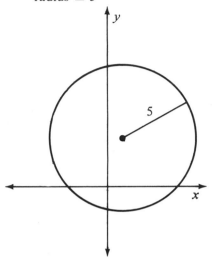

31. center = (−2, 4),
radius = $2\sqrt{2}$

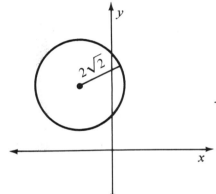

33. center = (−1, −1),
radius = 1

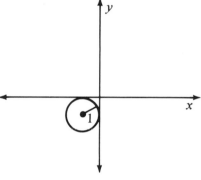

35. center = (0, 3)
radius = 4

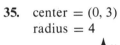

37. center = (−1, 0),
radius = $\sqrt{2}$

39. center = (2, 3),
radius = 1

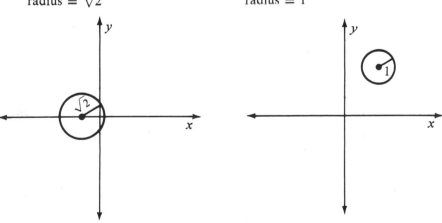

41. center $= (-1, -\frac{1}{2})$,
radius $= \sqrt{17}/2$

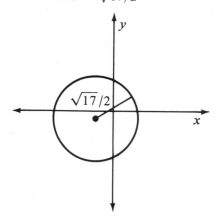

43. center $= (\frac{1}{2}, \frac{3}{2})$,
radius $= 3\sqrt{2}/2$

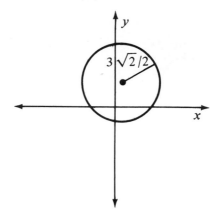

45. $x^2 + y^2 = 25$ **47.** $x^2 + y^2 = 16$ **49.** $(x - 2)^2 + (y - 5)^2 = 10$
51. $y = \sqrt{9 - x^2}$ corresponds to the top half; $y = -\sqrt{9 - x^2}$ is the equation of the bottom half.
53. 10π meters   **55.** 3   **57.** 5   **59.** $\frac{2}{3}$

Problem Set 9.3

1.

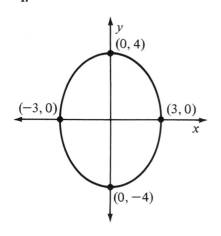

3.

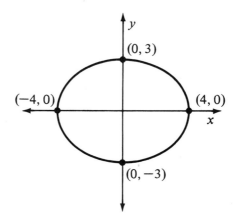

5.

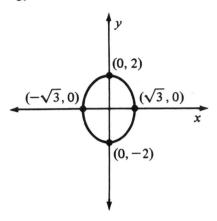

7.

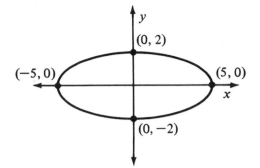

9.

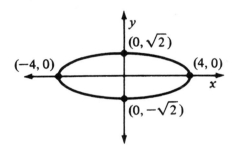

11.

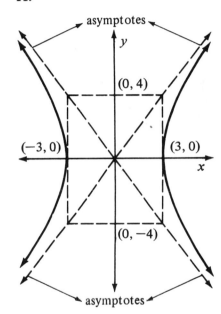

13.

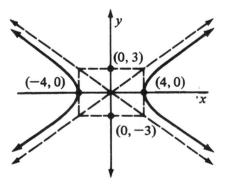

15.

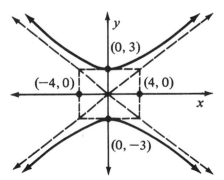

17.

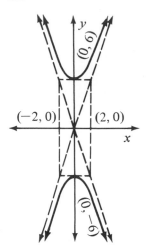

19.

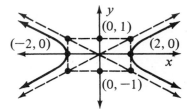

27.

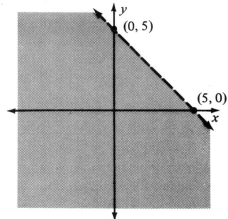

23. $y = \frac{3}{4}x;\ y = -\frac{3}{4}x$ **25.**   8

29.

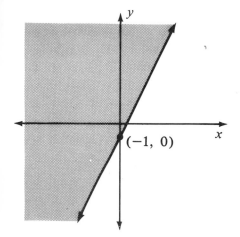

31.

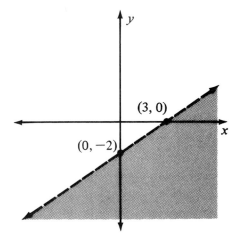

Problem Set 9.4

1.

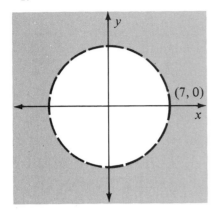

3.

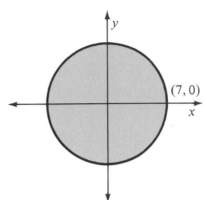

5.

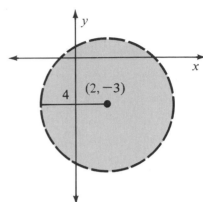

7.

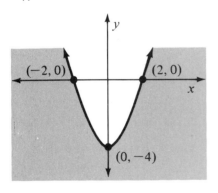

9.

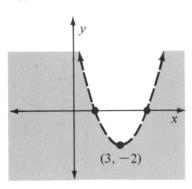

11.

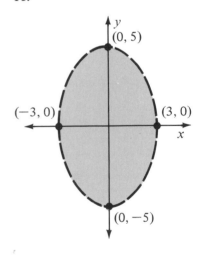

13.

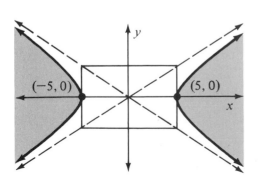

15.

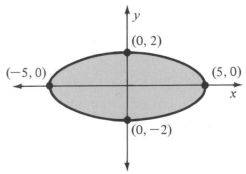

17.

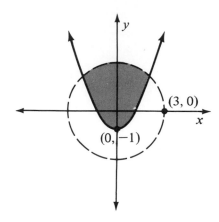

19.

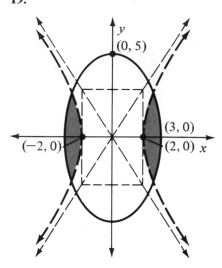

21.

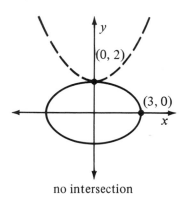

no intersection

23.

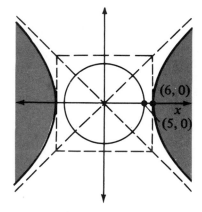

25. $(1, 2)$ **27.** $(0, 3)$ **29.** $(3, 4)$

Problem Set 9.5

1. $(0, 3)$, $(\frac{12}{5}, -\frac{9}{5})$ **3.** $(0, 4)$, $(\frac{16}{5}, \frac{12}{5})$ **5.** $(5, 0)$, $(-5, 0)$
7. $(0, -3)$, $(\sqrt{5}, 2)$, $(-\sqrt{5}, 2)$ **9.** $(0, -4)$, $(\sqrt{7}, 3)$, $(-\sqrt{7}, 3)$
11. $\left(\dfrac{-3 \pm 3\sqrt{5}}{2}, \ 3 \mp \sqrt{5}\right)$ **13.** $(-4, 11)$, $(\frac{5}{2}, \frac{5}{4})$ **15.** $(3, 0)$, $(-3, 0)$
17. $(\sqrt{7}, \sqrt{6}/2)$, $(-\sqrt{7}, \sqrt{6}/2)$, $(\sqrt{7}, -\sqrt{6}/2)$, $(-\sqrt{7}, -\sqrt{6}/2)$ **19.** $(4, 0)$, $(0, -4)$
21. $(8, 5)$ or $(-8, -5)$ **23.** $(6, 3)$ or $(13, -4)$ **25.** $y = -3x + 5$ **27.** $y = 5x - 17$
29. $y = \frac{2}{3}x - 2$

Chapter 9   Test

1.
(a)

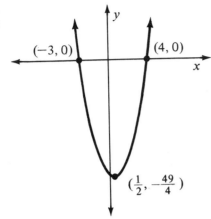

$(-3, 0)$ $(4, 0)$

$(\frac{1}{2}, -\frac{49}{4})$

(b)

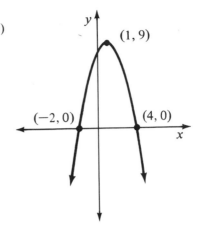

$(1, 9)$

$(-2, 0)$ $(4, 0)$

2. $\sqrt{82}$ **3.** -5 and 3 **4.** $(x + 2)^2 + (y - 4)^2 = 9$
5. $x^2 + y^2 = 25$

6.   center $= (5, -3)$
       radius $= \sqrt{39}$

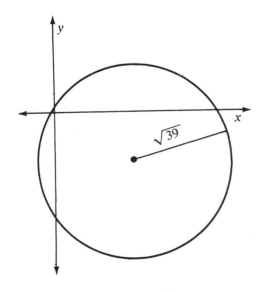

$\sqrt{39}$

7.

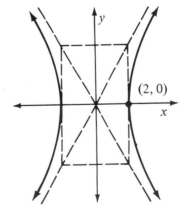

$(2, 0)$

8.

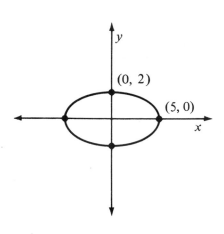

9. center $= (2, -1)$
   radius $= 3$

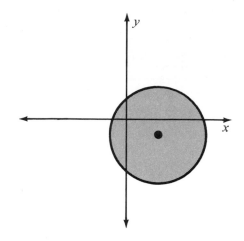

10.

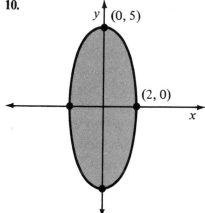

11. $(0, 5), (4, -3)$ **12.** $(0, -4), (\sqrt{7}, 3), (-\sqrt{7}, 3)$

CHAPTER 10

Problem Set 10.1

   1. $D = \{1, 2, 4\}$, $R = \{3, 5, 1\}$, yes   **3.** $D = \{-1, 1, 2\}$, $R = \{3, -5\}$, yes
   5. $D = \{7, 3\}$, $R = \{-1, 4\}$, no   **7.** $D = \{4, 3\}$, $R = \{3, 4, 5\}$, no
   9. $D = \{5, -3, 2\}$, $R = \{-3, 2\}$, yes   **11.** yes   **13.** no   **15.** no   **17.** yes
   19. yes   **21.** $\{x \mid x \geq -3\}$ **23.** $\{x \mid x \geq \frac{1}{2}\}$ **25.** $\{x \mid x \leq \frac{1}{4}\}$ **27.** $\{x \mid x \neq 5\}$
   29. $\{x \mid x \neq \frac{1}{2}, x \neq -3\}$ **31.** $\{x \mid x \neq -2, x \neq 3\}$ **33.** $\{x \mid x \neq -2, x \neq 2\}$

35.

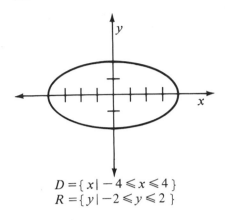

$D = \{x \mid -4 \leqslant x \leqslant 4\}$
$R = \{y \mid -2 \leqslant y \leqslant 2\}$

37.

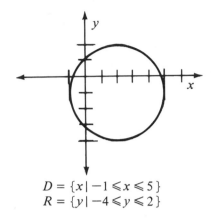

$D = \{x \mid -1 \leqslant x \leqslant 5\}$
$R = \{y \mid -4 \leqslant y \leqslant 2\}$

39.

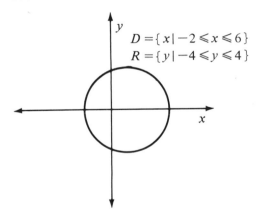

$D = \{x \mid -2 \leqslant x \leqslant 6\}$
$R = \{y \mid -4 \leqslant y \leqslant 4\}$

41.

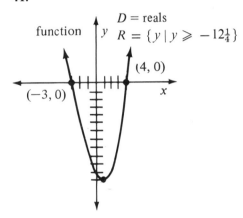

$D = \text{reals}$

function $R = \{y \mid y \geqslant -12\frac{1}{4}\}$

$(4, 0)$

$(-3, 0)$

43.

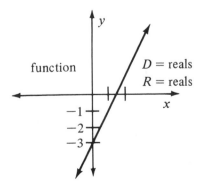

function

$D = \text{reals}$
$R = \text{reals}$

-1
-2
-3

45.

$D = \{x \mid x \leqslant -2 \text{ or } x \geqslant 2\}$

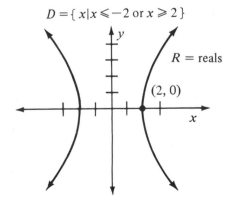

$R = \text{reals}$

$(2, 0)$

47.

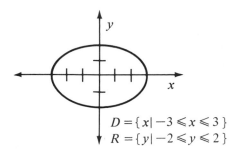

$$D = \{x \mid -3 \leqslant x \leqslant 3\}$$
$$R = \{y \mid -2 \leqslant y \leqslant 2\}$$

49.

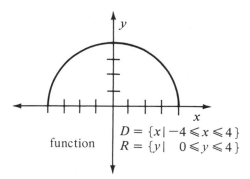

function | $D = \{x \mid -4 \leqslant x \leqslant 4\}$
$R = \{y \mid 0 \leqslant y \leqslant 4\}$

51.

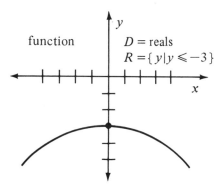

function | D = reals
$R = \{y \mid y \leqslant -3\}$

53.

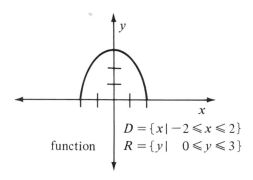

function | $D = \{x \mid -2 \leqslant x \leqslant 2\}$
$R = \{y \mid 0 \leqslant y \leqslant 3\}$

55. **a.** yes
b. $D = \{t \mid 0 \leq t \leq 6\}$
$R = \{h \mid 0 \leq h \leq 60\}$
c. 3 **d.** 60 **e.** 0 and 6

57. $(1, 2, 3)$ **59.** $(1, 3, 1)$

Problem Set 10.2
1. -1 **3.** -11 **5.** -5 **7.** 2 **9.** 4 **11.** 35 **13.** -13 **15.** 4
17. 0 **19.** 2 **21.** 2 **23.** $-\frac{1}{2}$ **25.** 1 **27.** -9 **29.** $3a^2 - 4a + 1$
31. $3a^2 + 14a + 16$ **33.** 16 **35.** 15 **37.** 1 **39.** $12x^2 - 20x + 8$ **41.** 2
43. $2x + h$ **45.** -4 **47.** $3(2x + h)$ **49.** $4x + 2h + 3$ **51.** 3 **53.** 4
55. $x + a$ **57.** 5 **59.** $x + a$ **61.** \$107, \$175, \$1000 **63.** 2 and 3 **65.** $-\frac{3}{2}$
67. -13 **69.** 31 **71.** -14

Problem Set 10.3
1. $6x + 2$ **3.** $-2x + 8$ **5.** $8x^2 + 14x - 15$ **7.** $(2x + 5)/(4x - 3)$
9. $4x - 7$ **11.** $3x^2 - 10x + 8$ **13.** $-2x + 3$ **15.** $3x^2 - 11x + 10$
17. $9x^3 - 48x^2 + 85x - 50$ **19.** $x - 2$ **21.** $1/(x - 2)$ **23.** $3x^2 - 7x + 3$
25. $6x^2 - 22x + 20$ **27.** 15 **29.** 98 **31.** $\frac{3}{2}$ **33.** 1 **35.** 40 **37.** 147
39. **a.** \$2.49 **b.** \$1.53 **c.** 5 min **43.** -2 and 3 **45.** $(2, 3)$
47. $(-\frac{15}{43}, -\frac{27}{43})$ **49.** $(1, 3, 1)$

Problem Set 10.4

1.

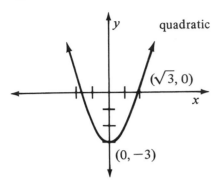

quadratic

$(\sqrt{3}, 0)$

$(0, -3)$

3.

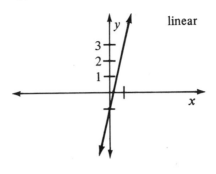

linear

5.

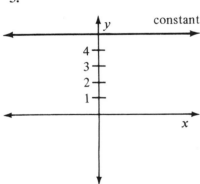

constant

7.

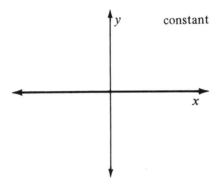

constant

9.

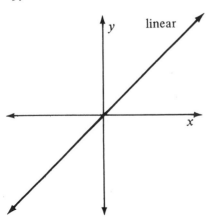

linear

11.

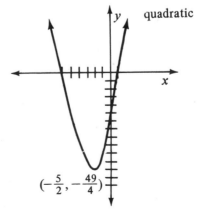

quadratic

$\left(-\frac{5}{2}, -\frac{49}{4}\right)$

13. $\frac{1}{9}$ **15.** $\frac{1}{4}$ **17.** 1   **19.** 2   **21.** $\frac{1}{27}$ **23.** 13   **25.** 17   **27.** 2   **29.** $\frac{17}{72}$

31.

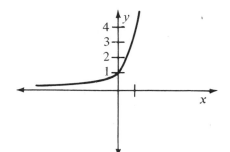

33.

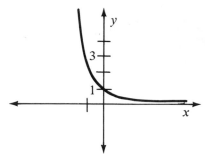

35.

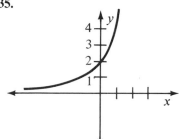

37.

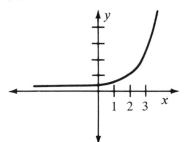

39.

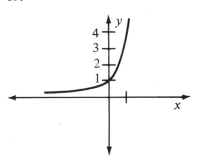

41.

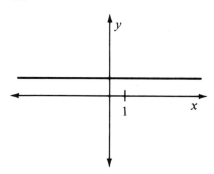

45. 200, 400, 800, 1600, 10 days   **47.** 5   **49.** 3 or -1 **51.** Center $(0, 0)$, $r = 5$
53. Center $(-3, 2)$, $r = 4$

Problem Set 10.5

1. $f^{-1}(x) = \dfrac{x + 1}{3}$ **3.** $f^{-1}(x) = \dfrac{1 - x}{3}$ **5.** $f^{-1}(x) = \pm\sqrt{x - 4}$

7. $f^{-1}(x) = 4x + 3$ **9.** $f^{-1}(x) = 2(x + 3)$ **11.** $f^{-1}(x) = \pm\sqrt{3 - x}$

13.

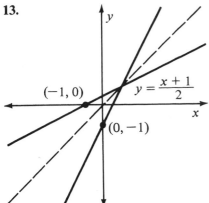

$(-1, 0)$

$y = \dfrac{x + 1}{2}$

$(0, -1)$

15.

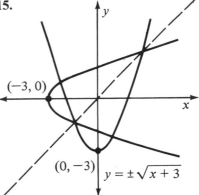

$(-3, 0)$

$(0, -3)$ $y = \pm\sqrt{x + 3}$

17.

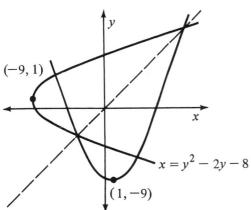

$(-9, 1)$

$x = y^2 - 2y - 8$

$(1, -9)$

19.

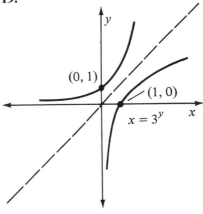

$(0, 1)$

$(1, 0)$

$x = 3^y$

21.

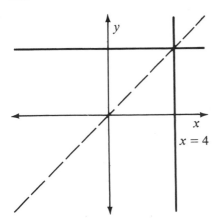

$x = 4$

23.

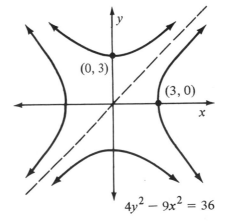

$(0, 3)$

$(3, 0)$

$4y^2 - 9x^2 = 36$

25.

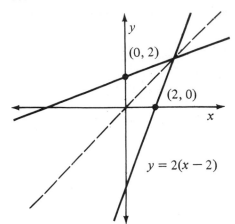

27.

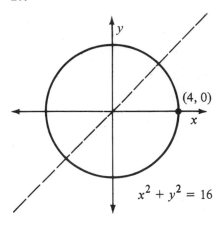

29.

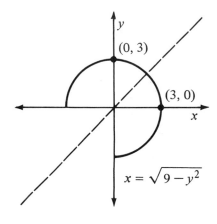

31.

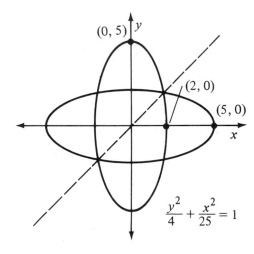

33. a. 4   **b.** $\frac{4}{3}$ **c.** 2   **d.** 2   **35.** $f^{-1}(x) = \dfrac{1}{x}$ **37.** $(3, 0)$, $\left(-\frac{9}{5}, -\frac{12}{5}\right)$

39. $(0, 4)$, $(0, -4)$ **41.** $(0, -2)$, $(\sqrt{3}, 1)$, $(-\sqrt{3}, 1)$

Chapter 10   Test

1. $D = \{-3, 2, 3, 0\}$
$R = \{1, 3, 5\}$, yes

2. $D = \{-2, -3\}$
$R = \{0, 1\}$, no

3. $D =$ all real numbers
$R = \{y \mid y \geq -9\}$, yes

4. $D = \{x \mid -3 \leq x \leq 3\}$
$R = \{y \mid -2 \leq y \leq 2\}$, no

5. $x \geq 4$ **6.** $x < 3$ **7.** $x \neq -1$ **8.** $x \neq 4$, $x \neq -2$ **9.** 11   **10.** -4

11. $4x + 2$ **12.** $x - 2$ **13.** 15   **14.** -14

15.

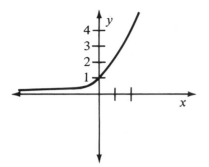

16.

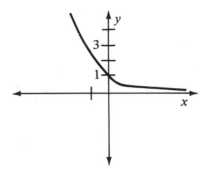

17.

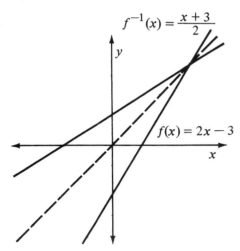

18.

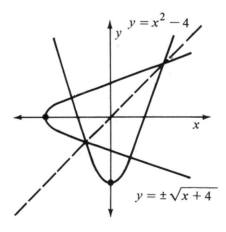

CHAPTER 11

Problem Set 11.1

1. $\log_2 16 = 4$ **3.** $\log_5 125 = 3$ **5.** $\log_{10} .01 = -2$ **7.** $\log_2 \frac{1}{32} = -5$
9. $\log_{1/2} 8 = -3$ **11.** $\log_3 27 = 3$ **13.** $10^2 = 100$ **15.** $2^6 = 64$ **17.** $8^0 = 1$
19. $10^{-3} = .001$ **21.** $6^2 = 36$ **23.** $5^{-2} = \frac{1}{25}$ **25.** 9   **27.** $\frac{1}{125}$ **29.** 4
31. $\frac{1}{3}$ **33.** 2   **35.** $\sqrt[3]{5}$

37.

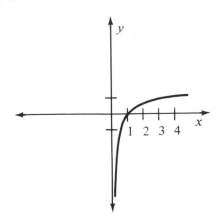

39.

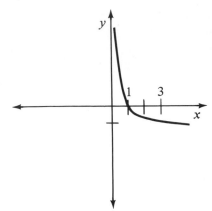

41.

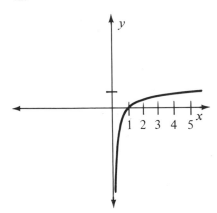

43.

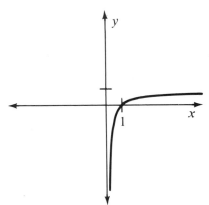

45. 4   **47.** $\frac{3}{2}$ **49.** 3   **51.** 1   **53.** 0   **55.** 0   **57.** $\frac{1}{2}$ **59.** 2.52 min

61. $2^y = -4$ can never be true   **63.** -2 **65.** -5 **67.** 2, -4 **69.** $\frac{1}{2}$, -3

Problem Set 11.2

1. $\log_3 4 + \log_3 x$ **3.** $\log_6 5 - \log_6 x$ **5.** $5 \log_2 y$ **7.** $\frac{1}{3}\log_9 z$

9. $2\log_6 x + 3\log_6 y$ **11.** $\frac{1}{2}\log_5 x + 4\log_5 y$ **13.** $\log_b x + \log_b y - \log_b z$

15. $\log_{10} 4 - \log_{10} x - \log_{10} y$ **17.** $2\log_{10} x + \log_{10} y - \frac{1}{2}\log_{10} z$

19. $3\log_{10} x + \frac{1}{2}\log_{10} y - 4\log_{10} z$ **21.** $\frac{2}{3}\log_b x + \frac{1}{3}\log_b y - \frac{4}{3}\log_b z$ **23.** $\log_b xz$

25. $\log_3 \dfrac{x^2}{y^3}$ **27.** $\log_{10} \sqrt{x}\ \sqrt[3]{y}$ **29.** $\log_2 \dfrac{x^3\sqrt{y}}{z}$ **31.** $\log_2 \dfrac{\sqrt{x}}{y^3 z^4}$ **33.** $\log_{10} \dfrac{x^{3/2}}{y^{3/4}z^{4/5}}$

35. $\frac{2}{3}$ **37.** 18   **39.** Possible solutions -1 and 3; only 3 checks   **41.** 3   **43.** Possible solutions -2 and 4; only 4 checks   **45.** Possible solutions -1 and 4; only 4 checks

47. Possible solutions $-\frac{5}{2}$ and $\frac{5}{3}$; only $\frac{5}{3}$ checks   **49.** 3.94×10^8 **51.** 3.91×10^{-2}

53. 0.00523   **55.** 50,300

Problem Set 11.3
1. 2.5775 **3.** 1.5775 **5.** 3.5775 **7.** 8.5775 − 10 **9.** 4.5775
11. 2.7782 **13.** 3.3032 **15.** 7.9872 − 10 **17.** 8.4969 − 10 **19.** 9.6010 − 10
21. 759 **23.** .00759 **25.** 1430 **27.** .00000447 **29.** .0000000918 **31.** 9260
33. 1.27 **35.** 20 **37.** 10.1 **39.** 386 **41.** 40,200,000 **43.** 24,800
45. .0000075 **47.** 258,000,000 **49.** 42 lb **51.** 3.3×10^4 **53.** $x > -7$
55. $x \le 1$ **57.** $x < -4$ or $x > \frac{3}{2}$ **59.** $3 \le x \le 5$

Problem Set 11.4
1. 1.4651 **3.** .6825 **5.** −1.5439 **7.** −.6477 **9.** −.3333 **11.** 2.0000
13. −.1846 **15.** .1846 **17.** 1.6168 **19.** 2.1132 **21.** 1.3333 **23.** .7500
25. 1.3917 **27.** .7186 **29.** .9650 **31.** 1.0363 **33.** 2.6356 **35.** 4.1629

37 $n = \dfrac{\log A - \log P}{\log(1 + r)}$ **39.** 60 geese, 48 ducks

41. Let x = number of oranges and y = number of apples

$\dfrac{10x}{3} + \dfrac{5y}{4} = 680$ He bought 150 oranges and 144 apples.

$\dfrac{50x}{3} + \dfrac{5y}{16} = 2545$

Problem Set 11.5
1. 2.40 **3.** 5.30 **5.** 4.38 **7.** 1.07 **9.** 3.98×10^{-4} **11.** 3.16×10^{-7}
13. **a.** 1.62 **b.** .87 **c.** .00293 **d.** 2.86×10^{-6} **15.** 5600 **17.** $8950
19. $4120 **21.** 14.2 years **23.** 11.9 years **25.** $\frac{7}{3}$, 1 **27.** ∅
29. $x < -1$ or $x > 9$ **31.** $-1 \le x \le 2$

Chapter 11 Test
1. 64 **2.** $\sqrt{5}$

3.

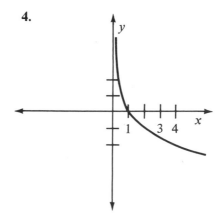

4.

5. $\frac{2}{3}$ **6.** 1.5645 **7.** 4.3692 **8.** 8.0899 − 10 **9.** 14,200 **10.** 15.6 **11.** 2.19
12. 49,200 **13.** 1.4651 **14.** 1.25 **15.** 15 **16.** 8 **17.** $5110 **18.** 3.52

CHAPTER 12

Problem Set 12.1

1. 4, 7, 10, 13, 16   **3.** 3, 7, 11, 15, 19   **5.** 1, 2, 3, 4, 5   **7.** 4, 7, 12, 19, 28
9. $\frac{1}{4}, \frac{2}{5}, \frac{3}{6}, \frac{4}{7}, \frac{5}{8}$ **11.** $\frac{2}{3}, \frac{3}{4}, \frac{4}{5}, \frac{5}{6}, \frac{6}{7}$ **13.** $1, \frac{1}{4}, \frac{1}{9}, \frac{1}{16}, \frac{1}{25}$ **15.** 2, 4, 8, 16, 32
17. $\frac{1}{3}, \frac{1}{9}, \frac{1}{27}, \frac{1}{81}, \frac{1}{243}$ **19.** $2, \frac{3}{2}, \frac{4}{3}, \frac{5}{4}, \frac{6}{5}$ **21.** $0, \frac{3}{2}, \frac{8}{3}, \frac{15}{4}, \frac{24}{5}$ **23.** $-2, 4, -8, 16, -32$
25. $n + 1$ **27.** $4n$ **29.** $3n + 4$ **31.** n^2 **33.** $3n^2$ **35.** 2^{n+1} **37.** $(-2)^n$

39. $\dfrac{1}{2^{n+1}}$ **41.** $\dfrac{n}{(n + 1)^2}$ **43.** \$50,500; \$51,005; \$51,515; \$52,030; \$52,550; \$53,076;

$a_n = 50,000(1.01)^n; \; a_{24} = 50,000(1.01)^{24} = 63,487$ **45.** $-8x^6$ **47.** $\frac{7}{12}$ **49.** $\dfrac{1}{x^9}$

51. $\dfrac{y^9}{8x^{21}}$

Problem Set 12.2

1. 36   **3.** 9   **5.** 50   **7.** $\frac{163}{60}$ **9.** $\frac{62}{15}$ **11.** 60   **13.** 40
15. $5x + 15$ **17.** $(x + 1)^2 + (x + 1)^3 + (x + 1)^4 + (x + 1)^5 + (x + 1)^6 + (x + 1)^7$
19. $\dfrac{x + 1}{x - 1} + \dfrac{x + 2}{x - 1} + \dfrac{x + 3}{x - 1} + \dfrac{x + 4}{x - 1} + \dfrac{x + 5}{x - 1}$
21. $(x + 3)^3 + (x + 4)^4 + (x + 5)^5 + (x + 6)^6 + (x + 7)^7 + (x + 8)^8$
23. $(x + 1)^2 + (x + 2)^3 + (x + 3)^4 + (x + 4)^5 + (x + 5)^6$ **25.** $\displaystyle\sum_{i=1}^{4} 2^i$ **27.** $\displaystyle\sum_{i=2}^{6} 2^i$

29. $\displaystyle\sum_{i=2}^{6} (i^2 + 1)$ **31.** $\displaystyle\sum_{i=3}^{7} \dfrac{i}{i + 1}$ **33.** $\displaystyle\sum_{i=1}^{4} \dfrac{i}{2i + 1}$ **35.** $\displaystyle\sum_{i=3}^{6} (x - i)$ **37.** $\displaystyle\sum_{i=3}^{5} \dfrac{x}{x + i}$

39. $\displaystyle\sum_{i=2}^{4} x^i(x + i)$ **41.** 208 ft, 784 ft   **43.** $(2x + 3)^2$ **45.** $(x^2 + 9)(x + 3)(x - 3)$
47. $(x + 2)(x^2 - 2x + 4)$ **49.** $(x - 3 + y)(x - 3 - y)$

Problem Set 12.3

1. 1   **3.** not an arithmetic progression   **5.** -5
7. not an arithmetic progression   **9.** $\frac{2}{3}$ **11. a.** $a_n = 4n - 1$ **b.** 39   **c.** 95
d. 210   **e.** 1176   **13. a.** $a_n = -2n + 8$ **b.** -12 **c.** -40 **d.** -30 **e.** -408
15. a. $a_n = 4n + 1$ **b.** 41   **c.** 97   **d.** 230   **e.** 1224   **17. a.** $a_n = -5n + 17$
b. -33 **c.** -103 **d.** -105 **e.** -1092 **19. a.** $a_n = n/2$ **b.** 5   **c.** 12
d. $\frac{55}{2}$ **e.** 150   **21. a.** $a_n = 6n - 2$ **b.** 58   **c.** 142   **d.** 310   **e.** 1752
23. a. $a_n = 2n + 5$ **b.** 25   **c.** 53   **d.** 160   **e.** 720
25. a. \$15,000; \$15,850; \$16,700; \$17,550; \$18,400   **b.** $a_n = 850n + 14,510$ **c.** \$22,650
27. $\dfrac{3(x - 4)}{x + 2}$ **29.** $\dfrac{x + 4}{x + 3}$ **31.** $\frac{5}{3}$ **33.** $\dfrac{x + 4}{x - 4}$

Problem Set 12.4

1. 5   **3.** $\frac{1}{3}$ **5.** not geometric   **7.** -2 **9.** not geometric
11. a. $a_n = 4 \cdot 3^{n-1}$ **b.** $4 \cdot 3^5$ **c.** $4 \cdot 3^{19}$ **d.** $2(3^{10} - 1)$ **e.** $2(3^{20} - 1)$

13.   **a.** $a_n = -2(-5)^{n-1}$ **b.** $-2(-5)^5$ **c.** $-2(-5)^{19}$ **d.** $(5^{10} - 1)/3$
e. $(5^{20} - 1)/3$ **15.**   **a.** $a_n = -4(\frac{1}{2})^{n-1}$ **b.** $-4(\frac{1}{2})^5$ **c.** $-4(\frac{1}{2})^{19}$ **d.** $8[(\frac{1}{2})^{10} - 1]$
e. $8[(\frac{1}{2})^{20} - 1]$ **17.**   **a.** $a_n = 2 \cdot 6^{n-1}$ **b.** $2 \cdot 6^5$ **c.** $2 \cdot 6^{19}$ **d.** $2(6^{10} - 1)/5$
e. $2(6^{20} - 1)/5$ **19.**   **a.** $a_n = 10 \cdot 2^{n-1}$ **b.** $10 \cdot 2^5 = 320$ **c.** $10 \cdot 2^{19}$
d. $10(2^{10} - 1) = 10230$ **e.** $10(2^{20} - 1)$ **21.**   **a.** $a_n = (\sqrt{2})^n$ **b.** 8   **c.** 1024
d. $\sqrt{2}[(\sqrt{2})^{10} - 1]/(\sqrt{2} - 1)$ **e.** $\sqrt{2}[(\sqrt{2})^{20} - 1]/(\sqrt{2} - 1)$ **23.**   **a.** $a_n = 5 \cdot 2^{n-1}$
b. 160   **c.** $5 \cdot 2^{19}$ **d.** $5(2^{10} - 1) = 5115$ **e.** $5(2^{20} - 1)$ **25.**   **a.** $10,600
b. $10,000; \$10,600; \$11,236; \$11,910.16; \$12,624.77   **c.** $a_n = 10,000(1.06)^{n-1}$
27. $x^2 + 10x + 25$ **29.** $x^3 + 3x^2y + 3xy^2 + y^3$ **31.** $x^4 + 4x^3y + 6x^2y^2 + 4xy^3 + y^4$

Problem Set 12.5

1. $x^4 + 8x^3 + 24x^2 + 32x + 16$
3. $x^6 + 6x^5y + 15x^4y^2 + 20x^3y^3 + 15x^2y^4 + 6xy^5 + y^6$
5. $32x^5 + 80x^4 + 80x^3 + 40x^2 + 10x + 1$
7. $x^5 - 10x^4y + 40x^3y^2 - 80x^2y^3 + 80xy^4 - 32y^5$
9. $81x^4 - 216x^3 + 216x^2 - 96x + 16$ **11.** $64x^3 - 144x^2y + 108xy^2 - 27y^3$
13. $x^8 + 8x^6 + 24x^4 + 32x^2 + 16$ **15.** $x^6 + 3x^4y^2 + 3x^2y^4 + y^6$
17. $x^3/8 - 3x^2 + 24x - 64$ **19.** $x^4/81 + 2x^3y/27 + x^2y^2/6 + xy^3/6 + y^4/16$
21. $x^9 + 18x^8 + 144x^7 + 672x^6$ **23.** $x^{10} - 10x^9y + 45x^8y^2 - 120x^7y^3$
25. $x^{10} + 20x^9y + 180x^8y^2 + 960x^7y^3$ **27.** $x^{15} + 15x^{14} + 105x^{13}$
29. $x^{12} - 12x^{11}y + 66x^{10}y^2$ **31.** $x^{20} + 40x^{19} + 760x^{18}$ **33.** $x^{100} + 200x^{99}$
35. $x^{50} + 50x^{49}y$ **37.** $\frac{21}{128}$ **39.** $4 - 10i$ **41.** $6 + 12i$ **43.** $-5 + 12i$
45. $-\dfrac{10}{13} + \dfrac{15}{13}i$

Chapter 12   Test

1. $-2, 1, 4, 7, 10$ **2.** $3, 7, 11, 15, 19$ **3.** $2, 5, 10, 17, 26$
4. $2, 16, 54, 128, 250$ **5.** $2, \frac{3}{4}, \frac{4}{9}, \frac{5}{16}, \frac{6}{25}$ **6.** $4, -8, 16, -32, 64$
7. $a_n = 4n + 2$ **8.** $a_n = 2^{n-1}$ **9.** $a_n = (\frac{1}{2})^n = 1/2^n$ **10.** $a_n = (-3)^n$
11. **a.** 90   **b.** 53   **c.** 130   **12.** 3   **13.** 6   **14.** **a.** 320   **b.** 25
15. **a.** $S_{50} = 3(2^{50} - 1)$ **b.** $5(4^{50} - 1)/3$ **16.** **a.** $x^4 - 12x^3 + 54x^2 - 108x + 81$
b. $32x^5 - 80x^4 + 80x^3 - 40x^2 + 10x - 1$ **c.** $27x^3 - 54x^2y + 36xy^2 - 8y^3$

Problem Set 13.1

1. Acute, complement $= 80°$, supplement $= 170°$
3. Acute, complement $= 45°$, supplement $= 135°$
5. Obtuse, complement $= -30°$, supplement $= 60°$
7. Complement $= 90° - x$, supplement $= 180° - x$ **9.** $\dfrac{\pi}{6}$ **11.** $\dfrac{\pi}{2}$ **13.** $\dfrac{2\pi}{3}$
15. $\dfrac{-5\pi}{4}$ **17.** $\dfrac{-5\pi}{6}$ **19.** $\dfrac{7\pi}{3}$ **21.** $60°$ **23.** $120°$ **25.** $-210°$
27. $-270°$ **29.** $330°$ **31.** $720°$ **33.** $45°$ or $\dfrac{\pi}{4}$ **35.** $120°$ **37.** $180°$
39. 5   **41.** 24   **43.** 5   **45.** $2, \sqrt{3}$ **47.** $4, 4\sqrt{3}$ **49.** $\dfrac{1}{3}, \dfrac{\sqrt{3}}{3}$

51. 3, 6   **53.** $2\sqrt{3}, 4\sqrt{3}$ **55.** $\sqrt{2}$ **57.** $\dfrac{4\sqrt{2}}{5}$ **59.** 8   **61.** $2\sqrt{2}$

63. $\dfrac{\sqrt{2}}{2}$ **65.** $\dfrac{3\sqrt{13}}{13}$ **67.** $\sqrt{6}+2$ **69.** $5+2\sqrt{6}$

Problem Set 13.2

	$\sin\theta$	$\cos\theta$	$\tan\theta$	$\cot\theta$	$\sec\theta$	$\csc\theta$
1.	$\frac{4}{5}$	$\frac{3}{5}$	$\frac{4}{3}$	$\frac{3}{4}$	$\frac{5}{3}$	$\frac{5}{4}$
3.	$\frac{12}{13}$	$-\frac{5}{13}$	$-\frac{12}{5}$	$-\frac{5}{12}$	$-\frac{13}{5}$	$\frac{13}{12}$
5.	$-\dfrac{3}{\sqrt{10}}$	$-\dfrac{1}{\sqrt{10}}$	3	$\dfrac{1}{3}$	$-\sqrt{10}$	$-\dfrac{\sqrt{10}}{3}$
7.	$\dfrac{b}{\sqrt{a^2+b^2}}$	$\dfrac{a}{\sqrt{a^2+b^2}}$	$\dfrac{b}{a}$	$\dfrac{a}{b}$	$\dfrac{\sqrt{a^2+b^2}}{a}$	$\dfrac{\sqrt{a^2+b^2}}{b}$

	sine	cosine	tangent
9.	$\dfrac{1}{2}$	$\dfrac{\sqrt{3}}{2}$	$\dfrac{1}{\sqrt{3}}$
11.	$\dfrac{1}{\sqrt{2}}$	$-\dfrac{1}{\sqrt{2}}$	-1
13.	$\dfrac{1}{2}$	$-\dfrac{\sqrt{3}}{2}$	$-\dfrac{1}{\sqrt{3}}$
15.	0	1	0

17.

θ	$\sin\theta$	$\cos\theta$	$\tan\theta$
$30°$ or $\dfrac{\pi}{6}$	$\dfrac{1}{2}$	$\dfrac{\sqrt{3}}{2}$	$\dfrac{1}{\sqrt{3}}$
$45°$ or $\dfrac{\pi}{4}$	$\dfrac{1}{\sqrt{2}}$	$\dfrac{1}{\sqrt{2}}$	1
$60°$ or $\dfrac{\pi}{3}$	$\dfrac{\sqrt{3}}{2}$	$\dfrac{1}{2}$	$\sqrt{3}$

19. I, IV   **21.** III, IV   **23.** III   **25.** I, IV   **27.** $\cos\theta=\frac{5}{13}$, $\tan\theta=\frac{12}{5}$, $\cot\theta=\frac{5}{12}$, $\sec\theta=\frac{13}{5}$, $\csc\theta=\frac{13}{12}$ **29.** $\sin\theta=\dfrac{\sqrt{3}}{2}$, $\tan\theta=-\sqrt{3}$, $\cot\theta=-\dfrac{1}{\sqrt{3}}$, $\sec\theta=-2$, $\csc\theta=\dfrac{2}{\sqrt{3}}$ **31.** $\sin\theta=-\dfrac{4}{5}$, $\tan\theta=\dfrac{4}{3}$, $\cot\theta=\dfrac{3}{4}$, $\sec\theta=-\dfrac{5}{3}$,

$\csc \theta = -\dfrac{5}{4}$ **33.** $\sin \theta = \dfrac{2}{\sqrt{5}}$, $\cos \theta = \dfrac{1}{\sqrt{5}}$ **35.** $\cos(-45°) = \dfrac{1}{\sqrt{2}} = \cos 45°$

37. 0.4362   **39.** 8.4362 − 10   **41.** 67.6

Problem Set 13.3

1. 0.7242   **3.** 0.9494   **5.** 0.9598   **7.** 19.7403   **9.** 34°   **11.** 20.8°
13. 37°   **15.** −0.1908   **17.** −0.7427   **19.** −0.7467   **21.** 0.5774

23. −0.8151   **25.** 198°   **27.** 132.5°   **29.** 217.7°   **31.** 303.2°   **33.** $\dfrac{\sqrt{3}}{2}$

35. −1   **37.** $-\frac{1}{2}$ **39.** $\frac{1}{2}$ **41.** $\frac{1}{2}$ **43.** 240°   **45.** 135°   **47.** 300°
49. 1   **51.** $\csc \theta \geq 1$ **53.** $\sin 90° = 1$, $\cos 90° = 0$, $\tan 90°$ is undefined
55. $\sin 180° = 0$, $\cos 180° = -1$, $\tan 180° = 0$
57. $\sin \theta = -\frac{12}{13}$, $\tan \theta = \frac{12}{5}$, $\cot \theta = \frac{5}{12}$, $\sec \theta = -\frac{13}{5}$, $\csc \theta = -\frac{13}{12}$

Problem Set 13.4

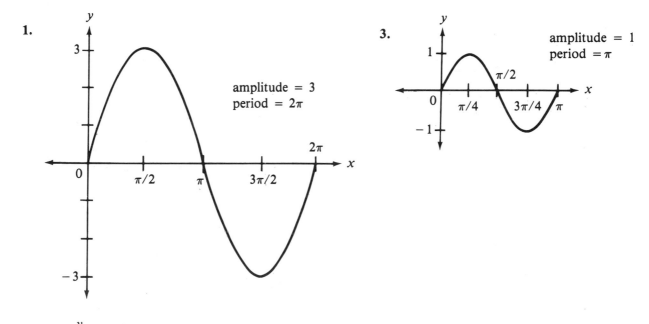

1.   amplitude = 3   period = 2π

3.   amplitude = 1   period = π

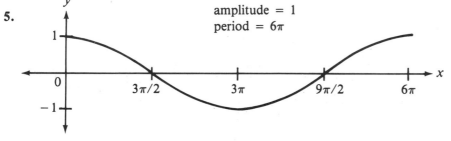

5.   amplitude = 1   period = 6π

7.

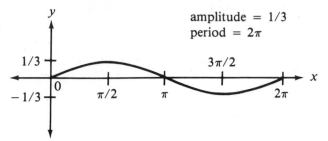

amplitude = 1/3
period = 2π

9.

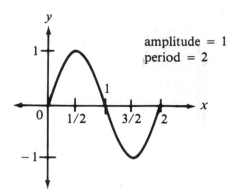

amplitude = 1
period = 2

11.

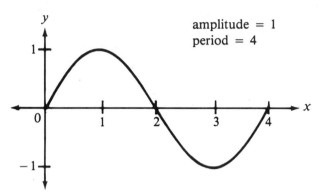

amplitude = 1
period = 4

13.

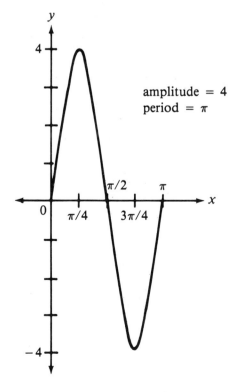

amplitude = 4
period = π

15.

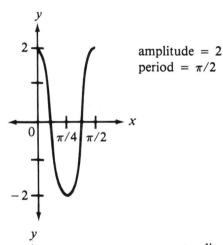

amplitude = 2
period = $\pi/2$

17.

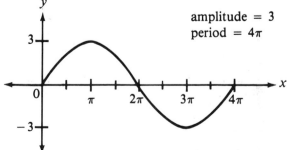

amplitude = 3
period = 4π

19.

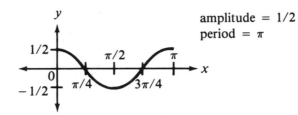

amplitude = 1/2
period = π

21.

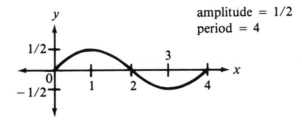

amplitude = 1/2
period = 4

23.

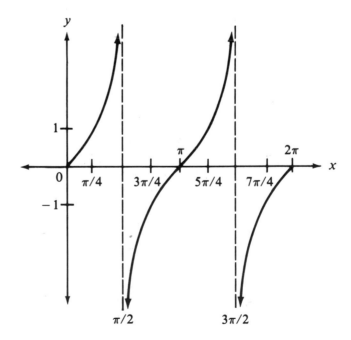

25.

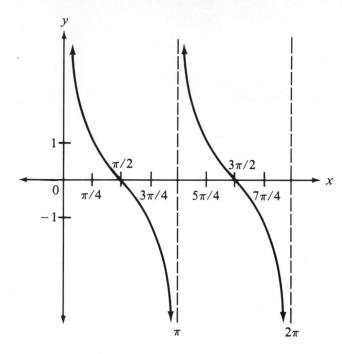

27.

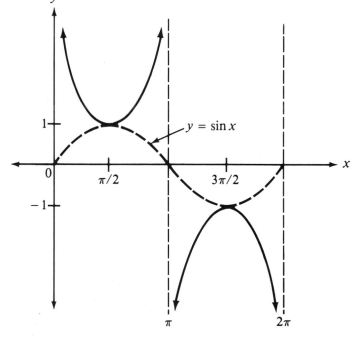

29. $\sin \theta = \dfrac{1}{\sqrt{5}}, \; \cos \theta = -\dfrac{2}{\sqrt{5}}, \; \tan \theta = -\frac{1}{2}$

31. $\sin \theta = -\dfrac{1}{\sqrt{2}}, \; \cos \theta = -\dfrac{1}{\sqrt{2}}, \; \tan \theta = 1$

Problem Set 13.5

1.

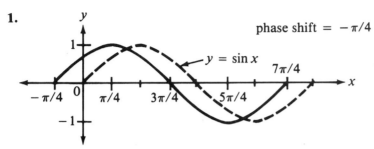

phase shift $= -\pi/4$

3.

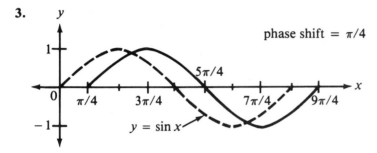

phase shift $= \pi/4$

5.

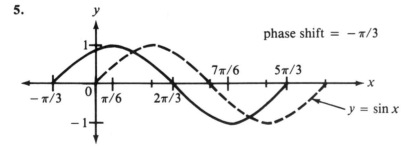

phase shift $= -\pi/3$

7.

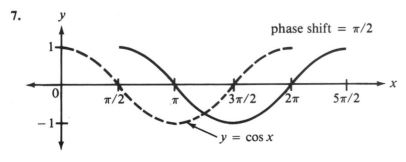

phase shift $= \pi/2$

9.

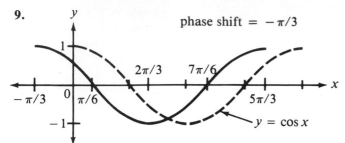

phase shift $= -\pi/3$

$y = \cos x$

11.

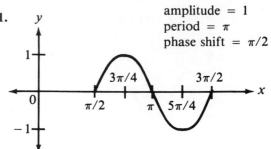

amplitude $= 1$
period $= \pi$
phase shift $= \pi/2$

13.

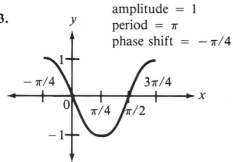

amplitude $= 1$
period $= \pi$
phase shift $= -\pi/4$

15.

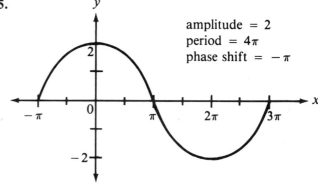

amplitude $= 2$
period $= 4\pi$
phase shift $= -\pi$

17.

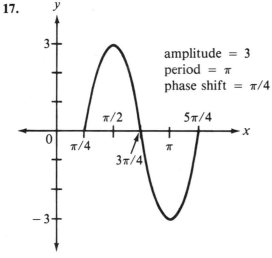

amplitude $= 3$
period $= \pi$
phase shift $= \pi/4$

19.

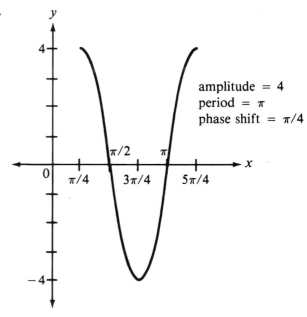

amplitude $= 4$
period $= \pi$
phase shift $= \pi/4$

21. amplitude = 1/2
period = $2\pi/3$
phase shift = $\pi/6$

23. amplitude = 2/3
period = $2\pi/3$
phase shift = $-\pi/6$

25. 0.6046   **27.** -0.3730 **29.** -1.1918

Chapter 13   Test

1. $\dfrac{5\pi}{12}$ **2.** $\dfrac{2\pi}{3}$ **3.** $\dfrac{25\pi}{9}$ **4.** $\dfrac{61\pi}{36}$ **5.** $200°$ **6.** $330°$ **7.** $120°$

8. $810°$ **9.** $5\sqrt{3}$, 10   **10.** $6\sqrt{2}$ **11.** $\sin\theta = -\frac{3}{5}$, $\cos\theta = \frac{4}{5}$, $\tan\theta = -\frac{3}{4}$,

$\cot\theta = -\frac{4}{3}$, $\sec\theta = \frac{5}{4}$, $\csc\theta = -\frac{5}{3}$ **12.** $\sin\theta = \dfrac{1}{\sqrt{5}}$, $\cos\theta = -\dfrac{2}{\sqrt{5}}$, $\tan\theta = -\frac{1}{2}$,

$\cot\theta = -2$, $\sec\theta = -\dfrac{\sqrt{5}}{2}$, $\csc\theta = \sqrt{5}$ **13.** III   **14.** II   **15.** $\cos\theta = -\dfrac{\sqrt{3}}{2}$,

$\tan\theta = -\dfrac{1}{\sqrt{3}}$, $\cot\theta = -\sqrt{3}$, $\sec\theta = -\dfrac{2}{\sqrt{3}}$, $\csc\theta = 2$ **16.** $\sin\theta = -\frac{3}{5}$, $\cos\theta = \frac{4}{5}$,

$\cot\theta = -\frac{4}{3}$, $\sec\theta = \frac{5}{4}$, $\csc\theta = -\frac{5}{3}$ **17.** 0.6060   **18.** 0.5030   **19.** 3.3759
20. 3.0415   **21.** -0.9641 **22.** 0.7536   **23.** -0.6678 **24.** -0.9191
25. 0.8391   **26.** 0.2588   **27.** $20.8°$ **28.** $69.2°$ **29.** $292.6°$ **30.** $247.4°$
31. $240°$ **32.** $135°$ **33.** $\dfrac{1}{\sqrt{2}}$ **34.** $\dfrac{1}{\sqrt{2}}$ **35.** $-\frac{1}{2}$ **36.** $-\frac{1}{2}$ **37.** $\frac{1}{2}$

38. $\frac{1}{2}$

39.

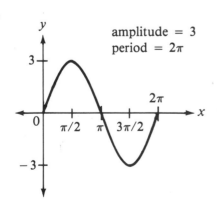

amplitude = 3
period = 2π

40.

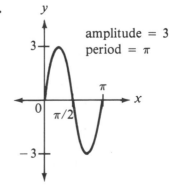

amplitude = 3
period = π

41.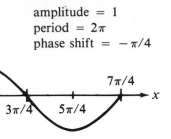

amplitude = 1
period = 2π
phase shift = $-\pi/4$

42.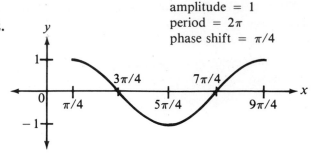

amplitude = 1
period = 2π
phase shift = $\pi/4$

43.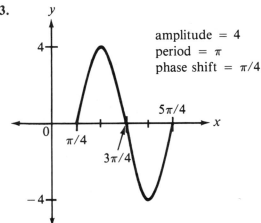

amplitude = 4
period = π
phase shift = $\pi/4$

44.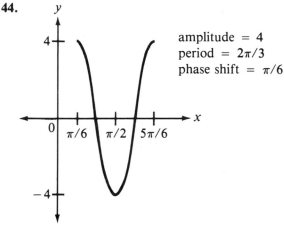

amplitude = 4
period = $2\pi/3$
phase shift = $\pi/6$

Problem Set 14.1

1. $\frac{5}{4}$ **3.** $-\frac{1}{2}$ **5.** $\dfrac{1}{a}$ $(a \neq 0)$ **7.** $\sqrt{2}$ **9.** $\frac{12}{5}$ **11.** $-\frac{13}{5}$ **13.** $-\frac{3}{5}$

15. $\frac{1}{2}$ **17.** $\cos\theta = \dfrac{\sqrt{3}}{2}$, $\tan\theta = \dfrac{1}{\sqrt{3}}$, $\cot\theta = \sqrt{3}$, $\csc\theta = 2$, $\sec\theta = \dfrac{2}{\sqrt{3}}$

Instead of answers for problems 19 through 39, we will give some hints that may be helpful in getting the proofs started.

19. Change $\tan\theta$ to $\dfrac{\sin\theta}{\cos\theta}$ and then simplify the left side.

21. Change everything to sines and cosines and then simplify the left side.
23. Change to sines and cosines and simplify.
25. Multiply out the left side, change to sines and cosines, then simplify.
27. Multiply out the left side and look for Pythagorean identity.
29. Multiply out the left side and look for Pythagorean identity.
31. Factor the numerator on the left side.
33. Multiply numerator and denominator on the right side by $1 + \sin\theta$.
35. Change to sines and cosines.

37. Change left side to sines and cosines, find the LCD, then add fractions on the left side.

39. See Example 11.

41. $\sin (30° + 60°) = \sin 90° = 1$, $\sin 30° + \sin 60° = \dfrac{1 + \sqrt{3}}{2} \neq 1$

43. $\dfrac{\sqrt{3}}{2}$ **45.** $\dfrac{\sqrt{3}}{2}$ **47.** $15°$

Problem Set 14.2

1. $\dfrac{\sqrt{6} - \sqrt{2}}{4}$ **3.** $\dfrac{\sqrt{6} + \sqrt{2}}{4}$ **5.** $\dfrac{\sqrt{6} - \sqrt{2}}{\sqrt{6} + \sqrt{2}}$ **7.** $\dfrac{\sqrt{6} + \sqrt{2}}{4}$ **9.** $\dfrac{-\sqrt{6} - \sqrt{2}}{4}$

For problems 11–19, expand the left side of each using the appropriate sum or difference formula and then simplify.

21. $\sin 5x$ **23.** $\cos 6x$ **25.** $\sin (45° + \theta)$ **27.** $\sin (30° + \theta)$
29. $\cos 90° = 0$ **31.** $-\frac{16}{65}, \frac{63}{65}, -\frac{16}{63}$ **33.** $2, \frac{1}{2}$ **35.** 1
37. $\sin 2x = 2 \sin x \cos x$ **39.** 0.7581 **41.** 1.1626 **43.** $34.7°$ **45.** $24.2°$

Problem Set 14.3

1. $\frac{24}{25}$ **3.** $\frac{24}{7}$ **5.** $-\frac{4}{5}$ **7.** $\frac{4}{3}$ **9.** $\frac{1}{2}$ **11.** 2 **13.** $-\dfrac{1}{\sqrt{10}}$ **15.** $-\frac{1}{3}$

17. $-\sqrt{10}$ **19.** $\dfrac{2}{\sqrt{5}}$ **21.** $-\frac{7}{25}$ **23.** $-\frac{25}{7}$ **25.** $\dfrac{3}{\sqrt{10}}$ **27.** $\frac{7}{25}$ **29.** 0

45.

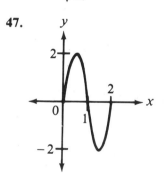

47.

Problem Set 14.4

1. $30°, 150°$ **3.** $30°, 330°$ **5.** $\dfrac{\pi}{3}, \dfrac{2\pi}{3}$ **7.** $\dfrac{\pi}{2}$ **9.** $30°, 90°, 150°$ **11.** $0°,$

$60°, 180°, 300°, 360°$ **13.** $\dfrac{\pi}{2}, \dfrac{7\pi}{6}, \dfrac{11\pi}{6}$ **15.** $\dfrac{\pi}{3}, \dfrac{5\pi}{3}$ **17.** $\varnothing$ **19.** $53.1°, 306.9°$

21. 201.5°, 338.5° **23.** 30°, 330° **25.** 45°, 135°, 225°, 315°
27. 30°, 90°, 150°, 270° **29.** 210°, 270°, 330° **31.** 90°, 210°, 330°
33. $h = -16t^2 + 750t$

37.

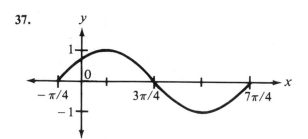

39.

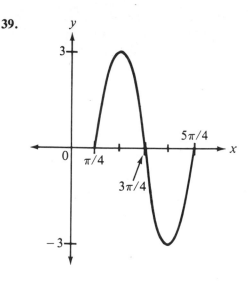

Chapter 14   Test

13. $\frac{63}{65}$ **14.** $-\frac{56}{65}$ **15.** $-\frac{119}{169}$ **16.** $-\frac{120}{169}$ **17.** $\frac{120}{119}$ **18.** $\frac{119}{120}$ **19.** $\dfrac{1}{\sqrt{10}}$

20. $-\dfrac{3}{\sqrt{10}}$ **21.** $\dfrac{\sqrt{6} + \sqrt{2}}{4}$ **22.** $\dfrac{\sqrt{6} + \sqrt{2}}{4}$ **23.** $\dfrac{\sqrt{6} - \sqrt{2}}{\sqrt{6} + \sqrt{2}}$ **24.** $\dfrac{\sqrt{6} + \sqrt{2}}{\sqrt{6} - \sqrt{2}}$

25. $\cos 9x$ **26.** $\sin 90° = 1$ **27.** $\dfrac{3}{5}, -\sqrt{\dfrac{5 - 2\sqrt{5}}{10}}$ **28.** $\dfrac{3}{5}, \sqrt{\dfrac{10 - \sqrt{10}}{20}}$

29. 30°, 150° **30.** 150°, 330° **31.** 30°, 90°, 150°, 270° **32.** 0°, 60°, 180°,

300°, 360° **33.** 30°, 150° **34.** 0°, 120°, 240°, 360° **35.** $\dfrac{\pi}{4}, \dfrac{3\pi}{4}, \dfrac{5\pi}{4}, \dfrac{7\pi}{4}$

36. $\dfrac{\pi}{2}, \dfrac{7\pi}{6}, \dfrac{11\pi}{6}$ **37.** $\dfrac{\pi}{2}, \dfrac{7\pi}{6}, \dfrac{3\pi}{2}, \dfrac{11\pi}{6}$ **38.** $0, \dfrac{2\pi}{3}, \dfrac{4\pi}{3}, 2\pi$ **39.** 35.9°, 144.1°

40. 63.3°, 296.7°

Problem Set 15.1

	$\sin A$	$\cos A$	$\tan A$	$\cot A$	$\sec A$	$\csc A$
1.	$\frac{4}{5}$	$\frac{3}{5}$	$\frac{4}{3}$	$\frac{3}{4}$	$\frac{5}{3}$	$\frac{5}{4}$
3.	$\dfrac{2}{\sqrt{5}}$	$\dfrac{1}{\sqrt{5}}$	2	$\dfrac{1}{2}$	$\sqrt{5}$	$\dfrac{\sqrt{5}}{2}$
5.	$\dfrac{1}{2}$	$\dfrac{\sqrt{3}}{2}$	$\dfrac{1}{\sqrt{3}}$	$\sqrt{3}$	$\dfrac{2}{\sqrt{3}}$	2

7. $\sin A = \dfrac{2}{3} = \cos B$, $\cos A = \dfrac{\sqrt{5}}{3} = \sin B$ **9.** $\sin A = \dfrac{1}{\sqrt{2}} = \cos B$,

$\cos A = \dfrac{1}{\sqrt{2}} = \sin B$ **11.** $\sin A = \dfrac{\sqrt{3}}{2} = \cos B$, $\cos A = \dfrac{1}{2} = \sin B$ **13.** 10

15. 33   **17.** 26.6°   **19.** 53.1°   **21.** $B = 70°$, $a = 8.2$, $b = 23$ **23.** $A = 14°$,
$a = 1.4$, $b = 5.6$ **25.** $A = 19.8°$, $B = 70.2°$, $c = 103$ **27.** $A = 41.4°$, $B = 48.6°$,
$a = 132$ **29.** 45°   **31.** 43°   **33.** height = 39.2 cm, base angles = 69.1°
35. 38.6 ft   **37.** 36.6°   **39.** 55.1°

Problem Set 15.2

1. 16 cm   **3.** 71 in   **5.** 68 yd   **7.** $C = 70°$, $c = 44$ km   **9.** $C = 80°$,
$a = 11$ cm   **11.** $C = 66.1°$, $b = 295$ in, $c = 284$ in   **13.** $C = 39°$, $a = 7.8$ m,
$b = 10.8$ m   **15.** $A = 141.8°$, $b = 118$ cm, $c = 214$ cm   **17.** $\sin B = 5$, which is
impossible   **19.** 209 ft   **21.** $\dfrac{\sqrt{6} + \sqrt{2}}{4}$
23. $\sin (\theta + 360°) = \sin \theta \cos 360° + \cos \theta \sin 360° = \sin \theta$ **25.** $\frac{56}{65}$ **27.** $-\frac{56}{33}$

Problem Set 15.3

1. $\sin B = 2$ is impossible   **3.** $B = 35.3$ is the only possibility   **5.** $B = 77°$ or
$B' = 103°$ **7.** $B = 54°$, $c = 88°$, $c = 67$ ft or $B' = 126°$, $C' = 16°$, $c' = 18$ ft
9. $B = 28.3°$, $C = 39.5°$, $c = 30$ cm   **11.** no solution   **13.** no solution

15. $B = 26.8°$, $A = 126.4°$, $a = 65.7$ km   **17.** 15 ft or 38 ft   **19.** $\frac{24}{25}$ **21.** $\dfrac{1}{\sqrt{5}}$

23. $-\frac{24}{7}$ **25.** $\sin 15° = \sqrt{\dfrac{1 - (\sqrt{3}/2)}{2}} = \dfrac{\sqrt{2 - \sqrt{3}}}{2}$

Problem Set 15.4

1. 87 in   **3.** $C = 92.9°$ **5.** 9.4 m   **7.** $A = 128.2°$ **9.** $A = 44°$, $B = 76°$,
$c = 62$ cm   **11.** $A = 29°$, $B = 46.5°$, $C = 104.5°$ **13.** $A = 15.6°$, $C = 12.9°$,
$b = 727$ m   **15.** $A = 39.5°$, $B = 56.7°$, $C = 83.8°$ **17.** $A = 55.4°$, $B = 45.5°$,
$C = 79.1°$ **19.** $a^2 = b^2 + c^2 - 2bc \cos 90° = b^2 + c^2 - 2bc(0) = b^2 + c^2$
21. 24 in   **23.** 30°, 150°   **25.** 90°, 210°, 330°   **27.** 0°, 60°, 180°, 300°, 360°
29. 45°, 135°, 225°, 315°

Chapter 15   Test

1. $\sin A = \frac{5}{13} = \cos B$, $\cos A = \frac{12}{13} = \sin B$ **2.** $\sin A = \frac{4}{5}$, $\tan A = \frac{4}{3}$, $\csc A = \frac{5}{4}$,
$\cot A = \frac{3}{4}$ **3.** 14 cm   **4.** 169 cm   **5.** 65.2°   **6.** 37.8°   **7.** $B = 63.7°$,
$b = 97$ yd, $c = 108$ yd   **8.** $A = 71.9°$, $a = 68$ yd, $b = 22$ yd   **9.** 6.7 in
10. 4.3 in   **11.** $C = 78.4°$, $a = 27$ cm, $b = 38$ cm   **12.** $B = 49.2°$, $a = 19$ cm,
$c = 43$ cm   **13.** $\sin B = 3.0311$, which is impossible   **14.** $B = 29°$ is the only
possibility   **15.** $B = 71°$, $C = 58°$, $c = 7.1$ ft or $B' = 109°$, $C' = 20°$, $c' = 2.9$ ft

16. $B = 59°$, $C = 95°$, $c = 10.9$ ft or $B' = 121°$, $C' = 33°$, $c' = 6.0$ ft   **17.** 11 cm
18. 19 cm   **19.** $C = 95.7°$ **20.** $B = 69.5°$ **21.** $A = 43.8°$, $B = 17.3°$,
$c = 8.1$ m   **22.** $B = 33.7°$, $C = 111.6°$, $a = 3.8$ m   **23.** $50.8°$ **24.** 59 ft
25. 411 ft   **26.** 14.2 m

APPENDIX A

1. $x - 7 + \dfrac{20}{x + 2}$ **3.** $3x - 1$ **5.** $x^2 + 4x + 11 + \dfrac{26}{x - 2}$

7. $3x^2 + 8x + 26 + \dfrac{83}{x - 3}$ **9.** $2x^2 + 2x + 3$ **11.** $x^3 - 4x^2 + 18x - 72 + \dfrac{289}{x + 4}$

13. $x^4 + x^2 - x - 1 - \dfrac{1}{x - 2}$ **15.** $x + 2 + \dfrac{3}{x - 1}$ **17.** $x^3 - x^2 + x - 1$

19. $x^2 + x + 1$

APPENDIX B

1. $D = 16$, two rational   **3.** $D = 0$, one rational   **5.** $D = 5$, two irrational
7. $D = 17$, two irrational   **9.** $D = 36$, two rational   **11.** $D = 116$, two irrational
13. ± 10 **15.** ± 12 **17.** 9   **19.** -16 **21.** $\pm 2\sqrt{6}$

Index

I1

                    4
B   5
C   6
D   7
E   8
F   9
G   0
H   1
I   2
J   3